GRAND CANYON
GEOLOGY

GRAND CANYON
GEOLOGY

Second Edition

Edited by
Stanley S. Beus
Michael Morales

New York • Oxford
OXFORD UNIVERSITY PRESS
2003

DEDICATION

This book is dedicated to the memory of Dr. Edwin Dinwoodie McKee (1906–1985), who stands as one of the premier scientists of the Grand Canyon in this century. Eddie began his career in 1929 as a park naturalist at Grand Canyon where for eleven years he made the natural history of this marvelous place a labor of love. In that time, he made extensive collections and observations in all areas of natural history, increased the interpretive program of the park, and became an expert on the biology, archaeology, and geology of the Grand Canyon. He also initiated studies of Paleozoic rock units that would occupy him, at least part time, for the rest of his career and would result in the publication of seven books on Grand Canyon geological history and four other monographs, together with more than 200 shorter articles, on geology or other topics of natural history.

His hiking exploits in the Grand Canyon became legendary—especially his frequent weekend journeys on foot from the south rim to the north rim and back (42 miles round trip) to woo and win a young biology student, Barbara Hastings. Barbara and Eddie were married in 1929 to begin a union that would last more than 50 years.

Following his years with the Grand Canyon National Park, Eddie served for a time as Chairman of the Geology Department at the University of Arizona, Tucson, as well as assistant director of the Museum of North Arizona, Flagstaff. In 1953 he began a 30-year career with the U.S. Geological Survey in Denver. From there, he conducted worldwide and world-class scientific investigations in stratigraphy, sedimentation, and depositional processes. These investigations included frequent returns to the Grand Canyon, where he continued his work on the Paleozoic strata.

An entire generation of geologists who have studied the Grand Canyon, including the authors of this book, have benefited from the stimulation, encouragement, and inspiration of Edwin D. McKee. A number of us have spent long-remembered days with him in the field. His tireless efforts to develop new insights, stimulate discussion, ask the right questions, and share the excitement and curiosity of scientific investigation serve as a model for all who would pursue the hidden secrets of the Grand Canyon.

Oxford University Press

Oxford   New York
Athens   Auckland   Bangkok   Bogotá   Buenos Aires   Calcutta
Cape Town   Chennai   Dar es Salaam   Delhi   Florence   Hong Kong   Istanbul
Karachi   Kuala Lumpur   Madrid   Melbourne   Mexico City   Mumbai
Nairobi   Paris   São Paulo   Shanghai   Singapore   Taipei   Tokyo   Toronto   Warsaw

and associated company in Berlin

Published by Oxford University Press, Inc.
198 Madison Avenue, New York, New York, 10016
http://www.oup-usa.org

Oxford is a registered trademark of Oxford University Press

ISBN 0-19-512298-4 (p)

ISBN 0-19-512299-2 (pb)

9 8 7 6 5 4 3 2 1

Printed in the United States of America
on acid-free paper

CONTENTS

Foreword / ix

1 Introducing the Grand Canyon / 1
Stanley S. Beus and Michael Morales

2 Paleoproterozoic Rocks of the Granite Gorges / 9
K. E. Karlstrom, B. R. Ilg, M. L. Williams, D. P. Hawkins, S. A. Bowring, and S. J. Seaman

3 Grand Canyon Supergroup: Unkar Group / 39
J. D. Hendricks and G. M. Stevenson

4 Grand Canyon Supergroup: Nankoweap Formation, Chuar Group, and Sixtymile Formation / 53
Trevor D. Ford and Carol M. Dehler

5 Geologic Structure of the Grand Canyon Supergroup / 76
J. Michael Timmons, Karl E. Karlstrom, and James W. Sears

6 Tonto Group / 90
Larry T. Middleton and David K. Elliot

7 Temple Butte Formation / 107
Stanley S. Beus

8 Redwall Limestone and Surprise Canyon Formation / 115
Stanley S. Beus

9 Supai Group and Hermit Formation / 136
Ronald C. Blakey

10 Coconino Sandstone / 163
Larry T. Middleton, David K. Elliott, and Michael Morales

11 Toroweap Formation / 180
Christine E. Turner

12 Kaibab Formation / 196
Ralph Lee Hopkins and Kelcy L. Thompson

13 Mesozoic and Cenozoic Strata of the Colorado Plateau Near the Grand Canyon / 212
Michael Morales

14 Post-Precambrian Tectonism in the Grand Canyon Region / 222
   Peter W. Huntoon

15 History of the Grand Canyon and of the Colorado River in
   Arizona / 260
   Ivo Lucchitta

16 Hydraulics and Geomorphology of the Colorado River in the Grand
   Canyon / 275
   Susan Werner Kieffer

17 Late Cenozoic Lava Dams in the Western Grand Canyon / 313
   W. K. Hamblin

18 Earthquakes and Seismicity of the Grand Canyon Region / 346
   David S. Brumbaugh

19 Holocene Terraces, Sand Dunes, and Debris Fans along the Colorado
   River in Grand Canyon / 352
   *Kelly J. Burke, Helen C. Fairley, Richard Hereford, and
   Kathryn S. Thompson*

20 Debris Flows and the Colorado River / 371
   Robert H. Webb, Theodore S. Melis, and Peter G. Griffiths

21 Side Canyons of the Colorado River in Grand Canyon / 391
   Andre R. Potochnik and Stephen J. Reynolds

Bibliography / 407

Index / 428

FOREWORD

It is strangely ironic that the Grand Canyon of the Colorado River, one of the most frequently visited natural wonders of the world, was for three centuries ignored by the Europeans who explored and exploited western North America. Of course, the Native Americans of the Southwest had known the canyon for thousands of years before some of them led a little band from the Coronado expedition to the south rim of the canyon in 1540. The Spaniards were impressed, and said so, but that was about the extent of their interest in this magnificent natural feature. During the long interval from that fateful day, four hundred and fifty years ago, when Don García López de Cárdenas and his companions gazed across the canyon until the third decade of the nineteenth century, very few people of European heritage visited or bothered to describe the canyon.

Then, in 1831 there appeared a description of the canyon by mountain man James Ohio Pattie, the first account to be written by an American. Yet his published portrayal of the canyon did not bring a rush of people to see it. It was hard to reach, and its breathtaking proportions made it a barrier to travel along the southwestern lines of latitude. Indeed, it was as a barrier that those Americans who knew anything about the canyon were prone to view it.

In the middle years of the nineteenth century, there were several United States Army surveying expeditions through the Southwest, in part to search for a feasible railroad route to the West Coast. One of the expeditions, led by Lieutenant Joseph Christmas Ives, reached the floor of the canyon near the mouth of Diamond Creek in 1858. The canyon seemed anything but inviting to Ives, who wrote that "the increasing magnitude of the colossal piles that blocked the end of the vista, and the corresponding depth and gloom of the gaping chasms into which we were plunging, imparted an earthly character to a way that might have resembled the portals of the infernal regions."

Fortunately, Ives had with him Dr. John Strong Newberry, one of the great pioneer American geologists, and it was Newberry who first looked at the canyon with a geologist's eye. To him the canyon did not look in the least like the gates of hell, for he wrote that it was "the most splendid exposure of stratified rocks that there is in the world."

Then, in 1869 and 1871 John Wesley Powell led his exploring parties down the Colorado River in their frail, wooden boats, penetrating the canyon from Green River to the Grand Wash Cliffs. Suddenly, people throughout the world were made aware of this great natural wonder. From then until the present day, the canyon has been scientifically explored and studied, so that there now exists a vast body of literature on all aspects of the canyon. All of this has taken place within two lifetimes.

This I know in a personal way, for one evening in the thirties I sat on a couch in the Explorers' Club, New York, with none other than Frederick Dellenbaugh, one of the members of Powell's second expedition. For an hour or more, I enjoyed a firsthand account of Powell and his party and of their adventures.

Grand Canyon Geology is dedicated to Edwin D. McKee—"Eddie" to all of us who knew him. No more appropriate dedication could be made. Eddie was the foremost modern authority on the geology of the Grand Canyon. Geologists today and in years to come will benefit from the profound studies of canyon geology that have come from Eddie's pen.

This book is not intended for the casual reader; he who turns its pages will need a certain amount of geological sophistication to appreciate the twenty chapters that describe and interpret Grand Canyon geology. Because each chapter is written by one of more authorities in their respective fields, there is here a storehouse of geological knowledge, much of it new. For geologists and those interested in geology, this is a book to be read, consulted, and treasured for years to come.

Edwin H. Colbert

GRAND CANYON
GEOLOGY

• 1 •

Introducing the Grand Canyon
Stanley S. Beus and Michael Morales

The Grand Canyon contains a marvelous record of geologic and paleontologic events spanning most of the last two billion years, nearly one half of the life span of this planet. Although it is neither the deepest nor the longest canyon in the world, it is one of the few places where so many chapters of earth history are clearly displayed. The igneous and metamorphic rocks of the canyon's inner gorge are part of the basement of the North American continent. Two great packages of sedimentary and volcanic rocks—one of Middle Proterozoic age and one of Paleozoic age—make up most of the canyon's walls. Beyond the rim to the north and east are the giant stairsteps of Mesozoic and lower Cenozoic strata whose cliffs have been retreating from the canyon's edge for millions of years. In the central part of the canyon, upper Cenozoic volcanic rocks have poured intermittently into the gorge, temporarily blocking the Colorado River and altering its cutting action. During the past two billion years, many dramatic, earth-changing events have occurred in western North America. We are incredibly fortunate that the rocks exposed in the Grand Canyon record this time period like no other place on the continent. This great chasm is truly a unique "window into the past."

The Grand Canyon (Fig. 1.1) lies entirely in the northwestern part of Arizona. It extends nearly 278 miles (448 kilometers) between Lake Powell on its eastern end and Lake Mead to the west. In 1919, the region became a national park, which today encompasses approximately 1900 square miles (4921 square kilometers) of land. This most famous of all canyons was formed by swiftly flowing waters of the Colorado River cutting into rock layers of the southwestern Colorado Plateau, a vast uplifted tableland that includes a large portion of the Four Corners states: Arizona, Colorado, New Mexico, and Utah. The land surrounding the Grand Canyon includes six local plateaus and one low-lying platform, all of which are bounded by faults or monoclines (Fig. 1.2). In a west-to-east cross section to the north of the canyon (Fig. 1.3), you can see how each of these blocks of land has been uplifted, downdropped, or tilted relative to its neighbor.

At its narrowest, the Grand Canyon is a little less than a mile (1.6 kilometers) across. Along the canyon's north and south rims, the relief is relatively gentle, except for incisions made by runoff waters flowing into the gorge. Within the canyon itself, however, the topography is quite varied and spectacular. The maximum depth of the canyon at any single place is about 6000 feet (1829 meters) from the rim to the floor. The maximum drop in elevation of the canyon as a whole, however, is approximately 6600 feet (2012 meters) between Point

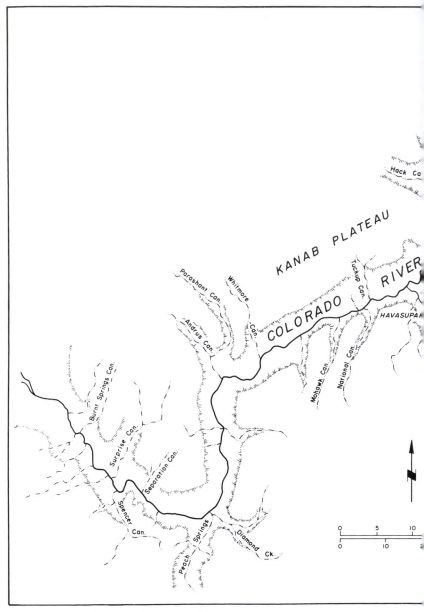

FIGURE 1.1. Map of the Grand Canyon.

Imperial on the north rim (8803 feet [2683 meters]) and the floor of the canyon near Lake Mead (1200 feet [366 meters]).

Many people are surprised to learn that there is a difference in elevation between the north and south rims. On the canyon's southern edge, the altitude ranges from 6000 to 7500 feet (1829 to 2286 meters) above sea level. The north-

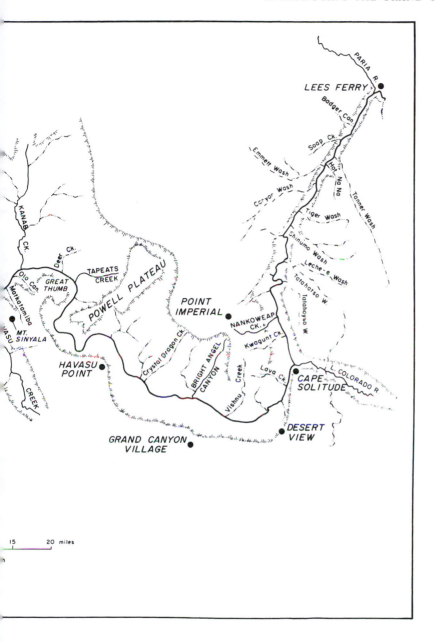

ern edge, however, is 1000 to 1200 feet (305 to 366 meters) higher even though both rims are capped by the same rock unit, the Kaibab Formation. The difference in height occurs because the various rock layers into which the canyon has been cut do not lie completely flat in this region. Instead, they arch or dome upward. The Colorado River carved the canyon through the southern flank of the dome, where the rock layers are tilted gently down to the south. Layers on

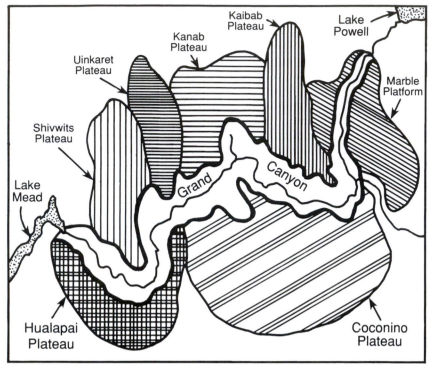

FIGURE 1.2. Generalized map of the area surrounding the Grand Canyon.

the north rim, therefore, have been uplifted higher than the same layers on the south rim (Fig. 1.4).

Water that flows into and through the Grand Canyon comes primarily from four merging rivers—the Green, San Juan, Little Colorado, and Colorado—that drain hundreds of square miles of the Four Corners states (Fig. 1.5). As it courses through the canyon, the Colorado River drops about 2000 feet (610 meters) in elevation (Fig. 1.6). This steep gradient allows the river to continue its erosion of the chasm's floor.

Humans have been in the Grand Canyon for at least four thousand years, but the early records are sparse. Rare artifacts, remains of prehistoric dwellings, and the numerous mescal pits all attest to the early exploration and habitation

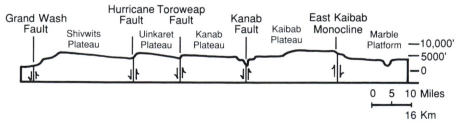

FIGURE 1.3. Generalized west-to-east cross section of the area just north of the Grand Canyon.

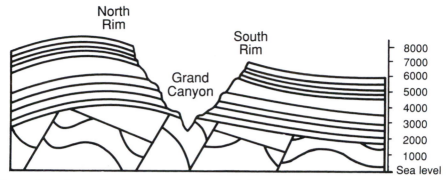

FIGURE 1.4. Generalized cross section through the Grand Canyon from the north rim to the south rim.

FIGURE 1.5. Drainage area for runoff water that flows into and through the Grand Canyon.

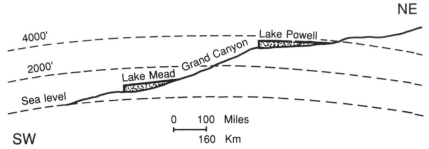

FIGURE 1.6. Diagrammatic profile of the Colorado River through the Grand Canyon, from Lake Powell to Lake Mead.

by American Indians such as the Anasazi and the Cohonina. The first Europeans to see the canyon were a Spanish party of thirteen in search of the fabled lost cities of gold, under the command of Captain Don García López de Cárdenas. In 1540, Hopi Indian guides led them to the south rim in the eastern part of the Grand Canyon, but they were unable to reach the river below. In the next three centuries, only two visits to the region have been reliably recorded. In 1776, Father Francisco Tomás Garcés, a Spanish missionary, explored Havasu Canyon in the south-central part of the Grand Canyon. In the same year, two Spanish priests, Father Silvestre Vélez de Escalante and Father Francisco Domíngues, led an expedition to the region and discovered a ford across the Colorado River (Crossing of the Fathers).

Among the earliest geological reports of the Grand Canyon country are those of Jules Marcou (1856) and John Strong Newberry (1861), who described the region's Paleozoic stratigraphy from their explorations of the canyon and the land to the south. Newberry was a geologist in the War Department-sponsored Ives expedition of 1857–1858. Ives' rather discouraging and, as it turned out, unprophetic statement about the Grand Canyon was:

> Ours has been the first, and will doubtless be the last, party of whites to visit this profitless locality. It seems intended by nature that the Colorado River, along the greater portion of its lonely and majestic way, shall be forever unvisited and undisturbed.

In 1869, John Wesley Powell led a party of ten men (reduced later to nine and finally only six) on an epic journey by boat down a thousand miles of the Colorado River from Green River, Wyoming, across Utah, and finally through the Grand Canyon to the mouth of the Virgin River at what is today the north end of Lake Mead. Powell's work was followed up by a small group of outstanding scientists through the turn of the nineteenth century. These included G.K. Gilbert, who was the first to apply formal rock unit names to Grand Canyon rocks; C.E. Dutton, who wrote the first monograph on the geology and geologic history of the Grand Canyon; A.R. Marvine, who participated in the U.S. Geographical Survey West of the 100th Meridian; and C.D. Walcott, who described both Paleozoic and Precambrian rocks in the canyon's central and eastern parts. These pioneer studies laid the groundwork for all subsequent research in the Grand Canyon.

Edwin D. McKee, to whom this book is dedicated, stands out as the premier research scientist of Grand Canyon geology in the twentieth century. Be-

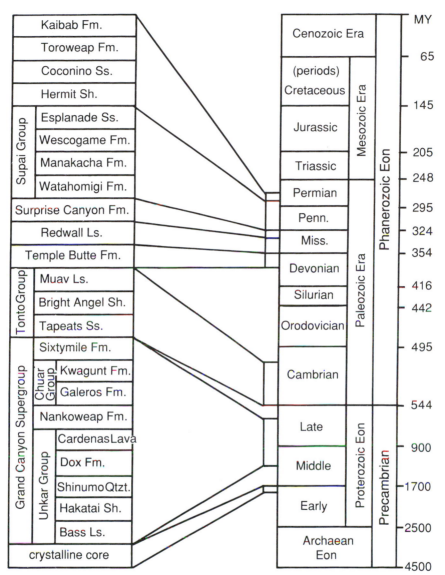

FIGURE 1.7. Comparison of the geologic column of the Grand Canyon with the Geologic Time Scale. (After Haq and Van Eysinga 1987.)

tween 1933 and 1982, McKee either authored or coauthored five monographic publications on various Paleozoic rock units of the canyon—including the Redwall Limestone, Supai Group, Coconino Sandstone, the Toroweap and Kaibab Formations, and Cambrian units. McKee also organized a special symposium in 1964 at the Museum of Northern Arizona to summarize the data and interpretations on the origin and evolution of the canyon. Much of our present understanding of the stratigraphy and age of Grand Canyon rocks (Fig. 1.7) is based on McKee's seminal work.

For more than a century, then, the Grand Canyon has attracted the extraordinary interest and effort of geologists to map its course, examine its rocks and fossils, and understand its record of earth history. In 1870, Powell wrote of the canyon country of the Colorado River:

> . . . the thought grew in to my mind that the canyons of this region would be a Book of Revelations in the rock-leaved Bible of geology. The thought fructified and I determined to read the book.

Since then, several generations of earth scientists have searched the river and its canyon for answers to questions of when and how this part of our planet's crust developed. Although some questions have been answered, new ones are raised, and so the search goes on. This book is written for those who inquire about the revelations of the Grand Canyon. It is a compilation of the best efforts of a number of authors, all experts in their field, to summarize what is now known about the geology of the canyon. Much of the information presented here is an update and refinement of earlier works by pioneer scientists of the past. All of us who have written for this volume owe a great debt to those who preceded us in the search for the geologic secrets of the Grand Canyon.

· 2 ·

PALEOPROTEROZOIC ROCKS OF THE GRANITE GORGES

K. E. Karlstrom, B. R. Ilg, M. L. Williams, D. P. Hawkins, S. A. Bowring, and S. J. Seaman

INTRODUCTION

The Precambrian Eons represent eight-ninths of earth history, from the time of formation of our planet about 4.6 Ga (billion years ago), to the rapid diversification of life on earth starting in the Cambrian 545 Ma (million years ago). North America records a rich Precambrian geologic record, from the oldest rocks on earth, 4.0–3.5 Ga in the Slave and Wyoming provinces, through younger belts that were sequentially added to the south (Fig. 2.1). The Archean nucleus of the continent had a complex history, including the development of a rifted shoreline in southern Wyoming about 2 billion years ago (Karlstrom et al. 1983). However, there was no continental lithosphere in the Grand Canyon region before ~1.85 billion years ago, when volcanic arcs began to form on oceanic crust.

The Grand Canyon region provides a spectacular record of the growth of the North American continent in the Paleoproterozoic Era (2.5–1.6 Ga), particularly within a 200-million-year interval from 1.8 to 1.6 Ga. To understand the history of this continental growth, we rely on detailed studies of the age and character of ancient rocks, analyzed in the context of modern plate tectonic analogues. The best modern analogue for growth of continents may be the Indonesian region, where the Australian plate is moving north and colliding with volcanic arcs and back arc basins across a very complex series of subduction zones and transcurrent faults. We envision that the Grand Canyon region resembled the Indonesian region at about 1.75 Ga, with tectonic blocks (terranes) consisting of volcanic island arcs, arc basins, and older continental fragments poised to the "south" and ready to become welded to the continental nucleus during progressive arc-continent collisions (Condie 1984; Karlstrom and Bowring 1988).

This plate tectonic model (Fig. 2.2) suggests that Proterozoic continental crust of the southwestern Unites States was differentiated from the mantle in magmatic arcs above subduction zones. In the Grand Canyon, these arcs range in age from 1.84 to 1.71 Ga. We infer that oceanic island arcs approached each other, and the older Archean part of the North American continent, as a result of subduction of the dense intervening oceanic crust and back-arc basins (Fig. 2.2a). This collision process resulted in the welding together of less dense materials (sediments, volcanics, and plutonic rocks) into new continent (Fig. 2.2b). The main orogeny, or "mountain building" episode, lasted from 1.74 to 1.65 Ga in the

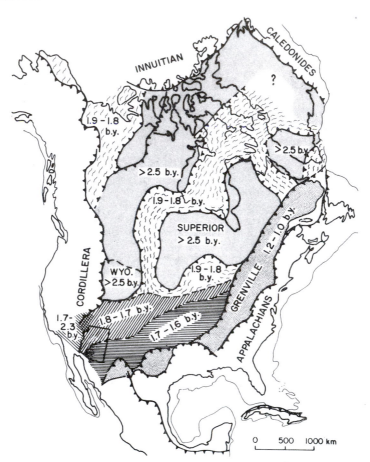

Figure 2.1. North American age provinces. The wide belt of Proterozoic crust in the southwestern United States was added to the Archean nucleus of the continent from 1.8 to 1.6 Ga. (Adapted from Hoffman 1988.)

Grand Canyon region, and it culminated with intense contractional deformation at 1.7–1.68 Ga. By 1.65 Ga, rocks were complexly deformed, metamorphosed, and beginning to cool off at depths of 10 km in the middle crust of the thickened orogen (Fig. 2.2b). This type of orogeny did not create high mountains like the Himalayas. Instead, thickening of originally thin crustal fragments appears to have resulted in a relatively low-elevation orogen, like the Indonesian region today (Bowring and Karlstrom 1990).

A continuing goal of our work in the Grand Canyon region is to identify the once-separate arc terranes, the tectonic boundaries between terranes (sutures), and the timing and geometry of collisions. Beyond tectonic models, our research is also designed to simply map, describe, and analyze the rocks in ways that will allow us to understand deep crustal processes, such as the symbiotic interactions between deformation, metamorphism, and plutonism (Williams and Karlstrom 1996; Karlstrom and Williams 1998). In working on such complexly

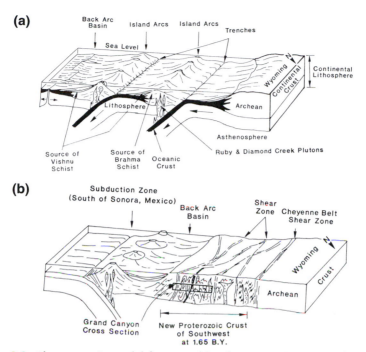

FIGURE 2.2. Plate tectonic model for assembly of Proterozoic crust in the Southwest. (a) Hypothetical geometry of island arcs south of the Archean nucleus at about 1.74 Ga. (b) The continent grew southward by assembly of arcs and accompanying compressional deformation between 1.74 and 1.65 Ga. Rocks of the Granite Gorge were at depths of 10–20 km in the middle crust (box) during assembly and stabilization of the continent.

deformed and metamorphosed rocks, it is important to realize that we cannot directly witness processes that operate at the 10- to 20-km depths from which these rocks have come, nor can we simulate in the laboratory the slow rates and interacting processes that went on during the 100-million-year plate collision period. Thus, many questions about the middle crust can only be answered by detailed collaborative studies such as this one. We approach the metamorphic rocks with a hope that while modern analogues may help us understand ancient mountain belts, the reverse is also true and the old eroded orogens offer a chance to understand tectonic processes that operate deep in the earth today.

PREVIOUS WORK AND NOMENCLATURE

The east–west trending Grand Canyon transect presents spectacular exposures of Paleoproterozoic rocks for 200 km across the ancient orogenic belt and our only window into the nature of the Precambrian rocks under the Colorado Plateau (inset to Fig. 2.3). In the Upper Granite Gorge, these rocks are continuously exposed from river mile 78 to 120 (Figs. 2.3 and 2.4). The Middle Granite Gorge has discontinuous exposures from mile 127 to mile 137. An isolated outcrop oc-

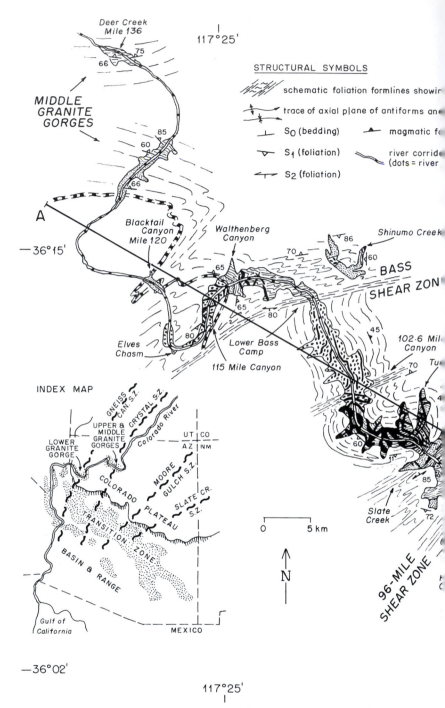

FIGURE 2.3. Geologic map of Paleoproterozoic rocks of the Upper and Middle Granite Gorges showing rock types, foliation patterns, and major shear zones. Index map shows location of Upper + Middle and Lower Granite Gorge transects, major physiographic provinces, Proterozoic rocks (stipple), and regional shear zones (S.Z.). (Adapted from Ilg et al. 1996).

112°00'

EXPLANATION

 Biotite ± muscovite granite and pegmatite:
1698 to 1662 Ma.

 Hornblende-biotite granodiorite, tonalite,
diorite, and gabbro: 1840 to 1713 Ma.

Ultramafic rocks: coarse-grained relict
cumulate textures.

Vishnu Formation: biotite-muscovite-quartz schist
and pelitic schist; local graded beds.

Rama and Brahma Formations: interlayered felsic
to mafic composition metavolcanic rocks: Rama
Fm. contains quartzofeldspathic gneiss and felsic
metavolcanics; Brahma Fm. contains amphibolite
and biotite schist. Orthoamphibole-bearing
gneisses (━━━) occur at base of Brahma Fm.
locally.

36°15' —

f and S$_2$

nforms

ion

s)

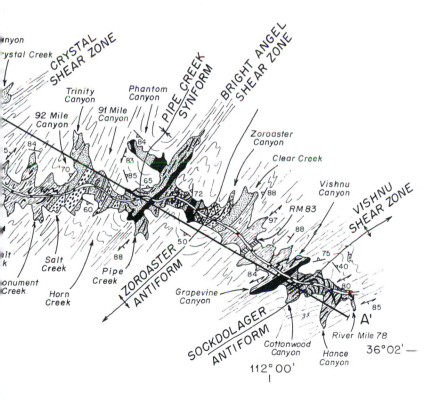

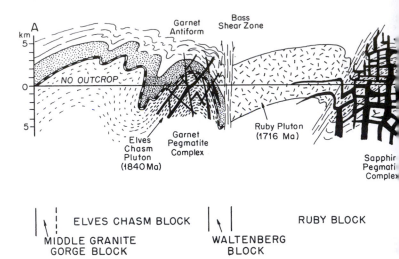

FIGURE 2.4. Cross section of Paleoproterozoic rocks of the Upper and Middle Granite Gorges showing rock types, foliation patterns, major shear zones and folds, and tectonic blocks; see Fig. 2.3 for explanation of rock units and Table 2.2 for description of blocks and shear zones. (Adapted from Ilg et al. 1996).

curs near the Hurricane Fault at mile 190–191. The Lower Granite Gorge contains near-continuous outcrops from mile 207 to mile 261 (Figs. 2.5 and 2.6).

To John Wesley Powell (1876), the Precambrian "granite" and "Grand Canyon schists" were dreaded because these harder rocks were associated with a narrower river and more difficult rapids. Walcott (1889) identified the Vishnu "terrane" as a complex of schist and gneiss. Since then, workers have continued to refine the subdivisions of the Precambrian rocks. Noble and Hunter (1916) provided the first detailed petrologic work by descending into side canyons from the south Tonto Platform (from Garnet Canyon east to Red Canyon). They identified domains of contrasting rock packages and recognized that differences reflected the presence of both metasedimentary and intrusive igneous rocks and that some of the gneisses possibly were a basement on which the metasedimentary units were deposited. Campbell and Maxson (1938), following six years of work and a three-boat expedition, identified different mappable units: Vishnu "Series" and Brahma "Series" (Maxson 1961). However, Campbell and Maxson underestimated structural complexities and probably overestimated stratigraphic thickness when they proposed that the combined stratigraphic sequence of metasedimentary and metavolcanic rocks was 8–16 km thick. This stratigraphic approach was called into question by Ragan and Sheridan (1970), who noted complex folding and interfingering of schist and amphibolite in the Phantom Ranch area. Subsequently, Brown et al. (1979) also emphasized the complex deformational features and lumped all of the metasedimentary and metavolcanic rocks under the name "Vishnu Complex."

Our approach (Ilg et al. 1996) recognizes the need to simultaneously pursue both tectonic and stratigraphic subdivisions of the Proterozoic rocks. The

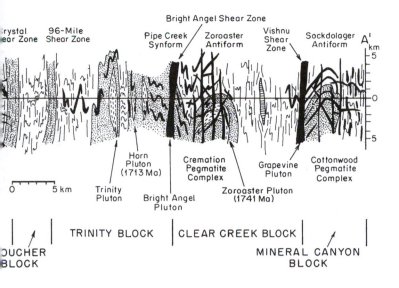

Bright Angel Shear Zone

Crystal Shear Zone 96-Mile Shear Zone Pipe Creek Synform Zoroaster Antiform Vishnu Shear Zone Sockdolager Antiform

A'
km
-5
-0
-5

Horn Pluton (1713 Ma) Cremation Pegmatite Complex Grapevine Pluton Cottonwood Pegmatite Complex

Trinity Pluton Bright Angel Pluton Zoroaster Pluton (1741 Ma)

0 5 km

BOUCHER BLOCK   TRINITY BLOCK   CLEAR CREEK BLOCK   MINERAL CANYON BLOCK

tectonic perspective emphasizes that the complex deformation and metamorphism in these rocks make it impossible to measure thickness and to confidently reconstruct regional stratigraphy. Furthermore, there are important shear zones boundaries of unknown displacement, but some possibly representing suturing of separate tectonic blocks that were once hundreds to thousands of kilometers apart (e.g., Crystal shear zone). Thus, we cannot be certain that all metasedimentary schists are strictly correlative (same age, same basin, same depositional sequence). Nevertheless, our work to date permits a stratigraphic interpretation (Noble and Hunter 1916) that metasedimentary rocks across the transect are broadly of similar rock type and age (1.75–1.73 Ga). Thus, the entire metasedimentary–metavolcanic package could have been deposited within 10–20 m.y. in a single arc basin or in similar, but tectonically separate, basins.

We proposed a new name, the Granite Gorge Metamorphic Suite, for metasedimentary and metavolcanic units in the Grand Canyon (Ilg et al. 1996). According to the U.S. Code of Stratigraphic nomenclature (Henderson et al. 1980), the term "metamorphic suite" can have the stratigraphic significance of "group" or "supergroup" and hence can be subdivided into mappable formations or groups. However, the term also implies complex structures, high-grade metamorphism, and difficulty in unraveling original stratigraphic relationships. We assign names for major mappable rock types in the Upper Granite Gorge: Brahma Schist for mafic metavolcanic rocks (after the Brahma "series" of Campbell and Maxson 1938) and Rama Schist for felsic metavolcanic rocks. We reserve the term Vishnu Schist for metamorphosed sedimentary rocks, as probably intended by Walcott (1894), recommended by Noble and Hunter (1916, their Vishnu schist), and proposed by Campbell and Maxson (1938, their Vishnu "Series"). Rocks in the Lower Granite Gorge also fall into these three basic lithologic groups.

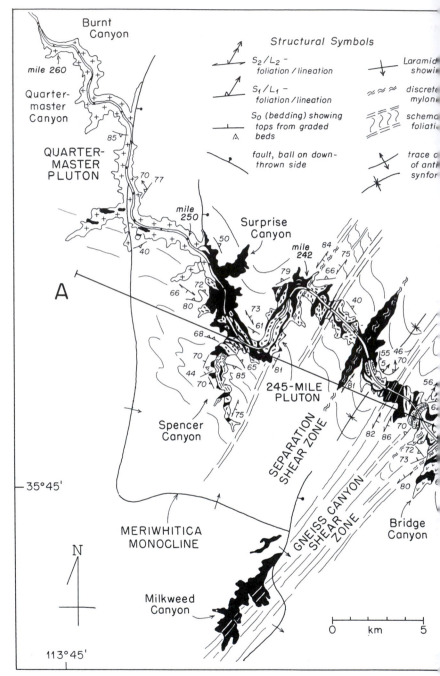

FIGURE 2.5. Geologic Map of the Proterozoic rocks of the Lower Granite Gorge showing rock types, foliation patterns, and major shear zones.

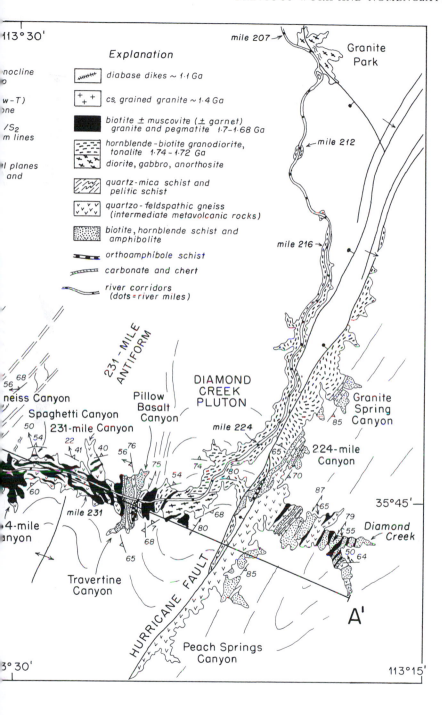

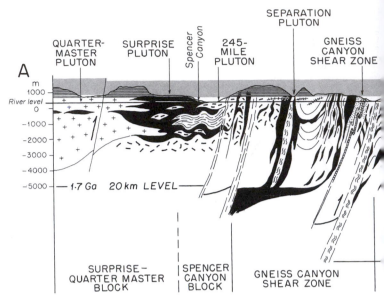

FIGURE 2.6. Cross section of the Lower Granite Gorge; see Fig. 2.5 for explanation of rock units and structural symbols, and see Table 2.2 for description of blocks and shear zones.

OLDER BASEMENT—1.84-GA ELVES CHASM PLUTON

All sedimentary and volcanic rocks must be deposited on some older substrate or "basement," but "basement" and "sedimentary cover" often get detached from each other and tectonically interlayered during deformation. High-grade metamorphism also obscures the nature of the original sedimentary and volcanic protoliths and the original contact relationships. Noble and Hunter (1916) posed this problem and speculated that some of the gneisses of the Grand Canyon might be basement for the schists. Subsequent workers recognized that gneisses are deformed intrusive rocks (Campbell and Maxson 1938; Brown et al. 1979; Babcock 1990), but there was still debate on whether the Elves Chasm and Trinity gneisses are possible basement for the Vishnu schist.

New mapping and geochronology indicate that the "gneisses" have different ages in different areas. The Trinity "gneiss" (Babcock et al. 1979) and the Zoroaster "gneiss" (Lingley 1973) are deformed and metamorphosed granodiorites that intruded into the Granite Gorge Metamorphic Suite. U-Pb zircon geochronology by Hawkins et al. (1996) shows that the Trinity pluton is 1.73 Ga and that the Zoroaster pluton is 1.74 Ga.

In contrast, the Elves Chasm pluton is 1.84 Ga, the oldest rock known in the southwestern United States and apparently the basement for the turbidites of the Vishnu Schist. The contact zone between the Elves Chasm pluton and the overlying Granite Gorge Metamorphic Suite is exposed in several areas—notably at Walthenberg, 113-mile, and Blacktail canyons—and several places in the Middle Granite Gorge (Fig. 2.3). The contact is gradational over an interval of several meters between the foliated pluton and a distinctive orthoamphibole-bear-

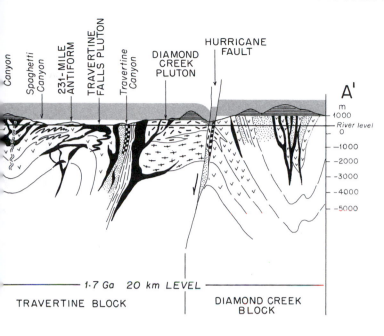

ing gneiss. The composition of this gneiss is unusual (high Ca/K ratio; Clark 1979). Babcock (1990) suggested this alteration took place during weathering, and new mapping suggests that the several occurrences of orthoamphibole gneiss in the Upper and Middle Granite Gorges may represent the same weathering zone repeated by folding (Figs. 2.3 and 2.4). This zone separates the 1.84-Ga pluton from Vishnu Schist that contains 1.75-Ga detrital zircon (D.P. Hawkins, unpublished data from Walthenberg Canyon), indicating that the Vishnu sediments were probably deposited on the older basement.

GRANITE GORGE METAMORPHIC SUITE
(1.75–1.73 GA)

Metasedimentary and metavolcanic rocks of the Granite Gorge Metamorphic Suite make up about half of the exposed rocks in the Grand Canyon, the rest being intrusive rocks (discussed below). Overall, the Granite Gorge Metamorphic Suite consists of a variety of schists containing quartz, feldspar, micas, and other metamorphic minerals. In the following discussions, we use descriptive metamorphic rock names for rocks we see in outcrop or thin section. We also infer the original sedimentary or volcanic "protolith" based on rock composition and a limited number of primary structures that survived the deformation and metamorphism.

The Rama Schist includes quartzofelspathic schist and gneiss with locally preserved phenocrysts of quartz and feldspar and possible relict lapilli (e.g., in Shinumo and Diamond Creek) that suggest a felsic to intermediate volcanic origin. Felsite layers within Brahma amphibolite at Clear Creek yield a zircon date of 1.75 Ga, quartzofelspathic gneiss in the Sockdologer antiform give a date of 1.74 Ga, and an intermediate-composition dike in "Pillow Basalt Canyon" (mile

229.5, Fig. 2.5) gives a date of 1.73 Ga. Thus, available geochronology suggests that there may be more than one age of felsic to intermediate volcanic rocks within the 1.75- to 1.73-Ga interval.

The Brahma Schist consists mainly of hornblende–biotite schists and amphibolites. Orthoamphibole-bearing schists are interbedded with amphibolites in several places (Bright Angel, Travertine Canyon) and are also included in the Brahma Schist. Geochemistry suggests that the massive amphibolites have tholeiitic character, compatible with an origin as island arc basalts (Clark 1979). Primary features that indicate the nature of the depositional setting are preserved in several places. For example, pillow structures in amphibolites indicate that mafic lava flows erupted under water. These are found in Clear Creek (Campbell and Maxson 1938), Horn Creek, 92-Mile Canyon, Crystal Creek, Slate Creek, Shinumo Creek, near Blacktail Canyon, and "Pillow Basalt Canyon" (mile 229.5). The latter locality also contains volcanic breccias that were fragmented in the presence of seawater.

The Rama and Brahma metavolcanic schists can be complexly interlayered. We lump them together in some generalized maps (Fig. 2.3), although they are generally separable at 1:24,000 scale. None of the mafic rocks are directly dated, but they are intruded or interlayered with 1.75-Ga felsic rocks near Clear Creek and intruded by a 1.73-Ga dacite dike in Pillow Basalt Canyon. Contact relationships support variable relative ages between mafic and intermediate metavolcanic rocks. In the Upper Granite Gorge, Rama Schist is underneath and older than the Brahma Schist in the Sockdolager antiform and Shinumo areas. In the Diamond Creek and Travertine areas of the Lower Granite Gorges, intermediate volcanic rocks and dikes appear younger than the amphibolites based on younging of pillows and intrusion of the 1.73-Ga dacite dikes into amphibolites.

The Vishnu Schist consists of thick sections of quartz-mica schist and pelitic schist that are interpreted to be metamorphosed sandstones and mudstones that were deposited in submarine conditions on the flanks of the eroding oceanic islands. Thick (several kilometers) sections show rhythmic bedding and graded bedding suggesting deposition as submarine turbidites. The absence of conglomerates and general fine grain size suggests a lack of high-energy proximal facies. Calc silicate pods are numerous and are interpreted to be concretions. Preserved graded bedding indicates that it was deposited stratigraphically above the Brahma Schist in the Walthenberg area. Detrital zircon crystals from the Vishnu Schist of the Upper Granite Gorge range from 3.3 Ga to 1.736 Ga. The older grains were probably transported from the Archean Wyoming province; the younger grains (1.74 Ga) are likely close to the depositional age of the Vishnu Schist, because these rocks overlie the 1.74- to 1.75-Ga Brahma Schist and were cross cut by the 1.74-Ga arc plutons (Trinity and Zoroaster plutons) that contain schist and calc-silicate xenoliths.

INTRUSIVE ROCKS

Intrusive rocks make up the other half of the crystalline rocks of the Grand Canyon. Campbell and Maxson (1933) thought there was a single major period of igneous "invasion," and this led to the convention of lumping all plutonic rocks of the Grand Canyon under a single name such as "Zoroaster Gneiss" (Campbell and Maxson 1933), "Zoroaster Granite" (Maxson 1968), and "Zoroaster Plutonic Complex" (Babcock et al. 1979). However, new mapping (Ilg et al. 1996) and geochronology (Hawkins et al. 1996) show that plutonic rocks record a long and complex evolution of the crust. For example, magma crystallized at numerous times between 1.84 and 1.37 Ga, and petrology and geochemistry indicate a wide variation in types of intrusions. Thus, a single rock name is misleading, and we prefer to use the names of individual plutons or dike swarms (Table 2.1). We subdivide the intrusive rocks into four groups of plutons, based on age and tectonic groups (modified from Babcock 1990; Table 2.1).

Older Basement

The 1.84 Elves Chasm pluton is part of an older basement terrane. This pluton is dominantly hornblende–biotite tonalite to quartz diorite. It represents a part of an older arc terrane upon which younger arcs and arc sediments were deposited. The Elves Chasm pluton is distinguished geochemically from other plutons in the Grand Canyon by its lower concentration of large ion lithophile elements and its lower concentration of light rare earth elements relative to heavy rare earth elements. These characteristics may suggest that the Elves Chasm pluton bears a less direct genetic relationship to slab subduction than that shown by younger arc plutons.

Arc Plutons

The 1.74- to 1.71-Ga granodiorite (plus gabbro-diorite) complexes are interpreted as arc plutons, compatible with petrologic and geochemical data (Babcock 1990). These plutons formed from melting above the subducting plate, then melts rose to form large magma chambers that, in turn, fed volcanic eruptions within the island arcs. We infer that they were emplaced at shallow levels in arcs because of an absence of contact metamorphic aureoles, but their original shape is not preserved. Some are now large folded sheet-like plutons (Zoroaster, Trinity, and Ruby plutons), whereas others are massive differentiated plutons (Diamond Creek pluton) or smaller stock-like bodies (Pipe Creek, Horn Creek, Boucher, and Crystal plutons). One characteristic of many of these plutons is the presence of enclaves of a range of compositions (gabbro to granodiorite) that record co-mingling of magmas within the arc magma chamber (e.g., Ruby and Diamond Creek plutons). Cumulate-textured ultramafic rocks found in several places (mile 81, 83, 91, Salt Creek, Crystal, Granite Park, Diamond Creek), typically as tectonic slivers, are interpreted to be variably dismembered parts of the base of these arc plutons, rather than as ophiolite fragments. This interpretation is based upon the abundance of phlogopite and light rare earth elements in wehrlite successions, implying a geochemical contribution from subducting slab material. Compositional relations of the mile 83, 91, and Crystal ultramafic bodies are consistent with their origin as a single cumulate succession related by fractional crystallization (Seaman et al. 1997). The arc plutons are relatively rich in the feldspar minerals and hence were relatively strong during deformations. Thus, some of these plutons are not as obviously deformed (foliated) as the schists. However, they were all intruded before the period of intense 1.70- to 1.68-Ga deformation.

Syncollisional Granites

The 1.71- to 1.66-Ga granites and granitic pegmatites have a different composition, intrusive style, and deformational character than the arc plutons. They are

TABLE 2–1. Plutons and Dike Swarms

Name	Mile	Age-Ma
Cottonwood Pegmatite Complex	77–82	1685 ± 1; 1680 ± 1
Grapevine Camp Pluton	**81.5**	**1737**
83-mile ultra-mafic	83	Undated
Zoroaster Pluton	**84.6–85.5**	**1740 ± 2**
Cremation Pegmatite Complex	84–88	1698 ± 1, with older inherited monazite cores
Bright Angel Pluton	88 and Bright Angel Creek	Undated
Phantom Pluton	Phantom Canyon	1662 ± 1
Pipe Creek Pluton	89	Pb–Pb dates of 1.74–1.69 Ga
Horn Creek Pluton	Near 90.6	1713 ± 2
91-mile ultramafics	**91**	**Undated**
Trinity Pluton	**91.5**	**1730 ± 93**
Boucher Pluton	**96.2**	**1714 ± 1**
Crystal Pluton	**97**	**Undated**
Tuna Pluton	**99**	**1750–1710 Ga, with inherited zircons >2.0 Ga**
Sapphire Pegmatite Complex	99–104	Undated
Ruby Pluton	**102–108**	**1716 ± 0.5**
Garnet Pegmatite Complex	111–115	1697 ± 1

Composition	Deformation	Interpretation
Biotite and biotite–muscovite granite and granitic pegmatite dikes and sills	Dated dikes cross-cut S2 and are weakly deformed	Late syn-D2
Medium-grained foliated biotite granite	**Contains S1; alligned along S2 Vishnu shear zone**	**Age and fabric suggests pre-D1 arc pluton, but composition similar to peraluminous plutons**
Layered ultramafic containing cumulate layers of olivine and pyroxene	Layering parallel to S1, truncated by S2-parallel margins	Tectonic slice of cumulate rocks from an arc pluton
Foliated medium-grained biotite granite to granodiorite orthogneiss	**Contains S1, folded by F2**	**Pre- or syn-D1**
Fine- to medium-grained biotite–muscovite granite and granite pegmatite	Cross-cuts S1, contains weak S2	Late syn-D2
Coarse-grained friable granite and pegmatitic granite	Intruded along Bright Angel shear zone; contains weak S2	Late syn-D2
Fine- to medium-grained biotite–muscovite granite	Cross-cuts S2, contains synmagmatic shear zones	Syn-D3
Granite to granodiorite	Contains S1, folded by S2	Pre-D1
Foliated medium-grained hornblende quartz diorite to tonalite	Contains magmatic S1, solid-state S2	Syn-D1
Cumulate layered ultramafic	**Contains S1, boudinaged by D2**	**Pre-D2 tectonic sliver of arc pluton cumulate rock**
Medium- to coarse-grained biotite granodiorite to granite orthogneiss	**Contains S1 and S2**	**Pre-D1**
Granodiorite to tonalite	**Weakly foliated**	**Pre- or syn-D1**
Granite to granodiorite	**Weakly foliated**	**Syn-D1**
Medium-grained foliated granodiorite	**Contains S1 and S2**	**Pre- to syn-D1**
Granite and granitic pegmatite dikes and sills	Cross-cut S1, contain S2	Syn-D2
Hornblende–biotite granodiorite, diorite, and **gabbro**	**Contains S1 as** magmatic **layering**	**Syn-D1**
Biotite and biotite–muscovite granite and granite pegmatite dikes and sills	Cross-cuts S1, weak S2	Syn-D2

(*continued*)

TABLE 2–1. Plutons and Dike Swarms (*Continued*)

Name	Mile	Age-Ma
Elves Chasm Pluton	**112–118**	**1840 ± 1**
Granite Park Mafic Complex	207–209	**Undated**
Diamond Creek Pluton	216–227	**1736 ± 1**
229-mile granite	228.7	Undated
Travertine Falls Pluton and dike complex	230–231	1704 ± 1
232-mile Pluton and dike complex	231.7–232.4	Undated
234-mile Pluton and dike complex	233.5–235	Undated
237-mile Pluton	236.7–237	Undated
Separation Pluton	239.3–239.8	Undated
245-mile pluton	**242–246**	**1720 ± 5**
Spencer Pluton and dike complex	242–245 and in upper Spencer Canyon	Undated
Surprise Pluton	246–252	Undated
Quartermaster Pluton	Injects into Surprise Pluton from 250–261	1375

Data from Clark 1976; Babcock et al. 1979; Babcock 1990; Ilg et al. 1996; Hawkins et al. 1996; Hawkins 1996. **Bold = metaluminous arc plutons**; light = peraluminous granites; * = 1.4 Ga granite.

Composition	Deformation	Interpretation
Lineated and foliated hornblende–biotite tonalite to quartz diorite and granodiorite, shows minging textures between intermediate and mafic units	**Contains S1 and S2**	**Pre-D1, older arc basement**
Alternating layers of gabbro, anorthosite, and granodiorite with gabbroic pegmatite	**Contains S1-related, top to SW, thrust-sense shear zones**	**Pre-D1 layered Mafic, may be base of Diamond Creek pluton**
Granodiorite, tonalite, diorite and gabbro, shows mingling textures and cumulate ultramafic	**Contains magmatic S1 and solid-state S2 shear zones**	**Syn-D1**
Fine- to medium-grained biotite granite	Contains strong S2 foliation and S2-parallel psuedotachylite	Syn-D2
Medium-grained biotite granite	Injects into and cross-cuts S1, contains variable S2	Syn-D2
Medium-grained biotite granite	Injects into and cross-cuts S1, contains variable S2	May be same as Travetine Falls Pluton, syn-D2
Medium-grained biotite granite	Cuts S1, contains variable S2	May be same as Travetine Falls Pluton, syn-D2
Medium-grained biotite granite	Cuts S1, contains weak and magmatic S2	Syn-D2
Coarse-grained to megacrystic biotite, (+muscovite) granite	Contains strong S2	Syn-D2
Granodiorite, tonalite, to diorite mingled Mafic Pluton intruded by Spencer Pluton	**Contains S1 and S2**	**Pre-D2**
Medium-grained biotite (muscovite) granite	Contains variable S2	Syn-D2
Medium grained biotite + muscovite (garnet) granite	Contains weak S2	Syn-D2, same as separation Pluton
Coarse-grained granite	Cross-cuts S2	Post-D2

biotite–granite, biotite–muscovite (garnet) granite, and granitic pegmatite that probably formed by partial melting of the lower crust during deformation. Melts rose opportunistically along cracks and shear zones and froze to form dikes or coalesced as small stocks and plutons. It is common for a network of granite dikes to locally make up more than half of the exposed rock volume, blurring the distinction between dike swarms and plutons and indicating that these plutons coalesced from dikes. In the Upper Granite Gorge, intrusive complexes are made up of nearly equal proportions of medium-grained granite and granitic pegmatite. These are named the Cottonwood, Cremation, Sapphire, and Garnet Canyon complexes (Fig. 2.3). In the Lower Granite Gorge, intrusives are more massive and pluton-like (e.g., Travertine Falls, Separation, and Surprise plutons; Fig. 2.5) and commonly are composed of sheets concordant to foliation.

Granites are commonly stretched (boudinaged), folded, and foliated, but also occur as undeformed cross-cutting tabular dikes and orthogonal dike networks. These relationships suggest that they were intruded during regional deformation. Plutonism was also synchronous with peak metamorphism; the heat from the molten rock combined with high ambient temperatures at depth caused chemical reactions and growth of metamorphic minerals in the schists. Highest metamorphic temperatures are recorded in areas of most voluminous granites indicating that granites created zones that were >200°C hotter than surrounding areas (see below). Based on detailed studies, the range of ages of these dikes (1.70–1.68 Ga) is thus interpreted to also be the time of peak metamorphism and the main contractional deformation. Plutonism continued until 1.66 Ga (Phantom Pluton), and local deformation also continued. Cessation of magmatism, deformation, and metamorphism by ~1.65 Ga is interpreted to be the beginning of a long period (200 m.y.) of tectonic stability of this newly formed part of continental North America.

Post-Orogenic Granites

The youngest plutonic rock in the Grand Canyon is the 1.35-Ga Quartermaster pluton (S.A. Bowring, unpublished U/Pb zircon data) and related pegmatites (Fig. 2.5). These represent part of a regionally important period of intracratonic magmatism in the southwestern United States. The Quartermaster pluton is a coarse-grained (rarely megacrystic) friable granite that intrudes and is intermixed with older biotite granite of the Surprise pluton from mile 250–261. A pegmatite dike of similar age was dated in Diamond Creek, and several other subhorizontal dikes of pegmatite in the Lower Granite Gorge may also be this age. These plutons and dikes postdate the main contractional deformation as shown by cross-cutting relations and a lack of fabric. There are no 1.4- to 1.3-Ga plutonic rocks yet known from the Upper Granite Gorge.

DEFORMATIONAL HISTORY

Rocks of the Granite Gorge Metamorphic Suite were deposited at the earth's surface (1.75–1.74 Ga) and then buried to depths of 20–25 km and complexly squeezed by thrusting and folding (Fig. 2.2b). Metamorphic data suggest they were exhumed part way back to the surface (to 10-km depths) during the main deformation at 1.7–1.68 Ga. The tectonic fabrics that give rocks of the Granite Gorge Metamorphic Suite their distinctive vertically layered appearance were formed in the middle crust during a progression of tectonic events from 1.74 to

1.66 Ga. Deformations took place throughout this time interval whenever the stresses generated by arc collisions exceeded the strength of the crust. Thus, deformation was accentuated when rocks were hot and ductile—for example, in areas heated by magmas. The main deformational features are folds and foliations that record NW–SE squeezing and shear zones that record shearing along boundaries of tectonic domains. All of these structures formed when the rocks were solid, but had consistencies something like stiff Silly Putty. The final structures record deformation over a long time interval, variations in temperature and rock strengths from place to place and time to time, and changing stress fields involving combinations of squeezing and shearing. There is also evidence for an interaction of brittle and ductile deformation; for example, dike swarms filled a complex network of cracks that developed even as the rocks were deforming ductilely. The result is a very complex set of structures. We begin to understand the history of this deformation by recognizing a number of generations of structures that can be distinguished based on their style and overprinting relationships (hence relative age; Fig. 2.7).

The most obvious result of cumulative deformation is a profound verticality of tectonic layering below the Cambrian unconformity. This NE-striking vertical fabric is a second-generation composite (S2) fabric that records two effects: (1) the rotation and transposition of original bedding and an earlier tectonic layering (S1) to a subvertical orientation during development of tight folds (such as the Sockdologer and Zoroaster antiforms (Fig. 2.3) and (2) alignment of micas and other inequant grains perpendicular to tectonic shortening. This fabric formed synchronously with metamorphism and granite intrusion mainly during a 15-m.y. period from 1.7- to 1.685-Ga. Evidence for this is that melt pods that give U-Pb dates of 1.7–1.69 Ga are folded and boudinaged, then cross-cut by unfolded dikes of granite that give U-Pb dates of 1.685–1.68 Ga (Hawkins et al. 1996).

The main northeast-trending subvertical foliation and its associated folds are second-generation structures (S2 and F2, respectively) that fold an earlier tectonic layering (as well as bedding). This early S1 fabric can be seen in areas that escaped intense shortening (Sockdologer synform; Walthenburg area) and as aligned mineral trails preserved within the protective housing of metamorphic minerals such as garnet (Fig. 2.8). In these areas, the early (S1) foliation is commonly subparallel to bedding and has a northwest strike and shallow dip. However, in other areas (e.g., Phantom Creek) S1 is also subvertical, presumably due to its rotation during D2. The regional foliation pattern in both the Upper and Lower Granite Gorges (Figs. 2.3 and 2.5) involves domains of northwest-striking S1 that are folded and variably reoriented into the northeast-striking subvertical S2 orientation. We speculate that early (D1) tectonism involved thrusting and subhorizontal foliations, but the detailed configuration of early structures, such as direction of movement of thrusts, has not yet been worked out, mainly due to the intensity of the later (D2) shortening.

D3 structures are defined as ones that fold or deflect the main S2 subvertical layering. These include northwest-striking crenulations, late synplutonic mylonite zones (e.g., in the cores of 1.68-Ga dikes and in the 1.66 Phantom pluton), and low-temperature shear zones that reactivated older zones and may have allowed differential uplift of blocks (e.g., in the Gneiss Canyon shear zone). This group of structures represents several styles and periods of deformation, including both continued 1.68- to 1.65-Ga shortening and possible 1.4-Ga deformation. For the low-T (greenschist grade, <450°C) shearing in the 96-mile shear zone and in several discrete zones within the Gneiss Canyon shear zone (Fig. 2.5), kinematic indicators suggest west-side-up and dextral oblique shear.

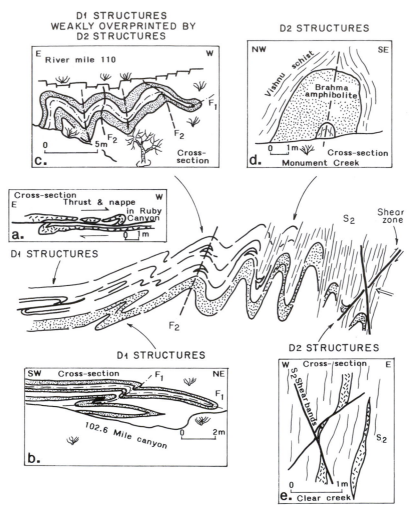

FIGURE 2.7. Overprinting of folds and fabrics and nomenclature for unraveling polyphase deformation. D1 involved thrusting, F1 recumbent folding, and development of S1 foliation; D2 involved NW–SE shortening, development of upright F2 folds, and dominantly west-side-up shear on shear zones; D3 involved minor crenulations and adjustments. Examples from outcrop photos: (a) and (b) recumbent F1 folds in Ruby Canyon and 102.6 Mile Canyon, respectively; (c) F1 recumbent fold refolded by F2 upright folds in area between Shinumo and Walthenberg canyons; (d) F2 fold with axial plane cleavage developed in metasedimentary rocks but not amphibolite, Monument Creek; (e) intense S2 and conjugate shear bands developed in Clear Creek Canyon. (Adapted from Ilg et al. 1996.)

TECTONIC BLOCKS

As is true throughout the southwestern United States, the Proterozoic crust in the Grand Canyon is segmented into blocks of different character by shear zones (Karlstrom and Williams 1998). Figures 2.3 and 2.5 show the main blocks and shear zones of the Grand Canyon transect; Table 2.2 summarizes tectonic evo-

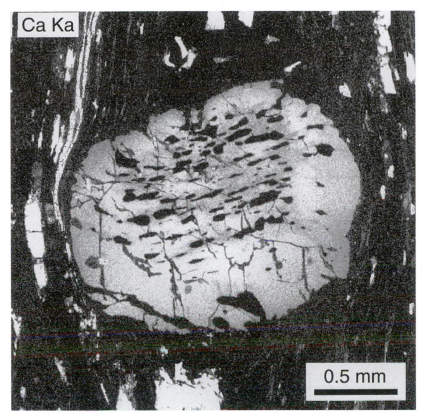

FIGURE 2.8. Electron microprobe, Ca Ka map showing compositional zoning in a single garnet crystal from mile 82. The thin-section view is approximately vertical and perpendicular to NE-striking S2 foliation. Note that the core of the garnet overgrew a slightly folded (S1) fabric. This early fabric was folded and transposed into the vertical S2 orientation now seen in the garnet rim and outside of the crystal. The increasing Ca content of the garnet, along with the composition of other associated minerals, suggests that the garnet core grew as temperature and pressure increased. Peak conditions of approximately 550°C and 6 kbar (20-km depths) were reached during the growth of the outer part the garnet, but the compositions of the extreme garnet rims indicate that pressures decreased during the final increments of growth. The composition of matrix minerals indicate that the final matrix recrystallization occurred at conditions near 550°C and 3 kbar (10-km depths). Thus, this one garnet crystal and its matrix record increasing pressures and the subsequent decompression from 20- to 10-km depths all during D2 deformation. The compositional map was collected on a Cameca SX-50 electron microprobe at the University of Massachusetts.

lution of blocks and shear zones. As discussed earlier, there is a general similarity of metasedimentary and metavolcanic rock types, similar types and ages of intrusive rocks, and similar styles and generations of deformational fabric across the transect. However, there are also important differences between blocks that might possibly indicate juxtaposition of once-separate tectonic terranes. The next sections discuss two ways that blocks differ: their isotopic compositions and metamorphic P–T histories.

TABLE 2–2. Blocks and Shear Zones

Name	Mile	Rocks and Structures
Mineral Canyon block	Mile 77–81	Migmatitic Rama–Brahma–Vishnu schists, intimately injected by the Cottonwood granite/pegmatite swarm, all folded into kilometer-scale moderately NE-plunging F2 Sockdologer antiform.
Vishnu shear zone	Mile 81	Narrow NE-striking subvertical fault zone with quartz veins marks juxtaposition of Grapevine Camp pluton abruptly against Vishnu Schist; kinematic history poorly known due to brittle overprint and likely multiple movements at high and low temperature.
Clear Creek block	Mile 81–88	Vishnu Schist with tectonic sliver of ultramafic rock at mile 83; Vishnu is structurally underlain by Brahma Schist then Zoroaster pluton in core of SW-plunging F2 Zoroaster antiform.
Bright Angel shear zone	Mile 88	2-km-wide zone of NE-striking subvertical S2 fabric that transposes NW-striking subvertical S1 fabric. The shear zone contains tectonically intermixed Rama, Brahma, and Vishnu schists, complexly injected and deformed peraluminous granite and pegmatite of the Cremation swarm, and the Bright Angel pluton that is elongated parallel to the zone.
Trinity block	Mile 88–96	Large refolded folds of Brahma and Vishnu schist, the Trinity orthogneiss, and an ultramafic sliver at mile 91.
96-mile shear zone	Mile 96	300-m-wide NE-trending subvertical mylonite zone in Vishnu and Brahma schist and metadacite. Shear sense is west-side-up and dextral oblique slip.
Boucher block	Mile 96–98	Vishnu Schist intruded at the east and west by arc plutons; S2 dominates, but earlier fabrics evident in Boucher Canyon.
Crystal shear zone	Mile 98	1-km-wide, NE-striking subvertical zone of intense S2 fabric. Heterogeneous lithologies include various schists, granodiorite, ultramafics; anastomozing foliation and lenticuar boudins of diverse lithologies in chlorite schist form tectonic melange.
Ruby block	Mile 98–108	Vishnu Schist intruded by plutonic rocks of the Tuna Creek and Ruby plutons, all intruded at the east of the block by the Sapphire Pegmatite Complex. Moderately south-dipping S1 foliation folded into kilometer-scale F2 folds with S2 cleavage best developed in schist.

Metamorphism and Cooling History

Upper amphibolite grade conditions of 600–720°C, 6 kbar were reached syn-D2 at 1.72–1.695 Ga; rapid cooling to 600–550°C by 1690 Ma; slower cooling to 550–500°C with continued fluid flux until 1660; cooling through 350°C at 1.4 Ga.

Migmatites on the east are juxtaposed with amphibolite-facies rocks on west so zone marks a ~100°C temperature discontinuity; presence of sillimanite in schist near the fault may reflect heating during juxtaposition in D2 time.

Except for higher-grade rocks against the Vishnu fault, peak metamorphic grade increases toward the west, from 500 to >600°C, all at 6 kbar. Volume of peraluminous granite also increases westward, and a 1698-Ma date on the granite is interpreted to be the time of peak metamorphism and D2 shortening, in agreement with U-Pb sphene dates of 1.7–1.695 Ga at mile 84. Late andalusite in Clear Creek Canyon (mile 84) and garnet zoning indicate decompression from 6 to 3 kbar during D2.

Upper amphibolite grade, with ubiquitous sillimanite and migmatites showing evidence for partial melting of schist; parallel magmatic and solid-state fabrics suggests that movement was synchronous with granite emplacement.

Upper amphibolite grade, 650–725°C, 6 kbar; 1662 Phantom Creek Pluton is the youngest peraluminous granite in the Canyon. Metamorphic zircon, monazite, xenotime, and sphene give dates of 1660–1550 Ma with a systematic decrease in metamorphic age from east to west, apparently unrelated to the density of peraluminous dikes; the block may have been tilted 1.55–1.4 Ga; muscovite Ar–Ar dates of 1.4 Ga suggest cooling through 350°C at 1.4 Ga.

Extreme grain size reduction and greenschist grade indicate low-T (~450°C) movement; 1.4-Ga muscovite Ar dates to east versus older dates to west suggest post-1.4-Ga movement.

Greenschist to lower amphibolite amphibolite grade, $T \sim 500°C$, $P \sim 6$ kbar. U-Pb sphene dates yield the same age as the Boucher Pluton (1714 Ma), suggesting that the pluton cooled rapidly below ~600°C. Ar–Ar muscovite dates of >1.6 Ga indicate that rocks cooled through 350°C by 1.65 Ga and remained cold while rocks east of the 96-mile shear zone were at high grade from 1.66 to 1.55 Ga and >350°C until 1.4 Ga.

Movement sense and timing poorly constrained; presence of melange and abrupt change in Pb isotopes across the zone suggests it may be a paleosuture zone.

Metamorphic grade highest in area of pegmatite complex; about 600–700°C and 6–8 kbar. Metamorphic sphene from the ~1.72-Ga Tuna Pluton yields U-Pb dates of about 1.70 Ga and apatites from the pluton yield U-Pb dates of about 1.65 Ga. These data suggest that peak metamorphism occurred at about 1.70 Ga and the rocks cooled through about 500°C by 1.65 Ga, in agreement with >1600 Ma Ar–Ar dates on micas

(continued)

TABLE 2–2. Blocks and Shear Zones (*Continued*)

Name	Mile	Rocks and Structures
Bass shear zone	Mile 107.8–108.2	0.5-km-wide ENE-striking subvertical zone within Vishnu Schist on west margin of Ruby Pluton; pluton itself is internally undeformed; anastomosing shear bands and foliation fish give dextral shear sense.
Walthenberg–Shinumo block	Mile 108–112 and in Shinumo Creek	Brahma and Rama schists overlain stratigraphically by Vishnu Schist near Walthenberg and in Shinumo Creek; kilometer-scale nappe-like F1 folds with moderately north-dipping axial planes are nearly E–W against Bass shear zone.
Contact Zone of Elves Pluton	Mile 112–113, Waltenberg Canyon area, in 115-mile Canyon, and Middle Granite Gorge	Concordant contact between the Elves Chasm Pluton and overlying Granite Gorge Metamorphic Suite is a 0.5-km-wide transitional contact marked by distinctive orthoamphibole-bearing gneisses and evidence for in situ partial melting. Contact is folded into large-scale F2 folds with a generally northwest-striking fold envelope. The contact may be a paleosol indicating deposition of Vishnu Schist on older basement, or a high-temperature shear zone, or both.
Elves Chasm block	Mile 113–127	1.84-Ga Elves Chasm Pluton contains mafic and intermediate plutonic units, including tabular amphibolite bodies that are interpreted to be dikes; all units are strongly foliated and lineated by S1 and L1; voluminous dikes of the Garnet Canyon Pegmatite Complex cross-cut S1; contact zone is exposed near Blacktail Canyon.
Middle Granite Gorge block	Mile 127–139	Massive amphibolites are overlain by felsic volcanics, then turbidites near Bedrock rapid; section probably correlates with Walthenburg block, numerous granite dikes intrude metasedimentary schists.
Granite Park block	Mile 209	Cumulate layered gabbro, anorthosite, amphibolite, and gabbroic pegmatite; rocks are strongly foliated and contain thrust sense top-to-west shear zones.
Diamond Creek block	East of Hurricane fault, in Diamond Creek	Thick sections of intermediate volcanic schists interlayered with amphibolite (with local pillows); east side of the block cross-cut by peraluminous granite dikes; intensely developed NE-striking subvertical foliation may mark a ductile high strain zone parallel to the Hurricane fault.
Travertine block	Mile 212–234	Diamond Creek Pluton exposed from mile 212–228.5, mingled magma textures in area 227–228. Travertine Canyon area contains Vishnu Schist overlain on the west by pillow basalts, then intermediate volcanics. Area from 230–235 is Travertine Falls, 232-mile and 234-mile plutons, and related dikes and sills intruding metavolcanic rocks.

Metamorphism and Cooling History

Shearing took place at high grade as shown by abundant sillimanite and K-spar and leucosomes that may be partial melt pods. Lower T shearing is indicated by chlorite and sericite alteration.

A 1697 U-Pb date on a syn-D2 pegmatite suggests D2 accompanied metamorphism (6–8 kbar, 550°C) at 1.7–1.69 Ga.

Peak metamorphic grade was approximately 600°C, 8 kbar, but garnet shows evidence for polyphase growth. Metamorphic monazite from pelitic schist in the contact zone yield U-Pb dates of 1700 Ma, which are interpreted as the time of peak metamorphism. Garnet with coronas of granitic leucosome suggest in situ partial melting of the Elves Chasm contact zone, but melting is interpreted to record channelization of heat and fluids in the contact zone at 1.7 Ga rather than contact metamorphism at 1.84 Ga. Kyanite is abundant, as is gedrite, with late-stage growth of sillimanite and cordierite indicating syn-S2 decompression to midcrustal levels.

Metamorphism was high grade as suggested by migmatitic gneisses in pluton margins; U-Pb data from sphene within the pluton suggest that rocks cooled through about 600°C 1675–1655 Ma.

Metamorphism amphibolite grade.

Metamorphic grade is lower amphibolite, 500°C, 3 kbar.

Metamorphic grade at Travertine Canyon is 550°C, 3 kbar; grade increases toward the plutons in the western part of the block; sphene in the Travertine area gives a U-Pb date of 1698 ± 7 Ma; hornblende in shear zones in the Diamond Creek Pluton gives an Ar age of 1.67 Ga. The former gives the time of peak metamorphism; the latter gives the time of cooling through about 500°C.

(continued)

TABLE 2–2. Blocks and Shear Zones (*Continued*)

Name	Mile	Rocks and Structures
Gneiss Canyon shear zone	Mile 234– 242.2	Migmatitic gneisses of mixed metasedimentary and metavolcanic protolith intimately injected with granites. Carbonate at mile 236.5 may mark important tectonic boundary. Structures consist of NE-striking subvertical S2 foliation zones separating domains of NW-striking folded S1 fabric; shear sense in the shear zones is NW-side up and dextral oblique slip. Discrete retrograde mylonites occur at mile 234, 235, 236, 239.5, and 242.2 and show dextral, west-side-up shear. 1.1-Ga diabase dikes are restricted to this zone
Spencer Canyon block	Mile 242.2–247	Spencer Canyon metasedimentary and metavolcanic gneisses are intruded by the 246-mile mafic arc pluton (1720 Ma) and by granite dikes of the Spencer pluton.
Surprise–ton; Quartermaster block	Mile 247–261	From mile 246–252, rock is mostly Surprise Pluton; from 252–261, 1375-Ga Quartermaster pluton intrudes the Surprise Pluton. Foliation is mainly weak magmatic layering that is NW striking and moderately dipping.

Data from Ilg et al. 1996; Hawkins et al. 1996; Williams, unpublished data; Hawkins, 1996; Hawkins and Bowring, in press; Hawkins and Bowring, unpublished data.

ISOTOPIC BOUNDARIES

Although rocks are broadly similar across the transect, different isotopic "signatures" suggest that blocks were derived from crustal domains of different age and composition. On a regional scale, Bennett and DePaolo (1987) and Wooden and Dewitt (1991) showed that rocks of the Mojave province of northwestern Arizona and California had isotopic signatures that indicated that there was a larger percentage of older crustal material at depth than beneath the Yavapai province of central Arizona. The nature of older material is not well known but probably includes Archean basement fragments at depth, more voluminous 1.84-Ga arc basement like the Elves Chasm gneiss at depth, or an increase in Archean detrital zircon grains deposited in the schists.

In the Grand Canyon transect, there are two shear zone boundaries that mark changes in isotopic character of the crust. Rocks in the eastern Grand Canyon (east of the Crystal shear zone—mile 98) have Pb isotopes similar to the Yavapai province of central Arizona that suggest that arc rocks were derived from the mantle at ~1.75 Ga without contamination from older crustal rocks. In contrast, rocks from mile 98–115 show elevated Pb signatures indicative of older material similar to the Mojave province (Hawkins et al. 1996). This may mean that the two crusts are fundamentally different composition terranes and were juxtaposed across the Crystal shear zone. Alternatively, the 1.75- to 1.73-Ga Granite Gorge Metamorphic Suite and related arc plutons may have developed across a deep crustal transition from oceanic crust to 1.84-Ga Elves Chasm arc crust, just as the present-day Aleutian volcanic arc climbs from oceanic basement to older Alaskan crustal basement. The Crystal shear zone is narrow (500 m) but contains highly tectonized rock called melange (Table 2.2) that could mark the

Metamorphism and Cooling History

Metamorphism is upper amphibolite facies, $T = 650°C$, $P = 6$ kbar, with evidence for partial melting; monazite in syntectonic granite are 1701 Ma; monazites in late tectonic cross-cutting dikes are 1678 Ma.

Metamorphic grade is upper amphibolite with evidence for partial melting, 600°C, 4–6 kbar in Spencer Canyon.

Metamorphism of metasedimentary screens reaches partial melting within the granites.

site where a whole ocean basin was subducted before the different sides collided and were sutured. Alternatively, this and other shear zones may mark simply a zone of relatively minor (kilometer-scale) slip that developed as crust was squeezed and thickened.

A similar, more subtle, isotopic transition occurs across the Gneiss Canyon shear zone in the Lower Granite Gorge. Here, Pb 207/204 values also increase to the west (J.L. Wooden, unpublished data). This shear zone is 10 km wide and contains a tectonic sliver of carbonate at mile 236.5 (Fig. 2.5), the only such outcrop of this rock type in the transect and hence an exotic lithology that may mark an important boundary. Like the Crystal shear zone, isotopic data show a similar though less dramatic increase in older (Mojave) crust on the west side. Structural and metamorphic studies suggest that the Gneiss Canyon shear zone is a thrust that brings deeper rocks up on the west (see below). If the Aleutian island analogue is valid, "Mojave province" may consist of older arc and Archean fragments at depth, on which the 1.74–1.71 arcs were built. The Yavapai arcs of the same age were built on oceanic crust, with a wide transitional boundary between the provinces, but an increase in the Mojave signature on the west side of major thrusts that mark the transition. The isotopic data do not prove that there are tectonic sutures at either shear zone, but they do indicate a change in bulk composition of the crust across the shear zones and suggest that these two boundaries are among the best candidates for suture zones.

METAMORPHISM

The Granite Gorge Metamorphic Suite contains a variety of metamorphic minerals that can be used to estimate the pressure (P) and temperature (T) history that rocks underwent. In particular, pelitic layers in the Vishnu Schist contain various combinations of garnet, sillimanite, staurolite, chloritoid, cordierite, kyanite, and andalusite (rare). Different combinations (assemblages) of minerals are indicative of specific P–T conditions. Similarly, garnet, biotite, and plagioclase change composition depending on the P–T conditions of their growth, so we can estimate P–T conditions by making compositional maps and calculating P and T (Fig. 2.8). Available P–T estimates for the different blocks are summarized in Table 2.2.

One remarkable finding is that rocks of the Upper Granite Gorge all record metamorphic pressures equivalent to depths of metamorphism of about 20 km; slightly greater depths are indicated between miles 98 and 115. This uniformity suggests either (1) that there are no structures (faults) that brought very deep rocks (e.g., 30–40 km) against shallow rocks, as one sees in continent–continent collision zones, or (2) that the metamorphic minerals that locked in the pressures grew after suturing such that any higher-pressure minerals were not preserved. Like the late intense contractional deformation, the peak 1.7- to 1.68-Ga metamorphism could have erased much of the evidence for earlier suturing.

In contrast, the Lower Granite Gorge records juxtaposition of different crustal levels. Rocks west of the Gneiss Canyon shear zone (Spencer Canyon block) record pressures that correspond to 15- to 20-km depths. These were thrust east onto rocks that were never deeper than about 10 km (Travertine block). This variation in exposed crustal depths, along with the large areas of uniform depth in the Upper Granite Gorge, allows us to make some generalizations about variations in the character of deformation and metamorphism with depth (Karlstrom and Williams 1998), but does not resolve our uncertainty about which boundaries may be suture zones.

Metamorphic temperature estimates tell another surprising story. Whereas depths remain relatively constant in the Upper Granite Gorge, minerals indicate that metamorphic temperatures varied from a low of 500°C (Boucher, Diamond Creek, and Travertine Canyons) to nearly 700°C (mile 78, near Horn Creek, 113–115 mile and Spencer Canyon). The hottest areas are generally associated with swarms of granite dikes and sills (Cottonwood, Cremation, and Garnet complexes), documenting the interaction of plutonism and metamorphism. These lateral temperature gradients of >200°C take place either abruptly across shear zones (e.g., 96-mile shear zone) or within distances of 5–10 km, as dike and granite complexes are approached (e.g., Clear Creek block—Table 2.2). We conclude that the temperature gradients were caused by additions of heat by the molten rock traveling through the dike complexes (Ilg et al. 1996). Similar temperature variability at 10 km in central Arizona (Karlstrom and Williams 1995) seem to suggest that the temperature distribution was laterally heterogeneous throughout the 10- to 20-km depths during 1.7- to 1.68-Ga tectonism (Karlstrom and Williams 1998).

In addition to the marked variation in temperature from place to place, metamorphic P–T data can also be used to document changes in the pressure and temperature of an individual rock through time. Figure 2.8 shows a compositionally zoned garnet mineral. These compositional changes, when compared to changes in plagioclase and biotite minerals in the same rock, suggest that this mineral decompressed from 20- to 10-km depths during its growth. Textures can document the relative timing of metamorphism and deformation. In Figure 2.8, the textures suggest that the core overgrew S1 and that the core and rim grew synchronously with progressive intensification of the S2 foliation seen in the matrix. Thus, we infer that this rock (and in fact most of the Upper Granite Gorge) passed from 20- to 10-km depths during the main 1.7- to 1.685-Ga deformational and metamorphic event. Similarly, the Spencer Canyon block of the Lower Granite Gorge was uplifted and thrust eastward along the Gneiss Canyon shear zone and stabilized at depths of about 10 km.

U-Pb dating of the metamorphic minerals monazite, xenotime, titanite, and apatite provides constraints on the thermal history of rocks during and immediately following the peak of metamorphism (Hawkins et al. 1996; Hawkins 1996; D.P. Hawkins and S.A. Bowring, in press; D.P. Hawkins and S.A. Bowring, unpublished data). The thermal history of the Mineral Canyon block (Fig. 2.4) is a model for most of the blocks in the Upper Granite Gorge. Vishnu Schist began to melt and was intruded by peraluminous granite melt at about 1710 Ma and continued to melt throughout the peak of metamorphism until about 1690 Ma. During this 20-million-year time interval, rocks decompressed from about 6 kbars to about 3 kbars and were rapidly cooled (at a rate of 15–30°C/ m.y.) to about 550°C. Then they cooled more slowly (at a rate of <2°C/ m.y.) for several hundred million years as the rocks resided at about 10- to 12-km depth.

The only block that does not seem to cool in this manner is the Trinity block, where metamorphic mineral growth took place between 1665 Ma and 1550 Ma (D.P. Hawkins and S.A. Bowring, in press). These ages become systematically younger from east to west for all of the dated minerals, and the minerals preserve little evidence for cooling. Perhaps this block represents a west-tilted crustal section for which the western end remained deeper and at higher temperatures and pressures than both the eastern end of the same block and the adjacent tectonic block to the west. This age pattern of the Trinity block emphasizes the fundamental importance of the block architecture for Proterozoic crust in the Grand Canyon. It also reveals how complicated the history of high-grade terrains can be and how difficult it is to correlate structures, fabrics, and metamorphic mineral reactions across such a terrain.

THE GREAT UNCONFORMITY

By 1.65 Ga, rocks throughout the Grand Canyon transect were at depths of about 10 km. For the Upper Granite Gorge, about 10 km of rock had been eroded during the 1.7- to 1.65-Ga deformation, reduced to particles, and transported and deposited elsewhere. At 1.35 Ga, rocks were probably still residing at near 10-km depths when they were intruded by granites like the Quartermaster granite. Thus, an additional 10 km of rock was eroded off the region between 1.35 and about 1.25 Ga to create a broad plain that was ready to receive sediments of the Grand Canyon Supergroup. The history of this unroofing of the middle crustal rocks is part of the development of the Great Unconformity (Powell 1876). Details are difficult to unravel because erosion has removed most of the record, and we do not know where the eroded sediment was deposited. Nevertheless, we are beginning to unravel parts of this history by examining the cooling history of the middle crustal rocks as they made their way toward the surface, as well as the sedimentary record of the Grand Canyon Supergroup (Chapter 5).

Ar isotope studies can date the last time a mineral underwent diffusion of ^{40}Ar. Diffusion ceases below about 500°C for hornblende and 300°C for muscovite. Data from across the transect show that micas become closed to diffusion at different times. Rocks between mile 96 and 115 cooled relatively quickly from 700°C at 1.7 Ga to below 300°C by 1.6 Ga, then stayed below 300°C. In contrast, rocks east of the Crystal shear zone did not cool through 300°C until approximately 1.4 Ga (Matt Heizler, unpublished data). This suggests differential cooling of the middle crust and continued importance of older shear zone boundaries during differential uplift at 1.4 Ga (Karlstrom and Humphreys 1998).

SYNTHESIS

Paleoproterozoic rocks of the Grand Canyon provide a rich laboratory for understanding the formation, stabilization, and reactivation of continental lithosphere. Crust formation took place in several steps: (1) differentiation of arc plutons at 1.84 and (mainly) at 1.75–1.71 above subduction zones, (2) deposition of 1.75- to 1.73-Ga volcanic and sedimentary rocks of the Granite Gorge Metamorphic Suite on the flanks of developing island arc volcanoes, and (3) collision of arcs with each other (1.74–1.71 Ga) and with the North American protocontinent at 1.7–1.68 Ga and resulting ductile deformation, melting of the lower crust and emplacement of syncollisional granites, and middle crustal metamorphism at depths of 10–20 km.

Stabilization of the new continental crust took place during and after the collisional orogeny as crust thickened and cooled. From 1.6 to 1.4 Ga, this part of North America remained little affected by tectonism, although Grand Canyon rocks remained in the middle crust and cooled fairly slowly. At about 1.4 Ga, renewed magmatism occurred; this may have driven differential uplift of blocks, erosion of about 10 km of crust, and development of a broad erosional surface that exposed middle crustal rocks and was to become the Great Unconformity. The northeast-trending basement shear zones and fabric of the Granite Gorge Metamorphic Suite were repeatedly reactivated in later Precambrian and Phanerozoic times to form dominant structures such as the Bright Angel and Hurricane faults.

ACKNOWLEDGMENTS

This chapter is based on an ongoing collaborative project that was initiated in the late 1980s. Funding was provided by NSF grants EAR-920645 and 9508096 (to Karlstrom), EAR-9507984 (to Williams), and EAR-34005696 (to Bowring), as well as by the UNM Kelley Silver Fellowship, Geological Society of America, and Colorado Scientific Society (for Ilg's PhD). We thank the Glen Canyon Environmental Studies project for logistical help and thank the Grand Canyon National Park for granting a research permit. This chapter also includes data from MS theses done by Katey Robinson (U. Mass) and Jimmie Hutchison (U.N.M.).

• 3 •

GRAND CANYON SUPERGROUP: UNKAR GROUP

J. D. Hendricks and G. M. Stevenson

INTRODUCTION

In Arizona's Grand Canyon, a series of gently tilted sedimentary and igneous rocks are exposed in isolated outcrops along the Colorado River and its tributary canyons. The rocks overlie the schists and granites of the inner gorge and occur below the flat-lying Paleozoic sedimentary units. These wedges of rock are quite noticeable both from the rim and from the river because of the angular difference of the beds (compared to those above and below) and the striking color and topographic variations within the sequence. Major outcrops of these rocks occur in seven separate locations within the Grand Canyon National Park (Fig. 3.1).

NOMENCLATURE

John Wesley Powell was the first person to note the geology of these rocks during his historic traverses of the Grand Canyon in 1869 and 1872. In his reports, he described their stratigraphic position and ascribed a tentative Silurian age to the sequence.

Charles D. Walcott conducted extensive field studies in the eastern Grand Canyon in 1882–1883 and reported his findings some years later. In his 1894 report, Walcott divided this sequence of rocks into two terranes: the Chuar (upper) and Unkar (lower). The combined sequence was named the Grand Canyon Series. In the same report, Walcott provided the first geologic map of the eastern Grand Canyon and the measured stratigraphic thickness of the Grand Canyon Series. Walcott (1894) reported the series to be approximately 12,000 feet (3660 m) in thickness, with the Unkar terrane being 6800 feet (2073 m) and the Chuar terrane being 5200 feet (1587 m). A Precambrian age was assigned to the Grand Canyon Series by comparison with the "Keweenawan Series" of the north-central United States.

From studies conducted in the Shinumo area (Fig. 3.1), Noble (1914) subdivided the Unkar terrane into five formations and assigned group status to the Chuar and Unkar. The names applied to the formations of the Unkar Group, in ascending order, were (1) Hotauta Conglomerate, (2) Bass Limestone, (3) Hakatai

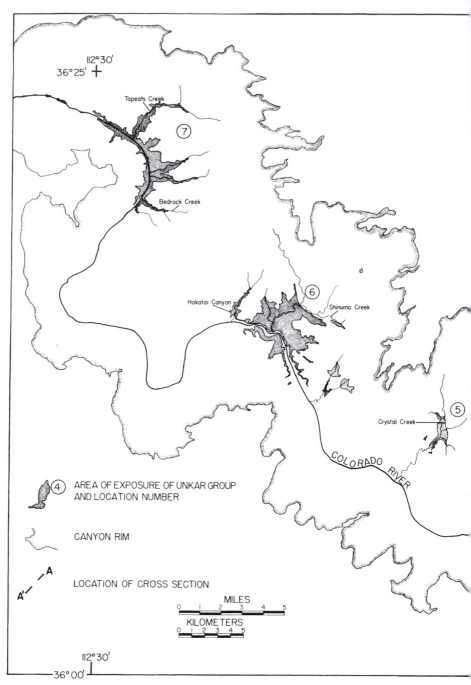

FIGURE 3.1. Location of exposures of the Unkar Group. (1) "Big Bend" region of the eastern Grand Canyon; (2) Clear Creek; (3) Bright Angel Creek; (4) Phantom Creek–Phantom Ranch; (5) Crystal Creek; (6) Shinumo Creek; (7) Tapeats Creek.

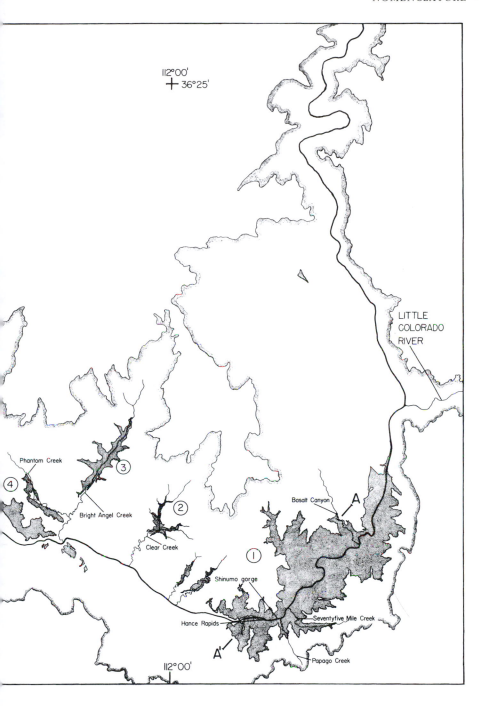

112°00'
+ 36°25'

LITTLE
COLORADO
RIVER

Phantom Creek

③

④

Bright Angel Creek

②

Clear Creek

Basalt Canyon

A

①

Shinumo gorge

Seventyfive Mile Creek

Hance Rapids

A'

Papago Creek

112°00'

Shale, (4) Shinumo Quartzite, and (5) Dox Sandstone. These names were all derived from local geographic features. Because of Precambrian erosional removal
of the Grand Canyon Series above the middle of the Dox Formation in the
Shinumo region, the upper part of the Unkar Group and all of the Chuar Group
were not named by Noble. He did provide, however, a geologic map and structural description of the Shinumo area and detailed petrologic descriptions of the
Unkar Group.

Van Gundy (1937, 1951) recognized two unconformities in the upper part
of the Unkar Group from work done in the eastern Grand Canyon. These breaks
were separated by approximately 330 feet (100 m) of sandstone and shale. He
applied the name Nankoweap Group to this unit, thus removing it from Walcott's (1894) Unkar terrane. The Nankoweap overlies a series of basaltic flows
and unconformably underlies sediments of the Chuar Group. Because the
Nankoweap had not been subdivided into individual formations, Maxson (1968)
classified this whole unit as the Nankoweap Formation.

Keyes (1938) used the name "Cardenas lava series" for the basaltic flow sequence at the top of the Unkar Group. Maxson (1968), however, in his geologic
maps of the Grand Canyon, designated these flows as the Rama Formation and
included it with intrusive rocks of similar composition found lower in the Unkar Group. Ford et al. (1972) formally named the basaltic flows at the top of the
Unkar Group the "Cardenas Lavas." This term has been modified to the Cardenas Lava by Lucchitta and Beus (1987). Because the term lava applies to a fluid
and not a rock, this designation is not favored nomenclature, but the name Cardenas Lava is used commonly for these rocks in the current literature and is retained herein.

Dalton (1972) studied the Bass Limestone and Hotauta Conglomerate of Noble (1914) and suggested that the Hotauta be included as a member of the Bass.
This designation is adopted in this discussion. Dalton (1972) also suggested that
the Bass Limestone should be reclassified as the Bass Formation because of the
variety of rock types within the unit and the fact that limestone is a minor lithology. Stevenson and Beus (1982) suggested that the Dox Sandstone of Noble
(1914) should be renamed the Dox Formation also because of the lithologic diversity. For the purpose of continuity of nomenclature, the original names of Noble, with the exception of the re-ranking of the Hotauta Conglomerate, naming
of the Cardenas Lava by Ford et al. (1972) and Lucchitta (1984), and the designation of the Dox Sandstone as the Dox Formation, will be used in the discussion of the individual formations of the Unkar.

As a result of these studies, the Unkar Group currently is subdivided into
five formations: (1) Bass Limestone, (2) Hakatai Shale, (3) Shinumo Quartzite,
(4) Dox Formation, and (5) Cardenas Lava. A stratigraphic section is presented
as Figure 3.2, which depicts this nomenclature, the average thicknesses of the
formations and members, and the general lithologies of the Unkar Group. Following the current code of nomenclature, the Unkar, Nankoweap, and Chuar
comprise the Grand Canyon Supergroup.

All of the formations comprising the Unkar Group, except the Cardenas Lava,
were named for localities in the Shinumo quadrangle. They occur in small, rotated, downfaulted blocks or slivers and commonly are partially exposed. Of the
five recognized formations comprising the Unkar Group, only the Hakatai Shale
is well-represented in the Shinumo Creek vicinity without further consideration
for an "alternate" area. However, in the case of the three other sedimentary units,
sections should be selected where they are best preserved. In the case of the
Bass Limestone, a much more complete marine section is present to the west at
Tapeats Creek (Fig. 3.1). Both the Shinumo and Dox formations have good sec-

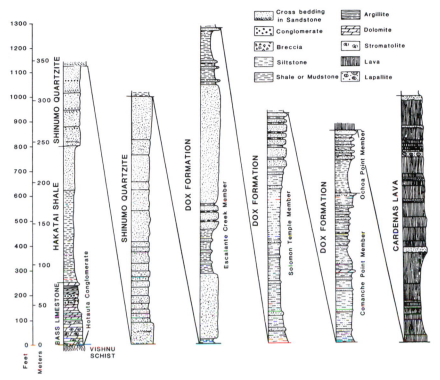

FIGURE 3.2. Columnar section of the Unkar Group.

tions eastward of Shinumo quadrangle in the Vishnu quadrangle where complete stratigraphic intervals are preserved and well-exposed. Perhaps the best locality for the Shinumo is in "Shinumo Gorge" near Hance Rapids, and the most revealing look at the Dox can be found in the "Big Bend" area (area 1 of Fig. 3.1).

AGE OF THE UNKAR

The Grand Canyon Supergroup overlies the metamorphic and granitic basement complex of Early Proterozoic age (1700 million years ago [m.y.a.] and underlies the middle Cambrian Tapeats Sandstone (550 m.y.a.). Thus, the supergroup occupies a portion of the 1150-million-year interval between 550 and 1700 m.y.a. Two methods of age determination have been applied to establish the time of the Unkar Group's formation. The Cardenas Lava provides the only stratigraphically controlled lithology of the supergroup, discovered to date, suitable for radiometric age determinations. McKee and Noble (1974) obtained an age of 1100 m.y.a. for the Cardenas Lava using the Rb–Sr method. The K–Ar method has produced Cardenas ages that are considerably younger than 1100 m.y.a. Ford et al. (1972) presented a single K–Ar age of 845 ± 20 m.y.a. for samples of the Cardenas Lava, and McKee and Noble (1974) obtained ages of 810 ± 20, 790 ± 20, and 781 ± 20 m.y.a. (K–Ar) for samples of the Cardenas Lava. The lower ages obtained using the K–Ar method may reflect an episode of heating and resetting of the K–Ar clock about 800 million years ago (McKee and Noble 1974).

Extensive study of the paleomagnetic pole positions and polar wandering paths by Elston (summarized 1986) has led to the conclusion that the Unkar Group accumulated in the time interval 1250 to 1070 m.y.a. These results are in agreement with the Rb–Sr age of the Cardenas Lava (McKee and Noble 1974). It appears, therefore, that the unconformity ("greatest angular unconformity"; Noble 1914) between the Early Proterozoic basement complex and the Unkar Group represents a time period of about 450 million years, whereas the unconformity between the Cardenas Lava and Nankoweap probably reflects a relatively short period of geologic time.

DESCRIPTIONS OF THE FORMATIONS
OF THE UNKAR GROUP

The sedimentary sequence of the Unkar Group records a major west-to-east transgression of the sea. During the nearly 250-million-year time span postulated for Unkar deposition, the region apparently was at (or very near) sea level. Only one unconformity is documented within the Unkar: between the Hakatai Shale and Shinumo Quartzite. Minor fluctuations of sea level or sediment surface elevation is recorded by features suggesting both subaerial and marine deposition throughout the sequence. Apparently, the Unkar Group was deposited in a basin in which the rate of subsidence was approximately the same as the rate of deposition. The only suggestion of relatively deep-water deposition is noted by textural features in dolomites and mudstones in the middle part of the Bass Limestone in the western Grand Canyon (Dalton 1972).

The Bass Limestone and Hotauta
Conglomerate Member

The Vishnu surface, over which the Unkar sea advanced from the west, was smooth with a local relief of probably no more than 150 feet (45 m). The Hotauta Conglomerate Member, the lowermost unit of the Bass Limestone, was deposited in low areas of the Vishnu terrane. This conglomerate consists of rounded, gravel-sized clasts of chert, granite, quartz, plagioclase crystals, and micropegmatites in a quartz sand matrix. It is found in the eastern Grand Canyon. In the Unkar exposures of the western Grand Canyon, the lowermost part of the Bass contains intraformational breccias and small pebble conglomerates, suggesting that the source of these clasts was toward the east.

The lithology of the Bass Limestone is predominantly dolomite with subordinate amounts of arkose and sandy dolomite with intercalated shale and argillite. Intraformational breccias and conglomerates also are found throughout the sequence (Dalton 1972). One feature of note within the Bass is the presence of biscuit-form and biohermal stromatolite beds (Nitecki 1971). The thickness of the Bass Limestone shows a general increase to the northwest ranging from 330 feet (100 m) at Phantom Creek (Fig. 3.1) to 187 feet (57 m) at Crystal Creek. The anomalously thin section at Crystal Creek probably reflects the presence of a Vishnu topographic high in this area during deposition. The Bass generally forms cliffs or stair-stepped cliffs: The more resistant dolomites make the risers, and the shale and argillite form steep treads.

Sedimentary features common to all exposures of the Bass Limestone include symmetrical ripple marks, desiccation cracks, interformational breccias/conglomerates, and both normal and reversed small-scale, graded beds (associated with stromatolites). All of these features suggest a relatively low-energy intertidal to supratidal environment of deposition. Although no evaporites presently are recognized in the Bass, some of the interformational breccias may be the result of collapse of earlier-formed gypsum. Dalton (1972) noted monoclinic crystal clasts in chert layers of the Bass Limestone. This is suggestive of a dolomitic replacement of gypsum.

The lithology and sedimentary structures observed in the Bass Limestone suggest deposition in an easterly transgressing sea. During the maximum incursion of the sea, carbonates and deep-water mudstones accumulated in the western Grand Canyon. In the east, stromatolites were forming, and shallow-water mudstones were being deposited. Following this period of transgression, the sea slowly regressed. Evidence for this includes ripple marks, mudcracks, and deposits of oxidized shales in the upper part of the Bass—all suggesting periods of subaerial exposure. Evaporite-forming conditions probably existed also during this regressive phase (Dalton 1972). Eventually, deltaic conditions predominated, which marked the beginning of Hakatai Shale deposition. The contact between the Bass Limestone and Hakatai Shale is gradational in the east and sharp, though conformable, in the west.

Hakatai Shale

The Hakatai Shale probably is the most colorful formation in the Grand Canyon, with colors that vary from purple to red to brilliant orange on outcrop. The colors result from the oxidation state of the iron-bearing minerals in the formation. The Hakatai is subdivided into three informal members based on lithology (Beus et al. 1974; Reed 1974. The lower two units are highly fractured argillaceous mudstones and shales that weather to gentle-to-moderate, granular slopes. The upper unit consists of cliff-forming beds of medium-grained quartz sandstone (Fig. 3.2). The Hakatai varies in thickness from about 445 feet (135 m) at Hance Rapids to nearly 985 feet (300 m) at the type section in Hakatai Canyon in the Shinumo Creek area.

Sedimentary structures, such as mudcracks, ripple marks, and tabular-planar cross bedding, suggest that the Hakatai was deposited in a marginal marine environment. Mudstones and shales of the lower two members probably were deposited in a low-energy, mud-flat environment. The upper sandstones suggest a higher-energy, shallow-marine environment (Reed 1974).

Although the contact between the Hakatai Shale and Bass Limestone is gradational in the eastern Grand Canyon and sharp and easily located in the western exposures, the contact between the Hakatai and the overlying Shinumo Quartzite is evident in all exposures. It is marked by an unconformity that truncates cross beds and channel deposits of the Hakatai. From observations made in the canyon of Bright Angel Creek, Sears (1973) indicates that Hakatai deposition in the area ended in conjunction with tectonic activity along a series of northwest-trending, high-angle, reverse faults.

Shinumo Quartzite

In contrast to the slope-forming argillaceous beds of the Hakatai Shale below and the Dox Formation above, the Shinumo Quartzite is a series of massive, cliff-forming sandstones and quartzites. The color of the Shinumo ranges from muted red, brown, and purple to white.

Four or possibly five poorly defined members have been recognized within the Shinumo Quartzite (Elston 1986; Daneker 1974). The lower units, in ascending order, consist of conglomeratic subarkose and submature quartz sandstone; to mature quartz sandstone; to brown quartz sandstone with abundant cross beds, clay galls, and mudcracks.

Near Shinumo Creek, the uppermost unit is the thickest member. It consists of fine-grained, well-sorted, and rounded quartz grains in a siliceous cement. Beds in the upper part of the upper member are contorted by fluid evulsion, which suggests that there might have been tectonic activity during this period.

The thickness of the Shinumo Quartzite shows a general increase to the west and ranges from 1132 feet (345 m) at Papago Creek in the east to 1328 feet (405 m) (Noble 1914) in Shinumo Creek. Because the Shinumo is such a resistant unit, it formed hills where exposed during the pre-Tapeats erosional event.

This feature, the pinching out of the Tapeats Sandstone against Shinumo Quartzite highs, can be seen today in exposures near the bottom of the Grand Canyon. Analysis of the lithology and the sedimentary structures of the Shinumo suggest that the environment of deposition was near-shore, very shallow, marginal marine, and part fluvial, part deltaic (Daneker 1974). The contact between the Shinumo and the overlying Dox Formation appears to be conformable in most locations and is marked by the lowermost shaley interval of Dox lithology.

Dox Formation

The Dox Formation is the thickest unit of the Unkar Group. The only complete section presently exposed, however, is in the eastern Grand Canyon (area 1 of Fig. 3.3), with thicknesses variously reported to be 3020 feet (921 m), 3115 feet

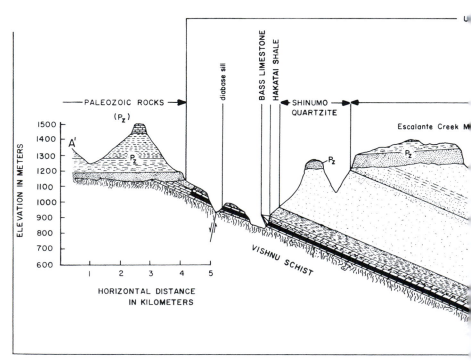

FIGURE 3.3. Cross section A-A' (eastern Grand Canyon). See Fig. 1.1 for location of section.

(950 m), and 3230 feet (985 m). The Dox consists of four members: (in ascending order) Escalante Creek, Solomon Temple, Comanche Point, and Ochoa Point. West of 75-mile Creek (western part of area 1, Fig. 3.1), only the Escalante Creek and Solomon Temple members are preserved; the Comanche Point and Ochoa Point have been removed by pre-Tapeats erosion.

The Escalante Creek Member is reported by Stevenson (1973) to be 1280 feet (390 m) thick where exposed in the eastern Grand Canyon. Here, it is a light-tan to greenish brown, siliceous quartz sandstone and calcareous lithic and arkosic sandstone that is 800+ feet (244+ m) thick, combined with an overlying 400-feet (122-m)-thick sequence of dark-brown-to-green shale and mudstone. The tan to brownish color of this lower member is in marked contrast to the characteristic red and red-brown color of the rest of the Dox Formation.

Sedimentary structures observed in the sandstones of the Escalante Creek included contorted bedding (within 100 feet [30 m] of the base), small-scale, tabular-planar cross beds, and graded beds (with shale interclasts at the base). The contacts between members of the Dox Formation are gradational and are based mainly on topographic expression, depositional environments, and color changes. The Escalante Creek is tan to brown and forms a cliff-slope topography, as opposed to the more red-orange overlying Solomon Temple Member, which forms rounded-hill topography that is more characteristic of the remainder of the Dox.

The Solomon Temple Member is a cyclical sequence of red mudstone, siltstone, and quartz sandstone. It is 920 feet (280 m) thick in the eastern Grand Canyon. The lower 700 feet (213 m) is a slope-forming series of predominantly

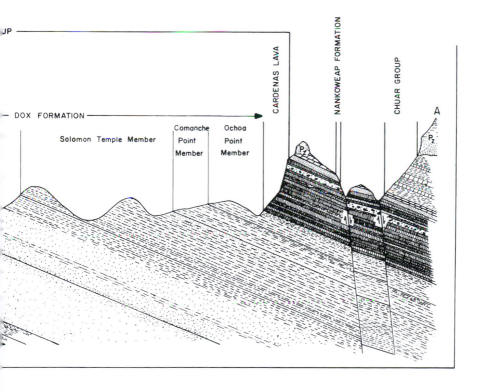

red-to-maroon shaley siltstone and mudstone with subordinate quartz sandstone. The upper 220 feet (67 m) of the member is primarily maroon quartz sandstone with numerous channel features. These channels, along with common low-angle, tabular cross beds, suggest a floodplain environment.

The Comanche Point Member occupies more than half of the Dox outcrop area and is distinguished from enclosing members by its slope-forming and color-variegated character. This member varies in thickness from 425 to 617 feet (130 to 188 m) in the eastern Grand Canyon. The lithology is primarily shaley siltstone and mudstone with minor amounts of sandstone. Five pale green-to-white, leached red beds, some up to 40 feet (12 m) in thickness, provide the variegated appearance of this unit. Stromatolitic dolomite layers are found within or directly adjacent to the leached beds. Sedimentary features found in this member include ripple marks; mudcracks and curls; salt casts; and wavy, irregular bedding.

The Ochoa Point Member is 175 to 300 feet (53 to 92 m) thick and forms steep slopes and cliffs below the Cardenas Lava. It consists of micaceous mudstone that grades upward into a predominantly red quartzose, silty sandstone. Sedimentary structures of this member also include salt crystal casts in the mudstone and asymmetrical ripple marks and small-scale cross beds in the sandstones.

The lithology and sedimentary structures found within the Dox Formation suggest, in ascending order, a subaqueous delta, floodplain, and tidal flat environment during deposition. According to Stevenson and Beus (1982), features of the lowermost member, the Escalante Creek, record a rapid transgression of the Dox sea followed by gradual basin filling.

This basin was filled by the close of Escalante Creek time, and the region was at or very near sea level for the remainder of Dox time. Stevenson and Beus (1982) also suggest a possible westerly source for some of the sediments of the Dox, which is opposite to the inferred source direction for other formations of the Unkar Group.

The Cardenas Lava

Cardenas Lava is the name given to a series of basalt and basaltic andesite flows and sandstone interbeds that are stratigraphically above the Dox Formation and below the Nankoweap Formation (Ford et al. 1972). This sequence of rocks is exposed only in the eastern Grand Canyon, where the thickness of the formation ranges from about 785 (240 m) to nearly 985 feet (300 m). The contact between the Cardenas Lava and the overlying Nankoweap Formation is unconformable, with an unknown amount of the Cardenas being removed prior to Nankoweap deposition.

The contact between the Dox Formation and the Cardenas is conformable and interfingering. This is highlighted by the presence of a thin, discontinuous basaltic flow in the upper Dox a few meters below the top of the formation. The Dox near the contact is mildly baked and locally displays small folds and convolutions that are suggestive of soft sediment deformation. At one location, the basalt of the lowermost flow sequence contains rounded masses of igneous rock, up to 3.3 feet (1 m) in diameter, completely surrounded by a thin layer of siltstone of Dox lithology. The uppermost Dox is fine-grained sandstone and siltstone deposited in a tidal flat environment. The region at the time of initial Cardenas eruption was at or very near sea level (Stevenson and Beus 1982). One interpretation is that the lava flowed over unconsolidated sandy and silty Dox sediments that were wet at the time. Whether these sediments were slightly above or slightly below water level is unknown.

Strictly on the basis of topography, the Cardenas Lava can be divided into two units (Hendricks and Lucchitta 1974). The lower unit forms granular slopes and is from 245 to 295 feet (75 to 90 m) thick. Although this unit is highly altered and weathered, many of the primary features are preserved. This "bottle-green member" (Lucchitta and Hendricks 1983) is a composite of many thin, discontinuous flows and sandstone interbeds. The basalt of this unit is broken and weathers into nodules typically 3 to 10 inches (10 to 30 cm) in diameter. Near the top of this unit, some 230 feet (70 m) above the base, the basalt is more massive and less altered. Petrographically, the lower unit is an olivine-rich basalt with a subophitic texture. Lower in the bottle-green member, the rock is highly altered but has a texture that suggests it may have been quite glassy originally.

Chemically, rocks from the lower unit are high in sodium and magnesium and depleted in potassium, suggesting spilitic alteration. The nodular character, glassy texture, and anomalous chemistry of rocks of the lower unit suggest rapid quenching in sea or brackish water. The thin, discontinuous sandstone interbeds also indicate that water flowed over or was ponded on the lavas during periods of volcanic quiescence.

Approximately 328 feet (100 m) above the base of the Cardenas is a continuous sandstone bed some 16 feet (5 m) in thickness that overlies the bottle-green member. This sandstone is laminated and forms vertical cliffs. In a number of locations, the sandstone occupies channels in the upper surface of the bottle-green flow member. The petrology of this sandstone suggests quiet-water deposition; the channels, therefore, probably are lava channels that were left as extrusion temporarily ceased and basin subsidence continued, thereby lowering the lava surface below sea level.

The upper unit of the Cardenas is a series of cliff-forming basalt and basaltic andesite flows along with additional sandstone interbeds. Features preserved in the individual flow units suggest that the volcanic pile accumulated at a slightly greater rate than basin subsidence. In ascending order, these are: an autoclastic breccia directly above the 328-foot (100-m) sandstone, a fan-jointed unit, ropy lava, and, finally, a lapillite unit at the 754-foot (230-m) level. Following eruption of the lapillite, volcanic activity ceased for a short period of time. The pyroclastic surface was smoothed as subsidence continued until the surface again was lowered below sea level. This is noted by a planar upper surface on the lapillite unit with a continuous sandstone layer directly above. At least two more eruptive events followed deposition of this sandstone layer.

Following cessation of volcanic activity, the sediments and igneous rocks of the Unkar were tilted gently toward the northeast. An unknown amount of Cardenas Lava was eroded prior to Nankoweap deposition. Elston and Scott (1976) suggested that major tectonic movement occurred along the Butte fault in the eastern Grand Canyon during the post-Cardenas, pre-Nankoweap interval.

Intrusive Rocks of the Unkar Group

Diabase sills and dikes intrude all formations of the Unkar Group below the Cardenas Lava. Sills are restricted to the Bass Limestone and Hakatai Shale. Dikes intrude the Hakatai Shale, Shinumo Quartzite, and Dox Formation above the sills. Feeder dikes or vents for the sills are not exposed. Above the sills, dikes can be traced, discontinuously, to within a few meters of the base of the Cardenas Lava.

Unkar sills range in thickness from 75 feet (23 m) at Hance Rapids in the eastern Grand Canyon to 985 feet (300 m) in Hakatai Canyon (Shinumo Creek area). Fine-grained, chilled margins suggest that the magma was highly fluid at the time of intrusion. Only the sills of the Shinumo Creek area show extensive differentiation and segregation products. Here, a picritic layer of unknown thickness is present near the base of the sill, while the top of the intrusion is marked by a 20-foot (6-m)-thick granophyre layer.

Contact metamorphism caused by intrusion of the diabase resulted in the formation of chrysotile asbestos above the sills where the magma intruded the Bass Limestone. Adjacent to the sills, the Hakatai Shale is a knotted hornfels containing porphyroblasts of andalusite and cordierite that have been replaced by muscovite and green chlorite, respectively.

The relation among the sills, dikes, and Cardenas flows is not self-evident. The mineralogy of the sills is uniform throughout the region and is identical to that of the unaltered parts of the bottle-green member of the Cardenas. Chemical variation diagrams (Hendricks and Lucchitta 1974) indicate that the flows are more felsic than the sills, but the evidence does not preclude the possibility of a common parentage. Paleomagnetic evidence (Elston 1986) suggests that the majority of the sills were intruded at the same time that the Comanche Point Member of the Dox Formation was being deposited. The Hance sill, the easternmost of the Unkar sills, produces an anomalous paleomagnetic pole position that Elston (1986) interpreted as an indication of a slightly greater age for this intrusion. If the majority of the sills were emplaced during Comanche Point time, approximately 330 feet (100 m) of sediment of the Comanche Point and Ochoa Point members of the Dox was deposited during the time interval between intrusion of the sills and initial eruption of the Cardenas Lava. Some of the dikes in the eastern Grand Canyon have paleomagnetic pole positions similar to the Cardenas and may represent feeders. These dikes cannot be seen connecting to the nearby Hance sill.

The sills, dikes, and flows of the Unkar Group may represent a single volcanic episode. If so, the earliest phases were intrusions of diabase sills followed by a period of igneous quiescence long enough for accumulation of approximately 300 feet (100 m) of Dox sediment. Later phases were eruptions of basalt and basaltic andesite flows via a network of thin dikes. Elston and McKee (1982) indicated that initial strontium-87/strontium-86 ratios are slightly different for a sill in the Shinumo Creek area and the Cardenas Lava and concluded that either the sill and flows are not comagmatic or the magma acquired ^{87}Sr from passing solutions. These ratios [0.7042 ± 0.0007 (sill) and 0.7065 ± 0.0015 (flows)], do not preclude a common parentage for the intrusive and extrusive rocks. This might occur if the magma ponded in the crust long enough for derivation of the basaltic andesite of the Cardenas Lava from the basaltic magma of the sills, provided that the intruded crust was assimilated and that it had an $^{87}Sr/^{86}Sr$ ratio greater than 0.7065.

OVERVIEW OF GEOLOGIC HISTORY

Prior to the beginning of deposition of the Unkar sediments, the region of the Grand Canyon was along the southwestern margin of the North American craton. The surface was a smooth terrain composed of Vishnu granitic and metamorphic rocks. Pre-Unkar relief on this surface was undulating with a relief on the order 30 to 500+ feet (a few tens to a few hundreds of meters). Relatively high terrain may have existed to the east and northeast, with probable open oceans to the west. As the land surface subsided, the ocean advanced eastward, marking the beginning of Unkar deposition about 1250 million years ago.

The Hotauta Conglomerate Member of the Bass Limestone, lowermost unit of the Unkar Group, was deposited in relatively low areas of the Vishnu terrain. The clasts size suggests an easterly source area. During this initial transgression, the sea advanced at least as far as Hance Rapids in the eastern Grand Canyon— and probably much farther. Sediments in the middle of the Bass Limestone in western Grand Canyon suggest that this area was below wave base and away from any strong currents. At the same time, indicators in the east suggest an in-tertidal-to-supratidal environment.

The initial transgression of the sea was followed by a gradual regression, which resulted from sediment accumulation. Sedimentary features found in the upper part of the Bass Limestone and Hakatai Shale indicate that conditions var-ied from subaerial to subaqueous throughout the region during this of time. Tec-tonic activity along northeast-trending, high-angle reverse faults marked the end of deposition of the Hakatai Shale and resulted in a period of erosion prior to the accumulation of sands of the Shinumo Quartzite.

Near-shore fluvial and deltaic conditions predominated in the region during Shinumo and early Dox time, marking a further subsidence of the region and the second transgression of the sea; marine conditions returned by the close of this period. Following deposition of about 165 feet (50 m) of Dox sediments, rapid subsidence caused the sea to advance further eastward followed by a sus-tained period of basin filling. Subaerial conditions returned by the close of Solomon Temple time. The contact between the Solomon Temple and Comanche Point members of the Dox Formation marks another transition to marine condi-tions with the remainder of the Dox environment fluctuating between marine and nonmarine conditions.

Features found in the lowermost Cardenas Lava suggest the outpouring of the basalt onto wet, probably shallow-water Dox sediments. Sporadic accumu-lation of the lava pile and continued basin subsidence resulted in conditions that varied between marine and nonmarine, but, generally, the flows accumulated at a greater rate than that of regional subsidence. Following extrusion of more than 985 feet (300 m) of Cardenas lava, tectonic uplift raised and tilted gently the re-gion of the eastern Grand Canyon toward the northeast. Subsequent erosion re-moved an unknown amount of the lavas prior to deposition of sediments of the Nankoweap Formation.

INTER-REGIONAL CORRELATIONS

The only other unmetamorphosed sequence of rocks that is younger than the 1700-million-year basement complex and older than Cambrian in the region is found in central Arizona. These rocks, the Apache Group and Troy Quartzite, are similar to those of the Unkar Group.

Shride (1967) has suggested a possible correlation of the Unkar Group with the Apache Group. This is based on similarities in the age and lithology of the two units. Elston (1986) reviewed the correlation of various Middle and Late Pro-terozoic sequences on a paleomagnetic basis. He concluded that the Mescal Lime-stone of the Apache Group correlates with the middle part of the Dox Forma-tion and that sills in the Unkar Group are similar paleomagnetically to sills that intrude both the Apache Group and Troy Quartzite. Elston (1986) also provided possible correlations of the Unkar Group with other Middle Proterozoic sequences of North America. These include the Uinta Mountain Group of Utah and Colorado, the Belt Supergroup of Montana and Idaho, the Sibley Series of Ontario, and the Keweenawan Supergroup of the Lake Superior Region.

SUMMARY

Sediments and lavas of the Unkar Group of the Grand Canyon Supergroup were deposited in a basin along the western edge of the North American craton during the period 1250 to 1070 million years ago. Features preserved in the Unkar record a west-to-east transgression of the sea with minor sea-level variations resulting from basin filling and subsidence. About 5800 feet (1770 m) of sediment was deposited in the Unkar basin before the onset of volcanic activity in the area. Unkar sedimentary rocks presently are exposed in isolated outcrops in the Grand Canyon and have been subdivided into formations; in ascending order they are, the Bass Limestone, Hakatai Shale, Shinumo Quartzite, and Dox Formation. A volcanic episode marked the end of sedimentation in the Unkar basin. During this volcanic period, igneous material formed sills in the lower parts of the Unkar and dikes in the upper parts. Lava was erupted onto sediments of the Dox Formation and accumulated to a thickness of nearly 1000 feet (305 m). These extrusive rocks have been named the Cardenas Lava. Following cessation of volcanic activity, the Unkar Group rocks were tilted slightly and the top of the Cardenas Lava eroded before the onset of further sedimentation in the region.

• 4 •

GRAND CANYON SUPERGROUP: NANKOWEAP FORMATION, CHUAR GROUP, AND SIXTYMILE FORMATION

Trevor D. Ford and Carol M. Dehler

INTRODUCTION

The upper half of the younger Precambrian strata of the Grand Canyon presents a sequence of approximately 6800 feet (2100 m) of rocks not exposed anywhere else in the southwestern United States. They have not been metamorphosed, and the sedimentary rocks include an unparalleled assemblage of late Precambrian fossils. For these reasons, the Nankoweap Formation, Chuar Group, and Sixtymile Formation deserve a chapter devoted to themselves. Furthermore, they provide unique evidence of sedimentation during tectonic activity, with examples of intraformational faults and unconformities in the Nankoweap Formation and the Chuar Group, as well as slump features in the Sixtymile Formation.

The younger Precambrian rocks of the Grand Canyon first were recognized as "Algonkian" by Powell (1876), who named them the Grand Canyon Series. Now redesignated as the Grand Canyon Supergroup (Elston and Scott 1976), the upper half comprises the Nankoweap Formation, overlain by a thick Chuar Group and a thin Sixtymile Formation. Walcott (1894) presented the first detailed descriptions, but divided the Supergroup into the Unkar and Chuar "terranes." Because the Unkar Group is treated elsewhere in this volume, only the younger part of the younger Precambrian is covered in this chapter.

The Nankoweap Formation and Chuar Group are exposed in the wide eastern part of the Grand Canyon, clearly visible from the Desert View Tower overlook (Fig. 4.1). The former is present in cliffs overlooking Basalt, Tanner, and Comanche Canyons and is visible from the river. The Chuar Group is only visible from the river immediately north of Basalt Canyon and crops out dominantly in the upper parts of several right-bank tributaries—notably Nankoweap, Kwagunt, Carbon, Chuar, and Basalt Canyons. The outcrops of the Chuar Group are bounded on the east by the Butte fault, and on all other sides by Powell's Great Unconformity and the overlying Cambrian Tapeats Sandstone (Fig. 4.1). The Sixtymile Formation is only exposed in the infrequently visited Sixtymile and Awatubi canyons and also caps Nankoweap Butte, which makes up the divide between Nankoweap Creek to the north and Kwagunt Creek to the south (Fig. 4.1).

The north rim overlooks at Point Imperial and Cape Royal provide fine views of the Chuar Group outcrops in these tributary canyons. Observers at Cape Royal can clearly see that the Chuar Group has been folded into a prominent

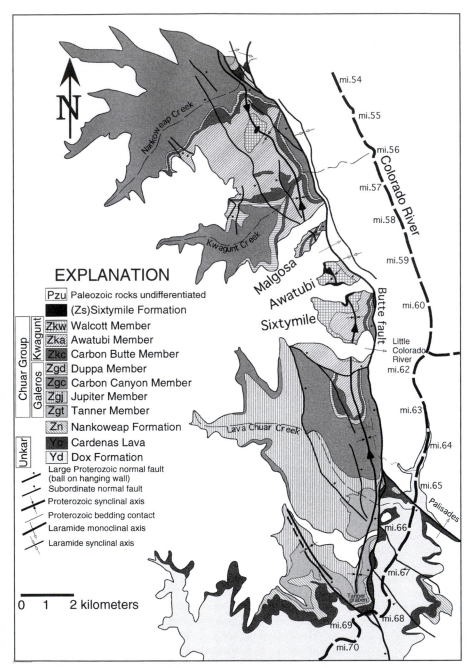

EXPLANATION

Pzu	Paleozoic rocks undifferentiated
(Zs)	Sixtymile Formation
Zkw	Walcott Member
Zka	Awatubi Member
Zkc	Carbon Butte Member
Zgd	Duppa Member
Zgc	Carbon Canyon Member
Zgj	Jupiter Member
Zgt	Tanner Member
Zn	Nankoweap Formation
Yc	Cardenas Lava
Yd	Dox Formation

Chuar Group — Kwagunt / Galeros
Unkar

Large Proterozoic normal fault (ball on hanging wall)
Subordinate normal fault
Proterozoic synclinal axis
Proterozoic bedding contact
Laramide monoclinal axis
Laramide synclinal axis

0 1 2 kilometers

FIGURE 4.1. Geologic map of the upper Unkar Group, Nankoweap Formation, Chuar Group, and Sixtymile Formation from Nankoweap Canyon southward to Basalt Canyon. (Timmons, unpublished data, 1998.)

north–south-trending syncline, west of and parallel to the Butte fault (Fig. 4.1). Displacement on the Butte fault was of the order of 10,500 feet (3200 m) down to the west in late Precambrian time, but this has been canceled out partly by west-side-up reactivation in Post-Paleozoic times of about 2660 feet (810 m).

THE NANKOWEAP FORMATION

In the middle part of the younger Precambrian sequence of the eastern Grand Canyon, a distinctive group of red-brown and tan sandstones, and subordinate siltstones and mudrocks, lies unconformably on top of the Cardenas Lava. The rocks crop out from just south of Carbon Canyon to Basalt Canyon on the west bank of the river and around Comanche Creek and Tanner canyon on the east bank. Splays of the Butte fault offset the Nankoweap Formation on both sides of the river. These sedimentary deposits were originally included partly in the top of the Unkar Group and partly in the basal Chuar Group by Walcott (1894), but were separated as a new unit, the Nankoweap Group, by Van Gundy (1937). Later, in 1951, Van Gundy gave a more complete description and noted the presence of unconformities at both the upper and lower contacts. Maxson (1968) introduced the term "Nankoweap Formation" on his geologic maps of the Bright Angel Quadrangle and eastern Grand Canyon, but made no comment on the unconformities. The name is taken from a small, fault-bounded block of the Nankoweap Formation in Nankoweap Canyon, although more extensive outcrops exist above and adjacent to the Cardenas Lava in Basalt Canyon (Fig. 4.3), Comanche Creek, and Tanner Canyon.

The Nankoweap Formation is divided into two informal members (Elston and Scott, 1976). The lower (ferruginous) member is 40 feet (13 m) thick and rests disconformably on the Cardenas Lava. It is composed of red, fine-grained quartzitic sandstones and siltstones with hematite laminae and lenses of volcanic detritus derived from the Cardenas Lava. The upper member, 330 feet (100 m) thick, disconformably overlies the lower member and is composed mainly of fine-grained quartzitic sandstones that are shaley and silty towards the top. These sandstones are thin-to-medium bedded, with cross-beds, ripplemarks, mudcracks, soft-sediment deformation, and rare salt pseudomorphs. Elston and Scott (1976) interpreted the lower member to have been deposited in quiet shallow water in a structurally controlled lake or pond. Sedimentary structures in the upper member suggest a moderate to low energy, shallow water, marine or lake environment.

Although exposures in the cliffs immediately north of Tanner Rapids show well-bedded, red-brown sandstone dipping evenly at about five degrees to the northeast, the upper parts of Basalt and Tanner Canyons reveal more complex stories, once post-Nankoweap faulting is removed. Substantial paleotopography cut into the Cardenas Lava can be observed in Tanner Canyon and Tanner Graben, as well as the disconformity that separates the lower and upper members. Some of this paleotopography is controlled by growth faults, which offset the Cardenas Lava and the lower member, yet not the overlying upper member (Elston and Scott, 1976). A growth fault was recently discovered within the upper member on the south-facing side of Basalt Canyon, signifying that syntectonic deposition did occur during upper member time.

No direct dating is possible on the Nankoweap Formation. The age of this formation can be bracketed between the 1070 ± 70-million-year age of the underlying Cardenas Lava and associated diabase dikes and sills (Elston and McKee 1982 and references therein) and the overlying Chuar Group, which Elston

(1979) has argued was terminated by the "Grand Canyon orogeny" approximately 823 million years ago. Limited paleomagnetic data (Elston and Grommé, 1974; Elston and McKee 1982; Elston 1989a) are consistent with an age of ~1070–1050 Ma for deposition of the lower member and ~950 Ma for deposition of the upper member.

Paleontology

A structure found in a sandstone bed of the Nankoweap Formation in Basalt Canyon was identified as a trace fossil impression of a stranded jellyfish (Van Gundy 1937, 1951). It comprises a series of lobes rounded at the extremities. Some lobes have a median groove radiating from a small, irregular hollow, and the whole structure is approximately five inches (12 cm) in diameter. Hinds (1938) and Bassler (1941) also considered this to be a jellyfish impression. It was named *Brooksella canyonensis* by Bassler as a new species of a genus well known in the Paleozoic Era, though the interpretation of the genus as a jellyfish is still in considerable doubt.

Cloud (1960, 1968) subsequently obtained a partial second specimen but claimed that the structures were of inorganic origin formed by "compaction of fine sands deposited over a compressible but otherwise unidentifiable structure, possibly a small gas blister." Glaessner (1969) was unconvinced by Cloud's explanation and drew comparisons with a Mesozoic stellate trace fossil, *Asterosoma*, deducing that it was of organic origin and that its "possible originator was a sediment feeder able to burrow into the sediment . . . worm-like in shape . . . probably an annelid."

Accordingly, Glaessner (1984) renamed the structure *Asterosoma? canyonensis* (Bassler, 1941) and apparently still accepts the trace fossil interpretation in spite of its age. In a 1981 study, Kauffman and Steidtmann (1981) supported Glaessner's interpretation as a burrow made by a sediment-feeding, worm-like organism.

An examination of both specimens revealed a similarity to small "sand-volcanoes" formed by the upward expulsion of gas or fluid from sediments as more sediment is loaded on top or as the sediment is shaken during seismic activity. These soft-sediment deformation features are ubiquitous in sandstone beds of the Grand Canyon Supergroup. In view of this, it is difficult to support Glaessner's interpretation of the phenomenon as trace fossils. Additionally, sinuous, high-relief ridges on the soles (undersides) of sandstone beds resemble infilled burrows, yet are more likely the partial casts of mudcracks, another ubiquitous sedimentary feature in the Grand Canyon Supergroup.

If the age of the Nankoweap Formation is ~1070–950 Ma, and if these specimens are truly fossils, whether burrows or jellyfish impressions, these specimens could be the earliest record of complex life on earth. A more intensive search of Nankoweap Formation outcrops for comparable structures obviously is needed.

THE CHUAR GROUP AND SIXTYMILE FORMATION

The Chuar Group has been subdivided by Ford and Breed (Ford et al. 1972a; Ford and Breed 1973b) into two formations and seven members. The Sixtymile Formation was originally included as the top member of the Chuar Group until it was recognized as a distinct coarse-grained unit and moved to formation sta-

TABLE 4–1. Stratigraphic Subdivisions of the Chuar Group and Sixtymile Formation Including Thickness Measurements by Various Workers[a]

			Thickness [in feet (meters)]			
Group	Formation	Member	Ford and Breed (1973b)	Elston (1989) and references therein[b]	Cook (1991)	Dehler and Timmons (unpublished data, 1998)
	Sixtymile Formation		118 (36)	194–210 (59–64)		
		Walcott Member	836 (255)	922 (281)	758–804 (231–245)	823 (251)[c]
	Kwagunt Formation	Awatubi Member	1128 (344)	987 (301)		653 (199)[c]
		Carbon Butte Member	249 (76)	164 (50)		112–223 (34–68)[c]
Chuar Group		Duppa Member	571 (174)	341 (104)		2050 (625)[c]
	Galeros Formation	Carbon Canyon Member	1545 (471)	1148 (350)		2918 (889.5)[c]
		Jupiter Member	1515 (462)	1424 (434)		868 (264.5)
		Tanner Member	640 (195)	512 (156)		604 (184)

[a]For locations of measured sections, see individual publications.

[b]Sixtymile Formation thickness from Elston (1979) and Elston and McKee (1982), and Chuar Group (member) thicknesses modified by Reynolds and Elston (1986) and Reynolds, written communication to Elston (1988).

[c]Member measured sections or intramember partial measured sections that show lateral thickness variations. Thickness variations observed by Dehler and Timmons (unpublished data, 1998) correspond with structural trends across the Chuar syncline that may explain the thickness variability amongst workers.

tus. The stratigraphic nomenclature is summarized in Table 4.1 and shown graphically in Figure 4.2.

Reynolds and Elston (1986) have revised the thickness because they found the Galeros Formation to be only 3000 feet (915 m) thick and the Kwagunt Formation to be 2083 feet (635 m). Ongoing work by Dehler and Timmons (unpublished data, 1998) has revealed thickness variations across the Chuar syncline (west–east) in stratigraphic intervals within the middle and upper Chuar Group (Table 4.1). These thickness variations may be attributed to, in part, lateral facies changes, along with changes in accommodation space controlled by movement of the syncline during deposition. The thickness variations have been documented in the upper Carbon Canyon Member and younger members, implying that the Chuar syncline was developing as these sediments were being deposited.

The Chuar Group shows an overall pattern of repeating carbonate–shale cycles on the order of hundreds of meters thick. In the middle of the Chuar Group,

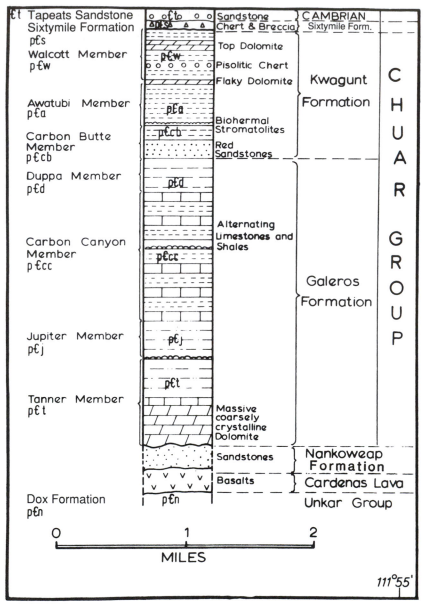

FIGURE 4.2. Schematic stratigraphic column of the Nankoweap Formation and Chuar Group.

the carbonate–shale cycles are interrupted by a distinctive red sandstone unit, the Carbon Butte Member, which marks the boundary between the Galeros Formation below and the Kwagunt Formation above. The Galeros Formation comprises the Tanner, Jupiter, Carbon Canyon, and Duppa Members; and the overlying Kwagunt Formation comprises the Carbon Butte, Awatubi, and Walcott Members. Research thus far suggests that contacts between members are gradational. These gradational contacts are also an attribute to the different thickness measurements attained by different workers (Table 4.1). Specifically, the contacts between the Jupiter, Carbon Canyon, and Duppa Members are very difficult to distinguish because there are no distinctive marker units to delineate these contacts, rather they are defined by the presence or absence of carbonate beds that are known to change lithologic character laterally.

GALEROS FORMATION
Tanner Member

The Tanner Member, which overlooks the Tanner Rapids on the Colorado River, consists of 20 to 80 feet (6 to 24 m) of thickly bedded, coarsely to finely crystalline dolomite at the base, and 580 feet (177 m) of almost entirely shales above. The basal Tanner dolomite disconformably fills in paleotopography cut into the upper member of the Nankoweap Formation. Sedimentary structures in the dolomite include parallel horizontal laminations and intraclast horizons. It forms a massive ledge capping the cliffs around Basalt Canyon (Fig. 4.3) and also crops out on the southern flank of Chuar Canyon. A fault-bounded sliver of Tanner dolomite exists on the north side of Nankoweap Canyon. The Tanner dolomite was included in the Unkar Group by Walcott, but it was transferred to the Chuar Group by Van Gundy (1951).

The overlying 177 meters of the Tanner Member is predominantly composed of shales, along with subordinate siltstones, sandstones, and dolomites. These strata are exposed throughout much of Basalt and Chuar Canyons. The shales are finely laminated to massive, are predominantly black, and weather ocherous yellow, orange, red-purple, pale green, and gray. The shales commonly have very thin to thin lenses and tabular beds of white to green siltstone and fine-grained sandstone. *Chuaria circularis* has been found in the upper 50 meters of black shale. Hematitic cement is common within the shales and sometimes weathers to goethite box-stones. Thicker sandstone and dolomite beds are present toward the top of the member. The sandstones are thinly to thickly bedded, green, and fine-grained, and they exhibit rare ripplemarks and mudcrack casts. The dolomite beds are massive and medium-bedded, and they are only found in the upper 2 meters of the member.

Jupiter Member

The Jupiter Member also consists of carbonates below and shales above. The basal division is about 40 feet (12 m) of stromatolitic limestones and dolomites, including undulating and broad-domed forms of algal laminate within a mass of dolomitized tufa-like rock, with flat-pebble conglomerates at the base. The upper part of the carbonate member has layers with abundant casts of gypsum crystals and a few, poorly defined and solitary stromatolite columns. These are similar to the form *Inzeria*, and to undulating stromatolites of the form *Stratifera*. The remainder of the 1516-foot (462-m)-thick member is predominantly shale, but commonly includes thin beds of sandstones and siltstones. The shales are variable in color, from red-purple to ocherous yellow to pale green to blue-black, and are commonly micaceous. Rare *Chuaria circularis* has been found in the black shales. The sandstones and siltstones are rarely more than a few inches thick and have abundant symmetric ripplemarks and mudcrack casts, as well as

FIGURE 4.3. Upper Basalt Canyon showing cliffs of Cardenas Lava in the bottom left corner, overlain by the flat-topped terrace of Nankoweap sandstones and Tanner Member dolomite, with shales up to the basal stromatolitic layers of the Jupiter Member, extending upwards toward the Great Unconformity beneath the Cambrian Tapeats Sandstone. Photograph by Parker Hamilton.

soft-sediment deformation features, ripple laminations, and rare raindrop prints and salt pseudomorphs. Hematitic cement is common in patches and sometimes weathers to goethite box-stones.

Carbon Canyon Member

The Carbon Canyon Member consists of alternating carbonate, shale, and sandstone beds [each a few feet thick (1 m)], attaining a total thickness of 1546 feet (471 m) (Fig. 4.4). The carbonates commonly are 3 to 6 feet (1–2 m) thick, and almost all are dolomicrite to dolosiltite with local chert nodules and common

FIGURE 4.4. The Chuar syncline looking north along its axis. Chuar Canyon in lower left; Carbon Canyon in the right center. The ridge in the middle foreground is of alternating dolomites, sandstones, and shales of the Carbon Canyon Member. The synclinal cuesta in the middle distance is made up of the Carbon Butte sandstone. The Butte fault trends north through the shadows on the right. (Photograph by John Shelton.)

laminations of quartz siltstone. In places, the carbonates grade into calcareous siltstones. Many of the carbonates have irregular (crinkly) laminations, which are interpreted to be of algal origin. The tops and bottoms of the carbonate beds commonly have symmetric ripplemarks and mudcracks. A dolomite marker bed in the upper part of this member exhibits a distinctive horizon of possible large-scale mudcracks (<0.5 m deep) that have undergone soft sediment deformation. These possible mudcracks form polygonal patterns on bedding planes, and probably indicate prolonged periods of exposure. Many carbonate beds grade from yellowish-tan crinkly laminated dolomites into reddish-orange massive dolomites, probably representing shallowing-upward cycles.

The interbedded fine-grained siliciclastics vary from blue-black, micaceous shale to red and green mudstones. Sandstones are fewer in number and rarely more than 2 feet thick (61 cm). They generally are green to gray to tan in color and have subangular to rounded quartz grains set in carbonate, silica, hematite, or chlorite cement, or clay matrix. Many of the fine-grained sandstones contain medium- to coarse-grained, well-rounded quartz sand that is randomly oriented in the fine-grained matrix. Mudcrack casts are common in the sandstones, and some laminae show truncated, incipient cracks that look misleadingly like worm tracks on weathered joint faces. Other sedimentary structures in the sandstones include symmetric ripplemarks, interference ripples, ripple laminations, soft-sediment deformation features, low-angle planar crossbeds, and trough crossbeds.

The stromatolites, located toward the top of the member, take the form of sharply widening, irregularly branching columns that show strongly convex laminae. These columns are clustered into domes, usually about 1.6 feet (0.5 m) in diameter and 1 foot tall (0.35 m), and taper at the base. Some clusters have been observed that are much as 6 feet (1.8 m) in diameter, yet still maintaining a height of 1 foot (0.5 m). The domes are commonly closely spaced, sometimes growing into one another. This stromatolitic horizon shows little lateral variation and is commonly interbedded with black shales and intraclastic dolomite. Dolomitization has destroyed internal detail, but the stromatolites appear to fall within the form *Baicalia* (probably *B.* aff. *rara*) Semikhatov.

Duppa Member

The Duppa Member is over 570 feet (174 m) thick and marks the return to a shaley character with only a few thin beds of dolomite. Calcareous siltstone beds up to 3 feet (1 m) thick are found throughout the member and contain well-rounded silt grains. The rest of the member is shale (generally micaceous) that grades into red mudstones and thinly bedded sandstones and siltstones toward the top. The contact with the overlying Kwagunt Formation is gradational.

KWAGUNT FORMATION

Carbon Butte Member

Because the Carbon Butte Member has at its base the only thick sandstone in the Chuar Group, it provides a distinctive marker for the base of the Kwagunt Formation. The member is 252 feet (76 m) thick at its type locality yet is found to vary in thickness across the Chuar syncline (Table 4.1). The basal 80 feet (24 m) comprises thickly bedded, tan to red, fine- to medium-grained sandstones

FIGURE 4.5. Mudcrack casts in the basal red sandstone of the Carbon Butte Member, Carbon Butte. Lens cap for scale. (Photo by Mike Timmons.)

with interbedded shales and siltstones which have weathered into a prominent cliff that surrounds Carbon Butte (Fig. 4.4). Sedimentary structures include 3- to 6-feet thick (1- to 2-m thick) cross-bed sets (epsilon?), symmetric ripplemarks, trough cross-beds, ripple laminations, soft-sediment deformation features, and mudcrack casts (Figs. 4.5 to 4.7). The overlying 170 feet (52 m) consists of predominantly red to purple mudstones and shales with subordinate thin to medium beds of fine- to medium-grained sandstone and siltstone. Some of these beds have mudcrack casts. In the upper part of this member is a 9-foot (3-m)-thick unit of white sandstone that exhibits very well preserved symmetric ripplemarks, interference ripplemarks, soft sediment deformation features, and trough cross-beds. Preliminary results from paleocurrent analyses indicate a dominant symmetric-ripple-crest trend of about north–south, similar to the trend of the Butte fault.

FIGURE 4.6. Symmetric ripplemarks in the basal red sandstone of the Carbon Butte Member, Carbon Butte. Preliminary paleocurrent data show that symmetric ripple crests are oriented north–south and parallel the Butte fault. Hammer for scale. (Photo by Mike Timmons).

FIGURE 4.7. Cross-bedding (epsilon?), and soft-sediment deformation in the basal red sandstone of the Carbon Butte Member, Carbon Butte. (Photo by Mike Timmons.)

Awatubi Member

The Awatubi Member comprises a basal stromatolitic carbonate unit overlain by dominantly shales and mudstones of varying colors. The basal 12 feet (3.7 m) of this 1128-foot (344-m)-thick member consists of biohermal domes, each 8–10 feet (2.5–3 m) in average height and width. These domes are made up of a complex of columns 2–3 inches (5–7.5 cm) in diameter, and are interbedded with confluent domes (Figs. 4.8 and 4.9). The columns show nearly flat laminae and generally are parallel-sided with rare bifurcation. Dolomitization has destroyed most of the internal detail, but a clearly defined wall is present, indicative of the form *Boxonia* Koroljuk. The matrix between columns generally is crystalline dolomite, whereas that between bioherms is coarsely granular dolomite of a highly porous nature. Flat-pebble conglomerates are present at the base of some bioherms.

The overlying variegated shales are interbedded with thin to very thin beds of sandstone and siltstone. These beds commonly exhibit ripple laminations, symmetric ripplemarks, interference ripplemarks, horizontal and low-angle planar laminations, and mudcrack casts. Approximately 30 feet (9 m) from the top of the member, black, finely fissile shales yield abundant *Chuaria circularis* on both the eastern and western slopes of Nankoweap Butte. The former is thought to be Walcott's type locality. These fossils have also been found in the same horizon in Sixtymile Canyon.

Walcott Member

The Walcott Member forms the topmost subdivision of the Kwagunt Formation and is much more diverse in character than those below it. It is 838 feet (255 m) thick and forms the upper part of Nankoweap Butte (Fig. 4.10). At the base

FIGURE 4.8. Stromatolite bioherm from the Awatubi Member, Kwagunt Canyon. Hammer for scale.

is a remarkable "flaky dolomite" that ranges in thickness from 12 to 31.5 feet (3.75 to 9.6 m) (Fig. 4.11; Cook 1991). This unit is composed of oolitic and intraclastic dolomite; folded and broken, crinkly, silicified and dolomitic laminations (stromatolite horizon); and wavy to horizontally laminated, silty intraclastic dolomicrite (Cook 1991).

Above the flaky dolomite are *Chuaria*-bearing black shales, silicified oolite and pisolite beds, and three distinctive carbonate units. The oolite and pisolite beds generally range in thickness from 6 inches to 1 foot (15 to 30 cm) and are found interbeddded with the black shale (Cook 1991). An exception is the white silicified oolitic bed, formerly classified as a pisolite by Ford and Breed (1973a), which is 4–5 feet (1.2–1.5 m) thick (Cook 1991). The pisolite beds are jet black and contain ooids and pisoids that are completely replaced by chert. The outer surfaces of some pisoliths contain a mat of algal filaments of at least two types, as well as spheroidal bodies (Schopf et al. 1973). Toward the top of the member are three dolomite units, the lower two are known as the "dolomite couplet," and the upper unit is known as the "karsted dolomite" (Cook 1991).

The lower dolomite couplet is 8.5 to 22 feet (3.5 to 7 m) thick and is composed of wavy to horizontally laminated dolomite (some contorted), trough–cross-bedded oolitic and intraclastic dolomite, a crinkly laminated, "cornflaky" algal stromatolite, and pink medium-grained quartz sandstone (Cook 1991). The upper dolomite couplet is 31–38 feet (9–12 m) thick and is a massive micrite to dolomicrite with rare mud chips, interbeds of black shale, and carbonate breccia zones (Cook 1991). The karsted dolomite is 40 feet (12 m) thick and is only present in Sixtymile Canyon. This unit comprises crystalline dolomite with vugs, cavities, dissolution features, and brecciated dolomite and sandstone clasts in some cavities (Cook 1991).

FIGURE 4.9. Internal columnar structure of a bioherm from the Awatubi Member, Carbon Canyon. Lens cap for scale. (Photo by Mike Timmons.)

Paleontology of the Chuar Group

During the past century, a number of fossils or possible fossils have been described from the Precambrian rocks of the Grand Canyon. In recent years, most of these either have been reassigned to biological groups other than those in which they first were placed, or their biologic affinities have been questioned. With the increasing interest in the early stages of life on our planet, several reviews of Precambrian life have been published, and these have presented conflicting interpretations of some of the fossils of the Grand Canyon. A review is presented here of the state of knowledge concerning fossils from the Chuar Group.

Chuaria circularis This small, disc-like, carbonaceous fossil first was found by Walcott during his pioneer work on the Grand Canyon Series (Precambrian) in

FIGURE 4.10. Nankoweap Butte, showing the upper Awatubi Member, the overlying Walcott Member with its ledges of pisolitic chert and flaky dolomite, capped by the Sixtymile Formation. Seen looking northeast toward the lower part of Nankoweap Canyon. The axis of the Chuar Syncline runs through Nankoweap Butte, and the east limb of the Chuar Syncline is in the upper right of center. The Butte fault is in the shadows in the upper part of the photo and runs right to left. (Photograph by John Shelton.)

1882–1883. It was formally named in 1899 and described as a primitive brachiopod allied with *Orbiculoidea* or *Discina*. As such, it was the first Precambrian brachiopod to be described, and it is somewhat surprising that it was largely overlooked by subsequent writers. The exceptions include Wenz (1938), who assigned *Chuaria* to the gastropods without giving very clear reason for so doing; Hantzschel (1962), who regarded *Chuaria* as inorganic; and both Glaessner (1966, 1984) and Cloud (1968), who tentatively regarded it as algal in nature, without going into details. A full description was given by Ford and Breed (Ford et al. 1972b; Ford and Breed 1973a).

Figure 4.11. The "Flaky dolomite" bed in the Walcott Member, Nankoweap Butte. Hammer for scale.

Others, including Vidal and Ford (1985), reexamined the problem and showed that this smooth, organic-walled, spheroidal microfossil was best classified with the Acritarcha (algal cysts of unknown affinity). It has been found at many horizons in the Chuar Group, where it is associated with assemblages of other late Riphean to early Vendian acritarchs.

The size range of *Chuaria circularis* revealed by palynological techniques was 70–712 μm; taken together with the megascopic specimens, this gives a total range of 70 μm to about 5 mm. The large forms are most common at what is presumed to be Walcott's-type locality, high in the Awatubi Member about 30 feet (9 m) below a prominent ledge of pisolitic chert on the eastern flank of Nankoweap Butte.

The large forms of *C. circularis* are crushed, carbonized, (originally) spheroidal objects commonly 2–3 mm in diameter, lying either alone or in small clusters (they are never seen to lie on top of another), indicating their globular shape at the time of deposition (Fig. 4.12). Specimens extracted from the shale with acids are hollow with a narrow marginal thickening. The central area is wrinkled like a crushed prune. No apertures have been seen and no ornamentation is recognized. It clearly is not a brachiopod or gastropod. Other tentative suggestions included the possibilities that *Chuaria* might be a trilobite egg or a non-calcareous foraminiferan; however, no evidence to support such suggestions can be found. The size range down to 70 μm, together with the composition and features noted above, makes it clear that *Chuaria* is an acritarch.

Microfossils Samples of the black shale beds throughout the Chuar Group have yielded organic detritus, and most horizons have yielded acritarchs. These are spheromorphic, cyst-like objects presumed to be the resting stage of some form of algae. Among the genera recovered are *Stictosphaeridium, Trachysphaeridium, Leiosphaeridia, Kildinosphaera, Cymatiosphaeroidies, Tasmanites*, and small *Chuaria*. A new form, *Vandalosphaeridium walcotti*, also was discovered.

Figure 4.12. *Chuaria circularis*—a mega-acritarch from the Walcott Member. Each carbonized disc is about 2 mm in diameter.

Assemblages with either megascopic *Chuaria* or the above microplankton have been recovered from many late Precambrian strata around the world, and they can be used for stratigraphic correlation. On these micropaleontological grounds, the Chuar Group can be correlated with sequences in Utah, northwest Canada, Greenland, Scandinavia, Svalbard, Russia, Iran, India, China, and Australia. In both Scandinavia and Russia, the sequences are earlier than the Vendian glacial deposits and are placed in the latest Riphean and earliest Vendian divisions of the Proterozoic—that is, around 800–700 million years ago. This is slightly younger that the date of ca. 823 million years deducted for the Grand Canyon disturbance by Elston and McKee (1982), but dating of Precambrian sedimentary deposits still has many uncertainties.

Vase-Shaped Microfossils A shale horizon in the Walcott Member, about 3 feet (1 m) below the prominent pisolitic chert, has yielded an assemblage of vase-shaped, organic-walled microfossils (from 32–170 μm in length) that Bloeser (1985) has named *Melanocyrillium*. Up to 10,000 specimens per cubic centimeter of shale have been found. The teardrop-shaped, vase-like, noncolonial microfossils have an ornamented aperture at the narrow end, and three species have been erected on the basis of ornamental detail. They are of uncertain biological affinity and are not chitinozoa, as first thought (Bloeser et al. 1977). Comparable forms have been found in rocks of late Precambrian age in Sweden, Greenland, and Brazil.

An assemblage of well-preserved filamentous and spheroidal plant microfossils have been found in a cherty pisolite bed of the Walcott Member on the slopes of Nankoweap Butte in eastern Grand Canyon (Schopf et al. 1973). The spheroidal forms probably are related to coccoid blue-green algae; the filaments bear a striking resemblance to *Eomycetopsis*, a thallophyte from bedded cherts in the Bitter Springs Formation of central Australia. Preliminary studies have revealed similar filamentous forms at other horizons of the Chuar Group.

Stromatolites Stromatolites are common in the Precambrian of the Chuar Group, where they first were noted as "concretionary limestones" by Walcott (1894). They were given the now obsolete name *Cryptozoon occidentale* by Dawson (1897), though their nature was not then understood. Ford and Breed (Ford et al. 1969, 1972a; Ford and Breed 1973a) have recorded three well-developed stromatolite horizons in the Chuar Group, as well as a large number of other carbonate beds with crinkly laminations of probable algal origin. They generally have shapes such as gentle undulations or low domes with confluent laminae and only rarely show the more distinctive columnar forms that Russian workers regard as diagnostic fossils. Forms regarded as *Inzeria, Baicalia,* aff. *B. rara,* and *Boxonia* have been recognized by comparing the external form with those illustrated by Cloud and Semikhatov (1969), though diagenetic alteration has obscured internal detail and made more specific identification impossible. Recognition of these genera allows correlation of the Chuar Group with the Russian Upper Riphean—that is, late Precambrian.

The three beds of stromatolites in the Chuar Group are well-exposed. The lowest horizon forms a prominent ledge and waterfall about a mile up Chuar Canyon from its mouth on the north side. It also is seen as a prominent ledge in the middle reaches of Basalt Canyon on the north side. The middle horizon is a single bed of limestone and dolomite within the upper part of Carbon Canyon Member—outcropping low on the flanks of Carbon Butte and also about 100 feet (33 m) above the creek in Nankoweap Canyon at the foot of Nankoweap Butte. The highest forms a bed of stromatolite "reefs," or bioherms, cropping out around the foot of Nankoweap Butte in Nankoweap and Kwagunt canyons (Figs. 4.9 and 4.10) and just above the prominent sandstone bench of Carbon Butte.

The paleoenvironment of stromatolites in the Chuar Group, judged by associated sedimentary rocks, is one of relatively low energy, shallow waters. Ripplemarks and mudcracks are common, suggesting a gentle current and intermittent desiccation. Intraclastic breccias associated with the stromatolites suggests periods of relatively higher energy current associated with possible dessication. The three major stromatolites and many other crinkly laminated carbonate beds are interbedded with black shales and maroon mudstones. These relationships suggest that algal growth may have been the most successful during times of less siliciclastic input, and when more siliciclastics came into the system, the water became too clouded for algae to survive.

Chuar Group Paleoenvironments

In the first and only study of the entire Chuar Group, Ford and Breed (Ford et al. 1972b; Ford and Breed 1973a) stated that the presence of megaplankton (*Chuaria*) and abundant microplankton, stromatolites, and filamentous algae indicate at least partial marine influences, and that sedimentologic evidence suggests a marine shoreline setting. Subsequent studies have resulted in a range of depositional interpretations—from deep and shallow marine, to alluvial and lacustrine (Horodyski 1986; Reynolds and Elston 1986; Cook 1991). Ongoing work on the sedimentology and stratigraphy of the Chuar Group is in general agreement with Ford and Breed's original interpretation (Dehler 1998). Knowing whether the Chuar Group sediments were deposited in a lake or in a coastal setting can reveal information about the paleogeography of the region at this time in the late Precambrian. If the Chuar Group is marine in origin, then it suggests that there was a continental margin in what is now northern Arizona during late Precambrian time. If the Chuar Group is lacustrine in origin, the Chuar basin was inboard of the continental edge.

Galeros Formation The basal Tanner dolomite is probably very similar to other overlying carbonate units, other than the fact that it has been severely diagenetically altered. The faintly laminated Tanner dolomite is likely similar to laminated dolomites in other parts of the Chuar Group that suggest a shallow subtidal or intertidal environment. The overlying black shales, sometimes *Chuaria*-bearing, represent a deeper water environment that occasionally received silts and sands during storms. Reynolds and Elston (1986) interpreted the Tanner Member to represent a sediment-starved basin, rich in organic material. The overlying Jupiter Member contains the basal *Inzeria* bed that also represents shallow subtidal to intertidal deposition. The overlying shales in the Jupiter are variegated and commonly display mudcracks, raindrop impressions, and salt crystal casts. These shales probably represent an intertidal to supratidal environment. Reynolds and Elston (1986) interpreted the Jupiter Member to represent a coastal or alluvial plain. The Carbon Canyon Member has a mixture of characteristics similar to the Tanner and Jupiter Members, and it represents fluctuating deeper subtidal to intertidal to supratidal environments. Mudcracked wave ripplemarks suggest fluctuations in currents that were tide- and wave-affected. Reynolds and Elston (1986) interpreted this member to represent a mixed coastal or paludal swamp. The only difference between the Duppa Member and the underlying Carbon Canyon Member is the lack of carbonate and sandstone beds. Therefore, the Duppa Member may also represent a deeper subtidal to supratidal environment. The Duppa Member is interpreted by Reynolds and Elston (1986) to represent an alluvial plain.

Kwagunt Formation The sedimentary structures in the Carbon Butte Member, such as mudcrack casts, wave ripplemarks, interference ripplemarkss, bidirectional trough crossbeds, and possibly epsilon crossbeds, indicate a tide- and wave-affected shoreline. However, some of these sedimentary structures could represent fluvial conditions. The Carbon Butte Member sandstones represent a dominance of relatively coarse, clastic sediments indicating an increase in sediment supply, a shift in environments (closer to the shoreline), or a significant change in energy conditions. The basal Awatubi Member bioherm suggests shallow subtidal to intertidal conditions. The upper Awatabi Member has symmetric ripplemarked sandstones with mudcrack casts similar to those in the Carbon Butte Member, and the uppermost Awatubi Member has *Chuaria*-bearing black shales. The Awatubi Member represents an intertidal to shallow subtidal environment that deepened through time. The Walcott Member has been interpreted by Cook (1991) to represent a carbonate ramp: The black shales are the deepest water environment, the oolite and pisolite beds are the shallow subtidal environments, and the carbonate beds represent shallow subtidal, intertidal, and supratidal environments.

The sedimentary rocks of the Chuar Group suggest that, in general, they were deposited in a quiet, nonturbulent embayment on a tide- and wave-affected marine platform fringing a continental mass. The outcrops are too limited, however, to indicate the direction of the coastline relative to this embayment, although ongoing basin analysis will result in more data regarding this problem (Dehler 1998; Timmons et al. 1998).

SIXTYMILE FORMATION

In contrast to the underlying Chuar Group, the Sixtymile Formation is composed largely of breccias and sandstones, with subordinate siltstones and mudstones. Slump folds are common, as are large boulders interpreted to be landslide blocks derived from the east limb of the Chuar syncline. About 200 feet (60 m) thick, the Sixtymile Formation caps Nankoweap Butte, where it has been misidentified in the past as basal Cambrian Tapeats Sandstone (Van Gundy 1951). However, magnificent exposures of this unusual formation are present in the little-visited and remote Awatubi and Sixtymile Canyons. The formation is named from the latter canyon, and its northwestern cliffs are well worthy of detailed study (Fig. 4.13). The Sixtymile Formation is beveled by the Great Unconformity and over-lain by the Cambrian Tapeats Sandstone in both Sixtymile and Awatubi Canyons.

Elston (1979) divided the Sixtymile Formation in Sixtymile Canyon into three unnamed members. He also provided a comprehensive and detailed description of the rock types, together with an analysis of the events resulting in this un-usual formation. The lower member, only present in Sixtymile Canyon, is as thick as 90 feet (27 m) and consists of coarse breccias full of many different types of clasts, including chert and dolomite derived from the underlying Chuar Group (Fig. 4.14). Among these are large blocks, the largest one measuring about 33 × 20 × 20 feet (10 × 6 × 6 m). A dolomitic limestone unit, 26 × 130 feet (8 × 40 m), was interpreted by Elston (1979) to be one of these landslide blocks. Cook (1991) studied this limestone unit in detail and concluded that it was *in situ*, and he named it the karsted upper dolomite (see Walcott Member discussion). Also in the lower member are red sandstones, some of which are thinly bedded and laminated.

FIGURE 4–13. Sixtymile Canyon: The Sixtymile Formation is the dark part of the cliffs on the right.

FIGURE 4–14. Intraformational breccia in the Sixtymile Formation, Nankoweap Butte. Hammer for scale.

The middle member is a thinly bedded and laminated, fine-grained quartzitic sandstone and siltstone 80 feet (25 m) thick, with common chert lenses, parting lineations, and massive white thin beds that may be tuffaceous. Slump fold axes in this member parallel the axis of the Chuar syncline.

The upper member is 40 feet (12 m) thick and comprises locally derived, fine- to coarse-grained sandstone, siltstone, conglomerate, and breccia. This member fills in paleotopography cut into the middle member, and the clasts are all derived from the middle member. These cut-and-fill features are as deep as 16 feet (5 m), and channel axes have northwest–southeast trends (parallel to the Butte fault). Other sedimentary features include contorted bedding, adhesion ripples, and small cut-and-fill structures filled with trough–cross-bedded sandstone showing paleoflow to the east–northeast.

Sixtymile Formation Paleoenvironments

A drastic change in environment is recorded in the Sixtymile Formation, yielding very coarse clastic sediments, probably under largely terrestrial(?) and syntectonic conditions. Landsliding and slumping in early Sixtymile time was followed by colluvial(?) and alluvial(?) sedimentation in later Sixtymile time, indicating that there was significant local relief throughout Sixtymile deposition.

Large boulders, breccias, and contorted bedding characterize the lower member and indicate mass-wasting processes. Elston (1979) interpreted these mass-wasting events to be controlled by folding of the Chuar syncline and uplift along the Butte fault. The lower member red sandstones were deposited in subaerial environments at or near sea level, and the red sandy siltstone represents mudflat deposition Elston (1979). The thinly bedded siltstones and sandstones of the middle member are probably representative of a low-energy fluvial environment such as a floodplain. Elston (1979) suggested that the middle member was deposited in a localized lake in the synclinal axis. Paleovalleys cut into the middle member and filled with coarse breccias and sandstones of locally derived, middle member clasts potentially indicate exhumation and deposition by debris flow, and fluvial processes. Elston (1979) interpreted these upper member deposits to be of fluvial origin that were eroding off the high-standing footwall of the Butte fault. Ongoing work on the Sixtymile Formation will contribute to our understanding of the depositional environment(s) of this enigmatic unit (Karl Karlstrom, personal communication, 1998).

CORRELATION OF THE NANKOWEAP FM, CHUAR GROUP, AND SIXTYMILE FM

Younger Precambrian strata are well- and widely exposed across central and southern Arizona, where they are known as the Apache Group. The Apache Group is correlated with the Unkar Group based on (a) lithologic similarities (Troy Quartzite with the Shinumo Quartzite, respectively) and (b) diabase sills of similar ages which cross-cut both groups (Link et al. 1993). There is no equivalent to the Nankoweap Formation, Chuar Group, or Sixtymile Formation in Arizona, yet comparisons can be made with strata in other parts of western North America.

In the Death Valley region of California, a thick infra-Cambrian sequence lies beneath the equivalent of the Tapeats Sandstone and rests on the Pahrump Group. This lower part of the Pahrump Group (Crystal Spring formation) is correlated with Unkar Group and the Apache Group based on similar texture and chemistry of crosscutting diabase sills (Link et al. 1993). The Chuar Group likely correlates with the Beck Springs Dolomite of the upper Pahrump Group, based on lithologic (shales, stromatolites, carbonate rocks), paleoenvironmental (fluvial-marine), and acritarch assemblage similarities. Unconformably above the Beck Springs Dolomite lies the Kingston Peak Formation, which contains tillite (Link et al. 1993). No glacial deposits have been identified in the Chuar Group, and based on microfossil global correlations, the Chuar Group predates the late Precambrian glacial events (Vidal and Ford 1985). The Red Pine Shale of the Uinta Mountain Group in Utah has yielded microfossils almost identical with those in the Chuar Group. Several geologists have proposed a correlation between these two groups (Hoffmann 1977; Vidal and Ford 1985).

The Belt Supergroup, predominantly located in Montana, was previously thought to be coeval with the Apache and Unkar Groups. Recent U-Pb ages of about 1400 Ma or older for the Belt Supergroup make this correlation unlikely (Aleinikoff 1996). The northern Canadian Windermere Supergroup unconformably overlies a dike of about 778 Ma and is cross-cut by a sill of about 723 Ma (Link et al. 1993). The presence of *Chuaria* in the Hector Formation of the Windermere Group provides support for a correlation with the Chuar Group (Ford and Breed 1973a, 1973b). The Little Dal Group of the Mackenzie Mountain Supergroup in northwest Canada is cross-cut by a diabase that is about 778 Ma, and it has yielded microplankton fossils similar to those in the Chuar Group (Hoffmann and Aitken 1979; Link et al. 1993). As noted earlier, the age of the Grand Canyon disturbance has been dated at about 823 Ma, which correlates reasonably well with the structural disturbance that terminated the Little Dal Group in northwestern Canada at about 778 Ma.

Support for these correlations comes from paleomagnetic studies of several of the above-mentioned sequences. Comparisons of paleomagnetic poles from Keweenawan igneous rocks and the Cardenas Lava are consistent with an age of about 1100 Ma (Link et al. 1993 and references therein). Limited paleomagnetic data (Elston and Grommé 1974; Elston and McKee 1982; Elston 1989a) are consistent with an age of less than about 1050 Ma for deposition of the upper member of the Nankoweap Formation. Paleomagnetic results from sills in the Little Dal Group yield paleomagnetic poles compatible with those for the Chuar Group strata (Don Elston, personal communication, 1997). There are strong indications that many of the Supergroup and Sixtymile Formation rocks contain primary magnetizations, but detailed and necessary field-based tests remain to be conducted to test this hypothesis (John Geissman, personal communication, 1998).

Although the ages of some of these sequences cannot yet be ascertained by isotopic dating, the microfossil evidence provided by acritarch assemblages strongly suggests intercontinental correlation with the Visingso Group of Sweden and, thus, with the latest Riphean of Russia. Isotopic age determinations in Sweden and Svalbard indicate the age of these beds to be around 800–700 million years (Vidal and Ford 1985). Applying a correlation by long-distance lithologic comparison, micropaleontologic assemblages, paleomagnetism, and very limited isotopic dating, it seems that the Nankoweap Formation, Chuar Group, and Sixtymile Formation were deposited within the period ~1000–700 million years ago.

SUMMARY

The Chuar Group represents deposition associated with a fluctuating shoreline in a tectonically active marine basin. Preliminary paleocurrent data on symmetric ripplemark crests show trends that parallel the Butte fault, suggesting the shoreline orientation was also north–south. Black shales throughout the Chuar Group, many *Chuaria*-bearing, represent the deepest subtidal environment. Maroon mudstones, pisolite and oolite beds, and carbonate beds (including stromatolites) were deposited in subtidal environments. Many sediments were exposed intermittently or for prolonged periods in intertidal and supratidal zones. Sandstones represent wave- and tide-affected shallow subtidal and estuarine(?) conditions.

The late Precambrian paleontological record from the Chuar Group consists of a variety of microfossils from filamentous algal sheets, coccoid algae, acritarchs (including the megascopic *Chuaria*), vase-shaped microfossils of unknown affinity, and stromatolites. The lobate markings from the Nankoweap Formation can be regarded as a dubio-fossil and may not be of organic origin at all.

The Nankoweap Formation has no known correlative sequences. The Chuar Group and Sixtymile Formation(?) are likely correlative with the Beck Springs Dolomite in Death Valley, the Red Pine Shale in Utah, and the Little Dal Group or the Windermere Group in northern Canada, based on lithologic, paleontologic, and geochronologic data. Global correlations can be made with 700- to 800-Ma successions in Svalbard and Sweden using archritarch assemblages.

• 5 •

Geologic Structure of the Grand Canyon Supergroup

J. Michael Timmons, Karl E. Karlstrom, and James W. Sears

INTRODUCTION

This chapter summarizes the geologic structure and tectonic history of the Grand Canyon Supergroup. This sequence of tilted Precambrian strata is approximately 4 km thick and is sandwiched between the overlying, subhorizontal Cambrian Tapeats Sandstone [500 Ma (million years ago)] and the underlying, highly contorted Granite Gorge Metamorphic Suite (1700 Ma) (Chapter 2, this volume; Ilg et al. 1996; Hawkins et al. 1996). The Grand Canyon Supergroup is unmetamorphosed, and its near pristine preservation provides a remarkable record of the formation of intracratonic extensional basins in the Late Precambrian.

John Wesley Powell first observed these tilted rocks in 1869 on his historic journey down the Colorado River and recognized that they record an important chapter in the tectonic history of the region. He concluded that the beds were folded, uplifted, and deeply eroded before the overlying Paleozoic rocks were "spread over their upturned edges" (Powell 1876). Charles Walcott outlined the geologic structure and stratigraphy of the sequence in a series of reports (e.g., Walcott 1883, 1890, 1894). Walcott recognized that the Grand Canyon Supergroup is Precambrian in age and suggested a correlation with other sequences in North America.

Since Walcott's pioneering geologic work, geologists have refined the stratigraphy and continue to debate how the Grand Canyon Supergroup correlates with other Precambrian strata of North America and other continents (see Chapters 3 and 4, this volume). Stratigraphic studies show there are major uncomformities within the Grand Canyon Supergroup that divide it into several sequences. From the base to the top, the major uncomformity-bounded sequences are as follows: (1) the Unkar Group and associated 1.1-Ga (billion years old) basaltic magmatism, (2) Nankoweap Formation, (3) Chuar Group, and (4) Sixtymile Formation. The groups are subdivided further into formations and, in most cases, members. Chapters 3 and 4 of this volume discuss the stratigraphy and depositional history of the Grand Canyon Supergroup.

Radiometric dating suggests that (a) the Unkar Group was deposited sometime between about 1.3 Ga and 1.1 Ga, in the Mesoproterozoic era, which lasted from 1600 to 1000 Ma, and (b) the Chuar Group was deposited sometime between about 1100 and 800 Ma, mostly in the Neoproterozoic Era, which lasted from 1000 to 540 Ma (Elston and McKee 1982; Bowring et al. 1996).

The Late Precambrian is an important yet still poorly understood period in the history of western North America, as well as globally. Recently, several researchers have proposed that during this time North America and other continental masses were assembled into a supercontinent called Rodinia. Following assembly at 1.0 Ga, Rodinia began to break apart in one or more episodes of rifting that took place between 750 to 550 Ma. The Grand Canyon Supergroup records part of this long history, and it is a key locality for unraveling the timing and processes of intracratonic rifting, the evolution in the character of global seawater, and the rapid diversification of life on earth leading to the Cambrian 'explosion' (Dalziel 1997).

Figure 5.1 shows the distribution of Mesoproterozoic sedimentary rock exposures in the eastern Grand Canyon. The Unkar Group occurs in isolated wedge-shaped remnants (grabens and half grabens) in areas along the river, between mile 65 and 137 (shown in Fig. 5.2). The Chuar Group is exposed in the Chuar Valley, a region west of the Colorado River and just west of the East Kaibab monocline. This monocline is a great step-like fold on the eastern edge of the Kaibab Plateau that formed in Laramide time by the reactivation of a Precambrian fault (see below). Farther west, Paleoproterozoic rocks of the Granite Gorge Metamorphic Suite (GGMS) emerge again near Granite Park (mile 207) and in the Lower Granite Gorge, but there are no more exposures of the Grand Canyon Supergroup beyond Deer Creek.

Early workers recognized that the Grand Canyon Supergroup is only preserved in fault-bounded, downdropped blocks, which protected it from pre-Tapeats erosion. We can piece these fragments together to decipher some aspects of the Meso and Neoproterozoic tectonic history of the Grand Canyon

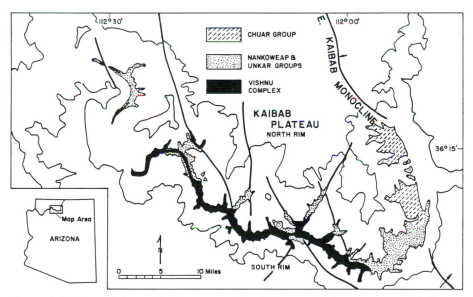

FIGURE 5.1. Precambrian rocks in eastern Grand Canyon. Major structures with Precambrian ancestry are also shown.

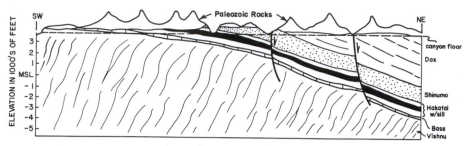

FIGURE 5.2. Cross section along the course of Bright Angel Creek, illustrating the wedge-like character of remnants of the Grand Canyon Supergroup.

region, but the original extent and distribution of the sediments and the shape of the basins remain unknown. Figure 5.3 provides a hypothetical view of the geology of the Grand Canyon region as it might have looked at the close of Proterozoic time, before Paleozoic sediments buried the region. The Proterozoic sedimentary rocks occupy a few large fault blocks bounded by north to northwest-trending, southwest-dipping normal faults with west-side-down sense of movement. Within the fault blocks, Proterozoic strata generally dip gently to the northeast, toward the southwest-dipping normal faults. Due to this regional tilt, basement metamorphic rocks emerge on the southwestern edges of the fault blocks and strata become younger toward the north. The youngest of the Proterozoic sedimentary rocks are found adjacent to the Butte fault, the eastern edge of a large asymmetric graben.

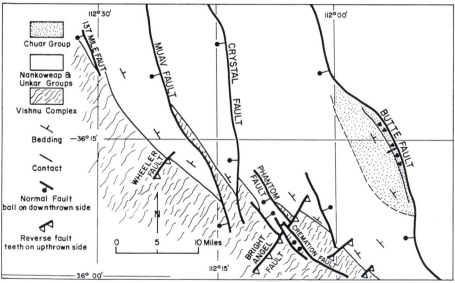

FIGURE 5.3. Schematic map of Precambrian structures and Supergroup rocks as it might have looked at the end of the Precambrian, prior to Paleozoic burial. This figure shows the same area as Fig. 5.1.

POST-1400-MA UPLIFT: EXHUMATION OF THE MIDDLE CRUST IN THE SOUTHWEST

The tale of the Grand Canyon Supergroup begins where the story of the Granite Gorge Metamorphic Suite (GGMS) ends. The GGMS experienced a complex interaction of deformation, metamorphism, and magmatism at 10- to 20-km depths between 1700 and 1680 Ma (Chapter 2, this volume; Hawkins et al. 1996). One wonders how and when the >10-km-thick section of crust was eroded prior to the Grand Canyon Supergroup deposition.

The Great Uncomformity developed by erosion of about 10 km of crust to expose middle crustal metamorphic and igneous rocks at sea level as a broad low-relief surface. This erosion took place mainly after the emplacement of voluminous granites at ca. 1.4 Ga (Anderson 1982; Nyman et al. 1994) like the Quartermaster pluton in the Lower Granite Gorge, and before the deposition of the Bass Limestone. The Bass Limestone was deposited between 1.3 Ga, a cooling age on micas in schists below the uncomformity (Bowring et al. 1996), and 1.1 Ga, the age of mafic intrusions into the limestone.

STRUCTURAL GEOLOGY OF THE UNKAR GROUP

The Unkar Group is a 2-km-thick section with carbonate at its base overlain by thick marine to fluvial siliciclastic rocks. The section is cross-cut by sills and dikes of diabase (ca. 1.1 Ga) and covered by the uppermost formation of the Unkar Group, the ca. 1.07-Ga Cardenas Lavas (Elston and McKee 1982). The Unkar Group contains both contractional and extensional faults, as discussed below, and apparently record a period of basin formation that overlapped in time with the early stages of Grenville collisions to the south (see regional implications below).

Contractional faults horizontally shorten (and vertically thicken) a section of rocks. A number of northeast trending, steeply southeast dipping contractional faults of Mesoproterozoic age cut the lower parts of the Grand Canyon Supergroup and the underlying Granite Gorge Metamorphic Suite. These brittle faults record southeast-side-up sense of movement (reverse sense), suggesting northwest–southeast-directed shortening. Faults die out up-section into step-like, monoclinal folds in Unkar strata, where slip was taken up along bedding planes. Figure 5.4 illustrates the evidence for Unkar-age slip on the Bright Angel fault and monocline. The monocline is very tight in the lower beds, which are vertical and locally overturned adjacent to the fault, but the fold opens steadily upwards in the Unkar section. Within the Dox Formation, the fold is a broad, subtle deflection. A thickness change of 60 m within the Shinumo Formation on the northwest side of the Bright Angel fault suggests that faulting and sedimentation were synchronous. Total displacement across the fault in Proterozoic time is about 240 m, southeast-side-up.

Other Proterozoic reverse faults and monoclines are exposed in Bass, Vishnu, and Red Canyons. Contractional faults tend to parallel the regional northeast trending metamorphic fabric (grain) of the Granite Gorge Metamorphic Suite. Apparently the faults follow older planes of weaknesses in the earth's crust. Proterozoic contractional structures are known only in the Unkar Group and older

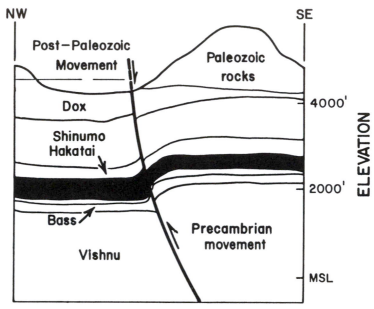

FIGURE 5.4. Cross section of the Bright Angel structure, northeastern Bright Angel Creek. Note that the Shinumo Quartzite is 60 m thicker on the northwest side of the fault, suggesting that faulting and sedimentation were synchronous.

rocks, suggesting that contractional brittle deformation is restricted to Unkar age. The timing of contractional faulting is important when we try to place the Unkar Group in a regional tectonic history (see regional implications below).

Extensional, or "normal" faults form during horizontal stretching of the earth's crust. Figure 5.5 shows five major extensional faults forming an array about 50 km wide from east to west in the depths of the Grand Canyon. Individual faults trend toward the north-northwest along curving traces, and they generally dip about 60° toward the west–southwest with west-side-down sense

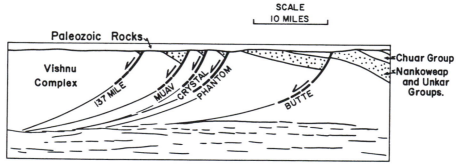

FIGURE 5.5. Schematic cross section showing major extensional faults of the Grand Canyon Supergroup. Rotation of the beds suggests that the faults flatten with depth.

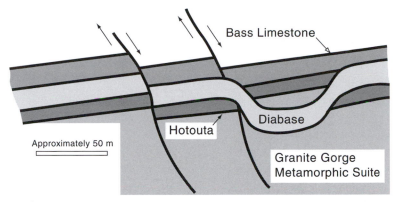

FIGURE 5.6. Fault relationship illustrating pre- and postdated 1.1-Ga faulting in the Unkar Group.

of movement. This geometry of tilted domino-style blocks is common in areas of crustal extension. Tilting of beds toward faults can also form if the faults, which are steep where exposed, curve and flatten with increasing depth. This listric geometry (Fig. 5.5) is also a pattern commonly documented in extensional terrains (Basin and Range province, Rio Grande rift, East African rift, and many others).

Documenting the timing of movement of extensional faults found within the Unkar Group is difficult. These faults are often truncated by the Cambrian Tapeats Sandstone and hence are Precambrian in age, but we have no way to test if those faults were truncated by younger Chuar sediments (indicating pre-Chuar age faulting). To further complicate timing interpretations, there is increasing evidence that extensional faults were active throughout deposition of the Grand Canyon Supergroup, suggesting that many of the faults had multiple movement histories. Figure 5.6 illustrates a fault relationship that indicates extensional faulting for both pre- and postdated 1100-Ma sill intrusions. There is also evidence for soft sediment deformation in the Dox Formation along the Butte fault, and the extensive convoluted bedding in the Shinumo Quartzite may suggest seismicity during Unkar deposition. Thus, available evidence suggests that Mesoproterozoic NW-trending extensional faults formed coevally with NE-trending contractional faulting and folding during Unkar deposition.

Nankoweap Formation

The Nankoweap Formation is a thin (100 m) section of red sandstones that is bounded above a below by uncomformities. Some workers have proposed that it represents a continuation of Unkar-like red bed sedimentation after the Cardenas Lavas, and noted a low angle angular uncomformity within the formation (Elston et al. 1993). Both the angular uncomformity and observed synsedimentary fault relationships suggest that extensional deformation was active during Nankoweap time. For example, in the Tanner graben we see a fault relationship that shows clear offset of the lower member of the Nankoweap Formation (Fig. 5.7). These offsets are filled and covered by the upper member of the Nankoweap Formation, clearly implying that faulting occurred during early Nankoweap time.

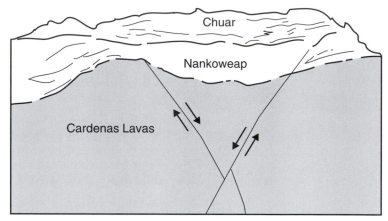

FIGURE 5.7. Graben in the Cardenas Lavas and lower Nankoweap Formation covered by upper Nankoweap and Chuar Group rocks.

STRUCTURAL GEOLOGY OF THE CHUAR GROUP AND SIXTYMILE FORMATION

Perhaps the preeminent structure in the Grand Canyon is the Butte fault system of the eastern Grand Canyon (Walcott 1890; Figs. 5.1 and 5.5). This structure was reactivated and now forms the main fault of the Laramide-age East Kaibab monocline, but its Precambrian movement history was longer and involved larger displacements. The Butte fault system is the easternmost structure of the array of exposed Precambrian normal faults (Fig. 5.5); it has the largest displacement and records a longer history of movement, warranting a separate discussion. Previous studies indicate the Butte fault may have been a basin-bounding normal fault, active during the deposition of the uppermost Chuar Group and Sixtymile Formation (Elston 1979; Ford and Breed 1973b; Cook 1991). Our new work has updated previous studies and suggests that most of the Chuar Group (Carbon Canyon, Duppa, Carbon Butte, Awatubi, Walcott Members) and Sixtymile Formation show thickness variations consistent with synchronous deposition, fault movement, and synclinal development throughout much of Chuar time.

The Chuar Group is a 2-km-thick section of shale and interbedded thin dolomite and sandstone that is spectacularly exposed in the eastern Grand Canyon. This section records relatively quiet water deposition, and the relatively abundant stromatolites and microfossils suggest an important increase in the diversity of organisms in the late Precambrian. The uppermost section also contains hydrocarbon-rich shales that are of interest as an oil source-rock. The sequence is not well-dated, but is post-1.1 Ga; uppermost units may be about 800 Ma, based on (a) reset K–Ar (potassium–argon) age determinations in the Cardenas Lavas (Elston and McKee, 1982) and (b) fossils present in the section (Andrew Knoll, personal communication, 1997).

BUTTE FAULT, CHUAR SYNCLINE, AND SUBORDINATE FAULT SYSTEM

The present Chuar Group exposures are bound on the east by the Butte fault, which is exposed for a length of 18 km before it is covered by Cambrian strata.

The continuation of the Butte fault, as expressed by the East Kaibab monocline, extends far to the north across the Utah border. The Butte fault trends to the north–northwest and dips moderately (60–70°) to the west. It presently has approximately 2900 m of Proterozoic west-side-down stratigraphic separation, but when Laramide reactivation (300 m west-side-up) is removed, it must have had on the order of 3200 m of west-side-down stratigraphic separation at the end of the Proterozoic (Elston and Mckee 1982; Elston 1989b). This magnitude of offset can only be seen in a traverse from the Colorado River to the Chuar Group outcrops along Carbon Creek. Movement was predominately dip-slip, as shown by slickensides oriented parallel to the dip of the fault.

Subordinate normal faults within the Chuar Group are parallel, and from the sedimentary record (see below) it apparently moved at the same time as the Butte fault. These faults include (a) other west-dipping faults that are parallel to the Butte fault and (b) a few east-dipping faults that form the sides of symmetrical grabens (as opposed to half grabens) such as the Basalt Canyon fault. Displacements on these faults are on the scale of meters to tens of meters and thus do not change the overall asymmetry of the Butte system.

The Chuar syncline is a broad, open, trough-shaped fold just west of and parallel to the north–south trace of the Butte fault (see Fig. 4.1, Chapter 4). The fold is asymmetrical with steeper dips on the east limb, adjacent to the Butte fault. The axial plane of the syncline dips 60–70° east, an orientation that is conjugate to the 60°-west-dipping Butte fault (Fig. 5.8). Parallelism of the synclinal axial plane and the Butte fault, its asymmetrical geometry, and tightening of the syncline at depth combine with intraformational faults and sedimentary evidence to show that the syncline is a growth structure that developed during progressive deepening of the basin on the downthrown side of the fault. The synclinal fold axis porpoises along the trace of the fault, suggesting differential extensional displacement along the fault during Chuar time. None of the synclinal features can be observed in the Paleozoic cover, indicating that the syncline is Proterozoic in age and unrelated to the reactivation (reverse sense) of the Butte fault in Laramide time.

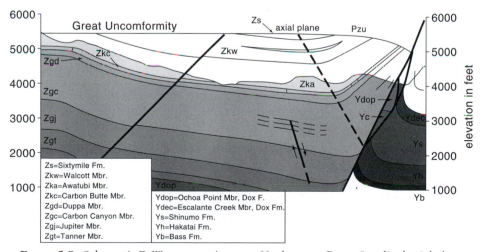

FIGURE 5.8. Schematic E–W cross section near Nankoweap Butte. Synclinal axial plane and Butte fault plane are parallel in strike (north) and are conjugate pairs (~50° angle between planes). Note thickness variation of upper members across syncline, along with intraformational fault in the Carbon Canyon Member.

SEDIMENTARY RECORD FOR PROLONGED INTERACTION OF SEDIMENTATION AND TECTONISM

Previous studies have proposed that there is evidence for synchronous deposition and fault movement in the uppermost Chuar Group and the Sixtymile Formation. For example, carbonate layers of the upper Walcott Member near the top of the Chuar Group pinch out toward the Butte fault, suggesting that deposition and synclinal development were synchronous (Fig. 5.8) (Elston 1979; Cook 1991). More dramatically, convolute bedding, bedding parallel slip surfaces, and intraformational breccias collectively indicate that synclinal development continued during deposition of the Sixtymile Formation. For more detail, see Chapter 4.

Our recent work shows that fault movement can be documented by sedimentation patterns and sedimentary structures within the Chuar Group. For ex-

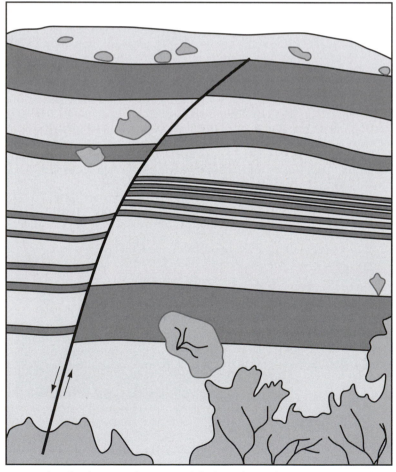

FIGURE 5.9. Sketch of an intraformational fault in the Carbon Canyon Member of the Galeros Formation, indicating that sedimentation and faulting were synchronous. View is to the southeast.

ample, measured sections between the basal sandstone of the Carbon Butte Member and the stromatolite at the base of the Awatubi Member are thicker in the synclinal axis and thinner on the east limb proximal to the fault. This is true for many different facies such that facies alone do not account for the thickness variation, and instead we suggest that the syncline was deepening and acting as a depocenter. Furthermore, the tightening of the syncline at depth requires there to be more sediment in the axis than on the east limb (Fig. 5.8). Abrupt thickness changes across subordinate faults can be observed in the Carbon Canyon Member (Fig. 5.9), and Carbon Butte Member. These structures tell us that the Butte fault system was responding to regional extensional stresses during sedimentation and synclinal growth. Wave-ripple crests in the sandstones of the Carbon Butte Member trend parallel to the syncline and Butte fault, suggesting that shorelines were parallel to structural trends. Widespread water-escape structures found in sandstone beds of the Carbon Butte Member may reflect recurrent seismic activity during deposition.

As detailed work continues in the Chuar Group, the model for Chuar basin evolution will become more refined. Combined with previous work in the Walcott Member of the Chuar Group and Sixtymile Formation, our new work suggests that basin deepening and synfaulting sedimentation took place during recurrent extensional slip on a regional scale fault system throughout Chuar Group deposition. The Sixtymile Formation marks a culmination of this syntectonic deposition, and possibly a change in the nature of deposition and tectonism at ca. 800 Ma.

Sixtymile Formation

The Sixtymile Formation marks a dramatic change in the style of deposition in the Chuar Group, from quiet-water marine shale to coarse sedimentary breccia and red sandstone indicating both higher energy deposition and probable tectonism on the Butte fault (Elston 1979). The Sixtymile Formation is exposed only in four outcrops in the hinge of the Chuar syncline. The lowermost member of the Sixtymile Formation is characterized by large (10-m scale) dismembered blocks surrounded by shale and olistostrome beds (slumped or gravity slide deposits) that record seismicity and fault movement during Sixtymile deposition. Several higher chert horizons and interbedded interclastic breccia in shale and tuffaceous shale mark continued episodic emergence and tectonism. The upper breccia on Nankoweap Butte fills 3-m-deep incised paleocanyons cut in shale that trend parallel to the Butte fault, suggesting continued influence of the Butte fault on sedimentary patterns.

POST-SIXTYMILE EROSION AND PHANEROZOIC REACTIVATION

There is no record preserved between the Sixtymile Formation and the uncomformity at the base of the Cambrian. Neoproterozoic faulting and tilting probably continued after Sixtymile deposition to bring the Cardenas Lavas and Dox Sandstone in the foot wall adjacent to upper Chuar Group shale. By Middle Cambrian time, erosion had again reduced the Grand Canyon Supergroup and underlying metamorphic rocks to a nearly smooth, flat plain. Ridges of Shinumo Sandstone rise up to 600 feet (180 m) above the generally level plain along the trends of the tilted blocks. The sea advanced over this plain from west to east and drowned the islands of Shinumo Sandstone in sand and mud of the Cambrian sea, about 510 million years ago.

After Paleozoic and Mesozoic strata buried the Grand Canyon region, many of the Precambrian Faults reactivated again and generated faults and folds in the younger beds. Nearly all of the major structures cutting Paleozoic rocks in the Grand Canyon region have a Proterozoic ancestry. The Butte fault was reactivated in a west-side-up sense and caused folding of the Paleozoic cover into a great east-facing monocline (Fig. 5.8). This monocline can be traced far to the north across the Utah border, and south to coincide with the Proterozoic Cherry Creek and Canyon Creek faults in the Apache Group of central Arizona. For more details on post-Paleozoic tectonism and reactivation of Proterozoic faults see Chapter 14.

REGIONAL IMPLICATIONS AND CONTINUED PROBLEMS FOR SUPERGROUP TECTONISM

This section highlights continued problems and uncertainties in understanding the long history recorded by the Grand Canyon Supergroup. Meso and Neoproterozoic sedimentary rocks of the Grand Canyon Supergroup, important uncomformities within this section, and the bounding uncomformities collectively record about 700 million years of earth history, a period longer than the entire Phanerozoic. We will never know what happened during this interval as well as we know the Phanerozoic because of incomplete outcrop and sparse fossil record, but ongoing studies are yielding new insights and questions about this important time in earth history.

The record as we now understand it involves a complex history of interaction between intracratonic basin formation (by faulting) and deposition to fill the basins. Was this a long progressive history or several discrete episodes? While it is possible that rifting took place more or less continuously from 1.3 Ga to 550 Ma, we favor the idea that rifting took place in two main pulses, recorded by sediments of the Unkar and Chuar Groups. The Unkar Group appears to contain evidence for synsedimentary NW contraction, NE extension, and mafic magmatism, and is possibly the result of far field stress associated with the Grenville collision in Texas (Fig. 5.10). Following a hiatus of unknown duration (Nankoweap uncomformities), the rocks of the Chuar Group and Sixtymile Formation record reactivation of preexisting faults that seem to be responding to regional E–W-directed extensional stresses around 800 Ma.

Much of North America was affected by the 1.3- to 1.1-Ga Grenville orogeny. This was a collisional suture between North America and another unknown continent, similar to the ongoing collision between Asia and India. The mountain belt with associated magmatic activity and metamorphism was concentrated in a belt that extended from Texas to New York. While this area was experiencing intense NW-directed contraction, large regions of the continental interior were trying to rift apart, extending in a NE–SW direction (Fig. 5.10). For example, the 1109- to 1086-Ma Midcontinent Rift System represents a large failed rift (Van Schumus and others 1992). Another 1100-Ma structure is the Central Basin Platform of northern Texas and southeastern New Mexico (Adams and Keller 1996). In Arizona, sedimentary rocks of the Apache Group and the Unkar Group were deposited between 1.4 Ga and 1.1 Ga and were intruded by 1.1-Ga diabase sills.

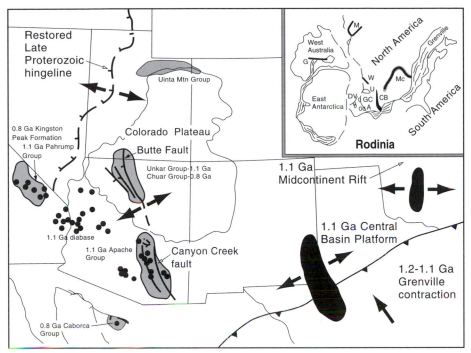

FIGURE 5.10. Mafic igneous rocks (black), sedimentary basins (stippled), and extensional structures of the Southwest from 1.1 to 0.8 Ga. Inset shows part of the Rodinian reconstruction after Borg and DePaulo (1994). (After Moores 1991.)

Both the Apache and Unkar sediments and their associated mafic sills may record the NW contraction and the NE extension caused by the Grenville orogeny. Although the shape of the basins are unknown (we do not know if the Unkar and Apache Groups were contiguous), we speculate that the development and filling of these basins may record the far-field effects of a continental scale collision.

There are several hints that the Chuar Group may record a second, separate, intracratonic rifting event at about 800 Ma. These rocks were deposited in a deepening fault-controlled basin or set of basins, with the N–S Butte fault as an important synsedimentary fault. Both fossils and preliminary paleomagnetic data suggest that the Chuar Group is less than 1 billion years old, and if the proposed 800-Ma age for the Sixtymile Formation is correct (Elston and McKee 1982), our new data suggests that this age is the culmination of Chuar rifting rather than the initiation of rifting as suggested by Elston.

There is also a record of Late Proterozoic extension in other ancient sedimentary basins of the western margin of North America (Fig. 5.10). In a zone from Death Valley to Northern Canada there is a major westward-thickening prism of sediment, the Cordilleran miogeocline, that marks the development of an Atlantic-style rift margin in Western North America in the Late Proterozoic. The immense length of this margin suggests that a major continental mass rifted away, but we do not know what continent it was, nor when the rift separation took place (Burchfiel et al. 1992). At the base of this thick prism of sedimentary rock are preserved patches of rock that are 1.45 Ga (Belt Supergroup), >1.1 Ga (lower Pahrump Group of Death Valley), and 800–700 Ma (Windemere Group, Uinta Mountain Group), suggesting that episodic rifting occurred over an extended period of time. Some workers suggest that the actual continental separation took place around 700 Ma (Ross et al. 1989), whereas others suggest that it took place about 550 Ma (Levy and Christie-Blick 1991). Evidence for long-lasting E–W extension in the Chuar Group may support the idea of intracratonic rifting in the 800- to 700-Ma time interval, but unfortunately does not clarify the timing of continental dismemberment.

What were the continental masses to the south and west of North America in the late Precambrian? Continental plate reconstructions based on paleomagnetic data and lithostratigraphic correlations of Proterozoic rocks remain a heated topic of debate. One model proposed by Sears and Price (1978) places Siberia off of western North America in the Late Proterozoic prior to rifting and development of the Western Cordillera. This plate reconstruction remains the preferred model of one of the coauthors. Other workers have postulated that West Australia and East Antarctica were our neighbors to the west and that portions of South America were to the south (Fig. 5.10, inset) (Moores 1991; Hoffman 1991; Dalziel 1991, 1997). This latter model suggests that North America was part of a supercontinental mass, named Rodinia, which was assembled around 1.0 Ga and broke apart between 750 and 550 Ma. If it existed, Rodinia would be the distant ancestor of the more widely accepted supercontinent Pangea, which assembled about 300 Ma and is still drifting apart today.

Ongoing work in the Grand Canyon Supergroup will continue to improve regional and global correlations as well as absolute ages for sedimentation and tectonism. Often, the subtle record preserved in sedimentary basins in continental interiors can provide clues to the nature of events at distant plate margins. It remains to be seen if the Grand Canyon Supergroup is recording the far-field effects of (a) 1100-Ma collision and coeval failed rifting (supercontinent assembly) and (b) ~800-Ma incipient rifting of a supercontinent.

SUMMARY

The Grand Canyon Supergroup offers a unique record that fills part of the gap in Powell's "Great Uncomformity," a time period we are just beginning to unravel. Emerging models point toward a prolonged history of tectonism in western North America that is cryptically recorded by an interaction of sedimentation and faulting in the Grand Canyon Supergroup. The Unkar Group and associated mafic magmatism appears to record an intracratonic basin that formed in response to NW contraction related to the Grenville collisions to the south (1.2–1.1 Ga). This was followed by a hiatus of unknown duration marked by uncomformities bounding the Nankoweap Formation. The Chuar Group records renewed intracratonic rifting, probably related to the early stages of supercontinent breakup. This rifting apparently failed in the Grand Canyon region, but eventually led to successful rifting and formation of the Cordilleran miogeocline between North America and the ancestral Pacific Ocean by 550 Ma (Link et al. 1993).

ACKNOWLEDGMENTS

Work came from a Masters thesis by Sears, and new data came from early stages of a collaborative NSF-funded project EAR-9706541 (to Karlstrom, Maya Elrick, John Geissman, Sam Bowring, and Andy Knoll). Timmons' work is part of a Masters project. We thank Carol Dehler who is doing concurrent sedimentology and stratigraphy in the Chuar Group.

• 6 •

TONTO GROUP

Larry T. Middleton and David K. Elliott

INTRODUCTION

The Cambrian of the Grand Canyon is, without question, one of the classic sequences of sedimentary rocks exposed in North America. These strata crop out along a prominent, essentially horizontal surface known as the Tonto Platform in the central part of the canyon and near the banks of the Colorado River in western areas of the canyon. The surface of the Tonto Platform roughly coincides with the top of the oldest Cambrian formation, the Tapeats Sandstone. Above the Tapeats, a series of small cliffs are separated by thicker intervals of slopes composed of finer-grained deposits of the Bright Angel Shale. These, in turn, are overlain by cliffs of resistant carbonate of the Muav Limestone, the youngest Cambrian formation of the Tonto Group.

This three-part system of sandstone, mudstone, and limestone is well known to most geologists and to a great number of Grand Canyon hikers and tourists. Despite this, research into the origin of these rocks has not kept pace with the developments in the last twenty years concerning the dynamics of nearshore and shelf depositional systems. The classic work of McKee and Resser (1945) endures as the most comprehensive study of the Cambrian system in the Grand Canyon.

To date, only a few studies have attempted to document carefully the lateral and vertical facies associations recorded in these strata (McKee and Resser 1945; Wanless 1973; Hereford 1977; Martin 1985; Middleton 1988). A major objective of this chapter is to present new data to enable recreation of the depositional systems that existed during the Cambrian in northern Arizona. Both sedimentologic and ichnologic data will be used to examine the depositional history of the Tonto Group.

Cambrian deposits in the Grand Canyon and throughout the Rocky Mountains long have been cited as representing a classic transgressive sequence of sandstone, mudstone, and limestone that accumulated on the slowly subsiding Cordilleran miogeocline and adjacent craton (McKee and Resser 1945; Lochman-Balk 1970, 1971; Stewart 1972; Stewart and Suczek 1977). During Early and Middle Cambrian time, a north–south trending strandline migrated progressively eastward across the craton. This shoreline was characterized by numerous embayments and offshore islands that affected sedimentation in nearshore areas. Shoreline migration for the most part was eastward, resulting in deposition of coarse clastics in shallow water areas to the east and finer clastics and carbonates in more offshore areas to the west. As will be discussed, numerous regressive phases interrupted this overall eastward transgression resulting in complicated facies interactions.

Cambrian strata in Arizona and along the entire western margin of North America thicken to the west, presumably reflecting more subsidence in the miogeoclinal or offshore shelf areas. Continued subsidence and/or sea-level rise, interrupted by a number of sea-level retreats or regressions, resulted in the complete submergence of the western cratonic margin by the Late Cambrian.

REGIONAL STRATIGRAPHIC RELATIONSHIPS

The Tonto Group comprises three formations (Fig. 6.1) that are, in ascending order: Tapeats Sandstone, Bright Angel Shale, and the Muav Limestone. The term "Tonto Group" was used first by G.K. Gilbert (1874) to describe this sandstone–shale–limestone sequence, though he considered these rocks to be of Silurian age. Subsequent stratigraphic and paleontologic work by Walcott (1890) established a Cambrian age for the Tonto Group, and Noble introduced the now-accepted formation names in 1914 during his mapping of the Shinumo Quadrangle in the Grand Canyon.

Strata of the Tonto Group also crop out along the Grand Wash Cliffs in western Arizona and to the east in the Juniper Mountains and the Black Hills in west-central Arizona. In these areas, the Tapeats Sandstone is overlain disconformably by the Devonian Martin Formation or the Chino Valley Formation, the age of which is uncertain (Hereford 1975). Presumably, the Bright Angle Shale and the Muav Limestone were removed during extensive pre-Devonian erosion. In central Arizona, scattered outcrops of the Tapeats Sandstone occur along the East Verde River and in the Sierra Ancha Range north of Young, Arizona. Tonto Group equivalents in southeastern Arizona include the Bolsa Quartzite and part of the overlying Abrigo Formation (Hayes and Cone 1975; Middleton 1988).

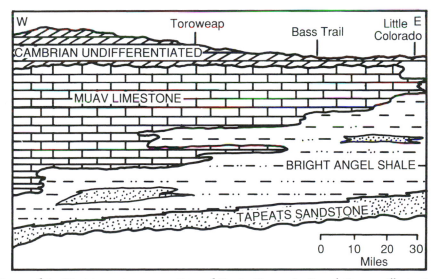

FIGURE 6.1. West-to-east cross section of Tonto Group in Grand Canyon illustrating stratigraphic relationships and eastward younging of the group. (From McKee and Resser 1945.)

Cambrian strata overlie a variety of Precambrian lithologies throughout the Grand Canyon. In the eastern part, the Tonto Group rests on tilted beds of the 1.4- to 1.1-billion-year-old Grand Canyon Supergroup, whereas in the western areas the Tonto Group nonconformably overlies older Precambrian (circa 1.7- to 1.6-billion-year-old) rocks of the Vishnu Group and other metamorphic units. This major unconformity between Precambrian and Tonto Group rocks, which has been recognized for a long time, obviously represents a considerable period of time during which the region was subjected to episodes of mountain building and extensive erosion. Walcott (1910) applied the name "Lipalian interval" to the period of time represented by this unconformity. Although dating of the igneous rocks within younger Precambrian strata establishes a minimum age, it is impossible to measure precisely the time from cessation of uplift to production of the nearly flat erosion surface onto which sediments of the Tapeats Sandstone were deposited. Clearly, we are dealing with a long period of time.

The surface upon which the Tonto Group accumulated was quite irregular. It was characterized by a rolling topography of resistant bedrock "hills" and lowlands. The Precambrian bedrock was weathered extensively in places and eroded during prolonged periods of subaerial exposure. Walcott (1880) and Noble (1914) were among the first to recognize that the Precambrian surface represented paleotopography and that sedimentation patterns were influenced by the relief and lithologies of these "hills." Other workers likewise have documented the influence of Precambrian topography on Cambrian sedimentation in other areas of the Rocky Mountains and in the midcontinent. There are numerous places in the canyon where the Tapeats Sandstone thins across or pinches out against these Precambrian highs. Where the Tapeats pinches out, the Bright Angel Shale overlies the Precambrian surface. The influence of these Precambrian highs will be discussed as the depositional environments of the Tonto Group are reconstructed.

A highly weathered horizon occurs on top of the Precambrian surface in several places in the canyon. The only effort to understand the genesis of this potentially very significant unit is that of Sharp (1940). His study suggested that extensive chemical weathering of Precambrian rocks occurred prior to deposition of Cambrian sediments. In places, this highly weathered surface, or regolith, is up to 50 feet (15.3 m) thick but generally is less than 10 feet (3.1 m) thick. Sharp speculated that where the Tapeats rests on unaltered basement, the regolith probably was removed by wave erosion associated with the initial Cambrian transgression. Sharp (1940) and McKee and Resser (1945) have suggested that the presence of such a thick, weathered horizon indicates that dominantly humid conditions existed during the early Paleozoic prior to deposition of the Tapeats Sandstone. Unfortunately, there have been no petrologic and geochemical studies that could substantiate this hypothesis. Considering that the time represented by the unconformity was likely several hundred million years, that the climate could have changed numerous times during this period, that this horizon was buried and exhumed numerous times prior to deposition of the Tapeats Sandstone, and that in the absence of terrestrial vegetation, weathering processes in soils would have been different (Basu 1981), a humid climate interpretation is quite tenuous. Obviously, considerable research needs to be done in this area.

STRATIGRAPHY OF THE TONTO GROUP

Tapeats Sandstone

The Tapeats Sandstone was named for exposure along Tapeats Creek in the western part of Grand Canyon National Park. For the most part, the formation is a medium- to coarse-grained feldspar and quartz-rich sandstone with granule and pebble-size, quartz-rich conglomerate present locally near the base. The percentage of feldspar is highest at the base and decreases upwards through the formation. The composition of the basal Tapeats reflects to varying degrees the mineralogy of the underlying Precambrian rocks. To date, however, there have been no petrologic studies of the Tapeats aimed at documenting the changes in mineralogy with respect to facies changes or evaluating the influence of basement lithology and paleotopography on the composition of the Tapeats.

The formation can be divided into two generalized packages. The majority of the Tapeats crops out as a cliff consisting of beds typically less than 3 feet (1 m) thick. Sedimentary structures include planar and trough cross-stratification and crudely developed horizontal stratification. Both the scale of the bedding and the cross-stratification decrease upwards. Overlying the main cliff is a thinner zone of interbedded fine- to medium-grained sandstone and mudstone. Stratification is of a smaller scale in these beds and is largely trough and ripple cross-stratification and horizontal stratification.

The significance of the upper unit is that it marks a major facies transition into the upper Bright Angel Shale (Fig. 6.2). An increase in fine-grained material indicates a reduction in the bedload to suspension load ratio. The concomitant changes in bedding thickness and scale of sedimentary structures are consistent with the above interpretation. The contact between the two formations, therefore, is arbitrary and probably should be placed at the top of the thickest sandstone bed within the transitional interval.

The Tapeats varies considerably in thickness throughout the Grand Canyon and also in areas to the south and west. A thickness of 393 feet (120 m) was reported by Noble (1922) along the Bass Trail. It is possible that this represents a maximum thickness in the canyon. Typically, the formation is between 100

Figure 6.2. Thin-bedded, horizontally and ripple-laminated upper portion of the Tapeats Sandstone. Overlying slope marks gradational contact with the Bright Angel Shale.

and 325 feet (30 and 100 m) thick. The thickness of the Tapeats clearly is controlled by relief of the underlying Precambrian surface. As previously stated, there are areas where the Tapeats thins across and/or pinches out against these Precambrian highs.

Except for trace fossils, which in places are quite common, body fossils are rare and only occur within the transition zone (McKee and Resser 1945). However, these fossils establish a late Early Cambrian age for the upper parts of the Tapeats Sandstone in the Grand Wash Cliffs in the western part of the canyon and an early Middle Cambrian age for the formation in the eastern canyon. These ages are based on trilobite assemblages (*Olenellus-Antagmus*) in the overlying Bright Angel Shale. This diachroneity is a reflection of the west-to-east sense of strandline migration.

Bright Angel Shale

The Bright Angel Shale is perhaps the least-studied formation in the Grand Canyon. It was named by Noble (1914) for exposures of slope-forming, interbedded, fine-grained sandstone, siltstone, and shale just above the Tonto Platform along Bright Angel Creek. Conglomerates and coarse-grained sandstones of the Bright Angel Shale contain quartz, minor amounts of potassium feldspars and sedimentary rock fragments, and glauconite. The latter is responsible for imparting the green color to many of the siltstones and sandstones. A number of the sandstones and siltstones contain a high percentage of hematitic ooids and iron oxide cements imparting a reddish brown coloration. The dominant lithology is greenish shale composed largely of illitic clay with varying amounts of chlorite and kaolinite. Inarticulate brachiopods, trilobites, and *Hyolithes* are locally abundant. Trace fossils are extremely abundant and varied.

McKee and Resser (1945) recognized one member in the Bright Angel Shale, which they termed the Flour Sack Member. This unit consists of shale, siltstone, and limestone and forms the uppermost part of the formation in the western canyon. Limestone decreases in abundance toward the east until the entire member is shale at its easternmost outcrop near Quartermaster Canyon. A number of rusty-brown dolomite tongues that occur in the Bright Angel represent carbonate extensions of the Muav Limestone (McKee and Resser 1945).

Sedimentary structures are numerous in the coarse-grained lithologies in the Bright Angel Shale throughout the Grand Canyon. These include horizontal laminations, small- to large-scale planar tabular and trough cross-stratification, and wavy and lenticular bedding (Fig. 6.3). Locally, structureless and crudely stratified conglomeratic sandstones typically overlie a scoured surface. Martin (1985) documented a number of coarsening-upward and fining-upward sequences in the Bright Angel in the central canyon.

The Bright Angel is over 450 feet (137 m) thick in the western Grand Canyon, only 270 feet (82 m) at Toroweap in the central canyon, and 325 feet (99 m) along Bright Angel Creek (McKee and Resser 1945). This variability in thickness is due to the complex intertonguing relationships with the Muav Limestone. The Bright Angel Shale thins toward the south and is only a few feet thick in the Juniper Mountains north of Prescott, Arizona. South and east of the Black Hills, the formation is absent—presumably the result of extensive erosion.

Like the rest of the Cambrian of the Grand Canyon, the Bright Angel crosses time lines, becoming younger toward the east. In the western part of Grand Canyon, the base of the formation lies below the *Olenellus-Antagmus* assemblage zone. It is, therefore, late Early Cambrian, whereas in the eastern part of

FIGURE 6.3. Wavy-bedded sandstone in the Bright Angel Shale along Pipe Creek. Internal structures suggest deposition by storm-enhanced currents.

the canyon, the lower third of the Bright Angel lies below the Middle Cambrian *Alokistocare-Glossopleura* assemblage zone.

Muav Limestone

The Muav Limestone is the youngest formation of the Tonto Group. It forms resistant cliffs above the Bright Angel Shale throughout the Grand Canyon. Noble (1914) named the formation for exposures in Muav Canyon. Contact with the Bright Angel Shale is gradational and characterized by complex intertonguing of the two formations. McKee and Resser (1945) defined seven members within the Muav. The members are differentiated on the basis of key marker horizons defined by fauna, intraformational conglomerate, or persistent beds of shale and/or thin-bedded limestone. The upper three members can be correlated throughout the entire canyon, whereas the lower four members are confined to areas west of Fossil Rapids.

The Muav consists of thin- to thick-bedded, commonly mottled, dolomitic, and calcareous mudstone and packstone, as well as beds of intraformational and flat-pebble conglomerate. Thin beds of micaceous shale and siltstone, minor amounts of fine-grained sandstone, and silty limestone occur at numerous horizons in the Muav, where they form small recesses and/or benches in the cliff-forming carbonate. The amount of siliciclastics increases toward the east, concomitant with a decrease in carbonates. Bedding thickness, in general, increases toward the west.

Most of the Muav comprises beds of structureless or horizontally laminated carbonate. Small-scale (less than 5 cm thick) trough, planar tabular, and low-angle cross-stratification occurs at many localities. Fenestral fabrics and desiccation cracks also are reported (Wanless 1975). Trace fossils, though not as abundant as in the Bright Angel Shale, are present throughout the canyon.

As a result of the intertonguing relationships between the Muav Limestone and the Bright Angel Shale, thickness trends within the Muav are variable. The unit as a whole thickens toward the west. McKee and Resser (1945) reported that the Muav is 827 feet (252 m) thick in the Grand Wash Cliffs near Lake Mead, 439 feet (134 m) thick at Toroweap in the central canyon, and only 136 feet (42 m) thick at the confluence of the Little Colorado and the Colorado rivers at the eastern end of the canyon.

In the western part of the canyon, the Muav lies above the *Alokistocare-Glossopleura* assemblage zone and is Middle Cambrian in age. In eastern Grand Canyon, the upper part of the Muav contains the *Bathyuriscus-Elrathina* zone and is late Middle Cambrian. The decrease in age of the Muav toward the east parallels the age trends of the Tapeats Sandstone and the Bright Angel Shale and reflects the west-to-east nature of the Cambrian transgression.

Cambrian Undifferentiated

In the western part of Grand Canyon, a thick (up to 426 feet [131 m]) sequence of dolostone overlies the Muav Limestone. McKee and Resser (1945) referred to this unit as the undifferentiated dolomites and considered it to be Upper Cambrian, though there is no paleontological evidence. Wood (1956) proposed the term "supra-Muav" for this unit, and Brathovde (1986) has suggested that these dolostones be named the Grand Wash Dolomite because the best exposures occur along the Grand Wash Cliffs in Western Arizona.

McKee and Resser (1945) recognized three lithofacies: white-to-buff massive dolomite; white-to-yellow, very fine-grained, thick-bedded dolomite; and gray, fine-grained, thick-bedded dolomite. Brathovde (1986) reported thick beds of oolitic grainstones and stromatolites that are interbedded with the fine-grained dolostones. Sedimentary structures include wavy and asymmetric ripple laminations and small-scale cross-stratification. Horizontal burrows and tracks are the dominant trace fossils.

PALEONTOLOGY

Since the early work of Resser (1946), there has been virtually no work done on the taxonomy and biostratigraphy of Cambrian strata in the Grand Canyon. Therefore, the systematics of the invertebrate fauna remain the same and will be reviewed only briefly here. Fossils have been described from the transition interval of the Tapeats Sandstone and from the Bright Angel Shale and Muav Limestone. There have been a few trace fossil studies of these strata, however, that have added to our understanding of the paleoecologic and environmental conditions during periods of deposition. These will be used in conjunction with the depositional environmental reconstructions to characterize the depositional systems.

Invertebrate Fossils

Despite the paucity of well-preserved invertebrate fossils, analysis of the fauna has provided some information concerning the paleoecology and certainly has facilitated the biostratigraphic zonation of the Tonto Group. Brachiopods and trilobites are the most common invertebrates reported from the Tonto Group, though preservation typically is poor. Fragments of sponges, primitive mollusks, echinoderms, and algae occur in the Bright Angel Shale and the Muav Limestone; however, these fossils are not very abundant. In addition to establishing the time-transgressive nature of these deposits, the biostratigraphic reconstructions have aided in documenting the numerous transgressive and regressive cycles that characterize the Cambrian of the Colorado Plateau and the Rocky Mountain regions (Lochman-Balk 1971; Aitken 1978).

Trilobites are the most abundant fossils in the Tonto Group, and common genera include *Olenellus, Antagmus, Zacanthoides, Albertella, Kootenia, Glossopleura,* and *Bolaspis.* Most specimens are poorly preserved and occur in the coarser-grained sandstones of the Bright Angel Shale and also in the mudstones of the Bright Angel Shale and the Muav Limestone. Resser (1946) reported 47 species of trilobites from the Tonto Group and suggested that these arthropods were, to some degree, facies-specific. More research needs to be done to evaluate this interesting relationship between depositional environment and trilobite distribution.

Brachiopods are locally abundant in the coarse-grained sandstones of the Bright Angel Shale. They also occur in some of the mixed siliciclastic–carbonate facies of the Muav Limestone. The most common genera in the Tonto Group are *Lingulella, Paterina,* and *Nisusia.* In general, the brachiopods tend to occur in beds containing few other invertebrate taxa.

Paleontologists have reported a number of species of primitive mollusks (Conchostraca) from the coarser-grained, hematitic sandstone of the Bright Angel Shale. The association of these fossils in coarse-grained sandstones led Resser (1946) to speculate that these mollusks occupied shallow-water habitats. Documentation of this environmental zonation, however, is unsubstantiated.

Resser (1946) reported sponge spicules from the Muav Limestone in the western Grand Canyon. These consist of thick, six-rayed spicules that Resser suggested were similar to purported sponge spicules of *Tholiastrella? hindei* that Walcott (1920) reported from the Cambrian of British Columbia. Elliott and Martin (1987) described six-rayed sclerites, which they assigned to the genus *Chancelloria,* from the Bright Angel Shale along Horn Creek in the Grand Canyon. Although Walcott (1920) considered *Chancelloria* to be a sponge (Phylum Porifera), Rigby (1976) and Elliott and Martin (1987) have questioned the assignment of this genus to this phylum and have suggested that *Chancelloria* represents a separate, yet unknown, phylum.

Algae, echinoderms, and gastropods have been described from the Tonto Group, though they certainly are the rarest taxa reported. Algal structures in the Muav Limestone consist of convex-upward laminae of calcite and/or dolomite and also small nodules composed of concentric laminations that have been termed *Girvanella* (McKee and Resser 1945). The environmental significance of the algae has yet to be established. Two well-preserved specimens of *Eocrinus* have been reported from the Bright Angel Shale. The excellent preservation of these echinoderms suggests relatively quiet water environments. Gastropods are represented by one species of *Scenella* from the Muav Limestone and several well-preserved species of *Hyolithes* from the Bright Angel Shale.

ICHNOLOGY

Trace fossils are common in all formations of the Tonto Group, particularly the Bright Angel Shale, and include a diverse array of tracks, trails, and burrows (Fig. 6.4 and Table 6.1). Despite this, the ichnofauna has been described in only a few studies (McKee 1932; McKee and Resser 1945; Seilacher 1970; Hereford 1977;

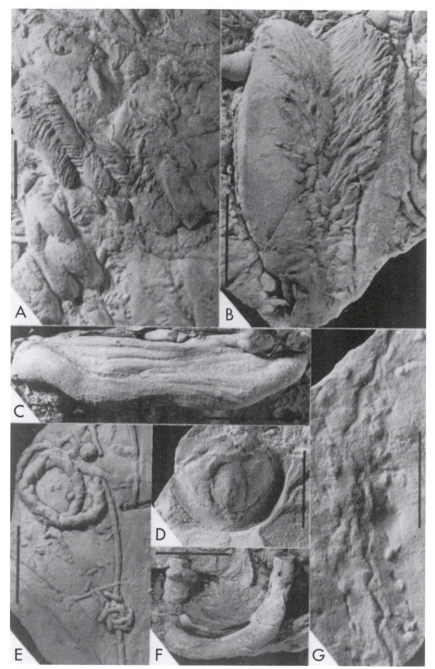

FIGURE 6.4. Trace fossils from the Bright Angel Shale in the central Grand Canyon. (a) *Cruziana* and *Rusophycus;* (b) *Rusophycus;* (c) *Teichichnus;* (d) *Glossopleura;* (e) *Phycodes pedum;* (f) *Diplocraterion;* (g) *Angulichnus alternipes.*

TABLE 6–1. Common Invertebrate
Trace Fossils

Ichnogenus	Formation
Angulichnus	Bright Angel Shale
Arenicoloides	Tapeats Sandstone
Corophioides	Tapeats Sandstone
Cruziana	Bright Angel Shale
Diplichnites	Bright Angel Shale
Diplocraterion	Bright Angel Shale
Palaeophycus	Bright Angel Shale
Phycodes	Bright Angel Shale
Rusophycus	Bright Angel Shale
Scalarituba	Bright Angel Shale
Scolicia	Bright Angel Shale
Skolithos	Tapeats Sandstone
	Bright Angel Shale
Teichichnus	Bright Angel Shale

Elliott and Martin 1987). Consequently, much remains to be done to establish the taxonomic affinities and the relationships between certain physical processes such as current strengths and substrate stability and the mode of infaunal and epifaunal behavior. Studies of other ancient shelf sequences (e.g., Crimes 1970), have demonstrated the benefits of integrating ichnologic and sedimentologic data.

Trace fossils are more abundant in the upper half of the Tapeats Sandstone, particularly in the transition interval into the Bright Angel Shale. These consist of single and paired vertical tubes and several types of horizontal traces.

Unbranched, straight vertical burrows assigned to the ichnogenus *Skolithos* are common at many localities. These sand-filled burrows occur near the top of beds. Burrows of this type probably functioned as dwellings and/or temporary resting structures of suspension-feeding organisms. Their occurrence in fine- to coarse-grained sandstones suggests an environment characterized by currents capable of active bedload transport. This is further substantiated by their occurrence in cross-bedded sandstones. Similar structures are common in many modern nearshore settings.

U-shaped burrows perpendicular to bedding also occur in the fine- to coarse-grained sandstones of the Tapeats Sandstone and the Bright Angel Shale. These tubes appear as paired holes on bedding planes or as concave-upward scours, where they have been eroded to the base of the burrow. These abundant traces, assigned to the ichnogenus *Corophioides*, occur in shallow-water deposits (Hereford 1977). The traces probably represent dwelling structures of suspension-feeding organisms, such as certain groups of annelids, and are common in many modern nearshore deposits.

Horizontal traces first were reported by Walcott (1918) from green shales in the Tonto Group and by McKee (1932) from the Tapeats Sandstone. These so-called "fucoides" are smooth-sided curving traces several inches in length that typically occur in large numbers covering entire bedding surfaces. Presumably, they were formed by the detritus-ingesting annelids moving through the sediment.

Trilobite crawling (*Cruziana*) and resting (*Rusophycus*) traces occur in the transition interval and throughout the Bright Angel Shale (Fig. 6.4a, b). Seilacher

(1970) provided the first detailed description of *Cruziana arizonensis* from the Tapeats Sandstone, and Martin (1985) reported trilobite trace fossils from the Bright Angel Shale. Martin (1985) noted the common occurrence of these traces in interbedded sandstones and mudstones in the Bright Angel Shale. Elliott and Martin (1987) suggested that the *Cruziana* traces were formed during fair-weather periods as these arthropods moved across the muddy shelf sediments. They also suggested that *Rusophycus* marks formed during storms.

Trace fossils are relatively uncommon in the coarser-grained, cross-stratified sandstones of the Bright Angel Shale (Martin 1985). Only the U-shaped trace *Diplocraterion* (Fig. 6.4f) is common, attesting to a relatively mobile substrate where the infauna frequently had to relocate their burrows (Martin 1985). *Diplocraterion* also occurs in upward-fining sequences where preservation of the complete burrow is common. In these sequences, the animals evidently recolonized the substrate and then evacuated the sediments as silt and clay were deposited from suspension following the passage of storms (Elliott and Martin 1987).

Interbedded sandstones and mudstones constitute the most abundant facies sequence in the Bright Angel Shale and also contain the most diverse trace fossil assemblage. Horizontal traces, which dominate these beds, include *Cruziana, Rusophycus, Palaeophycus, Diplichnites, Scalarituba, Scolicia, Angulichnus, Teichichnus,* and *Phycodes* (Elliott and Martin 1987). These traces were produced by organisms burrowing in mud, crawling and feeding on the sediment surface, or moving across the top of a sand bed covered with a thin layer of mud (Elliott and Martin 1987).

Trace fossils are not common in the Muav Limestone. Wanless (1975) described several horizons of burrowed, fine-grained carbonate, yet there has been no attempt to provide systematic descriptions of these trace fossils. Nor have we examined their potential environmental significance. Horizontal burrows, which appear to be the most abundant variety in the Muav, consist of relatively thin, sinuous traces.

DEPOSITIONAL SETTINGS OF THE TONTO GROUP

Cambrian strata in the Grand Canyon accumulated in environments ranging from braided streams to mid-shelf to offshore carbonate buildups. Environments range widely to include beach and intertidal flats; shallow, subtidal sand wave complexes (Tapeats Sandstone); offshore sand sheet deposits; open-shelf, fine-grained sandstones and mudstones (Bright Angel Shale); and subtidal and possibly intertidal carbonate buildups (Muav Limestone and Grand Wash Dolomite). The purpose of this section is to provide detail on the specifics of these depositional systems and to provide a brief review of previous sedimentologic and stratigraphic studies.

Tapeats Sandstone

Deposition of the Tapeats Sandstone was influenced by a variety of geomorphic factors, as well as processes inherent in fluvial and shallow marine depositional environments. The basal sediments of the Tapeats were deposited on a Precambrian surface that had been exposed for long periods of time. There was considerable relief on this surface. These "hills" have relief as great as 800 feet (244 m) (McKee and Resser 1945). Generally, however, the relief is considerably less. Large blocks of the younger Precambrian Shinumo Quartzite occur in the basal deposits of the Tapeats Sandstone near Bright Angel Canyon. Walcott (1883) reported large basement blocks mantling the sides of the areas of high relief. Dott (1974) has shown that similar deposits in Cambrian strata in Wisconsin almost certainly were eroded by storm waves.

McKee and Resser (1945) attempted to reconstruct the depositional environments of the Tapeats Sandstone based on texture of the sediments, paleocurrent trends, and, to some degree, the types of sedimentary structures. These authors concluded that most sedimentation occurred below the beach zone and seaward for tens of miles from the coast in water depths up to 100 feet (33 m). The prevalent west-to-southwest dip of the cross-bedding indicates net offshore transport of sediment. McKee and Resser believed that wide channels filled with cross-stratification and other large-scale scour-and-fill structures represented rip channels oriented perpendicular to the coast.

McKee and Resser (1945) also concluded that the monadnocks, or islands, had relatively little impact on sedimentation—other than serving as a local source of coarse, clastic sediment. They based their conclusions on the relatively consistent dip directions of the cross-stratification. The major influence of these islands, according to the authors, was in modifying sedimentation patterns in the inter-island embayments. They reported southeast-directed current trends in the major embayment that developed between the Shinumo Quartzite islands in Bright Angel Canyon.

A far more detailed facies analysis of the Tapeats Sandstone is that of Hereford (1977). This study was concentrated in Chino Valley and the Black Hills of north-central Arizona. Hereford recognized six environmentally specific lithofacies that he related to physical and biological processes operative on modern tidal flats, beaches, and braided river systems.

This study documented a continuum of tidal flat deposits ranging from lower tidal flats to upper tidal flats dissected by channels. Lower tidal flat sandstones are characterized by complex cross-stratification, reactivation surfaces and herringbone cross-stratification. The complex cross-bedding reflects the passage of smaller bedforms across the surface of larger dunes or sand waves. Reactivation surfaces are common in tidal systems where reversals of flow and/or erosion during periods of bedform immobility result in the scouring of lee-side avalanche deposits. The herringbone cross-stratification reflects bimodal–bipolar flow of the tidal currents. Despite the polymodality of current directions indicated in this study, there is a dominantly southwestern component of sediment transport that is in agreement with the data of McKee and Resser. Tidal systems typically are characterized by asymmetry in flow velocities and durations. In the case of the Tapeats, it is apparent that the ebb phase was the strongest and that it resulted in a preservational bias toward the structure produced during offshore flow.

Deposits of the high intertidal flats are characterized by interbedded sandstones and mudstones exhibiting a variety of features that attest to exposure and late-stage emergent runoff. Additionally, Hereford (1977) was able to document the presence of tidal channels that drained the flats.

Large channels occur near the top of the Tapeats at several localities in the central and western canyon (Fig. 6.5). The channels are up to 13 feet (4 m) deep and 60 feet (18 m) wide. At several localities, up to three laterally contiguous channels form a complex of southwest-oriented channel systems. Channel fill is variable and consists of thick sets of planar tabular cross-stratification, co-sets of planar tabular and trough cross-stratification, and/or simple vertical fills that conform to the shape of the channel. Flow tends to parallel the southwestern strike of the channel axis, and in some instances there is a well-developed bi-

FIGURE 6.5. Large subtidal channel complex near top of the Tapeats Sandstone. Transition zone is indicated by covered slope below first Bright Angel cliff.

modal–bipolar orientation to the cross-bed dip directions. Vertical trace fossils occur in the upper parts of the channel fills.

Although the geometry of these channels is similar to that of both fluvial and tidal flat channels, the internal stratification differs from that found in fluvial sequences in the Tapeats Sandstone (Middleton and Hereford 1981). The presence of trace fossils and bimodal–bipolar foreset dips, along with the absence of exposure features, suggests a subtidal channel complex dominated by off-shore flow, with minor preservation of flood-oriented structures. Although subtidal channels occur on modern tide-dominated coasts, comparatively little is known of their sedimentologic characteristics. In the lower intertidal and shallow subtidal zones, bedload transport and erosion can be intense, particularly along meso- and macrotidal coasts, because of the concentration of flow in these areas.

Johnson (1977) documented similar deposits in late Precambrian shallow-marine, quartz arenites in Norway. He demonstrated that these subtidal channels were oriented perpendicular to the coast and separated (dissected) subtidal sand bodies. To date, we have not gathered enough data to identify definitively the sandstone bodies that surround these channels as subtidal ridges or sandwaves. Nor have we established the fair-weather or storm-generated origin of these channels.

Two other facies reported by Hereford (1977) were not documented in the study of McKee and Resser (1945). One comprises low-angle, cross-laminated sandstone that likely formed on beaches. These tend to occur most frequently around Precambrian highs, where beach and upper foreshore sediments should have been common.

The second facies association represents braided stream deposits that grade into the marine units. Fluvial deposits in the Tapeats occur in the basal portions of the formation. Typically, they are less mature texturally and mineralogically than the associated marine deposits, which reflects a lack of extensive reworking that is common in the high-energy nearshore. These deposits are characterized by broad, shallow channels filled by horizontally stratified, coarse-grained sandstone and conglomerate that alternate with thick sets of planar-tabular and trough cross-stratified sandstone (Middleton and Hereford 1981). This sequence of structures, which is consistent with processes operative in coarse-grained, braided fluvial systems, has been reported from other pre-vegetation fluvial systems (Cotter 1978; Middleton et al. 1980; Cudzil and Driese 1987). Depositional occurred in wide, shallow streams where in-channel transport of sediment was accomplished by the movement of sheets of coarse-grained sediment along the bed and by migration of dunes and slightly sinuous transverse bars.

Bright Angel Shale

The Bright Angel Shale comprises a variety of lithologies and sedimentary and biogenic structures that indicate deposition in open shelf environments. McKee and Resser (1945) concluded that the Bright Angel Shale accumulated in waters below wave base at depths intermediate between the shallow water represented by the Tapeats Sandstone and the deeper waters of the Muav Limestone. More recent work by Wanless (1973), Martin et al. (1986), Elliott and Martin (1987), and Rose et al. (1998) have provided new data that permit more precise environmental reconstructions. Although generally supporting the conclusions of McKee and Resser, these workers have documented shallow-water deposits in the Bright Angel Shale, as well as providing important information concerning the roles of fair-weather and storm-related processes in controlling depositional patterns in the Bright Angel Shale.

Martin (1985) recognized eight facies in the Bright Angel Shale and grouped these into three genetically significant facies sequences. These include cross-bedded, upward-coarsening, and upward-fining sequences and a heterolithic sequence consisting of interbedded sandstone and mudstone. These facies sequences reflect deposition in subtidal areas influenced by tidal and meterologic processes. They also aid in tracking transgressive and regressive strandline movements.

Upward-coarsening sequences are up to 25 feet (8 m) thick and typically can be traced for several tens of kilometers. The lower parts of these sequences are characterized by laminated, bioturbated mudstones that were deposited during fair-weather suspension settling of silt and clay. The coarser-grained portions accumulated as sand waves, dunes, and ripples that migrated over sand sheets. These portions are characterized by thick sets of planer tabular cross-stratification (Middleton 1988). The presence of reactivation surfaces and abrupt changes in the dip of many foresets indicates periodic movement of these large bedforms, many of which are palimpset or relict. It also may indicate lee-side erosion during tidal reversals and/or storms. Deposition was entirely in subtidal areas, though the upper portions of many of these sequences were deposited in relatively shallow waters, as evidenced by eroded burrows of *Diplocraterion* (Elliott and Martin 1987).

Sequences that fine upwards are common and consist of a lower, normally graded small-pebble conglomerate or sandstone overlying an erosive base. This, in turn, is overlain by interbedded and fine-grained sandstone and mudstone. The basal coarse-grained facies represent deposition from high-energy, storm-induced currents that transported coarse materials from nearshore areas. The tops of these sequences contain symmetrical ripples, as well as appreciable amounts of laminated mudstone deposited from waning flows following the passage of storms (Fig. 6.3). These beds are very similar to those reported from both modern and ancient storm deposits. Complete vertical and horizontal traces occur at the top of many beds, indicating that the substrate was recolonized soon after deposition (Elliott and Martin 1987).

Lenticular beds of interbedded sandstone and mudstone constitute the majority of the Bright Angel Shale. This association comprises very fine-grained sandstone lenses and micaceous shale. Most beds are graded normally and contain an abundant and diverse trace fossil assemblage (Elliott and Martin 1987). These deposits represent post-storm suspension settling of muds and sands—and, possibly, remobilization during fair-weather periods. *Cruziana* and *Rusophycus* indicate a substrate inhabited by trilobites. Other trace fossils also indicate a relatively stable substrate colonized by a variety of infaunal and epifaunal organisms.

Muav Limestone

McKee and Resser (1945) considered the Muav Limestone to have been deposited in subtidal environments. The subtidal origin of much of the Muav is based on faunal and textural characteristics. These include an open-marine fauna, the very fine-grained nature of the mottled limestone and dolostone facies, and the fact that the Muav grades eastward into a shallow-water facies of the Bright Angel Shale. These authors also indicated that many of the flat-pebble conglomerates occurring throughout the formation were deposited in relatively deep water.

Intraformational or flat-pebble conglomerates are an extremely important facies in the Muav Limestone. These deposits, which are abundant from the Bass Trail eastward, consist of disc-like clasts of micrite and, occasionally, silt-size quartz and glauconite grains. The orientation of these clasts is variable. Some are oriented parallel with the bedding; some clasts are imbricated, and in some instances the clasts are vertical.

McKee and Resser (1945) described two associations of these conglomeratic beds. One variety consists of intraformational conglomerates that occur as scattered, discontinuous lenses within thinly bedded limestones. The other variety consists of one to several thin, conglomeratic beds that extend up to 45 miles. The great lateral persistence of these beds makes them ideal stratigraphic markers, and McKee and Resser used them to correlate over great distances in the canyon. These workers considered the widespread conglomerates to represent subtidal deposits formed during regressions.

The origin of the clasts obviously requires early lithification by cementation and/or compaction because they are derived from sediments within the basin. Where this induration takes place is controversial. Opinions range from ripups of carbonate muds exposed on tidal flats by storms and/or tidal channels to submarine lithification and subsequent erosion during storms.

Dew (1985) documented the occurrence of intraformational conglomerates similar to those reported from the Muav Limestone in the Upper Cambrian DuNoir Limestone in Wyoming. Based on facies associations, her study showed that both intertidal and subtidal limestone conglomerates can occur over a short stratigraphic interval. Sepkoski (1982) demonstrated that storm-induced currents were mostly responsible for the widespread distribution of flat-pebble conglomerates in Montana's Cambrian strata. In this case, the conglomerates are interbedded with shales that lack any evidence of subaerial exposure. Considering the stratigraphic importance that has been made of these conglomerates, it is clear that they represent key lithofacies and that documentation of their mode of origin (i.e., subaerial or subtidal) needs to be determined.

Regardless of their mode of origin, it is particularly interesting that they are abundant only in Cambrian and Ordovician rocks. Sepkoski (1982) speculated that with the proliferation of organisms living within the sediment during and following the Ordovician, the potential for early submarine cementation of carbonate shelf deposits was reduced substantially due to bioturbation processes. This hypothesis has yet to be tested and, of course, assumes a subtidal origin for these deposits.

Although many of the limestone and dolomite beds in the Muav are subtidal, Wanless (1973, 1975) reported intertidal and supratidal facies from outcrops in the western Grand Canyon. Many of the textures and structures reported by Wanless are similar to those found in modern tidal flats on Andros Island in the Bahamas. In particular, the laminated dolostones in the Muav have many characteristics in common with laminated dolomites that occur on supratidal levees adjacent to tidal channels on Andros Island. In these areas, fine-grained carbonate sediment is deposited during periods of overbank flooding following storms. Algae that inhabit the levees trap the sediment, resulting in the generation of continuous laminae of carbonate mud and pellets. Aitken (1967) referred to these laminated horizons as cryptalgal laminations because the evidence of algal binding had to be inferred. Discontinuous laminae also occur and are produced by traction transport of pellets and other grains over the algal-bound sediment. Wanless (1975) reported that these units are up to 66 feet (20 m) thick in the Muav Limestone. This suggests that there were prolonged periods of supratidal sedimentation far offshore from the Cambrian strandline.

It is clear that the Muav Limestone records episodes of both subtidal and peritidal deposition. A reasonable depositional model, therefore, might be one of offshore shoals surrounded by deeper water areas. Pratt and James (1986) proposed a tidal flat island model for Lower Ordovician shelf carbonates of Newfoundland (Fig. 6.6). In this model, small, localized carbonate islands occurred far offshore and were separated by subtidal areas. Middleton et al. (1980) documented similar facies distributions from Cambrian strata in Wyoming.

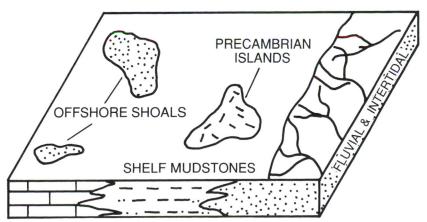

FIGURE 6.6. Block diagram illustrating the distribution of depositional environments represented by the Tapeats Sandstone, Bright Angel Shale, and Muav Limestone.

Undifferentiated Dolomites

The undifferentiated dolomites that overlie the Muav are not well understood in terms of their temporal and environmental significance. Brathovde (1986), however, has documented thick beds of oolitic grainstones and stromatolites interbedded with fine-grained carbonates. This association clearly indicates shallow subtidal and, possibly, intertidal environments.

SUMMARY

Facies analyses of the Tonto Group indicate deposition in a variety of fluvial, nearshore, and shallow shelf environments. Braided stream and intertidal-to-shallow subtidal deposits of the Tapeats Sandstone grade seaward into a complex array of shallow shelf sands and muds of the Bright Angel Shale. Shelf sedimentation was influenced by both tidal and storm currents. Sand ridges, sand waves, and broad areas where fine-grained siliciclastics were deposited from suspension settling—following storms and during fair-weather periods— characterized the shelf. Farther offshore, carbonate islands dotted the shelf. Here, the carbonate buildups were characterized by intertidal and possible supratidal zones separated by deeper water areas where tidal currents were active and where finer-grained carbonate sediments were deposited.

A number of transgressions and regressions resulted not only in the intertonguing of the formations of the Tonto Group but also in the vertical juxtaposition of facies belts that probably were not laterally adjacent. As Curray pointed out in his 1964 study, rapid migration of the strandline can result in the overstepping of offshore facies over nearshore deposits with no record of intervening environments. Numerous examples of such transitions occur in the Tonto Group, but most are poorly documented.

More detailed facies mapping and, in particular, documentation of lateral facies changes are needed. Only through such studies can the nature of the transgressive and regressive stratigraphies preserved in the strata of the Tonto Group be reevaluated and related to regional paleogeography.

7

TEMPLE BUTTE FORMATION

Stanley S. Beus

INTRODUCTION

Strata of the Temple Butte Formation of early Late Devonian and possibly late Middle Devonian age are exposed through most of the Grand Canyon but are relatively inconspicuous. Outcrops in the east are thin, discontinuous lenses, and in central and western Grand Canyon the exposures, though continuous, tend to merge with cliffs of the much thicker overlying Redwall Limestone. Temple Butte lithology is predominantly dolomite or sandy dolomite with minor sandstone and limestone beds.

NOMENCLATURE

Rocks of Devonian age in the Grand Canyon were first reported by Walcott (1880, 1883), who recognized Devonian strata beneath the Redwall Limestone in Kanab and Nankoweap canyons. He applied the name Temple Butte Limestone (Walcott 1889) to a thin band of dolomite along Temple Butte (on the west side of the Colorado River in eastern Grand Canyon). Although no specific type section has been designated, the Temple Butte site (Fig. 7.1) has gained universal acceptance. McKee (1939) and others have extended the name, as Temple Butte Formation, to the much thicker and more extensive outcrops in western Grand Canyon. West and north of the Grand Canyon, Devonian strata equivalent to the Temple Butte are designated as the Muddy Peak Limestone (Longwell et al. 1965), a name derived from the Muddy Mountains of southern Nevada (Longwell 1921). South of the Grand Canyon, along the Mogollon Rim of central Arizona, equivalent Devonian rocks are recognized as the Martin Formation (Teichert 1965), a name originally applied to Devonian strata in the Bisbee area of southeastern Arizona (Ransome 1904).

DISTRIBUTION

In eastern Grand Canyon and upstream in Marble Canyon the Temple Butte crops out as scattered, lens-shaped exposures that fill channels eroded into the upper surface of the Muav Limestone or the Cambrian undifferentiated dolomite. These channel-fill lenses commonly are less than 100 feet (30 m) thick but may

Figure 7.1. Approximate location of the type section of the Temple Butte Formation, on west side near south end of Temple Butte. Cu, Cambrian undifferentiated dolomite; Dtb, Temple Butte Formation; Mr, Mississippian Redwall Limestone.

be up to 400 feet (120 m) wide. Numerous exposures of these lenses are visible near river level in Marble Canyon beginning just below mile 37 (see Fig. 7.2) and also in the Little Colorado gorge. From Hermit Creek (about mile 95) westward throughout central and western Grand Canyon, the Temple Butte forms a continuous band of dolomite above local channel-fill deposits at the

Figure 7.2. Temple Butte channel-fill outcrop on right bank of Marble Canyon at approximately mile 38.4. Dtb, Temple Butte Formation.

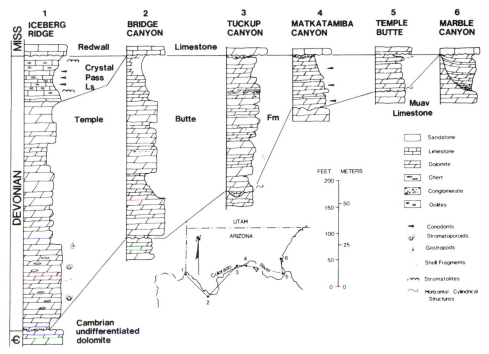

FIGURE 7.3. Selected stratigraphic sections of the Temple Butte Formation.

base. It gradually thickens to more than 450 feet (220 m) at Iceberg Ridge, five miles west of the mouth of the Grand Canyon (Fig. 7.3). Earlier descriptions of an Iceberg Ridge Devonian section more than 1200 feet (365 m) mistakenly included several hundred feet (220 m) of unnamed Cambrian dolomite beds [the undifferentiated Cambrian beds of McKee and Resser (1945)] as part of the Temple Butte.

CAMBRIAN–DEVONIAN UNCONFORMITY

In a broad, regional sense, Devonian strata truncate successively older rocks from west to east across northern Arizona (Fig. 7.4). The unconformity at the base of the Temple Butte is one of the major stratigraphic breaks in the Paleozoic sequence of the Grand Canyon. It probably represents latest Cambrian, all of the Ordovician and Silurian, and most of Early and Middle Devonian time. The time gap involved is probably somewhat less in western Grand Canyon because the uppermost Cambrian strata were deposited in a sea regressing westward and the Temple Butte was formed in a sea transgressing eastward.

In western Grand Canyon a sequence of light-gray dolomite beds up to 500 feet (150 m) thick overlies, and appears conformable with, the Muav Limestone. These strata were referred to as undifferentiated Cambrian dolomite by McKee and Resser (1945). The beds are truncated gradually eastward by the Temple Butte but are present near Lava Canyon in easternmost Grand Canyon, where McKee and Resser (p. 141) reported 163 feet (50 m) of dolomite and siltstone. Although no diagnostic fossils are known from these unnamed beds, they are

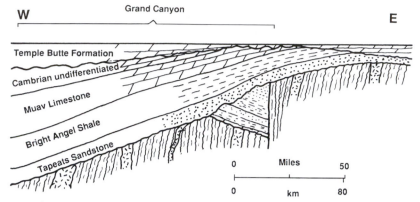

FIGURE 7.4. Regional pattern of Devonian–pre-Devonian unconformity and angular discordance of beds across northern Arizona. Vertical scale greatly exaggerated, relative thickness of units only approximate.

conformable with the Muav Limestone below (Brathovde 1986, p. 1). Korolev and Rowland (1993) have demonstrated on stratigraphic criteria a correlation of the undifferentiated dolomite in the Grand Canyon with the Banded Mountain Member of the Bonanza King Formation (probable late Middle Cambrian age) of Nevada.

The unconformable surface at the base of the Temple Butte is marked locally by considerable relief in the form of channels and depressions cut into the underlying Cambrian strata. These channels, up to 100 feet (30 m) deep in Marble Canyon, occur throughout most of the Grand Canyon. Where present, they clearly mark the Devonian base. They were first observed in Kanab Canyon by Walcott (1880) and were studied and described in detail between Garnet and Cottonwood Creek in eastern Grand Canyon by Noble (1922, pp. 49–51). In parts of the Grand Canyon, including the type section on Temple Butte (where the channels are absent), the Cambrian–Devonian strata appear in local exposures to be without angular discordance, and the contact is planar, with gray dolomite beds below and above. Here, the unconformity, even though representing more than 100 million years, may be difficult to locate.

STRATIGRAPHY

The strata forming the basal channel-fill part of the Temple Butte are commonly a distinct pale, reddish purple dolomite or sandy dolomite. The bedding generally is irregular, and it is gnarly in places. At some localities the beds are horizontal, but elsewhere they may conform to the walls of the channel they fill and be truncated with angular discordance by the overlying beds (Fig. 7.5). Rarely, there are basal conglomerate beds composed of subrounded dolomite pebbles. The upper beds in the channel-fill illustrated in Figure 7.2 contain about 20 percent insoluble residue consisting of detrital quartz grains, clay, and hematite. They also exhibit a peculiar columnar pattern of pale gray and purple dolomite—perhaps the result of leaching or weathering.

The more continuous strata above the basal channel-fill beds of the Temple Butte are exposed as uniformly medium to thick blocky ledges outlined by

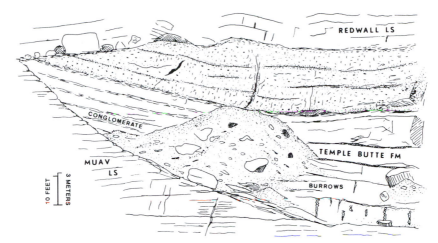

FIGURE 7.5. Field sketch by James Obey of portion of Temple Butte channel-fill lens in Marble Canyon on right bank at approximately mile 38.4.

parting planes or thin recesses (Figs. 7.6 and 7.7). The predominant lithology is dark-to-light olive gray, fine to medium crystalline dolomite that commonly weathers to a sugary texture. A subordinate, though extensive, second carbonate lithology is that of thin-bedded, very fine grained or aphanitic dolomite that is grayish-orange-pink. It weathers to a very light gray or yellow, exhibits a concoidal fracture, and commonly appears porcelaneous. Rounded, frosted quartz sand grains occur either scattered or as thin lenses or laminae in some of the

FIGURE 7.6. Basal part of Temple Butte Formation at Tuckup Canyon near river mile 164.5 showing channel cut into underlying Cambrian undifferentiated dolomite beds. Arrow indicates a 6-ft figure.

FIGURE 7.7. Paleozoic strata in western Grand Canyon near the mouth of Quartermaster Canyon, about river mile 265. Mr, Redwall Limestone; Dtb, Temple Butte Formation; Cu, Cambrian undifferentiated dolomite.

aphanitic dolomite beds. Locally they form marker beds of quartz sandstone up to 7 feet (2 m) thick, as at Havasu and Matkatamiba Canyon (Fig. 7.3). The dolomite beds commonly crop out as a uniform series of steep, receding ledges, but in central Grand Canyon they may form part of a vertical 1550-foot (460-m) carbonate wall that appears unbroken from the Muav Limestone up through the Redwall Limestone.

At Iceberg Ridge, five miles west of the Grand Canyon mouth, the upper 85 feet (26 m) of the Devonian carbonate section, above the Temple Butte Formation, consists of light gray lime mudstone to oolitic wackestone with a basal conglomerate or breccia. These beds previously were treated as part of the basal Whitmore Wash Member of the Redwall Limestone. Ritter (1983, p. 6) has recovered diagnostic latest Devonian (Famennian) conodonts from these strata and assigned them to the Late Devonian Crystal Pass Limestone. This unit apparently pinches out eastward and thus has not been recognized in the Grand Canyon.

PALEONTOLOGY AND AGE

Although marine fossils are abundant in the Upper Devonian Jerome Member of the Martin Formation in central Arizona (Teichert 1965; Beus 1978), the Temple Butte formation has yielded surprisingly few identifiable organic remains. Walcott (1883, p. 221) reported indeterminate brachiopods, gastropods, corals and "placoganoid" fish from the walls of lower Kanab Canyon. Noble (1922) reported fish plates identified as *Bothreolepis*, a fresh or brackish water form, from Sapphire Canyon. Denison (1951) confirmed the identification and the Late Devonian age assignment for the fish plates. Rare silicified corals, gastropods, crinoid plates, and massive stromatoporoids occur in the Temple Butte section at Iceberg Ridge, just west of the Grand Canyon mouth, but none are identifiable to the generic level.

Peculiar cylindrical trace fossil forms, somewhat resembling *Paleophycus*, occur in dolomite beds near the base of the Temple Butte at the type section and at Tuckup Canyon. These forms are subhorizontal, are straight to gently arcuate, and have a micritized central core within a cylindrical sleeve about 3 mm in diameter (Fig. 7.8). They may be trace fossils or possibly some sort of algal or stromatoporoid structure, but to date are indeterminate.

Conodont microfossils reported from the Temple Butte Formation at Matkatamiba Canyon (about 148 miles on the Colorado River in Grand Canyon) by Elston and Bressler (1977) are the most diagnostic and significant fossils yielded by the Temple Butte. These conodonts, identified by D. Schumacher (1978, written communication), indicate possibly a late Givetian lowermost *Polygnathus as-*

FIGURE 7.8. Subhorizontal cylindrical structures near the base of the Temple Butte type section on Temple Butte. Scale is in inches.

symertricus Zone assemblage occur at the base of the section, an early Frasnian assemblage between 20 and 40 feet (6 to 12 m) above the base of the Temple Butte, and probable early late Frasnian forms 20 feet (6 m) below the top of the formation. The uppermost 20 feet (6 m) of the Temple Butte at the Matkatamiba Canyon section are devoid of fossils. The conodonts suggest a latest Givetian to late Frasnian age (latest Middle Devonian through early Late Devonian) for most of the formation in central Grand Canyon. The Temple Butte is thus the approximate age equivalent of the Jerome Member of the Martin Formation in central Arizona, the Muddy Peak Formation of southern Nevada, and the Elbert Formation (Knight and Cooper 1995) of the subsurface in northeastern Arizona.

DEPOSITIONAL ENVIRONMENT

Devonian strata in the Grand Canyon are perhaps the least understood of the Paleozoic units studied there. The dolomitization of the original limestone and rarity of recognizable fossils have made refined environmental interpretations difficult. The carbonate facies and the few known fossils—crinoids, corals, stromatoporoids, and conodonts typical of nearshore biofacies (D. Schumacher 1984, written communication)—indicate accumulation in shallow, subtidal, open marine conditions for most of the Temple Butte in central and western Grand Canyon. The aphanitic dolomite beds in the Temple Butte may record local supratidal conditions. Similar modern, supratidal dolomites have been formed through evaporative pumping of magnesium-enriched sea water moving through porous supratidal sediments in the Persian Gulf as described by Illing et al. (1965) and Shinn et al. (1965).

The thinner and discontinuous channel-fill deposits of the Temple Butte in eastern Grand Canyon may have been deposited in tidal channels in an intertidal environment.

The regional paleogeography of the Frasnian (early Late Devonian) appears to have been shallow (but generally open-circulation) marine conditions in northwestern and central Arizona; intertidal-to-very-shallow subtidal conditions in central and eastern Grand Canyon; and shallow marine, restricted circulation conditions in northeastern Arizona (Beus 1980).

SUMMARY

The Temple Butte Formation records minor deposition of a thin carbonate sequence that gradually thickens westward across the Grand Canyon region. The easternmost outcrops and some of the basal strata to the west were probably deposited in narrow tidal channels. The more laterally extensive dolomite beds in the central and western portions of Grand Canyon accumulated in more subtidal but probably very shallow marine conditions across a gently submerged continental shelf.

8

REDWALL LIMESTONE AND SURPRISE CANYON FORMATION

Stanley S. Beus

INTRODUCTION

Two formations of Mississippian age are recognized in the Grand Canyon. The oldest of these, the Redwall Limestone, of Early and early Late Mississippian age, is one of most prominent topographically displayed units in the canyon wall. Recently, a second unit, The Surprise Canyon Formation of latest Mississippian age, has been recognized as the result of detailed mapping in the more remote part of western Grand Canyon. It occurs as isolated patches and lenses and occupies stream valleys, caves, and collapse structures developed in the top of the Redwall Limestone.

REDWALL LIMESTONE

Nomenclature

The Redwall Limestone consistently forms massive, vertical cliffs 500 to 800 feet (150 to 250 m) high about midway up the canyon wall (Figs. 8.1 and 8.2). The cliff face generally is stained red by iron oxide material washed down from the redbeds in the overlying Supai Group. The name Red Wall Limestone was applied first by Gilbert (1875, p. 177) in a report for one of the early surveys west of the 100th Meridian directed by John Wesley Powell. A type locality was later established by Darton (1910) when he introduced the name Rewall Canyon to a Grand Canyon tributary in the Shinumo quadrangle and thus provided a geographic place name for the Redwall. Four distinct stratigraphic units within the Redwall were recognized by Darton (1910). Later, they were described in more detail by Gutschick (1943) and given formal names by McKee (1963). These units, in ascending order, are: the Whitmore Wash, Thunder Springs, Mooney Falls, and Horseshoe Mesa members. All four have their type locality in the Grand Canyon or its tributaries, and all four can be traced throughout the Grand Canyon and beyond (McKee and Gutschick 1969, p. 3).

Distribution

The Redwall Limestone originally was deposited across virtually all of northern Arizona except the Defiance positive area of east central Arizona It is recognized

FIGURE 8.1. View of Redwall Limestone cliffs along the South Kaibab Trail in eastern Grand Canyon. Lowermost sharp "V" in trail switchbacks marks approximate Redwall Limestone-undifferentiated Cambrian(?) contact.

as far south as the Gold Gulch area just north of globe in south central Arizona (Racey 1974). Within the Grand Canyon, it is exposed in almost continuous outcrop on both canyon walls from about river mile 22 in Marble Canyon, where it first appears at river level, to the mouth of the canyon at the Grand Wash Cliffs (mile 277). Thickness of the formation gradually increases northwestward. It is just over 400 feet (120 m) thick at the Tanner Trail section in easternmost Grand Canyon and about 800 feet (245 m) at Iceberg Ridge, 5 miles (8 km) west of the Grand Canyon's mouth (Fig. 8.3) (McKee and Gutschick 1969, p. 3).

FIGURE 8.2. View of Redwall Limestone cliffs in western Grand Canyon near Separation Canyon, mile 240. H, Horseshoe Mesa Member; M, Mooney Falls Member; T, Thunder Springs Member; W, Whitmore Wash Member; Dt, Temple Butte Formation.

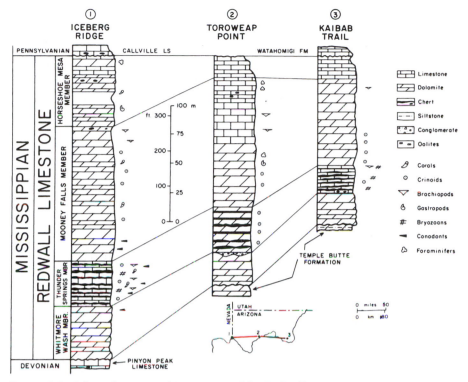

FIGURE 8.3. Selected stratigraphic sections of the Redwall Limestone in Grand Canyon.

PRE-REDWALL UNCONFORMITY

Throughout most the Grand Canyon, the Redwall Limestone rests without angular discordance upon Devonian strata or, where the Temple Butte Formation is missing, upon rocks of Cambrian age. The unconformity at the base of the Redwall spans all of latest Devonian (Famennian) time and the earliest part of the Missisippian Period (early Kinderhookian) through central and western Grand Canyon. The magnitude of the unconformity increases eastward because of the transgressive nature of the basal Redwall (Fig. 8.2). Basal beds are Early Missisippian (Kinderhookian) age in westernmost Grand Canyon, as indicated by diagnostic foraminifers, and are of late Early Mississippian (Osagian) in eastern Grand Canyon. In western Grand Canyon, the Mississippian–Devonian contact generally is well-marked by an irregular surface of erosion having up to 10 ft (3 m) of relief in a lateral distance of 100 to 200 feet (30 to 60 m).

Locally, a basal conglomerate composed of angular dolomite or limestone blocks of the Devonian Temple Butte Formation occurs at the unconformity. In eastern Grand Canyon, the Redwall–Devonian contact most commonly is a nearly horizontal surface with little or no relief. It is more difficult to recognize the unconformity where the Cambrian or Devonian strata beneath the unconformity and the Mississippian strata above are both dolomite. The nature of the unconformity suggests only gentle uplift and mild, though perhaps locally prolonged, erosion of pre-Mississippian strata before the beginning of Redwall deposition.

Stratigraphy

Whitmore Wash Member The type section for the Whitmore Wash Member is at Whitmore Wash (Colorado River mile 187.5 below Lees Ferry). Along this northern tributary to central Grand Canyon, McKee and Gutschick (1969) measured 101 feet (30 m) of thickly bedded, fine-grained dolomite. This member is composed mainly of fine-grained limestone in western Grand Canyon, but this changes to mostly dolomite in central and eastern Grand Canyon. The Whitmore Wash is nearly pure carbonate, having less than 2 percent insoluble residue content of minor gypsum and iron oxides (McKee and Gutschick 1969, p. 27). Common textural carbonates are pelleted and locally skeletal or oolitic wackestones and packstones (Kent and Rawson 1980). Thickness in the Grand Canyon is from about 100 feet (30 m) in the east to nearly 200 feet (60 m) at Iceberg Ridge, 5 miles (8 km) beyond the western end of Grand Canyon (Fig. 8.3). Bedding generally is thick, ranging from 2 to 4 feet (0.6 to 1 m) and even thicker locally. It usually forms a resistant cliff overlying a narrow bench or series of ledges typical of the Temple Butte Formation beneath. The upper boundary with the overlying Thunder Springs Member is conformable but is easily recognized by the lowest appearance of thin, dark, chert beds alternating with lighter gray limestone or dolomite beds typical of the Thunder Springs.

Fossils are rare in the Whitmore Wash Member, probably owing to extensive dolomitization of the original lime mud and sand. The Whitmore Wash is late Kinderhookian and early Osagian (late Early Mississippian) to the east as interpreted from brachiopod and foraminiferid fossils. Racey (1974) reported late Kinderhookian conodonts from the lower part of the member in the Salt River Canyon area south of the Grand Canyon. Ritter (1983, p. 17) noted conodonts of possible earliest Osagian age in the upper part at Iceberg Ridge.

Thunder Springs Member The Thunder Springs type section is at the head of Thunder River about 2 miles (3 km) north of Colorado River mile 136 in central Grand Canyon. The Thunder Springs Member is the most distinctive member of the Redwall Limestone because of the light and dark banded appearance imparted by alternating chert and carbonate beds. It consists of thin beds of light gray limestone or dolomite alternating with thin beds of dark reddish brown or dark gray weathering beds or lenses of chert. Thickness of the member increases gradually from 100 feet (30 m) in eastern Grand Canyon to about 150 feet (46 m) in the west (McKee and Gutschick 1969, p. 41).

Most of the carbonate rock in the Thunder Springs is thin-bedded, crinoidal grainstone or packstone. The rock tends to be limestone in the west and dolomite in the east. Thin-section analyses of the chert beds in the member as exposed in central Arizona reveal them to be silicified former bryozoan wackestones and mudstones (Bremner 1986, p. 55).

Invertebrate marine fossils are especially abundant in the chert beds of the Thunder Springs Member. These include corals (particularly colonial *Syringopora*), bryozoans, brachiopods, crinoids, and a few gastropods, blastoids, and cephalopods. Similar forms occur in the carbonate beds, though they are less abundant and not so well-preserved, probably owing to dolomitization. Diagnostic conodonts of Osagian age are reported by Racey (1974) and Ritter (1983). The Thunder Springs is everywhere conformable with the underlying Whitmore Wash Member. It is disconformable with the overlying Mooney Falls Member except in the extreme western end of Grand Canyon (Fig. 8.4). Locally the contact with the Mooney Falls is a low-angle unconformity, as at Kanab Canyon and Marble Canyon. This indicates minor structural activity, as well as erosion between Thunder Springs and Mooney Falls deposition.

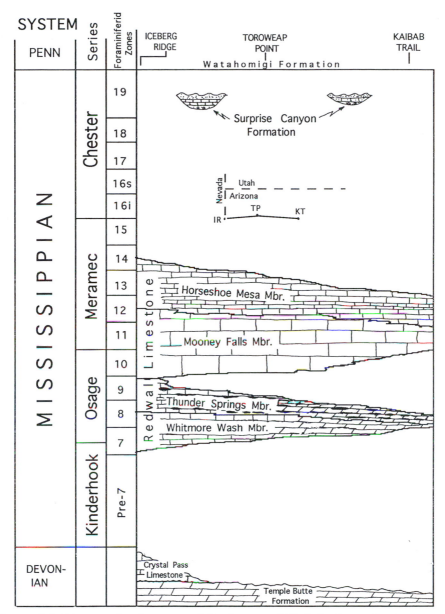

FIGURE 8.4. Diagram showing age and distribution of Mississippian rock units across northwestern Arizona. Foraminiferal zones after Mamet (Mamet and Skipp 1970). (Modified from Skipp 1979, Fig. 76.)

Mooney Falls Member The type section of the Mooney Falls Member is at Mooney Falls in Havasu Canyon (mile 153), about 4 miles (6.5 km) south of the Colorado River (McKee 1963). It is the thickest member of the Redwall, ranging from about 200 feet (60 m) in eastern Grand Canyon to nearly 400 feet (122 m) at the western end (McKee and Gutschick 1969, p. 56). It forms the major part of the sheer wall to which the Redwall name refers.

The Mooney Falls Member is predominantly pure limestone—except locally where it is dolomitized. Insoluble residue generally is less than 0.5 percent. The rock constitutes a favorable source for lime, which is being processed at two major plants (one south of the Grand Canyon near Peach Springs, and a second near Clarkdale). Carbonate grains include oolites, pellets, and a variety of skleletal fragments dominated by crinoid plates. One or two zones of thin beds or lenses of chert occur in the upper part of the member—generally near the contact with the overlying Horseshoe Mesa Member. Bedding is normally thick and appears massive in outcrop. Large-scale, tabular-planar cross-bedding is reported in the upper third of the member at several localities in central and eastern Grand Canyon, including the North Kaibab Trail, by McKee and Gutschick (1969, p. 61).

Invertebrate marine fossils are abundant throughout the member. They include solitary and colonial corals, spiriferid brachiopods, and crinoids. Diagnostic foraminifers (Skipp 1969; Mamet and Skipp 1970, p. 338) and conodonts (Racey 1974; Ritter 1983) indicate a late Osagian and Meramecian age.

The upper contact of the Mooney falls with the overlying Horseshoe Mesa Member is conformable and difficult to locate precisely in the field. The boundary is generally placed at the vertical change from medium- and coarse-grained limestone and thick or massive bedding of the Mooney Falls Member to the aphanitic and relatively thin-bedded, receding-ledge-forming limestone of the Horseshoe Mesa.

Horseshoe Mesa Member The type section for this member is on the south rim of Grand Canyon, along the Grandview Trail north of Horseshoe Mesa. The Horseshoe Mesa Member is the thinnest and least extensive member of the Redwall Limestone. Its thickness in the Grand Canyon varies from 45 to 125 feet (14 to 38 m), with the thinnest section in the east. Erosion causes the member to wedge out 30 to 40 miles (50 to 65 km) south of the Grand Canyon. It is also missing from the top of the Mooney Falls Member in most of central Arizona. The Horseshoe Mesa is composed of thin-bedded, light gray limestone with a mudstone to wackestone texture. Some chert lenses occur in the lower part. It typically forms weak receding ledges in contrast to the massive cliff of Mooney Falls below.

Well-preserved, invertebrate fossils are rare but present throughout the Horseshoe Mesa. Spiriferid brachiopods, bivalves, and corals are among the most abundant forms. At least 16 species of foraminifers are recognized in the member and indicate a Meramecian (early Late Mississippian) age (Skipp 1979, p. 298). The upper boundary of the Horseshoe Mesa Member is a major unconformity overlain by Early Pennsylvanian redbeds of the Watahomigi Formation, Supai Group. Locally it is overlain by lenses of the Surprise Canyon Formation of Late Mississippian age.

Paleontology and Age

Invertebrate fossils are common in certain lithofacies of the Redwall throughout the Grand Canyon (Fig. 8.5). Data from some 500 localities collected by McKee and Gutschick (1969) indicate that the most abundant megafossils are brachiopods and corals—followed by bryozoans, crinoids, bivalves, and cephalopods. Additional minor elements include blastoids, trilobites, ostracods, fish teeth, and algal remains. Foraminifers are abundant and were found in half of all samples selected for thin-sectioning.

Figure 8.5. Invertebrate fossils from the Redwall Limestone (1–9) and plant fossils from the Surprise Canyon Formation (10,11). (1) *Zaphrentites,* a solitary coral; (2) *Fenestella,* a bryozoan; (3) *Loxonema;* (4) *Bellerophon;* (5) *Orophocrinus saltensis* Macurda, a blastoid; (6) *Straporollus,* gastropods; (7,8) *Buxtonia viminolis* (White); (9) *Anthracospirifer,* brachiopods; (10,11) two forms of *Lepidodendron.* All figures 3 1.

Index fossils permit relatively accurate dating of most of the Redwall Limestone. As Fig. 8.4 shows, initial Redwall deposition began with the basal Whitmore Wash Member in western Grand Canyon during latest Kinderhookian (early Early Mississippian) time. Basal Redwall deposits become progressively younger as the sea transgressed eastward across northern Arizona and are no older than Osagian (late Early Mississippian) age in eastern Grand Canyon.

The Thunder Springs Member was formed by a regressing sea during middle Osagian to early Meramecian time. A second marine transgression deposited the Mooney Falls Member of the Redwall. The Horseshoe Mesa Member was formed during middle Meramecian (early Late Mississippian) time as a regressive deposit.

A single, 6.5-foot (2-m) limestone outcrop at the top of the Redwall near the Bright Angel Trail was considered to be Chesterian (late Late Mississippian) by McKee and Gutschick (1969, p. 74). The age assignment, based upon rare brachiopod and foraminiferid fossils, is considerably younger than any other Redwall Limestone outcrop. Billingsley and Beus (1985, p. 27) treated this occurrence as part of the overlying Surprise Canyon Formation of Chesterian age. However, recent examination of the fossils McKee and Gutschick collected from this outcrop confirms that they are from typical Redwall, not Surprise Canyon, lithology and are probably older than Chesterian age.

The age assignment of the Redwall places it as a correlative of the Escabrosa Limestone of southeastern Arizona, the Leadville Limestone of southwestern Colorado, and the Monte Cristo Group of southeastern Nevada. Four of the five formations in the Monte Cristo Group (excluding only the Arrowhead Limestone) are nearly identical in lithology and stratigraphic position with the four members of the Redwall in Grand Canyon (McKee and Gutschick 1969, p. 14). It is likely that the Redwall Limestone deposits were laterally continuous with all the above units at the end of the Mississippian Period.

Depositional Setting

Deposition of the Redwall Limestone sediments occurred in a shallow, epeiric sea that produced a submerged continental shelf across northern Arizona. Deposits formed during two major transgressive–regressive pulses, as demonstrated by McKee and Gutschick (1969). Detailed facies analysis by Kent and Rawson (1980) and Bremner (1986) have confirmed and refined this interpretation.

The basal part of the Whitmore Wash Member records initial deposition during the first transgression under nearshore, shallow subtidal conditions where high-energy currents produced oolitic shoals. As the transgression proceeded, more offshore deposits of skeletal grainstone and packstone accumulated under quieter water and more open marine conditions.

The Thunder Springs Member accumulated in increasingly shallower conditions as the sea regressed westward. The abundant chert layers (which exhibit a lack of sorting, an original lime mud texture, and a high proportion of delicate bryozoan fossil fragments) are considered by Bremner (1986, p. 62) to be preferentially silicified blue-green algal mats, which may have locally baffled marine currents. Alternating with the chert beds are abraded, sorted, crinoidal grainstone and packstone deposits that record a more vigorous current-washing of skeletal sand.

The Mooney Falls Member of the Redwall formed during a second marine transgression as crinoidal packstone and grainstone sediments developed widely across northern Arizona under generally open marine, offshore conditions. The Horseshoe Mesa Member formed under conditions of increasingly shallow and more restricted circulation during a final, slow regression of the sea.

SURPRISE CANYON FORMATION

Nomenclature

The Surprise Canyon Formation is a newly recognized rock unit in the Grand Canyon. It appears as isolated, lens-shaped exposures of clastic and carbonate rocks that fill erosional valleys and locally karsted topography and caves in the top of the Redwall Limestone. McKee and Gutschick (1969, p. 76) recognized conglomerate and gnarly mudstone beds filling local channels in the top of the Redwall Limestone but considered them part of the basal Supai Group. Subsequently these strata were recognized by Billingsley (1979) as a separate unit belonging to neither the Supai or the Redwall, and they were referred to as pre-Supai buried valley deposits by Billingsley and McKee (1982).

The name Surprise Canyon Formation was applied formally by Billingsley and Beus (1985, p. 27) and is taken from a large, northern tributary canyon in western Grand Canyon at mile 248. The type section (Fig. 8.6) is on the east-facing slope of a narrow ridge near the Bat Tower viewpoint in western Grand Canyon (mile 265)—about 12 miles (20 km) northwest of the mouth of Surprise Canyon. The Surprise Canyon Formation is probably the least visible rock unit in the Grand Canyon because of its discontinuous nature and the remoteness of the larger outcrops (Billingsley and Beus 1999).

FIGURE 8.6. Type section of the Surprise Canyon Formation, Bat Tower Section 2, is located about 1 mile (1.6 km) west of mile 263 on the Colorado River and 1.6 miles (2.6 km) southwest of the mouth of Tincanebits Canyon.

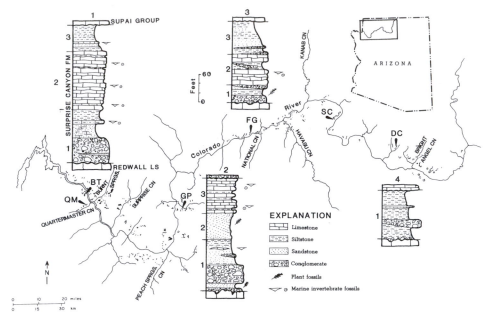

FIGURE 8.7. Index map showing major outcrops of the Surprise Canyon Formation (black patches) and selected stratigraphic sections. Section 1—type section near the Bat Tower (BT); section 2—Granite Park (GP); section 3—Fern Glen (FG); section 4—Dragon Creek (DC); the outcrop at Quartermaster Canyon (QM) is illustrated in Fig. 8.9.

Distribution

The Surprise Canyon Formation is nowhere a continuous stratum. Instead, it crops out as isolated, lens-shaped patches throughout much of the Grand Canyon and in parts of Marble Canyon to the east (Fig. 8.7). The valleys in which the formation occurs commonly are 150 to 200 feet (45 to 60 m) deep and up to 0.5 mile (1 km) wide in western Grand Canyon. They become shallower and relatively wider in central and eastern Grand Canyon (Fig. 8.8). Thickness of the formation corresponds to the depth of the valleys in which it occurs. The thickest section observed is at Quartermaster Canyon (Fig. 8.9), a southern tributary canyon in western Grand Canyon, where the formation is about 400 feet (122 m) thick (Billingsley and Beus, 1985, p. 27). Outcrops in central Grand Canyon are up to 50 feet (45 m) thick, whereas in eastern Grand Canyon and Marble Canyon they rarely are more than 60 or 70 feet (18 or 20 m) thick.

Redwall-Surprise Canyon Unconformity

The erosion surface that developed on the Redwall Limestone during the Late Mississippian Period, and on which the Surprise Canyon was deposited, must have been a relatively flat, resistant, limestone platform with considerable local relief. Most of the Surprise Canyon outcrops occupy gentle U-shaped or V-shaped notches cut into the top of the Redwall. By their nature and distribution these notches appear to have been part of a major dendritic drainage system that

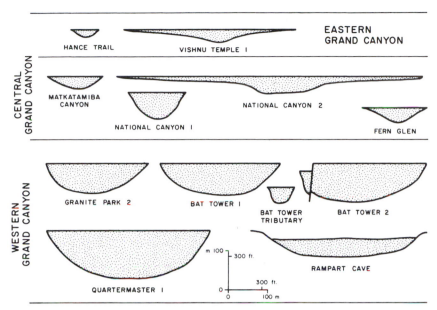

FIGURE 8.8. Cross sections of selected channel fill outcrops of the Surprise Canyon Formation, illustrating a general increase in thickness from eastern to western Grand Canyon.

FIGURE 8.9. Thickest known exposure of the Surprise Canyon Formation on the west wall of Quartermaster Canyon (QM in Fig. 8.7). Note distinct curved surface of the pre-Surprise Canyon valley wall cut into the top of the Redwall Limestone.

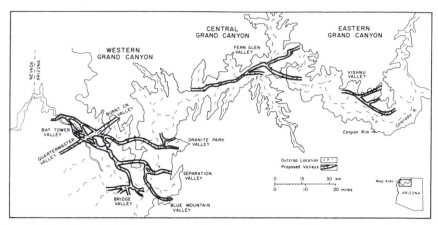

FIGURE 8.10. Hypothetical reconstruction of segments of the ancient valley system eroded into the Redwall Limestone in Late Mississippian time and onto which the Surprise Canyon Formation was deposited.

flowed generally from east to west. They were incised up to 400 feet (122 m) deep into the top of the Redwall Limestone in the western end of Grand Canyon. A preliminary reconstruction of the drainage pattern (Grover 1987) illustrates several major valleys that merge westward (Fig. 8.10). In addition, solution depressions and caves in the upper Redwall Limestone are filled locally with red mudstone of the Surprise Canyon Formation, indicating the development of this eroded and karsted topography prior to Surprise Canyon desposition (Fig. 8.11). The time available for the development of this topography on the top of the Redwall is just a few million years—the interval between the youngest Redwall (of middle Meramecian age) and the oldest Surprise Canyon (of late Chesterian age). The depth of the stream valleys cut into the top of the Redwall indicates either an uplift of the land surface or a drop in sea level of several hundred feet (120 m).

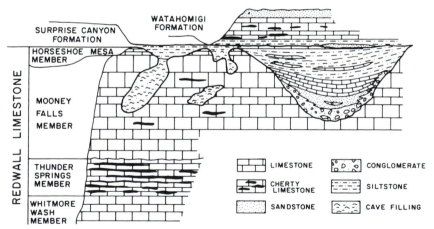

FIGURE 8.11. Cross section showing the stratigraphic relationship of the Surprise Canyon Formation to the underlying Redwall Limestone and overlying Watahomigi Formation of the Supai Group. (Modified from Billingsley and Beus 1985, Fig. 3.)

Stratigraphy and Lithology

The thicker sections of the Surprise Canyon Formation in western and central Grand Canyon can be divided into three major stratigraphic units: (1) a lower conglomerate and sandstone, mainly of terrestrial origin, that commonly forms a cliff and slope; (2) a middle unit of skeletal limestone of marine origin that commonly forms a cliff; and (3) an upper marine siltstone and silty or sandy limestone unit that typically forms a slope (Fig. 8.7). In the eastern Grand Canyon area the three units merge into a single red-brown, slope-forming conglomeratic sandstone and siltstone unit containing no limestone. The Surprise Canyon exhibits several lithofacies in each unit and has the most varied sedimentary lithology of any Paleozoic formation in Grand Canyon.

The basal part of unit 1 in most sections is a ferruginous pebble-to-cobble and local boulder conglomerate. Clasts are predominantly chert with minor limestone derived from the underlying Redwall Limestone. Some clasts contain typical Redwall Limestone fossils. The clasts commonly are grain supported and enclosed in a matrix of nearly pure quartz sand grains and some hematite. Locally the cobbles are sufficiently imbricated to indicate current directions at the time of deposition (Fig. 8.12). In most sections the conglomerate grades upward into a yellow to dark reddish brown or purple quartz sandstone or siltstone, or, in some sections, a dark, carbonaceous shale. The sandstone beds are commonly flat bedded, but some exhibit trough cross-strata or ripple laminations. *Lepidodendron* log impressions occur in the sandstone at numerous localities between Burnt Springs Canyon (mile 259.5) in western Grand Canyon and Cove Canyon (mile 169) in central Grand Canyon. Numerous plant fossils also occur in carbonaceous shale at Granite Park near mile 209. In eastern Grand Canyon, red-brown to purple mudstone beds, together with subordinate chert pebble conglomerate lenses typical of this lower unit, make up the entire formation. Trace fossils in sandstone beds of this unit are simple, vertical burrows and rare *Conostichus*, suggestive of the *Skolithus* ichnofacies of Crimes (1975).

Unit 2 is a coarse-grained, skeletal limestone that typically has a grainstone texture and is composed of whole or fragmented shells. Quartz sandstone beds

FIGURE 8.12. Imbricated pebbles and cobbles in conglomerate of the Surprise Canyon Formation, Dragon Creek section.

FIGURE 8.13. Algal stromatolites (oncolites) from near the top of the Surprise Canyon Formation at Quartermaster Canyon section 4. Centimeter scale.

up to 1.5 inches (3 or 4 cm) thick, alternating with skeletal limestone beds up to 4 inches (10 cm) thick, are common. The base of the limestone commonly truncates the unit 1 sandstone or siltstone beds occurring below on an erosion surface. The limestone unit typically forms resistant cliffs or ledges and weathers yellowish brown, rusty, or purple gray. Small-scale trough cross-strata exhibiting bimodal current directions are common. Marine invertebrate fossils are abundant and include crinoids and other echinoderms, brachiopods, bryozoans, corals, mollusks, and trilobites. Trace fossils typical of the shallow marine *Cruziana* ichnofacies of Crimes (1975) occur in sandy limestone beds. This unit is the thickest and topographically most prominent feature in many sections of central and western Grand Canyon. It is absent, however, in eastern Grand Canyon, east of Fossil Bay (about mile 130).

In western and central Grand Canyon, unit 3, the upper unit, is typically a dark red-brown to purple, ripple-laminated to flat-bedded calcareous siltstone of sandstone that forms weak slopes or receding ledges. Linguoid ripples are common. Resistant ledges of algal or ostracodal limestone are also common within the unit in western Grand Canyon. In at least three localities—Burnt Spring Canyon, Quartermaster Canyon, and National Canyon—nearly spherical algal stromatolites (oncolites) occur near the top of the unit (Fig. 8.13).

The boundary between the Surprise Canyon Formation and the overlying Watahomigi Formation of the Supai Group commonly is obscured by limestone rubble from above or by a covered slope developed on weak mudstone. Where well-exposed, the basal Watahomigi consists of: (1) a thin widespread, but locally discontinuous, limestone pebble conglomerate that contains minor chert clasts; or (2) where the conglomerate is absent, a purplish red calcareous siltstone and mudstone overlain generally by resistant gray limestone beds containing pale red-to-orange chert nodules (Billingsley and Beus 1985, p. 29). In a few localities a low-angle unconformity is recognizable at the contact (Fig. 8.14).

FIGURE 8.14. Low-angle unconformity at the Surprise Canyon–Watahomigi Contact (between the two arrows in the center of the photograph) about 0.3 mile (0.5 km) due west and on the opposite side of the ridge from the type section southeast of the Bat Tower.

Paleontology and Age

The fossil record of the Surprise Canyon Formation is one of the richest and most diverse of any Paleozoic unit in the Grand Canyon. It has yielded more than 60 species of marine invertebrates (Fig. 8.15) including foraminifers, conodonts, corals, bryozoans, brachiopods, echinoderms, mollusks, trilobites, and ostracodes (Beus 1995). Brachiopods are the most abundant forms preserved and are commonly associated with bryozoans, corals, and echinoderms. A reconstruction of a fossil community dominated by *Composita* and spiriferid brachiopods with associated echinoderms, bryozoans, corals, and mollusks typical of the unit 2 limestone beds in the Bat Tower area is shown in Figure 8.16.

Microfossil invertebrates are moderately abundant in the middle limestone unit, and some are present in the upper limestone beds of unit 3. Seven species of foraminifers were identified by Betty Skipp (Billingsley and McKee 1982, p. 144) Ten forms of conodonts have been identified from limestone beds in the middle and upper units of the Surprise Canyon (Martin 1992). Of these, *Adetognathus unicornis, Cavusgnathus unicornis, C. naviculus,* and *Gnathodus* spp. are the more abundant and significant forms.

The Surprise Canyon has also yielded a modest number of plant fossils (Fig. 8.5) mainly from sandstone and siltstone or shale beds of the lower unit in western Grand Canyon. Palynomorphs (spores) representing 22 species from the Granite Park area were identified by R. M. Kosanke (Billingsley and McKee 1982, p. 144). Several types of algal structures, and 12 species of plant megafossils, including three species of *Lepidodendron, Calamites,* and seed ferns (Tidwell et

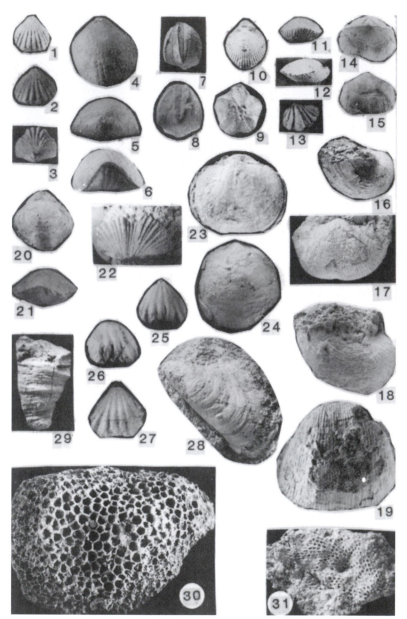

FIGURE 8.15. Invertebrate fossils from the Surprise Canyon Formation. (1–3) ?*Macropotamorhynchus* cf. *M. purduei* (Girty); (4–6) *Leiorhynchoidea carbonifera* (Girty), rhynchonellid brachiopods; (7–9) *Pentremites*, a blastoid; (10–12) *Eumetria*, (13) punctate spiriferid, brachiopods; (14,15) *Diaphragmus*, (16) *Inflatia*, (17–19) *Flexaria*, productid brachiopods; (20–21) *Composita*; (22) *Anthracospirifer*; (23) *Schizophoria*; (24) *Rhipidomella nevadensis* (Meek); (25–27) *Rotaia*, brachiopods; (28) *Septimyalina*, a bivalve; (29) *Barytchisma*, a solitary coral; (30) *Michelinia*, a colonial coral; (31) fenestellid bryozoan. All figures × 3/4.

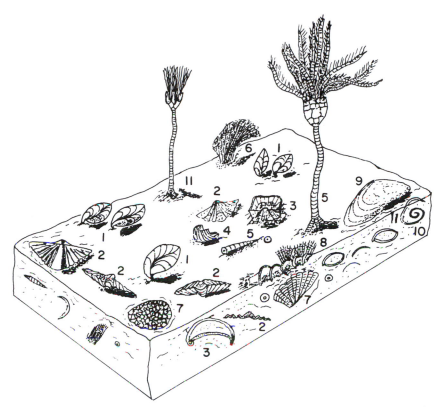

FIGURE 8.16. Reconstruction of a bottom-dwelling marine community typical of the middle limestone unit as it may have appeared in Late Mississippian time. Brachiopods are: (1) *Composita*; (2) punctate spiriferid; (3) *Inflatia*; and (4) *Ovatia*; (5) a crinoid; (6) fenestrate bryozoan; (7) the colonial coral *Michelinia*; (8) the bryozoan *Archimedes*; (9) the bivalve *Septimyalina*; (10) rare bellerophontid gastropod; (11) the blastoid *Pentremites*.

al. 1992), are known. Also bone fragments and some identifiable shark teeth are present in the middle unit.

Fossil evidence from spores, foraminifers, conodonts, brachiopods, and corals documents a Late Mississippian (Chesterian) age for the entire Surprise Canyon Formation. The brachiopod *Rhipidomella nevadensis* occurs through some 35 feet (10 m) of the middle limestone unit in the Blue Mountain Canyon area and marks the *Rhipidomella nevadensis* Assemblage Zone that Gordon (1984) considered coincident with foraminiferid Zone 19 in the western United States. Foraminiferids identified from limestone beds in the upper unit by Betty Skipp (Billingsley and McKee 1982, p. 144) include *Eosigmoilina explicata* that is restricted to foraminiferid Zone 19.

Conodonts provide the most precise data for age determination of the Surprise Canyon Formation. The upper Chesterian *Adetognathus unicornis* Zone is recognized in virtually all Surprise Canyon outcrops yielding conodonts. Although some zones barren of conodonts exist, taken collectively the occurrence of *Adetognathus unicornis* extends from the lowest limestone in the middle unit to the highest limestone in the upper unit of the formation. Thus the entire formation is considered equivalent to the *Adetognathus unicornis* Zone as defined by Baseman and Lane (1985) in the western United States. Strata containing the latest Mississippian *Rhachistognathus muricatus* conodont zone are not recognized in the Surprise Canyon Formation. The unconformity between the Surprise Canyon and the overlying Watahomigi Formation appears to represent the time interval of this missing zone (Fig. 8.4).

The Surprise Canyon Formation correlates well with other formations in the eastern Great Basin which contain the *Adetognathus unicornmis* Zone. These include the lower Indian Springs Formation of southern Nevada, the upper part of the Chainman Formation in western Utah, the lower part of the Ely Formation in west-central Nevada, and part of the Manning Canyon Formation in northern Utah. It may be the equivalent of the Log Springs Formation (Armstrong and Repetski 1980) of northern New Mexico. The Paradise Formation of southern Arizona is slightly older (foraminiferid zones 15–18) than the Surprise Canyon Formation (Armstrong et al. 1984).

Depositional Setting

The distribution and nature of the Surprise Canyon strata suggest that the entire formation was deposited within the confines of a broadly dendritic stream valley system and in the associated caves and collapsed depressions formed on a limestone platform (Fig. 8.17). The major shoreline must have been somewhere between the present western edge of the Colorado Plateau and Frenchman Mountain, located just east of Las Vegas, Nevada. Here, the outcrops of the Indian Springs Formation record an extensive shallow marine environment (Webster 1969). The marked lateral and vertical facies changes in the Surprise Canyon Formation are believed to record deposition in a major estuary system extending for at least 80 miles (130 km) east–west across northwestern Arizona to the eastern limit of marine fossils (Fig. 8.16). For the western and central Grand Canyon outcrops the depositional environment appears to have been fluvial for the ba-

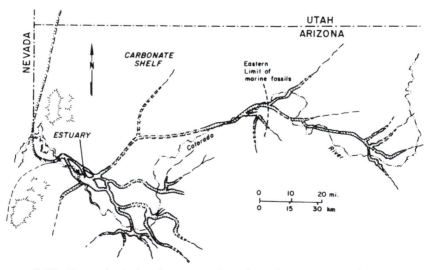

FIGURE 8.17. Hyopothetical paleogeography of northern Arizona during Surprise Canyon Formation deposition in Late Mississippian time.

sal unit. The strata become progressively more marine-dominated up section into the middle limestone and sandstone unit and end with the record of a restricted marine environment in the upper unit.

The basal conglomerate beds in unit 1 exhibit local imbrication that indicates strong, unidirectional, fluvial currents. The sandstone beds above the conglomerate commonly contain cut and fill structures and locally abundant plant remains. These beds grade upward into ripple-laminated or flat-bedded sandstones and siltstones. Grover (1987) has interpreted this sequence (especially well displayed in the Bat Tower area) as a record of continental and fluvial conditions changing to intertidal conditions as the sea transgressed eastward into the estuary and began to trap and rework clastic sediments within the tidal range.

The skeletal grainstone of the middle unit is well-developed in both the Bat Tower and Fern Glen paleovalley systems (Figs. 8.10 and 8.18). Abundant marine fossils attest to deposition in shallow marine conditions. The bimodal current directions (both upstream and downstream, but predominantly upstream or east), as recorded by prominent small-scale trough cross-strata in the Bat Tower area, suggest deposition in an estuary dominated by flood tides (Grover 1987) (Fig. 8.16).

In the Granite Park area, strata at the same stratigraphic position as the middle limestone unit are predominantly cross-stratified sandstone containing abundant plant fossils (section 2, Fig. 8.7; Fig. 8.19). Grover (1987) has interpreted this as deposition in a more fluvial- and ebb-tide-dominated valley where sand supply and deposition eclipsed minor marine limestone deposits.

For the most part, the upper unit (unit 3) of the Surprise Canyon Formation is ripple-laminated sandstone or siltstone, which alternates with algal and/or os-

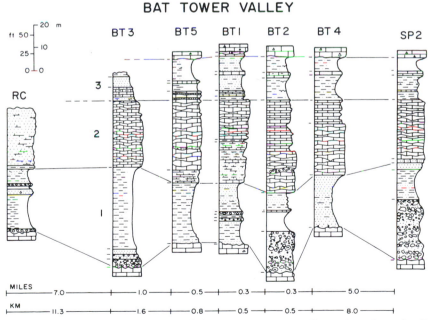

FIGURE 8.18. Stratigraphic sections of the Surprise Canyon formation along the Bat tower paleovalley portion of the estuary showing the maximum development of the unit 2 limestone. (From Grover 1987.)

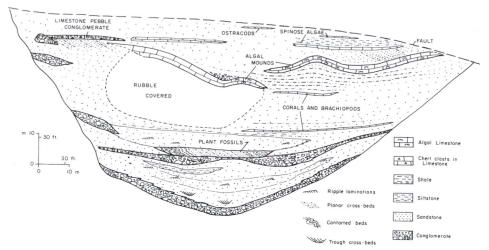

FIGURE 8.19. Sketch of a major part of the exposed lens of Surprise Canyon Formation at the Granite Park 2 section about 3.5 miles (5.6 km) southeast of the mouth of Granite Park Wash and mile 209.

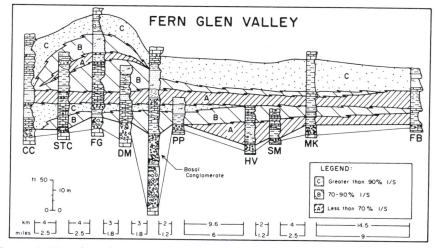

FIGURE 8.20. Selected stratigraphic sections of the Fern Glen paleovalley portion of the estuary, showing clay mineralogy facies distribution and environmental interpretations. Correlation is attempted only on fine-grained lithologic units. Facies A, having less than 70% illite/smectite clays, is considered to record nonmarine conditions; facies B, containing 70–90% illitic clays, is considered to record transitional or intermediate conditions; facies C, greater than 90% illitic clays, is considered to mark marine conditions. Two cycles of marine transgression and regression are indicated. (From Shirley 1987, Fig. 29.)

tracodal limestone beds that lack a normal, open marine fauna. Grover (1987) has interpreted these strata as a record of restricted-marine to upper tidal-flat environments developed during final infilling of the estuary. Detailed sedimentary and paleontological analysis of sections in the Fern Glen paleovalley indicate several fluctuations of marine and terrestrial or, at least, marginal marine conditions during deposition.

It is also possible that the original Surprise Canyon strata were part of a more widespread sheet formed in a shallow sea that covered much of northwestern Arizona. Subsequent uniform erosion then may have removed all but the lower valley-fill portion of the deposits. This would more easily explain the unusually great lateral extent of marine fossils [80 miles (130 km) east–west] within a narrow channel. However, in the absence of convincing evidence for marine deposition outside the confines of the narrow paleovalleys that presently contain the formation, deposition confined to a narrow estuary system seems the most reasonable interpretation.

Independent environmental interpretation based upon mineralogical and geochemical analysis of clays in the generally structureless and unfossiliferous mudstones of the Surprise Canyon permits confirmation and refinement of the above interpretation. Differential flocculation of clays in the zone of sea water/fresh water mixing within the Fern Glen paleovalley section of the estuary appears to have produced lateral distribution of kaolinite predominantly to the east (the landward direction) and illite to the west (seaward direction) (Shirley 1987, Fig. 23a–j) as in modern estuaries (Edzwald and O'Melia 1975). Using ratios of kaolinite/illite clays, Shirley (1987) has demonstrated a convincing record of two marine transgression–regression cycles in the Fern Glen paleovalley (Fig. 8.20).

The depositional environment of the fine-grained, red mudstones and local conglomerates of the Surprise Canyon Formation in easternmost Grand Canyon and Marble Canyon must have been mainly fluvial—in fresh, or perhaps brackish, water conditions. Almost totally absent are limestone beds and marine fossils, and only a few plant fossils have been recovered. Imbricated cobbles in the basal conglomerate of the Dragon Creek section (Fig. 8.12) indicate a vigorous westward-flowing current at the time of deposition.

ACKNOWLEDGMENTS

I am grateful for the encouragement and helpful suggestions offered during the early stages of this study by the late Edwin D. McKee. He had a continuing interest in the Surprise Canyon Formation, and his earlier work provided most of the data relating to the Redwall Limestone. Financial support for much of this work was provided by grants EAR 821743 and EAR 8618691 from the National Science Foundation and faculty grants from Northern Arizona University. The cooperation of the Hualapai Tribe and the Grand Canyon National Park in allowing access to remote areas of the Grand Canyon is appreciated. Helicopter transport to critical sites was provided by the U.S. Geological Survey through the efforts of Karen Wenrich and George Billingsley. The latter was the first to recognize the Surprise Canyon Formation, provided helpful comments and reviews of this chapter, and was a frequent companion in field excursions to Surprise Canyon outcrops.

• 9 •

SUPAI GROUP AND HERMIT FORMATION
Ronald C. Blakey

INTRODUCTION

The brilliant red cliffs and slopes present throughout the Grand Canyon comprise strata of continental, shoreline, and shallow-marine origin. Assigned to the Supai Group and Hermit Formation of Pennsylvanian and Early Permian age, these sedimentary rocks consist of a broad variety of lithologies—including sandstone, mudstone, limestone, conglomerate, and gypsum. Geologists have divided the rocks by their red-to-tan color and their weathering characteristics into five formations recognizable throughout the region. The five formations, in ascending order, are the Watahomigi Formation and Manakacha Formation (both Lower Pennsylvanian), Wescogame Formation (Upper Pennsylvanian), Esplanade Sandstone (Lower Permian), and Hermit Formation (Lower Permian). Geologists traditionally assign the first four formations to the Supai Group.

NOMENCLATURE AND DISTRIBUTION

Although the rocks now assigned to the Supai Group and Hermit Formation have a long and complicated nomenclature history (see McKee 1982a, page 3), three papers are responsible for the present terminology now in use in Grand Canyon. Darton (1910) proposed the name of Supai Formation for all the predominantly red strata between the Redwall Limestone and the Coconino Sandstone. Noble (1922) divided out the Hermit Shale from the top of Darton's Supai. McKee (1975) raised the Supai to group status and proposed four formations within the group: the Watahomigi, Manakacha, and Wescogame formations and the Esplanade Sandstone. The type section for each formation is near Supai in the Grand Canyon. Most publications within the last 35 years have used the term "Hermit Formation," in preference to shale, and this trend is followed here.

The Supai–Hermit terminology is sound throughout Marble Canyon and eastern and central Grand Canyon; however, several problems arise in western Grand Canyon. McNair (1951) assigned the lower part of the Supai interval to the Callville Limestone and recognized a Permian carbonate unit not present to the east, to which he assigned the term "Pakoon Limestone." McKee (1982a) did not recognize the Callville Limestone within the confines of Grand Canyon, and although he recognized the Pakoon Limestone, he did not include it within the Supai Group. A further complication exists in the western portions of the Grand Canyon where McNair (1951) assigned sandstone partly equivalent to the Esplanade and partly equivalent to the Hermit to the Queantoweap Sandstone.

The solution to this nomenclature problem is controversial and beyond the scope of this chapter, but the approach followed herein, pending further work, is to follow McKee's terminology in the Grand Canyon and Grand Wash Cliffs and to assign equivalent strata to the Callville and Pakoon limestones, Queantoweap Sandstone, and Hermit Formation (where applicable) in areas farther to the west, northwest, and north (Fig. 9.1).

To the southeast, along the southern margin of the Colorado Plateau, we can correlate the Supai–Hermit interval as far east as Sedona (Blakey 1979; Blakey, 1990; Blakey and Knepp 1988). Correlation to the southeast of Sedona is complicated by the addition of strata of Pennsylvanian age that is not directly time equivalent to the Supai. Termed the Naco Formation, these rocks were laid down in a basin separate from that in which the Supai was deposited. The fact that the Esplanade Sandstone is not present southeast of Sedona complicates the separation of the Supai and Hermit. Adding to the dilemma is the presence of a younger stratigraphic unit not present in the Grand Canyon, the Schnebly Hill formation (Blakey 1980, 1990).

To the east on the Defiance Plateau, Permian redbeds have been assigned to the Supai Formation (Read and Wanek 1961). Recent, Blakey (1990) demonstrated that only a portion of these redbeds are equivalent to part of the Supai in the Grand Canyon (see Fig. 9.2).

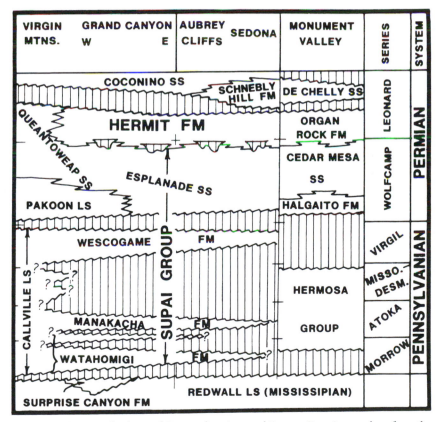

FIGURE 9.1. Time-rock chart of Pennsylvanian and Lower Permian rocks of northern Arizona. Vertical ruled lines show time represented by unconformities.

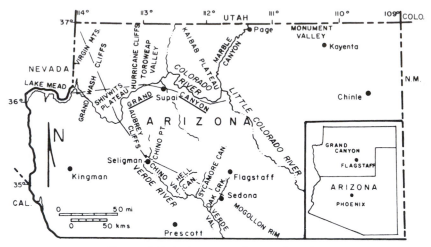

FIGURE 9.2. Index map of area of study.

LITHOLOGY AND STRATIGRAPHY

Watahomigi Formation

The Watahomigi Formation, the oldest formation in the Supai Group, consists chiefly of red mudstone and siltstone and gray limestone and dolomite (Fig. 9.3). The unit forms a broad, slightly westward-thickening sheet that ranges in thickness from 100 feet (30 m) in eastern Grand Canyon and at Sycamore Canyon on the east to 300 feet (90 m) in western Grand Canyon and along the Grand Wash Cliffs (Figs. 9.4 and 9.5). Carbonate content, which chiefly is very fine-grained (aphanitic) limestone to the east and granular (grainstone and packstone) limestone to the west, increases dramatically to the northwest. Grains include abraded fossil fragments, accretal grains, and pellets. Mudstone consists primarily of nondescript, slope-forming, poorly exposed units with thin, intercalated, bioturbated (disturbed by organisms), bright-orange, limey sandstone. Occasional bedding-plane exposures of the redbeds reveal various tracks, trails, and burrows. Most sections contain a basal chert–pebble conglomerate in which the clasts were derived from the underlying Redwall Limestone.

Throughout the Grand Canyon and the adjacent southern Colorado Plateau, the Watahomigi Formation can be divided informally into a lower redbed slope, middle carbonate ledge, and upper redbed slope (Blakey 1980; McKee 1982a). Based on included fossils and the presence of local conglomerate beneath the upper slope, McKee (1982a) assigned the lower slope and the middle ledge a Morrowan age and the upper slope an Atokan age (Fig. 9.2).

The lower contact of the Watahomigi Formation is everywhere sharp and unconformable with the underlying Redwall Limestone. Where the Surprise Canyon formation is present (see Chapter 8, this volume), geologists suspect an unconformity, but the contact is poorly exposed. The upper contact probably is conformable and was assigned by McKee to a zone of gray, jasper-bearing limestone and bright-orange sandstone and intercalated red mudstone. The contact apparently occurs at the change from primarily slope below to steep slope or cliff above and, therefore, may not be a consistent stratigraphic level across the region.

FIGURE 9.3. Typical outcrops of Watahomigi Formation. (a) Along Hermit Trail in Grand Canyon. (b) In western Mogollon Rim near Hell Canyon; arrow points to limestone marker bed in region. In both photos, dashed line marks top of formation.

The sharp increase in limestone west of a line paralleling the Hurricane Cliffs can be used to divide the Watahomigi Formation into an eastern redbed facies and a western carbonate facies (see McKee 1982a, Fig. 6). The westward increase in thickness, carbonate content, and marine fossils is typical of many Paleozoic rock units in the Grand Canyon region.

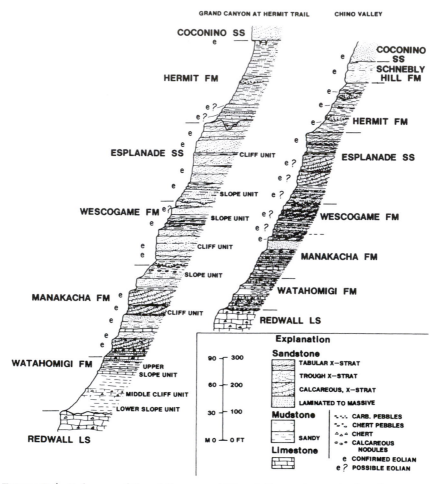

FIGURE 9.4. Columns of Supai Group and Hermit Formation and related strata showing distribution of known and suspected eolian strata.

Manakacha Formation

The Manakacha Formation marks an important change in the trend of Paleozoic depositional patterns in the Grand Canyon region. Following initial Cambrian sand deposition, the area was dominated by carbonates and minor mudstones from Middle Cambrian to Early Pennsylvanian time. The influx of quartz sand during the deposition of the Manakacha reflects a significant change across the western interior of the United States.

The Manakacha formation consists chiefly of quartz sandstone and intercalated red mudstone (Fig. 9.6). Unlike most other Paleozoic rock units, the formation is thicker in central Grand Canyon than it is in western Grand Canyon (McKee 1982a). The Manakacha forms a broad, sheetlike deposit across northwestern and central Arizona that averages about 300 feet (90 m) thick in Grand Canyon and 150 feet (45 m) thick across the Verde and Chino valleys.

Sandstone is the dominant lithology in the Manakacha Formation (Figs. 9.4 and 9.5). The composition ranges from very fine- to medium-grained quartz

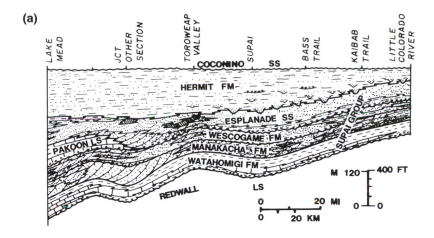

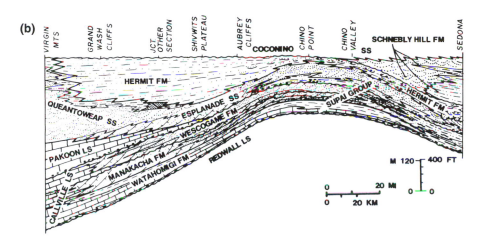

FIGURE 9.5. Restored stratigraphic cross section of Supai Group, Hermit Formation, and related strata. (a) West to east through Grand Canyon; (b) northwest to southeast from Virgin Mountains to Sedona. See Figure 9.4 for list of symbols.

grains to ooids, abraded fossils, and peloids. Most units, regardless of composition, are cemented with calcite. Jasper (red chert) is an accessory to many of the more limey units. Based on bedding types, geologists can recognize three kinds of sandstone. Cross-stratified sandstone comprises trough, planar, and compound sets that range in thickness from 1 foot (2.5 cm) to 30 feet (9 m). Careful examination of the strata within the sets reveals the ubiquitous presence of the climbing translatent strata described by Hunter (1977). Climbing translatent strata are thin laminae, generally less than several millimeters thick, that display reverse grading within each lamina. Each lamina displays the migration record of a single wind ripple (Hunter 1977) and, as such, is a powerful indicator of eolian deposition. The orientation of the strata reflects the nature and geometry of the surface on which they were deposited. Horizontal or very low-angle, climbing translatent strata form at the base of dunes,

Figure 9.6. Typical outcrops of Manakacha Formation. (a) Close-up showing two large eolian sets (each about 4–5 m thick). (b) Along Hermit Trail; dashed line marks top of formation; Manakacha is about 60 m thick.

between dunes, and on eolian sand sheets. Climbing translatent strata in cross-stratified sets form from the migration of eolian dunes. A broad range of both types of deposition is present within sandstone units of the Manakacha Formation.

A second type of sandstone consists of horizontally laminated to very low-angle, cross-stratified, fine-grained, calcareous, and silty units up to 20 feet (6 m) thick. Generally deeper red in color than cross-stratified units, these units also show rare ripple lamination and locally abundant tracks and trails on bedding planes. The existing evidence about the sandstone's origin suggests a subaqueous genesis, but few other clues are present.

The third type of sandstone typically is very fine-grained and grayish orange to bright reddish orange in color. It ranges from structureless (homogeneous) to extensively bioturbated. Actually, both the structureless and bioturbated sandstone are thought to be the result of bioturbation; however, the former is so extensively bioturbated that individual burrows or trackways cannot be distinguished. Because the latter is less thoroughly disturbed, we can see individual traces clearly.

Many sandstone units in the Manakacha Formation show additional alteration caused by diagenesis—chiefly the growth of calcite and dolomite crystals that alter or destroy the original fabric of the sediment. Such change is most prevalent in the upper portions of the sandstone units.

Mudstone and fine-grained limestone and dolomite are common throughout the Manakacha Formation. The mudstone is usually dark reddish brown in color, and it ranges from structureless to units that contain nodular carbonate concretions. Light gray carbonate units that bear jasper occur fairly commonly; they are seen with or without wispy lamination.

McKee (1982a) recognized a lower cliff unit and upper slope unit throughout Grand Canyon (Fig. 9.6). Sandstone dominates the former and is less abundant in the latter. These subdivisions are not as apparent in the Verde and Chino valleys; units assigned to the Manakacha Formation in these areas form a lower ledge and slope unit, a middle cliff unit, and an upper ledge unit. It should be emphasized that the lower level of relief along the southern margin of the Colorado Plateau (as opposed to the Grand Canyon) undoubtedly affects the topographic expression of all units in the Supai Group.

The Manakacha Formation displays a steady increase in carbonate content to the northwest. Some of these carbonates are cross-bedded, indicating deposition by fluid currents. In the Shivwits Plateau region, McKee (1982a, Fig. 9.8) classified the entire formation as limestone.

The upper contact with the overlying Wescogame Formation marks a regional unconformity, with most of the Middle Pennsylvanian absent. Conglomerate derived from the underlying Manakacha Formation, accompanied by scouring and channeling, marks the boundary. Poorer outcrops along the southern Colorado Plateau in the western Mogollon Rim region make this contact very difficult to locate.

Wescogame Formation

The Wescogame forms a sheetlike unit that ranges from 100 to 200 feet (30 to 60 m) thick across the Grand Canyon and the western Mogollon Rim region. Lithologic components are similar to those described above for the Manakacha Formation (Figs. 9.4, 9.5, and 9.7). Sandstone dominates the Wescogame throughout much of the central Grand Canyon and in western Chino Valley; to the east,

Figure 9.7. Typical outcrops of Wescogame Formation. (a) Close-up showing trough cross-bedding composed of eolian wind-ripple strata; about 2 m of section shown. (b) Along Kaibab Trail; dashed lines mark base and top of formation; Wescogame is about 40 m thick.

mudstone increases, whereas limestone increases west of Chino Valley (Blakey 1980; McKee 1982a).

McKee (1982a) informally divided the Wescogame Formation into a lower cliff unit and an upper slope unit, both traceable throughout much of the Grand Canyon. In the western part of the Mogollon Rim region, the Wescogame typically forms steep, ledgy slopes. Contrary to the implications of McKee's work, the location of the boundary between the Manakacha and Wescogame formations can be difficult to determine, both from a distance and from close range. The unconformity between the two formations varies, and even where noticeable relief is present, the boundary tends to be in a slope- or ledge-forming horizon, where debris from overlying units obscures the contact. The thickness and expression of associated conglomerate also varies and cannot be used to define the horizon in some areas. The topographic criteria provided by McKee (1982a) form the most reliable method of separating the two formations. The contact lies near the top of the slope that typically forms the upper part of the Manakacha Formation, a few feet below the overlying cliff unit of the Wescogame Formation.

The upper contact of the Wescogame Formation almost certainly marks the disconformity associated with the Pennsylvanian–Permian boundary (McKee 1982a). Although some of the problems discussed above also pertain to locating this contact, the top of the Wescogame Formation generally is easy to locate. Wherever exposures are clear, conglomerate and relief of up to 50 feet (15 m) mark the boundary. Based on regional patterns in the Grand Canyon and in the western Mogollon Rim region, geologists have defined three broad, north-northeast-trending facies belts. The eastern belt, the redbed facies, lies east of a line from near Hermit Basin to the western Verde Valley. It comprises subequal amounts of sandstone and red mudstone. The middle belt, the sandstone facies, extends west to a line running through the Shivwits Plateau region. The westernmost belt is chiefly cross-stratified limestone and limey sandstone. It forms the limestone facies.

Esplanade Sandstone (and Pakoon Limestone)

Sandwiched between the steep ledges and slopes of the underlying Wescogame Formation and the smooth slopes of the overlying Hermit Formation, the Esplanade Sandstone forms one of the most distinctive horizons in the Grand Canyon, especially in the central and western portions. The Esplanade Sandstone contains the highest percentage of sandstone of any of the units in the Supai Group (Figs. 9.4, 9.5, and 9.8). The formation consists of a steadily northwestward-thickening wedge that ranges from 200 to 250 feet (60 to 75 m) thick in the eastern Grand Canyon and western Mogollon Rim region to over 800 feet (240 m) in the western Grand Canyon and along the Grand Wash Cliffs (Blakey 1980; McKee 1982a). The thickness in the western Grand Canyon region includes the Pakoon Limestone.

Although the lithologic components are the same as those previously described in the Manakacha Formation, the dominance of cross-stratified sandstone in many sections distinguishes the Esplanade Sandstone. Most cross-stratified units are dominated by climbing translatent strata, which strongly suggests an eolian origin for the unit. The sandstone units forms beds generally 5 to 50 feet (1.5 to 15 m) thick that are separated by thin, red mudstone; by fine-grained carbonate units; or by prominent, irregular-bedding planes. These planes cause the Esplanade to weather to an irregular or ledgy cliff. Clean outcrops range in color from pale grayish orange to pale reddish orange, but staining from the overlying redbeds in the Hermit Formation causes the unit to appear dark reddish brown.

McKee (1982a) informally divided the Esplanade into a basal slope unit and main cliff unit. Typically, the upper part of the main cliff weathers back into a series of ledges, or ledges and slopes. In western portions of the Grand Canyon,

Figure 9.8. Typical outcrops of Esplanade Sandstone. (a) O'Niell Butte on Kaibab Trail. (b) On Hermit Trail. Most ledges are composed of eolian sandstone. Dashed lines mark base and top of Esplanade, which is about 90 m thick at both locations.

the basal slope unit grades westward into the dolomite and limestone of the Pakoon Limestone. The Pakoon thickens from east to west because of increased subsidence and the replacement of siliciclastic units by carbonate units in this direction. The Pakoon Limestone is approximately 300 feet (90 m) thick along the Grand Wash Cliffs. McKee chose not to include the Pakoon formally within the Supai Group, but I would suggest that future stratigraphic work in the area consider assigning the Pakoon Limestone to the Supai Group.

The nature of the upper contact of the Esplanade Sandstone with the overlying Hermit Formation is part of a complex regional problem. McKee (1982a, pp. 169–171, 202–203) examined the nature of the contact and concluded that overall evidence suggested the presence of a regional unconformity. At many locations, there is a definite transition between the Esplanade below and the Hermit above. At other locations, the Esplanade is overlain by an erosion surface with 30 to 50 feet (9 to 15 m) of local relief. The Hermit overlies that surface. Still other locations exhibit channeling into the Esplanade, but the channels originate from within the Hermit and not from the base. This suggests that the erosion surfaces formed after the deposition of the Hermit began. In a few places (such as near Toroweap, in the Shivwits area, and around Sedona), cross-stratified sandstone typical of the Esplanade occurs as much as 100 feet (30 m) above the Hermit–Esplanade contact. This suggests that the contact is variable from place to place and that it represents a zone of transition. The erosional relief also may be related to channel cut-and-fill sequences and not to a major regional unconformity. Obviously, there is a great deal that geologists do not know about this contact zone.

The most dramatic facies change in the Esplanade Sandstone occurs at the base, where the lower slope unit grades westward into the Pakoon Limestone in the Shivwits Plateau area (McKee 1982a). The upper part of the Esplanade displays a gradual change to increased carbonate content in the west. This change, however, is less sharp than that which occurs in the Manakacha and Westcogame formations. Bedded gypsum occurs near the top of the Esplanade in the Toroweap area (Fig. 9.5). The gypsum is intercalated with cross-stratified sandstone throughout a zone as much as 200 feet (60 m) thick.

Hermit Formation

Its overall fine-grained nature and relatively poor exposures, as well as a general lack of interest in the unit among geologists, contribute to making the Hermit Formation one of the poorest known units in Grand Canyon. White (1929) provided a comprehensive discussion of the formation, but it centered on the flora. Many other workers have provided local descriptions, but no major, recent, stratigraphic or sedimentologic study is available. This chapter will provide both sedimentologic and stratigraphic description—but primarily at the reconnaissance level.

For the most part, the Hermit Formation is composed of slope-forming, reddish brown siltstone, mudstone, and very fine-grained sandstone. The formation varies from approximately 100 feet (30 m) in thickness in the eastern portion of the Grand Canyon and near Seligman to over 900 feet (270 m) in the Toroweap and Shivwits Plateau areas (McNair 1951). Northward along the Hurricane Cliffs and into adjacent Utah, the sandstone content increases. Here, the Hermit interval generally is assigned to the Queantoweap Sandstone. In the western Mogollon Rim region east of Seligman, the Hermit can be traced eastward through discontinuous outcrops to the Sedona area, where it averages 300 feet (90 m) in thickness (Blakey 1980, 1990; Blakey and Knepp 1988).

McNair (1951) pointed out the misnomer of "shale" in the Hermit and proposed the word "formation" instead. Many subsequent workers have followed this suggestion. Overall, the Hermit Formation comprises a heterogeneous, chiefly fine-grained, siliciclastic sequence (Figs. 9.4, 9.5, and 9.9). Neither the vertical

FIGURE 9.9. Typical outcrops of Hermit Formation, Kaibab Trail. (a) Close-up of ledges (fluvial channels and slopes (overbank deposits). (b) Typical slope-forming nature.

nor the areal distribution of the various components defines any sharp trends, though regional variations are apparent. Most of the lithologic components of the Hermit are present to some extent in the underlying Supai Group. Silty sandstone and sandy mudstone are the most widespread and voluminous components in the Hermit. Silty sandstone ranges from structureless to ripple laminated to trough cross-stratified. Structureless units make up ledge-forming beds that av-

erage 3 feet (1 m) in thickness and may or may not contain limy, nodular concretions. Ripple-laminated units display subaqueous, faint-to-prominent ripple
cross-lamination. Trough cross-stratified sandstone consists of troughs up to several feet across (which may be organized within gently dipping master sets in
which the troughs are roughly perpendicular to the included master sets). Termed
epsilon cross-stratification, this structure is associated commonly with point-bar
deposition in meandering streams. Rare trough to planar–tabular sets of cross-
stratified sandstone, fine-grained and well-sorted, with climbing translatent strata,
occur near the base of the Hermit at several localities within Grand Canyon and
in the Sedona area.

Sandy mudstone forms slopes in the Hermit Formation. The mudstone commonly is featureless, though rare clean rock outcrops display fine ripple lamination and calcareous nodular concretions. A minor component overall, but locally abundant, is an intraformational conglomerate composed of sedimentary
pebbles. Most pebbles are carbonate grains obviously derived from the adjacent
carbonate concretions contained within intercalated sandstone and mudstone,
but some fine-grained, limey sandstone and siltstone pebbles also are present.
The conglomerate occurs both as beds and as an accessory to sandstone units.
McKee (1982a) reported conglomerate in the Hermit at 14 of his 35 localities in
Grand Canyon; it is most abundant in the Sedona area, where 10 to 15 horizons
are typical within the 300-foot (90-m)-thick section.

Most of the Hermit Formation consists of interbedded, silty sandstone and
sandy mudstone. At most locations, sandstone is more abundant near the base
of the formation, and mudstone increases upward. The two lithologies are interbedded rhythmically, typically with 15 or more cycles present in most sections. The extent or geometry of individual sandstone bodies is uncertain because of poor exposures. Reconnaissance work suggests an increase in sandstone
in the Shivwits Plateau area. This may reflect the increase in sandstone to the
northwest, where the Hermit is believed to grade into the Queantoweap Sandstone (Blakey and Knepp 1988). Conglomerate is most abundant in the Sedona
area. Current evidence indicates that it decreases in all directions from there.

The upper contact of the Hermit Formation is everywhere sharp and without gradation of any kind. Throughout Grand Canyon and into the Aubrey Cliffs,
it is overlain by the Coconino Sandstone. Cracks 20 or more feet deep at the
top of the Hermit frequently are filled with the overlying sandstone.

East of Seligman into the western Mogollon Rim region, a substantial sequence of strata not present in the Grand Canyon appears above the Hermit;
these strata are defined as the Schnebly Hill formation. The regional relations
shown in Fig. 9.5b are discussed in detail by Blakey (1980, 1990) and by Blakey
and Knepp (1988). Contact with the overlying Schnebly Hill Formation in this
area also is sharp, but it lacks the sandstone-filled cracks. Based on regional
stratigraphic relations (Blakey 1996), the sharp contact is interpreted as a major
regional unconformity, though sufficient paleontological evidence to confirm this
hypothesis is not yet available.

PALEONTOLOGY

Both White (1929) and McKee (1982a) have examined the faunal and floral content of the Supai Group and the Hermit Formation. Overall, both stratigraphic
units are sparsely and sporadically fossiliferous, but we have recorded a broad
variety of fossil types.

Trace Fossils

Trace fossils—the trackways, burrows, impressions, resting marks, and feeding marks formed by organisms in sediment and preserved in the rock record—are ubiquitous throughout the Supai and the Hermit. McKee reported their presence, but no comprehensive report on trace fossils has yet appeared. Nondescript bioturbation is the most common form of trace fossil in the Supai Group. The swirls, crinkles, and irregular patterns associated with distinct burrows attest to its abundance. Rocks showing disturbed sediment range from structureless (complete bioturbation) to distinctly burrowed. Burrowed structures are common both in the Supai and the Hermit, most frequently in silty sandstone and carbonate units. Many types of burrows are present, but those occurring most frequently are cylindrical, smooth burrows several millimeters in diameter that are parallel, perpendicular, or oblique to the bedding. McKee also described trackways, mostly attributed to vertebrates, from both the Wescogame Formation and the Esplanade Sandstone. Other miscellaneous trace fossils, including invertebrate trackways, resting marks, bioturbation caused by plant roots, and possible feeding marks, are distributed throughout the Supai Group and the Hermit Formation. Clearly, this is a field with strong future research potential.

Fossil Invertebrates

McKee (1982a, Chapters E and F) described and pictured many of the invertebrates in the Supai Group. He reported a fairly rich and varied brachiopod fauna from the Watahomigi Formation. Marine fossils, including locally abundant fusulinids, occur sporadically throughout the Supai—especially in western Grand Canyon. However, the presence of abraded fossils in well-sorted, cross-stratified sandstone or limestone should not be used as a criterion for marine deposition because skeletal grains can be, and were, blown inland and incorporated into eolian deposits.

Flora

Plant remains are widespread but sparsely distributed throughout the Supai Group and the Hermit Formation. The comprehensive reports by White (1929) and McKee (1982a) list the location, stratigraphic horizon, and plant taxa found throughout the units, and Blakey (1980) reported plants fossils from the Hermit formation in the Sedona area. McKee (1982a, p. 100) found that although most of the plant remains have little stratigraphic value, they do suggest the presence of broad floodplains developed during times of regressing seas and semiarid-to-arid climates.

AGE AND CORRELATION

Introduction

McKee (1975; 1982a) provided us with the first detailed regional account of the age of the Supai Group. He based his age determinations primarily on brachiopods in the Watahomigi Formation and fusulinids scattered throughout the group. Using this information, he assigned to the formations of the Supai Group the following ages: Watahomigi Formation—Morrowan and Atokan; Manakacha Formation—Atokan; Wescogame Formation—Virgilian; Pakoon Limestone—Wolfcampian; Esplanade Sandstone—Wolfcampian (lower part of formation). The age of the upper part of the Esplanade Sandstone is determined by its stratigraphic position.

White (1929, p. 40) assigned the Hermit formation an age of upper Lower Permian (Leonardian); McKee's work seems to confirm this. He used a fossil plant, *Callipteris arizonae*, found in the Bone Spring Limestone of Leonardian age in southeastern New Mexico, as his point of reference. If McKee is correct, the age of the Esplanade Sandstone must be no older than Wolfcampian and no younger than Leonardian.

In the Sedona area, the Esplanade Sandstone and the Hermit Formation lie several hundred feet below the Fort Apache Member of the Schnebly Hill formation. The Fort Apache Member has yielded conodonts of early late Leonardian age. This paleontologic evidence provides additional support for presuming an early Leonardian age for the Hermit Formation (Blakey and Knepp 1988).

Regional and Worldwide Correlation

In a 1988 study, Blakey and Knepp summarized the regional correlation of Pennsylvanian and Permian rocks of Arizona and adjacent areas. Based on age assignments presented above and using the worldwide geologic time scale of Van Eysinga (1975), they found that the Supai Group correlates with the Westphalian, Stephanian, and Sakmarian.

The Hermit Formation correlates with the Artinskian, which occurs in the standard Carboniferous and Permian units of Europe. In addition, both the Esplanade Formation and the Hermit Formation correlate with the eolian-bearing Rotliegendes of northern Europe.

ORIGIN OF DEPOSITS IN SUPAI AND HERMIT

Introduction

Interpretation of the depositional history of Pennsylvanian and Permian redbeds in the Grand Canyon region is plagued by a variety of problems—including sparsity of fossils, the complex cyclic nature of the strata, the fine-grained nature of some units, extensive diagenesis, inaccessibility of some outcrops, and some misconceptions concerning regional stratigraphy and depositional history.

Despite these problems, recent and detailed sedimentologic work (much of it previously unpublished) suggests the need to revise our interpretation of the depositional history of the Supai Group. Earlier studies suggested a fluvial, deltaic, beach, shallow-marine, or estuarine origin for Supai sandstone bodies. New evidence indicates that an eolian origin is much more likely. McKee (1982a) did not consider an eolian origin for any sections of the Supai Group. The presence of carbonate grains, including locally abundant fossil fragments, may have prevented his recognition of the eolian environment. Desert varnish and extensive diagenesis of some beds further obscure some important eolian characteristics.

Eolian Characteristics

Until fairly recently, eolian criteria consisted chiefly of the textural maturity of quartz sand, coupled with thick set size and a high-angle, cross-strata dip. Although many ancient eolian sandstones display these characteristics (the Coconino Sandstone, for example), these criteria are unreliable. Some eolian sand-stones lack large-scale, high-angle sets. In some cases, sandstones with these characteristics are noneolian. Hunter (1977) found that the most reliable eolian indicators are not set size, angle of dip, or even cross-stratification, but the na-ture of the individual laminae and strata that compose the sandstone. Migrating and climbing wind ripples, regardless of their orientation, form the inversely graded laminae (generally less than several millimeters thick) that Hunter termed "climbing translatent strata." This strata type can form and be preserved in many parts of the eolian environmental—including on most parts of dunes, between dunes (interedunes), and on duneless areas of blowing sand termed sandsheets.

Many dunes contain one or more active slipfaces that are characterized by the periodic avalanching of loose sand. Hunter (1977) described deposits formed by this process as structureless strata (as much as several centimeters thick) at or near the angle of repose of dry sand (30–34°, though generally 24–26° in an-cient sandstones because of compaction). He termed these "sandflow strata." Some small dunes, and most dunes without slip faces, may leave only climbing translatent strata in variously shaped sets with widely ranging angles of dip. Par-ticularly common are trough-shaped sets several feet thick and tens of feet wide that were produced by crescentic-shaped dunes or troughlike scours (blowouts) on dunes or sandsheets. Many dunes with slip faces produce avalanche strata that fade out down the dune, where they are replaced by climbing translatent strata. This toe of the avalanche, which is termed "sandflow toe," is an easily observed eolian indicator.

Eolian sandstones typically consist of one or more facies that relate to their type and position within the eolian sand sea (erg). The central portion of most sand seas contains dunes and larger bedforms called draas. Marginal sand seas include small, possibly scattered dunes and sandsheets, and the most distal ar-eas may consist of broad sandsheets, and the most distal areas may consist of broad sandsheets with few, if any, dunes. Sandsheets and small dune fields may filter into adjacent environments, such as fluvial plains, dry lake beds, and coastal plains. Facies formed in central sand–sea areas typically consist of cross-strati-fied sandstone comprised of sandflow strata and climbing translatent strata with interdune deposits of horizontal climbing translatent strata or various types of structureless to wavy-bedded sandstone or silty sandstone. Marginal sand–sea de-posits typically display smaller sets of cross-stratified sandstone and sandsheet deposits. These consist of horizontal to irregular beds of climbing translatent strata. The most complete discussion of the distribution of all known ancient eo-lian sandstone units and their facies can be found in Blakey et al. (1988).

Controls on Deposition

Controls on deposition during Supai and Hermit time were varied and complex. They included a complicated tectonic regime, rapid rise and fall in sea levels, and an influx of sand from the north.

The tectonic elements that affected sedimentation during the late Paleozoic have been addressed in several recent papers (Blakey 1980, 1988; Blakey and Knepp 1988). Areas in which the crust subsides relatively rapidly have thicker deposits than do areas of less subsidence. The former tend to have more ma-rine deposits and fewer redbeds. Areas of less subsidence are characterized by thin, red, terrigenous clastic deposits with few fossils and abundant unconfor-mities. During the late Paleozoic and early Mesozoic in the Southwest, such de-posits tended to form in fluvial, lacustrine (lake), and coastal-plain environments.

Areas of greater subsidence are characterized by deposits of limestone and thin sandstone and mudstone. They typically are grayish in color and fossiliferous. Marine and shoreline deposits dominate.

Geologists classify the resulting tectonic elements by their shape and their relation to adjacent elements. Circular to oblate areas with relatively high rates of subsidence are termed *basins*. Areas of intermediate subsidence rates covering broad areas are called *shelves*, and those more parabolic in shape are termed *embayments*. Arches are areas of little subsidence; typically, they interrupt brief periods of slight uplift. Areas dominated by uplift are referred to as upwarps or uplifts.

During the formative period of the Supai and Hermit, northern Arizona was bisected by a northeast-trending element of relatively little subsidence, the Sedona arch (Fig. 9.10). To the northwest, the Grand Canyon embayment domi-

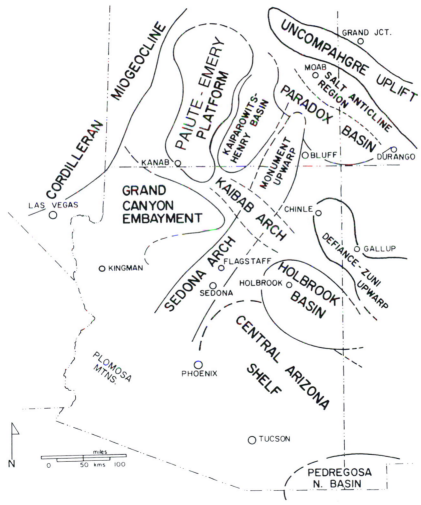

FIGURE 9.10. Map showing late Paleozoic tectonic elements of northern Arizona and vicinity.

nated the Grand Canyon region. East-central Arizona was the site of the Holbrook basin and the Mogollon shelf. Northeastern Arizona was dominated by the Defiance uplift during the Pennsylvanian or the Defiance arch in the Permian. To the north, in southeastern Utah, was the Paradox basin. During periods of relatively high worldwide sea level, the seas tended to encroach on Arizona from the west into the Grand Canyon embayment, from the east and south across the Mogollon shelf, and from the northeast into the Paradox basin. Although the uplifted areas (i.e., arches) received some sediment, much of it was nonmarine. Because great fluctuations of sea level (chiefly because of glaciation in the southern hemisphere) characterized the late Paleozoic, deposits tended to be cyclical in nature. Marine-dominated cycles occurred in basins, and continental-dominated cycles occurred across the arches. Movement on arches, uplifts, and basins altered a landscape already shaped by changes in sea level. Therefore, the resulting pattern of deposition is very complex.

The influx of quartz sand into the Colorado Plateau region from the north also complicates the setting. The sand tended to accumulate between the arches and adjacent shelves or basins as extensive eolian deposits (Blakey 1988). When sea level was relatively low, eolian deposits became widespread, often expanding into basins and across shelves; when the sea level was high, eolian deposits, if present, were confined to narrow belts along the flanks of arches (Chan and Kocurek 1988).

The Grand Canyon and the western Mogollon Rim region lay in a somewhat intermediate position between more negative areas to the northwest and higher ones to the southeast. Not unexpectedly, the resulting sedimentation is intermediate in character; cyclic sedimentation with more continental aspects is present in the east, whereas cyclic sedimentation with more marine aspects is present in the west.

The general structural grain of northwestern Arizona tended southwest–northeast during Supai and Hermit time. Shorelines and marine trends should have paralleled this pattern, and fluvial deposits should have flowed northwesterly, perpendicular to this grain. Indeed, marine and shoreline patterns do parallel the trend, but the sandstone bodies that McKee (1982a) suggested might be of fluvial origin show a paleocurrent reading to the south and southeast; this is as much as 180° from the expected trend. We have not documented the trends of Hermit Formation sandstone bodies.

Any one of the above controls on deposition was sufficient to produce a complicated depositional pattern. Together, they combined to produce a heterolithic record that varies abruptly, both vertically and laterally. The summary presented below must be considered preliminary. We have much to learn about the depositional history of the Supai Group and the Hermit Formation.

Depositional Environments

Recent studies tend to confirm the long-standing belief that the Supai Group was deposited on a broad coastal plain. Individual depositional settings ranged from shallow marine to continental. However, the recent identification of eolian-deposited sandstone units throughout the section necessitates a reinterpretation of Supai depositional history. Most of our knowledge of the Supai is based on the work of McKee (1982a, pp. 260, 261), who provided a general description of sandstone bodies in the Supai Group. He interpreted the large foreset planes in the Manakacha Formation as the fronts of large, subaqueous sand sheets or small Gilbert deltas in areas strongly affected by tides. Traditionally, geologists have interpreted sandstone bodies of the Wescogame Formation as having formed in a high-energy, fluvial environment. Marine bioclastic debris was thought to have been trapped in an estuarine setting.

McKee found the Esplanade the most difficult unit to interpret and offered a compromise of mixed fluvial, estuarine, and shallow-marine origins. It should be pointed out that although McKee's Supai publication appeared in 1982, the work actually started in the late 1930s and continued into the 1970s. Delays in publication created a large gap between the time when the work was written and the time it was published. Therefore, much of the state-of-the-art sedimentology of the late 1970s and the 1980s is not reflected in the publication; this is especially true of eolian sedimentology. McKee's monograph will continue to be a valuable source of information concerning the Grand Canyon Supai, but some of the interpretations, particularly concerning depositional environments, must be considered out of date.

Given the eolian criteria discussed earlier in this chapter, most cross-stratified sandstone units in the Manakacha and Wescogame formations and the Esplanade Sandstone now are regarded as eolian in origin. Evidence for this includes the fact that wind-ripple laminae (climbing translatent strata) are ubiquitous throughout these units in the eastern and central Grand Canyon and the western Mogollon Rim regions (Fig. 9.11). Most sets of cross-strata are composed chiefly or entirely of climbing translatent strata. There is little or no evidence, however, of avalanche-formed sand-flow strata. Studies of the eolian deposits in the Supai have not progressed enough to explain this, and several theories are possible. Kocurek (1986) reported that dunes with broad aprons (plinths) commonly do not preserve slipface deposits in the sedimentological record; longitudinal dunes, star dunes, and oblique dunes commonly have large aprons. Reversing dunes also tend to lack slipface deposits. Other late Paleozoic eolian deposits contain a high percentage of wind-ripple strata. They also contain at least locally abundant sand-flow strata (Blakey et al. 1988). But only in the Esplanade Sandstone of the Oak Creek Canyon area are sand-flow strata developed extensively in the Supai Group. Some cross-stratified sets (perhaps 40 percent in Grand Canyon and 60 percent in the western Mogollon Rim) have been so affected by diagenesis that the stratification type cannot be determined. A portion of these could contain sand-flow strata, and some sets could be noneolian in origin; however, close association with, and similarity to, sets with wind-ripple laminae strongly suggest that many or most unidentifiable sets formed as climbing wind-ripple deposits.

The geometry of eolian sandstone bodies in the Supai Group is somewhat unusual. Many well-known eolian sandstones, such as the Coconino Sandstone and the Navajo Sandstone, are mostly or entirely cross-stratified sandstone. Redbeds, conglomerate, and limestone are rare or absent. Blakey et al. (1988) have pointed out the problems of stereotyping eolian deposits as simple, pure sandstone bodies. Most are not. Individual eolian bodies range from irregular sheets to more common broad lenses, to wedges, and even small pods (Fig. 9.12). Bodies are separated from each other by red siltstone and mudstone, noneolian sandstone, limestone, and, rarely, conglomerate. Undoubtedly, this heterogeneous assemblage of lithologies and complex geometry is the chief reason that the Supai has not been interpreted previously as eolian in origin. Coupled with the high carbonate content and locally abundant sand-sized marine fossil grains in some eolian sandstone, the Supai eolian bodies are far removed from the "typical" eolian sandstone.

The origin of some noneolian deposits in the Supai Group is not clear. Bioturbated sandstone can form in any number of environments—including eolian,

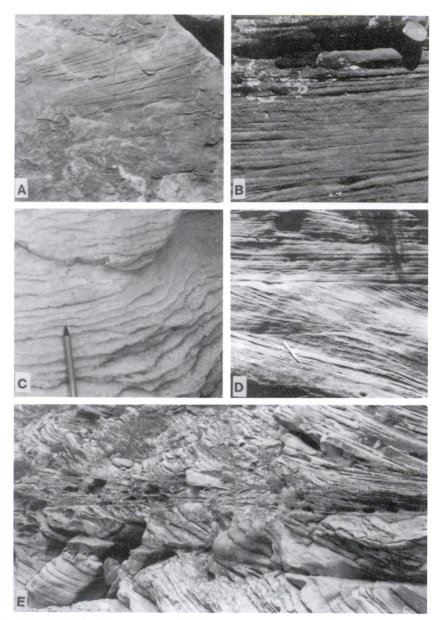

FIGURE 9.11. Eolian strata in Supai Group. (a–d) Eolian wind-ripple (climbing translatent) strata. (e) Eolian sand-flow (avalanche) strata.

fluvial, shoreline, and shallow marine. It is possible that this material has been reworked by organisms in a marine or continental setting. Horizontally laminated to low-angle, cross-stratified sandstone may have formed as plane-bed laminae in shallow streams or in beach swash zones, or as shallow-marine storm deposits. Fine-grained redbeds are particularly enigmatic. Their potential origin can range from fluvial flood plain to lacustrine and sabkha; through low-energy, shoreline (supratidal mudflats, lagoons); and into nearshore, shallow marine.

FIGURE 9.12. Eolian sandstone bodies in Supai Group. (a) Manakacha, Wescogame, and Esplanade along Hermit Trail. Dashed lines separate the formations. Most cliff units are composed of eolian sandstone. (b) Manakacha Formation on Hermit trail showing lens-like eolian sandstone bodies separated by thin sandstone and mudstone of uncertain origin. Thickest sand bodies are about 6 m thick.

Because the lower contacts of eolian sandstone bodies are sharp and probably disconformable (and upper contacts commonly are the same), stratigraphic positions an unreliable criteria for diagnosing the environmental of noneolian deposits. At most locations, a few sequences display the following cyclic pattern: sharp basal contact abruptly succeeded by cross-stratified, eolian sandstone; the eolian sandstone grades upward into bioturbated, micritic limestone that is overlain conformably by red mudstone, horizontally stratified sandstone, or a combination of these. Unfortunately, it cannot be determined whether these cycles are transgressive, regressive, or totally continental in origin. If transgressive, the most logical interpretation would be eolian, succeeded by low-energy shoreline, to local beach deposits. If regressive, the eolian likely would be succeeded by fluvial and lacustrine or sabkha deposits. If the cycle is continental (and not necessarily related to advance or retreat of the sea), the succession likely represents ingress and egress of fluvial and lacustrine environments into the eolian dune complex. Clearly, more work is needed on this topic.

Evaporite deposits in the Esplanade Sandstone, chiefly gypsum, could have formed in either coastal or continental sabkha environments. The evaporits are interbedded with redbed and eolian sandstone in the Toroweap area of western Grand Canyon.

We know now that not all carbonate units in the Supai Group are of marine origin; some arenaceous, cross-stratified limestones are of eolian origin. However, many Supai carbonates clearly are of shallow-marine origin. Micrite-rich, fossiliferous carbonates were deposited in open-marine, relatively low-energy environments (McKee 1982a, p. 345). Carbonate grainstone and packstone formed in higher-energy environments, such as on carbonate shoals and in tidal channels. Unfortunately, the detailed relations between marine limestone and eolian sandstone and arenaceous limestone are yet to be determined. McKee (1982a) has documented the abundance of marine carbonate in western facies of the Manakacha and Wescogame formations as well as Pakoon Limestone, but we have no complete study of their presumed interbedding with eolian sandstone. These relationships would be exposed best in the Shivwits Plateau region of the western Grand Canyon.

Documentation as to how far west eolian sandstone and arenaceous limestone extends also is nonexistent. Based on similarities to eolian strata that I have seen (photographs included in McKee 1982a), it seems likely that eolian strata exist in western Grand Canyon and northward along the Grand Wash Cliffs.

It is equally difficult to determine the depositional setting of the Hermit Formation. White (1929) used fossil flora, general lithology, and a regional setting to postulate deposition by sluggish streams on a broad, low-lying, arid coastal plain. Details of Hermit deposition in the Grand Canyon have yet to appear in the literature. Work currently in progress in the Sedona area is providing some detail of Hermit deposition for the western Mogollon Rim region. A brief examination of scattered Hermit outcrops in the Grand Canyon tends to confirm White's general conclusions.

This information, combined with more detailed work near Sedona, provides a few more specifics on the nature of Hermit streams. Several types of stream deposits have been identified. These include: narrow, channel-shaped bodies of limestone-pebble conglomerate that formed in local, arroyolike streams as well as broader sheets of conglomerate that probably formed in wider arroyos. The conglomerate was derived from caliche deposits on adjacent alluvial plains. Sheets of limy and nodular structureless-to-ripple-laminated sandstone were deposited as relatively unconfined stream deposits or as broad levee deposits. Trough, cross-stratified sandstone (with or without associated conglomerate) laid down in broad, gently dipping, irregular sheets (epsilon cross-stratification) represents the deposits of larger, more perennial, meandering streams. Red mudstone was deposited in low-lying areas as overbank deposits. Light-colored, large-scale, high-angle, cross-stratified sandstone (with climbing translatent strata) represents minor eolian dune deposits.

From the Sedona area westward to an area near Seligman, the types of deposits described above are distributed irregularly throughout the Hermit. In central and eastern Grand Canyon, geologists are able to identify all of these types. However, three general facies predominate.

The lower portion of the Hermit on both the Kaibab and Hermit trails consists of ledges and slopes of structureless and trough cross-stratified sandstone. Exposures there are not as good as some in the Sedona area, but the sand bodies are similar to the meandering stream deposits and unconfined stream deposits there. The bulk of the Hermit in eastern and central portions of the Grand Canyon consists of weak, ledge-forming, silty, faintly ripple-laminated sandstone and slope-forming mudstone. Generally, 10 to 15 cyclic alternations of these units are present. The ledges probably represent sluggish, shallow, perhaps unconfined stream deposits, and the slopes primarily are overbank deposits. Whether these cycles were formed by climatic fluctuation, by the migration of stream deposits, or by tectonic cycles is unknown.

Depositional Models

The high carbonate grain content, associated marine carbonates, and regional setting strongly suggest a coastal-plain setting for the eolian dune deposits in the Supai Group. Carbonate sand was derived locally by onshore winds from adjacent, bioclastic debris. The quartz sand was transported by wind and longshore currents into the Grand Canyon region from the north (Blakey et al. 1988). To the west was a broad, shallow, epicontinental sea, and to the east were low-lying coastal plains and low uplands (Fig. 9.13). The repetitive Pennsylvanian and Early Permian marine transgressions and regressions moved across this broad plain, but influxes of eolian sand formed obstacles to widespread transgression. During regression, the eolian dunes spread across the broad flats. The ensuing transgression interrupted eolian deposition and apparently prevented the development of large eolian sand seas (ergs).

The battle between dune and sea was repeated numerous times, and only a small fragment of the sedimentological record has been preserved. Sluggish and probably ephemeral streams continually washed fine-grained, terrigenous, and clastic debris into the area. The mud and fine-grained sand were trapped in lagoons, on tidal flats, and in river channels. The extremely arid climate provided a reflux action that pumped salty groundwater upward. This resulted in the precipitation of calcium carbonate and other minerals as deposits on the earth's surface. Some of these nodules were reworked later into conglomeretic deposits. Such a setting clearly was the site of a great deal of reworking of sediment.

Over time, the wind altered marine and fluvial deposits into dune deposits and eolian dust (loess). Rivers and advancing seas also reworked these eolian deposits. This intermixing and reworking of environments created an array of conditions that might seem incompatible within a given environment. For example, fluvial and tidal deposits might be better sorted than expected in one area because of the activity of the wind. Eolian deposits frequently display geometric patterns that are more typical of fluvial or shallow-marine deposits. The heterogeneity of much of the Supai is directly related to this complex intermixing of environments coupled with a broad range of available depositional material.

Hermit stream deposits do not fit readily into established fluvial models. The broad, low-gradient, arid plain was crossed by several kinds of streams, probably both perennial and ephemeral. Local coarse conglomerate indicates that

FIGURE 9.13. Map showing general depositional model and hypothetical paleogeography of eolian units in Supai Group. Map depicts region during regressive sequence. Carbonate sand is derived from exposed, older carbonate shoals; quartz sand is fed into the region from the north. Note state lines and abbreviations for Supai, Grand Canyon, Seligman, and Flagstaff.

streams occasionally had sufficient energy to transport larger clasts. Widespread and small dune fields were present on the floodplain, but the great eolian faucet from the north had been shut off temporarily.

STRATIGRAPHIC ANALYSIS

The incredible range of Supai depositional environments and the processes within them are only generally understood. Clearly, many problems remain and some may never be solved. Patterns of transgression and regression have yet to be documented. Hermit deposition is poorly understood, primarily because of a lack of regional and local data. Despite these problems, the ideas, models, and conclusions offered here permit a preliminary analysis of Supai and Hermit depositional history.

Morrowan History

The lower two-thirds of the Watahomigi formation was deposited in shallow-marine and low-energy shoreline environments (McKee 1982a). The Morrowan portion of the Watahomigi consists generally of a lower, fine-grained, siliciclastic sequence and an upper, fossil-bearing limestone. Poor exposures and a lack of reliable criteria prevent detailed analysis of the lower slope. A basal conglomerate suggests that continental or shoreline processes reworked weathered Redwall Limestone. It is possible that the overlying mudstone accumulated in low-energy, shoreline environments or on a broad coastal plain. A rise in sea level (or subsidence in the Grand Canyon region) permitted a period of widespread, clear-water, carbonate deposition. Local intercalated mudstone suggests that there were periods of influx by fine clastics. These may have been caused by fluctuating sea levels. A widespread erosion surface at or near the top of the limestone suggests a period of erosion near the close of Morrowan time (McKee 1982a, p. 160).

Atokan History

The upper Watahomigi Formation consists of thin limestone, redbeds, and minor sandstone. This suggests an advance of the sea into the Grand Canyon area. However, a strong influx of eolian material from the north initiated a trend of eolian deposition that would continue periodically for over 150 million years (Blakey et al. 1988). During a major regression, quartz and carbonate sand blew across the landscape. It was derived locally from exposed, recently deposited, shallow-marine materials. Quartz sand from the north was transported across the Paiute and Emery arches and was funneled into the Grand Canyon embayment (Fig. 9.13). Here, the sand was deposited in coastal dunefields. We are not certain whether the eolian sand was deposited in a broad coastal erg or in a series of small dunefields. However, eolian deposits were planed off by marine transgressions and perhaps minor fluvial events. The next regression saw a return to eolian deposition. Sharp bases to eolian sandstone units suggest a rapid change to eolian conditions.

DesMoinesian and Missourian History

No rocks of these ages have been reported from the Grand Canyon or the western Mogollon Rim regions. This period of time apparently is represented by the unconformity between the Manakacha and Wescogame formations. Whether sediments of Middle Pennsylvanian age were deposited and later removed or were never deposited extensively in the region is unknown.

Virgilian History

The pattern of deposition established during Atokan time was renewed during deposition of the Wescogame Formation. Repeated transgression and regression caused numerous depositional cycles. The widespread Wescogame cliff unit, believed to be mostly eolian in origin, may represent the development of a large erg, or of several ergs, across the region (Fig. 9.13). Redbeds at the top of the Wescogame Formation may have formed during a period of decreased eolian influx.

Another period of erosion followed. This is indicated by the unconformity at the presumed Pennsylvania–Permian boundary. Apparently, a lowering of the sea level caused an incision of arroyolike streams into the underlying Wescogame Formation.

Wolfcampian History

During this period, the sea advanced into the region at least as far east as the central Grand Canyon. The Pakoon Limestone formed in a variety of clearwater, shallow-marine environments. Farther east, coastal-plain and minor eolian deposits are represented in the lower slope unit of the Esplanade Sandstone. Later in the Wolfcampian, a vast blanket of eolian sand spread southward across the Colorado Plateau region. Forming the Cedar Mesa Sandstone in southeastern Utah, the Queantoweap Sandstone in southwestern Utah, and the Esplanade Sandstone in northern Arizona, eolian deposits of Wolfcampian age blanketed much of the Southwest (Blakey et al. 1988).

Interruptions in eolian deposition in the Esplanade Sandstone may have been caused by a rise in sea level, influxes of fluvial activity, or both. East of the Sedona arch, eolian deposits became sparse to absent, and a period of fluvial deposition dominated the landscape.

As eolian conditions waned, fluvial deposits spread westward into the Grand Canyon and the western Mogollon Rim region. Some fluvial channels were incised into the underlying eolian sandstone, but in many locations a general transition to fluvial conditions is evident.

Leonardian History

Most likely beginning in the Late Wolfcampian and continuing into the Leonardian, low-energy, fluvial conditions developed. Based on the distribution of the Hermit and coeval deposits in Utah, Colorado, New Mexico, and eastward into the High Plains, geologists have concluded that this marked a period of extremely widespread fluvial deposition. Deposits farther east were left behind by streams that had steeper gradients and carried coarser debris. As the streams crossed the broad, low-lying coastal plain, they carried fine material and locally derived carbonate clasts. A few scattered dune fields persisted into this period of time, but in general the great eolian episode was brought to a temporary halt.

Eolian deposition was renewed in the Leonardian, first with the development of eolian deposits in eastern and central Arizona in the De Chelly Sandstone and the Schnebly Hill Formation and then across most of the northern and central Arizona, including the Grand Canyon region in the Coconino Sandstone.

SUMMARY

Some of the conclusions concerning the depositional history of the Supai Group and Hermit formation are preliminary and probably will alter with further work. However, one aspect seems clear: Eolian depositional processes played an important role in the formation of the Supai Group and parts of the Hermit formation. The Atokan eolian deposits in the Manakacha formation may represent the very earliest record of the vast eolian systems that were to dominate the Southwest until the Late Jurassic.

• 10 •

COCONINO SANDSTONE

Larry T. Middleton, David K. Elliott, and Michael Morales

INTRODUCTION

The Early Permian Coconino Sandstone is one of the moist conspicuous formations in the Grand Canyon. The high-angle, sweeping sets of cross-stratification, which record the southerly advance of very large dunes, are visible at great distances throughout the area. Ironically, few geologists, except for McKee (1933b) and Reiche (1938), have studied the Coconino. These Sahara-like dunes were part of an enormous desert that once extended north into Montana.

During the last 10 years, a number of studies on modern and ancient sand seas, or ergs, have increased greatly our understanding of eolian bedform dynamics. At the same time, these studies have facilitated the development of facies models for eolian depositional systems. It is possible now to apply the results of these studies of stratification styles to the Coconino Sandstone and to discuss possible dune morphologies and migration paths. Additionally, we can describe trace fossils in the Coconino and relate them to the physical environments of deposition. This provides us with a more comprehensive reconstruction of the erg environments (e.g., substrate conditions). Although the cliff-forming nature of the formation can present certain logistical problems, the exceptional three-dimensional exposures in the Grand Canyon afford the sedimentologist and ichnologist the opportunity to examine the details of one of the more classic and interesting eolian deposits on the Colorado Plateau.

Because the results reported here are preliminary, it is our hope that they will spur more detailed studies of lateral and vertical facies relationships in the Coconino Sandstone, as well as initiate more effort at documenting the variety and distribution of the trace fossil assemblages.

REGIONAL STRATIGRAPHIC RELATIONSHIPS

The Middle Leonardian Coconino Sandstone crops out throughout much of the southern Colorado Plateau south of the Utah–Arizona border. The most exceptional exposures are in the Grand Canyon and the Marble Canyon areas and along the Mogollon Rim south and southeast of Flagstaff. Darton (1910) named the formation for exposures in Coconino County, Arizona. Although

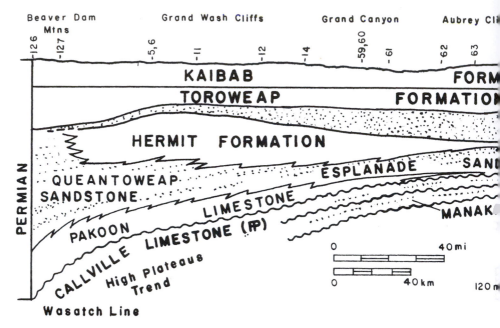

FIGURE 10.1. Regional stratigraphic relationships of Permian strata along the southern margin of the Colorado Plateau. (After Blakey and Knepp, 1988.)

there have been a number of controversies regarding Permian stratigraphy in northern Arizona, recent studies have established some regional stratigraphic relationships (Fig. 10.1) (Cheevers and Rawson 1979; Rawson and Turner-Peterson, 1980; Blakey and Middleton, 1983; Blakey and Knepp 1988; Blakey 1996).

In the Grand Canyon region, the Coconino Sandstone disconformably overlies the Permian (Wolfcampian–Leonardian) Hermit Formation. Throughout the canyon, the Coconino is overlain by, or intertongues with, the Permian Toroweap Formation. In easternmost localities where the Toroweap is not present, the Kaibab Formation conformably overlies the Coconino. The Coconino Sandstone grades eastward into the Glorieta Sandstone in western New Mexico. To the south, along the Mogollon Rim, it intertongues with the Schnebly Hill Formation and is transitional into the overlying Toroweap Formation or is sharply overlain by the Kaibab Formation.

Regional structural features (Fig. 10.2) control the marked thickness variations in the Coconino Sandstone. In the western part of the Grand Canyon near the Grand Wash Cliffs, the formation is 65 feet (20 m) thick, but it thins progressively to the west. The Coconino thickens in the central part of the canyon and is over 600 feet (183 m) thick near Cottonwood Creek. Eastward and northward, the Coconino thins to 57 feet (17 m) in Marble Canyon before pinching out in the vicinity of Monument Valley. The Coconino, likewise, wedges out in southwestern Utah.

McKee (1933b, 1974) reported a maximum thickness of nearly 1000 feet (305 m) near Pine, Arizona, along the Mogollon Rim (Fig. 10.2). This rapid southerly thickening probably is related to increased subsidence along the south-

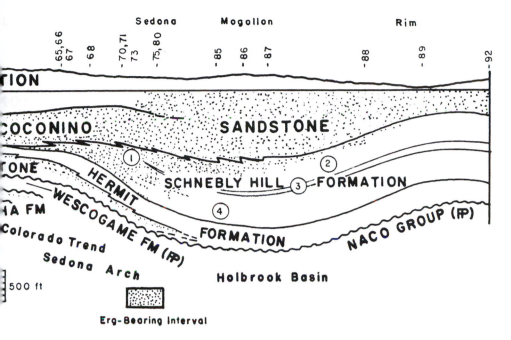

ern boundary of the Coconino depositional basin. The lithologies, textures, and sedimentary structures in the Toroweap Formation are similar to those of the Coconino toward the south. Hence, some of the thickening might reflect the presence of Toroweap equivalents in the upper part of the Coconino.

Blakey and Knepp (1988) proposed that the Sedona arch (Fig. 10.2) controlled facies distributions, as well as thickness variations, in several of the late Paleozoic units in northern Arizona. This seems likely because the Coconino thickens across and to the east of the Sedona arch, and it is in this area that the Toroweap Formation undergoes facies changes into the Coconino Sandstone.

ICHNOLOGY

Introduction and Preservation

Because body fossils have not been reported from the Coconino Sandstone, invertebrate and vertebrate trace fossils (Fig. 10.3) remain the only evidence of the organisms that lived in the Coconino desert. They are also the only data on which to base the studies of fossil diversity, morphology, and paleoecology. In particular, trace fossils have been important evidence in the debate on the depositional environment of the Coconino Sandstone. Work by Brady (1939, 1947), McKee (1944, 1947), and Brand (1979) has shed a considerable amount of light on the formation of the traces though differences in interpretation still remain.

Brady (1939, 1947) carried out a series of experiments with small vertebrates and invertebrates and pointed out similarities between the fossil traces and those formed by modern scorpions, millipedes, and isopods. In addition, he found that in wet or slightly moist sand, no trace was left by the scorpions he was working with but that they made clear impressions in dry sand. As will be discussed, work by Sadler (1993) has expanded on these studies.

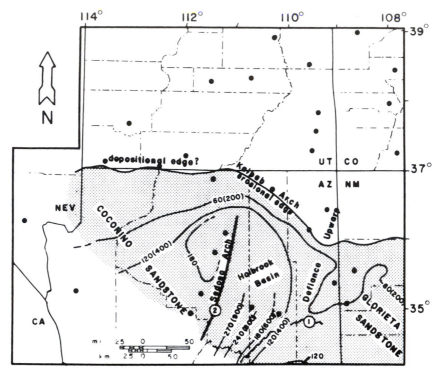

Figure 10.2. Isopach map of the Coconino Sandstone illustrating thinning away from the Sedona arch. (After Blakey and Knepp, 1988.)

In the 1940s, McKee conducted a series of experiments designed to duplicate the Coconino tracks. He filled a long trough with sand that formed a hill in the center. A variety of small vertebrates and invertebrates then were induced to walk along the sand and over the hill. By varying the slope of the hill and the moisture content of the sand, McKee was able to test the trace-forming capabilities of a variety of animals in a number of different environments. He tested the animals on a variety of surfaces—including dry sand, damp sand, saturated sand, and even sand that had been soaked and then allowed to dry. He found that only the largest animals tested (chuckwalla lizards) were able to make tracks in wet sand or crusted sand, and even then, the tracks were not as clear as those formed in dry sand. Smaller animals, such as millipedes and scorpions, were unable to make tracks in wet sediment and left clear traces only in dry, loose sand. He also found that at slopes below 27°, both uphill and downhill tracks were likely to be preserved. Avalanching, however, tended to destroy tracks on steep slopes.

Based on previous studies of conditions necessary to preserve surface features in dune sands, McKee (1945) suggested that the tracks in the Coconino Sandstone were formed initially in loose, dry sand that was dampened subsequently before being covered. The mists and fogs that intermittently are present in areas of coastal sand dunes would provide a suitable means of dampening the surface. Recent work by Brand (1978, 1979), however, suggests that clear traces with morphologies similar to those found in the Coconino Sandstone can be generated on sandy surfaces submerged in standing water. It may be, there-

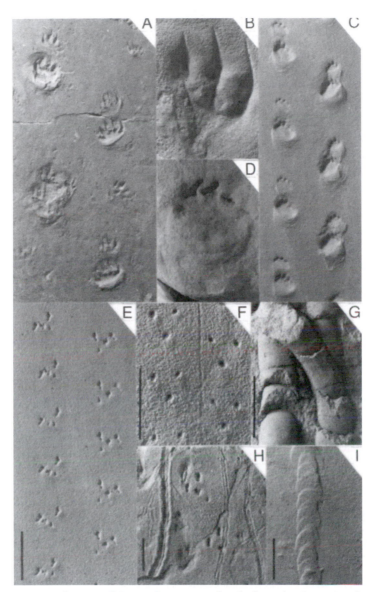

Figure 10.3. Vertebrate and invertebrate trace fossils from the Coconino Sandstone. (All specimens numbers refer to the Museum of Northern Arizona, Flagstaff, Geology Collection.) (a) *Laoporus* sp. (V3470); notice the four toes on the forefeet prints and the five toes on the hindfeet prints. (b) *Baryopus tridactylus* (V3360). (c) *Laoporus* sp. (V3472); notice the sand humps behind the tracks, indicating that the animanl walked uphill, pushing back loose dune sand. (d) *Baropus* sp. (V 3388). (e) Spider track (N3089). (f) *Paleohelcura dunbari* (N3666). (g) *Scolecocoprus cameronensis* (N3707). (h) *Diplodichnus biformis* (N3657). (i) *Scolecocoprus arizonae* (N3655).

fore, that a study of trace morphology alone is not sufficient to distinguish between sand surfaces exposed to subaqueous or subaerial conditions. Further experimentation may help to resolve this problem.

Invertebrate Trace Fossils

In 1918, Lull described trace fossils from the Coconino Sandstone that he collected near the Hermit Trail in the Grand Canyon (Lull 1918). Subsequently, C.W. Gilmore of the U.S. National Museum carried out more detailed work in the same area. Gilmore published his results in a series of papers (Gilmore 1926, 1927, 1928), describing both vertebrate and invertebrate tracks. In these studies, he described 10 genera and 17 species, of which the following five genera and species clearly were invertebrate: *Mesichnium benjamini,* a trail consisting of two parallel lines of footprints with a median row of suboval, regularly spaced depressions; *Octopodichnus didactylus,* a trail consisting of alternating sets of impressions in groups of four; *Paleohelcura tridactyla,* a trail consisting of alternating sets of three prints with a median tail drag; and *Triavestiga niningeri,* a trail composed of paired, longitudinal grooves. Gilmore ventured little opinion as to the identity of the trail-formers beyond suggesting that *Octopodichnus* and *Paleohelcura* showed some similarity to tracks made by modern crustaceans.

Through many years, Brady continued work on these tracks (Brady 1939, 1947, 1949, 1955, 1959, 1961). He described new traces, provided new insights into the conditions necessary for their formation, and identified the trace-formers. He pointed out (1939, 1947) that the *Paleohelcura* trails were very like those formed by the modern scorpion (*Centruroides*) in dry sand when the temperature was about 60°F (15°C). He also showed, however, that with variations in temperature and surface conditions, the same animal could leave a variety of traces—sometimes impressing two, three, or four feet on each side and leaving no tail drag, an intermittent one, or a complete impression of the tail. Based on this information, he showed that a variety of trails in the Coconino Sandstone were simply variations on *Paleohelcura.* It is clear from this work that two of the ichnogenera described by Gilmore (1926), *Triavestiga* and *Mesichnium,* represent variations on *Paleohelcura* and do not merit the status of separate ichnogenera. In addition, Brady recognized one new ichnospecies of *Octopodichnus,* *O. minor.*

Brady (1947) was able to show that some two- and three-grooved trails were very similar to those formed by modern millipedes as they move up and down sand slopes. He named these *Diplopodichnus biformis* (Fig 10.3H) and pointed out that the traces attributed by Gilmore (1926) to *Unisulcus* were identified incorrectly. The traces attributed to *Unisulcus* are slender, smooth, and wormlike and should be included in *Diplopodichnus.* Trails similar to those formed by the modern isopod *Oniscus* he named *Isopodichnus filiciformis.* However, he discovered subsequently that the name was already in use, and he renamed the ichnogenus *Oniscoidichnus* (Brady 1961). Traces that Brady attributed to oligochaete fecal pellets he named *Scolecocoprus cameronensis* and *S. arizonensis* (Fig. 10.3g and 10.3i). He described them as consisting of an overlapping series of pellets and attributed the sometimes grooved appearance of the lower surface to traces left by spines on the ventral surface of an oligochaete. It is more probable that these represent the infillings of burrows, the spreite representing what was thought to be the rounded ends of the fecal pellets. In 1961, Brady also described an additional species of *Paleohelcura,* *P. dunbari* (Fig. 10.3f).

TABLE 10.1. Invertebrate Ichnospecies from the Coconino Sandstone

---

Diplopodichnus biformis (Brady 1947)
Mesichnium benjamini (Gilmore 1926)
Octopodichnus didactylus (Gilmore 1926)
O. minor (Brady 1947)
Oniscoidichnus (Isopodichnus) filiciformis [Brady 1949 (1947)]
Paleohelcura tridactyla (Gilmore 1926)
P. dunbari (Brady 1961)
Scolecocoprus cameronensis (Brady 1947)
S. arizonensis (Brady 1947)
Triavestiga niningeri (Gilmore 1927)

Alf (1968) carried out the most recent work on invertebrate traces from the Coconino Sandstone, describing a trail composed of alternating sets of four prints (Fig. 10.3e). These clearly were different from *Octopodichnus*, which consists also of alternating sets of four prints, and Alf was able to show from experiments with modern tarantulas and wolf and trapdoor spiders that the most likely trace-former was a spider. He did not name the trace, however, though clearly it is a separate ichnogenus. The Coconino Sandstone's invertebrate trace fossils are listed in Table 10.1.

Traces similar to those reported from the Coconino Sandstone have been reported from other late Paleozoic and Mesozoic eolian sequences (Ekdale and Picard 1985; Sadler 1993). In a study of trace fossils from the Permian De Chelly Sandstone in northeastern Arizona, Sadler (1993) examined tracks similar to those in the Coconino Sandstone and conducted experiments using modern arthropods as a means of identifying the possible track makers. The De Chelly Sandstone is correlative to the lower part of the Coconino Sandstone, and like the Coconino is eolian in origin (Blakey and Knepp 1989). Sadler described four ichnospecies assigned to the ichnogenera *Paleohelcura* and *Octopodichnus* based on comparisons with similar traces in the Coconino Sandstone in Grand Canyon. Sadler's study involved experimental studies using scorpions and spiders to demonstrate that not only did these arachnids produce tracks similar to those found in both formations but that the trackways provided evidence of direction of movement and information concerning substrate moisture conditions.

Sense of movement in these two ichnogenera is indicated by a bifurcation of impressions that open in the direction of travel. In Figure 10.3 (f and h) the direction of movement is toward the top of the page. Sadler's study also showed that determination of true sense of movement requires more detailed and more quantitative analyses of the tracks because several trails exhibited a backward orientation to the bifurcation angle.

The consistency of the substrate as a requirement for preservation of tracks has long been a question. Early work by McKee (1947) indicated that preservation only was likely in dry cohesionless sand deposited on gentle slopes or horizontally. Sadler's work supported McKee's contention that preservation potential was greatest for tracks going up dune slip faces due to the problems of track obliteration by grain flows during down-slope movement of animals. Sadler (1993) demonstrated that track preservation was likely under these conditions but also showed that a damp sand surface also was conducive to preservation of tracks made by scorpions and spiders and that increased moisture content of dune sands can result in preservation of down-slope-directed tracks. Preservation of tracks made by arachnids in sands that are damp necessitates subsequent infilling by sand deposited by grainfall or wind-ripple migration.

Desert sands become damp and therefore somewhat cohesive for a variety of reasons. Moisture content of desert sands is of course low because most deserts typically receive less than 150 mm of mean annual rainfall. Modern settings such as the Namib Desert in western Africa have varying moisture contents as a function of their proximity to onshore coastal winds, to basin margin and basin-central fluvial systems, and also to the magnitude of water table rise during wetter seasons (Lancaster 1984, 1989, 1995). Studies by Hasiotis and Bown (1992) of coastal dune systems have shown that the abundance of insects and arachnids increases landward due to an increase in vegetative diversity and substrate stability.

As has been shown in several studies (Briggs et al. 1984; Sadler 1993), a single trackway can exhibit a myriad of track morphologies obviously generated by the same animal. More experimental studies utilizing insects and arachnids as well as more careful morphometric analyses of fossil trackways is clearly warranted and needed if definitive identification of the track maker is to be attained.

Vertebrate Trace Fossils

Lull (1918) produced the first description of tracks attributed to tetrapod vertebrates in the Coconino Sandstone. The tracks consist of forefoot prints with four toes and hindfoot impressions with five toes. Both of these tracks are relatively small, 0.5–1.5 inches (1.5–3.0 cm) in length, and show claw marks. Lull assigned the tracks to two ichnospecies, *Laoporus schucherti* and *L. noblei*. The rest of the vertebrate trace fossils found in the Coconino Sandstone were described by Gilmore in the late 1920s. On the basis of tracks similar to those described by Lull (though different in detail), Gilmore named the following ichnotaxa: *Agostopus matheri* and *A. medius*; *Allopus? arizonae*; *Amblyopus pachypodus*; *Baropezia eakini*; *Baropus coconinoensis* (Fig. 10.3d); *Barypodus palmatus*, *B. metszeri*, and *B. tridactylus* (Fig. 10.3b); *Dolichopodus tetradactylus*; *Nanopus merriami* and *N. maximus*; and *Palaeopus regularis*. Gilmore (1926) also redescribed *Linopus? coloradensis* and referred it to the ichnogenus *Laoporus*.

Baird (1952 and in Spamer 1984) and Haubold (1984) reevaluated the vertebrate tracks from the Coconino Sandstone. Baird (1952) decided that *Allopus? arizonae* and *Baropus coconinensis* are junior synonyms of *Baropezia eakini*. Later, he transferred *Nanopus maximus* to *Barypodus metszeri* and attributed the ichnospecies *N. merriami* and *Dolichopodus tetradactylus* to the ichnogenus *Laoporus*. However, Haubold (1984) maintained the generic distinction of *Dolichopodus*. He also had doubts about the taxonomic validity of *Amblyopus pachypodus* and considered *Agostopus, Barypodus, Nanopus*, and *Palaeopus* to be junior synonyms of *Laoporus* (Fig. 10.3a and 10.3c). This view may be a case of excessive taxonomic lumping, however, because Baird (1952 and in Spamer 1984) has not indicated such sweeping synonymies. To resolve this and related problems, a thorough taxonomic revision of the Coconino Sandstone's vertebrate tracks is needed. A list of all the original names given to vertebrate trace fossils from the Coconino Sandstone is given in Table 10.2.

TABLE 10.2. Original Names of
Vertebrate Ichnospecies from the
Coconino Sandstone

Agostopus:	*A. matheri* (Gilmore 1926)
	A. medius (Gilmore 1927)
Allopus?:	*A.? arizonae* (Gilmore 1926)
Amblyopus:	*A. pachypodus* (Gilmore 1926)
Baropezia	*B. eakini* (Gilmore 1926)
Baropus:	*B. coconinoensis* (Gilmore 1927)
Barypodus:	*B. palmatus* (Gilmore 1927)
	B. metszeri (Gilmore 1927)
	B. tridactylus (Gilmore 1927)
Dolichopodus:	*D. tetradactylus* (Gilmore 1926)
Laoporus:	*L. schucherti* (Lull 1918)
	L. coloradensis (Gilmore 1926)
	L. noblei (Lill 1918)
Nanopus:	*N. maximus* (Gilmore 1927)
	N. merriami (Gilmore 1929)
Palaeopus:	*P. regularis* (Gilmore 1926)

EOLIAN DEPOSITIONAL SYSTEMS

Although we have known about the eolian origin of the Coconino Sandstone for a long time, there are comparatively few data on the details of dune types and distribution. Nor is there documentation of small-scale stratification features within the larger co-sets of cross-stratification that characterize the formation. Within the last decade, a number of studies have examined ancient and modern eolian deposits. Geologists have developed criteria and facies models by which we can interpret these sediments. In the process, we have learned more about depositional processes and the geomorphic features of eolian sand seas, or ergs. When this information and these models are applied to the Coconino Sandstone, a better understanding of the genesis and evolution of the Coconino erg will emerge.

The Coconino Sandstone is composed of fine-grained, well-sorted, and rounded quartz grains and minor amounts of potassium feldspar. The cement is primarily silica in the form of quartz overgrowths. These textural and mineralogic characteristics are compatible with an eolian environment in which sediment transport involves numerous grain-to-grain collisions. These collisions result in the mechanical destruction of less stable grains and in winnowing by the wind. As McKee (1979) correctly pointed out, however, these characteristics do not substantiate conclusively a wind-blown origin. Although paleocurrent trends suggest a northern source for this sand, we cannot identify the source(s) of such a large quantity of quartz in the Coconino, as well as in correlative units to the north such as the Weber Sandstone in Utah and the Tensleep Sandstone in Wyoming and Montana.

Primary sedimentary structures in the Coconino Sandstone include small- to large-scale [up to 66 feet (20 m) thick] planar-tabular and planar-wedge cross-stratification (Fig. 10.4), compound cross-stratification, horizontal stratification, ripple marks (Fig. 10.5), and raindrop impressions (Fig. 10.6). Deformation features also occur and consist of small, pull-apart structures and a variety of slump-related features (Fig. 10.7). Small-scale stratification comprises wind-ripple lami-

FIGURE 10.4. Large-scale planar-tabular cross-stratification in the Coconino Sandstone in Hualapai Canyon. Contact with overlying Toroweap Formation is at slope above the large-scale cross-strata.

FIGURE 10.5. Ripple marks striking down large foreset of cross-strata. This orientation of ripple crests is common throughout the Coconino Sandstone and indicates wind flow and sediment transport across the surfaces of the larger bedforms. Lens cap for scale.

FIGURE 10.6. Well-preserved raindrop impressions on foreset on large set of cross-strata. The irregular margins of the circular pits are somewhat raised on the down dip side. Lens cap for scale.

FIGURE 10.7. Pull-apart (detachment structures on slipface of gently dipping foresets. These features probably represent small-scale avalanching of semicohesive sand. Lens cap for scale.

Figure 10.8. Thin, evenly continuous laminae that compose the bulk of the foreset deposits. These laminae represent climbing wind ripples.

nations and sandflow strata and minor grainfall laminae. Collectively, these features, together with facies geometries, support an eolian interpretation and can be used to characterize bedform morphologies as well as variations in depositional and substrate conditions.

The most obvious structures in the Coconino Sandstone are the thick sets of cross-stratified sandstone (Fig. 10.4). Internally, these strata exhibit thin (less than 1 cm thick) laminae that are continuous for considerable distances along the foreset of the cross-stratification (Fig. 10.8). These laminae conform to the shape of the foresets: planar where the foresets are straight and becoming concave upward on the tangential foresets. In most cases, these laminae contain no internal structures. Grain size, though quite uniform within the laminae, exhibits a slight inverse grading.

Recent studies, most notably that of Hunter (1977), demonstrate that these structures are the products of wind-ripple migration as they climb over the upwind and downwind sides of dunes and in interdune areas. For the most part, the wind ripples in the Coconino Sandstone moved transversely across the lee side of the dunes. This cross-dip path of migration is caused by secondary air currents that parallel the strike of the lee side of simple and complex dunes (or *draas*). The low height-to-wavelength ratio of the wind ripples as measured in plan view exposures of many foresets is consistent with those recorded from modern coastal and inland dunes (McKee 1979).

Thin, down-foreset, tapering laminae occur in many of the thicker cross-stratified sets. These deposits rarely exceed 3 inches (7.5 cm) in thickness and lack any noticeable grain-size grading and internal stratification. Wind-ripple laminae typically surround these wedge-shaped units. McKee et al. (1971) and Hunter (1977) have shown that these features form by the avalanching of loose sand on dune slipfaces. These "sandflow" strata form either by a loss of grain cohesion in a descending avalanche sheet or by scarp recession (Hunter 1977). In the latter case, sand moves down and away from the detachment area as the scarp itself migrates up and across the slipface. Whichever the mechanism, these strata represent the avalanching of loose, dry sand.

Other "detachment structures" in the Coconino Sandstone, however, suggest wetter substrate conditions. McKee et al. (1971) demonstrated that detachment and down-slope movement of moistened, somewhat cohesive sand produced steep-sided, pull-apart structures. Possible causes for the cohesiveness of the sand are difficult to verify, but wetting by rain and/or dew or early cementation are obvious possibilities. Our studies to date have demonstrated that cementation patterns generally are uniform throughout these strata, and it seems likely, therefore, that rainfall or dew caused the increase in substrate cohesiveness.

We can gain a great deal of information concerning the types of eolian bedforms in the Coconino erg by analyzing larger-scale features such as the thick sets and co-sets of cross-stratification and the truncation surfaces that bound or are contained within these sets. The Coconino Sandstone contains both simple sets of cross-strata and also complexly cross-stratified packages. Each is related to differences in dune types and/or flow processes associated with dune migration.

The most common types of cross-stratification in the Coconino Sandstone are planar-tabular and planar-wedge sets (Fig. 10.4). These consist of foresets with an average dip of 25 degrees and have a sharp or tangential basal contacts. McKee (1979) reported foreset lengths of up to 80 feet (24 m), but most are less than 40 feet (12 m) long. In general, the foresets do not contain any smaller cross-beds and are composed primarily of wind-ripple and subordinate sandflow laminae. Paleocurrent trends reported by Reiche (1938) and McKee (1979) are to the southeast and southwest. Our studies corroborate these earlier findings of a unimodal path of dune migration.

FIGURE 10.9. Stacked sets of planar tabular cross stratification separated by horizontal to gently dipping first order bounding surfaces. Concave-upward scouring of dune surface. Hualapai Trail, western Grand Canyon.

In most cases, these simple cross-bedded sets are underlain and overlain by horizontal to gently dipping planar surfaces (Figs. 10.4 and 10.9). These surfaces typically can be traced hundreds of meters along the outcrop. In all cases, these surfaces truncate the foresets of these thick, cross-bedded sets. Thin beds of horizontally laminated, fine-grained sandstone and coarse-grained siltstone overlie these surfaces in some localities. Typically, however, the surfaces separate large-scale cross-beds.

Although the origin of these features in ancient sandstone is controversial, the fact that planar surfaces truncate the cross-strata argues strongly that these are erosional features. Two major hypotheses put forth to explain the genesis of these laterally persistent surfaces involve the role of groundwater in desert basins and the phenomenon of migrating and climbing bedforms.

Stokes (1968) proposed that these extensive planes were the direct result of groundwater rise in dunes and subsequent wind erosion. According to this hypothesis, deflation of the dry, cohesionless sand above the saturation horizon resulted in the generation of a nearly planar surface. Subsequent migration of other dunes over these areas caused a vertical juxtaposition of cross-strata, such as that which occurs in the Coconino Sandstone and numerous other eolian sandstones on the Colorado Plateau. The net result of a repetition of this process during periods of basin subsidence were stacked sets of thick cross-strata separated by multiple truncation planes.

A major problem with the groundwater theory as applied to the Coconino Sandstone involves the nature of the water surface and the extensiveness of these surfaces. McKee and Moiola (1975) have shown that rather than being a horizontal surface, the upper level of the saturation horizon commonly is irregular and, in many instances, mimics the topography of the bedform. Deflation of the overlying dry sand in these cases would not result in a horizontal surface. It also seems rather improbable that this sand would cover a large enough area to create an extensive planation surface, notwithstanding the presence of interdune lows.

A more plausible hypothesis for the genesis of these first-order bounding surfaces in the Coconino Sandstone involves the downwind migration and climb of interdunal areas and dunes. Brookfield (1977) referred to these as first-order bounding surfaces and attributed their origin to the passage of large, complex eolian bedforms or draas. The planation surface most likely is caused by the truncation of downwind dunes or draas by migrating interdune areas. Interdune areas are zones of deflation that bevel the next downwind dune as the sand migrates downwind. Inherent in this model is the requirement that there is a net downwind climb to the bedforms. Rubin and Hunter (1982) have demonstrated the viability of bedform climb as a mechanism of generating these features. In a regional study of the Jurassic Entrada Sandstone, Kocurek (1981) proved that first-order surfaces can develop in this manner.

Another possibility that may explain the formation of these widespread surfaces involves climatic fluctuations. Talbot (1985) argues that basin-wide climatic changes can result in generation of these regional surfaces based on studies of geomorphic changes that have occurred along the southern margins of the Sahara Desert in northern Africa. In particular, this hypothesis suggests that periods of climatic shifts from arid to more humid periods during the Quaternary resulted in generation of broad areas where dunes were stabilized and eolian processes were relatively ineffective. In this model, stabilization of dunes would occur, erosion would ensue, and ultimately featureless areas would be common. Talbot documented this phenomenon in the southern Sahara where widespread areas are essentially devoid of large-scale dunes and planation surfaces are extensive. If preserved, these surfaces would represent regional bounding surfaces and are clearly indicative of widespread erosion.

There is scant evidence of fluvial and/or marine influences on development of the Coconino sand sea. These interactions would be expected around the margins of sand seas where, for example, rivers influence dune field buildup and migration (Middleton and Blakey, 1983; Langford, 1989; Langford and Chan, 1989; Herries, 1993) and strandline migrations modify coastal dune complexes (Chan and Kocurek, 1988). This is explicable if most of the Coconino Sandstone records deposition in interior ergs far removed from fluvial and coastal systems. There is no evidence of extradunal sand sheet deposition (Kocurek and Nielson, 1986).

The paucity of interdune deposits in the Coconino Sandstone is puzzling. Interdunes are common geomorphic elements of sand seas. A possible reason for their scarcity in the Coconino is that climatic fluctuations resulted in prolonged periods of stasis when bedform stabilization was protracted and interdune buildup was minimal. Also, it is possible that during bedform migration and climb, little deposition occurred in the interdunal corridors. Thus, the only record of interdunes would be the regional bounding surfaces.

Paleocurrent readings are strongly unimodal toward the south. This, coupled with the fact that the sets are internally simple, suggests that the bedforms that produced these cross-strata probably were relatively straight crested and that they were oriented transverse to the prevailing winds. McKee (1979) reported that in some localities the "curved shapes of small barchan types are recognizable." The majority of the dunes of the Coconino erg appear to have been of these two morphologies.

Although the aforementioned sets are characteristic of the Coconino, in many places the formation exhibits a complex system of bounding surfaces (Fig. 10.9) and extremely complicated sequences of internal erosional surfaces and stratification styles (Fig. 10.10). The concave-upward, bounding surfaces (Fig. 10.9) probably formed as the result of erosion—either by wind and the subsequent infilling of the eroded surface or by scouring associated with the migration of the next upwind dune–interdune couplet. Rubin and Hunter (1983) and Blakey et al. (1983) have documented the occurrence of these scoured surfaces in the late Paleozoic and Mesozoic sandstones on the Colorado Plateau.

Figure 10.10 illustrates some of the more complex geometries in the Coconino Sandstone. Low-angle surfaces truncate higher-angle sets of cross-stratification through this large set. These surfaces, which dip in the downwind direction, are over- and underlain by smaller sets. Brookfield (1977) has termed these low-angle, dipping surfaces second-order bounding surfaces "because they are truncated by the more extensive first-order surfaces."

These complex sets represent the deposits of large-scale, eolian draas (Wilson 1972a, b). Draas, which are common in most modern ergs, are characterized by smaller dunes migrating on both the stoss and lee sides of the larger bedform. Although it is possible that the second-order bounding surfaces represent periods of erosion of the lee face, it is more likely that these surfaces were formed by the migration of dunes down a draa that lacked a well-developed slipface.

Draas can be oriented either transverse or parallel to the prevailing wind direction. In the absence of plan view exposures that clearly show the geomorphic shape of the bedform, it is necessary to examine the orientation of the smaller-scale cross-stratification relative to the dip of the major bounding surface

FIGURE 10.10. Sets of complexly cross-stratified sandstone exhibiting both first- and second-order bounding surfaces and intraset cross-stratification. Hualapai Trail, western Grand Canyon.

(Rubin and Hunter 1985). In draas oriented parallel to the main wind direction (and, therefore, long-term sand-transport direction), the smaller-scale sets should exhibit current directions parallel to the draa set. In all sections of the Coconino Sandstone examined, the internal sets are oriented parallel to the dip of the master surfaces. Thus, it appears that flow-transverse bedforms also formed these sets.

Of minor importance in the Coconino Sandstone are sets of trough cross-stratification. These sets, typically less than a meter thick, are filled by wind-ripple laminae. Winds that scoured the surface probably formed the troughs. The pit itself then was filled by wind ripples migrating into the scour during periods of reduced wind strengths. Blakey et al. (1983) have proposed a similar origin for these units in the underlying Schnebly Hill Formation in Oak Creek Canyon.

SUMMARY

The sediments and trace fossils of the Coconino Sandstone record the advance and passage of a major eolian sand sea. Both invertebrate and vertebrate traces

indicate that the substrate was, for the most part, very dry, though it is clear that light rainfall or dew periodically moistened the dune surfaces. Although we know that the erg was characterized by both simple and complex bedforms that apparently were oriented approximately transverse to the main direction of sand transport, we have much to learn about the distribution of the bedforms and the dynamics of the Coconino erg.

• 11 •

TOROWEAP FORMATION

Christine E. Turner

INTRODUCTION

The Permian Toroweap Formation in the Grand Canyon region is one of the most intriguing, if not the most readily noticed, formation when viewed from the commonly visited overlooks in Grand Canyon National Park (Fig. 11.1). The Toroweap occupies the generally tree-covered interval between the overlying Kaibab Formation, which forms the rim of the Grand Canyon, and the underlying Coconino Sandstone, which forms a prominent light gray cliff that is visible from great distances. What the Toroweap Formation lacks in scenic splendor in the eastern Grand Canyon, it compensates for in its geologic diversity. The Toroweap exhibits some of the most striking lateral facies changes in the Grand Canyon sequence. McKee (1938) was the first to recognize the lateral facies changes in the Toroweap in his classic monograph on the Kaibab and Toroweap Formations. More recently, geologists have conducted stratigraphic and facies analyses of the Toroweap in the context of modern sedimentologic concepts. This work has permitted a reevaluation of the depositional environments and has resulted in a fairly complete paleogeographic reconstruction of the Toroweap Formation.

NOMENCLATURE AND DISTRIBUTION

The Toroweap Formation in Arizona, which covers an area of approximately 25,000 square miles (65,000 km^2), is best exposed in the Grand Canyon region and in outcrops along the Mogollon Rim. It pinches out in the subsurface to the east of the Grand Canyon, but extends northward in the subsurface into southern Utah, westward into eastern Nevada, and southward in the subsurface to the outcrop belt along the Mogollon Rim. Beyond this point, erosion has removed evidence of the formation.

In 1938, McKee described and named the Toroweap Formation, which he separated from the overlying Kaibab Formation (Fig. 11.2). The type locality for the Toroweap is in Brady Canyon, an eastern side canyon to Tuweep (Toroweap) Valley. McKee recognized three informal members: an upper evaporite and redbed interval—the alpha member; a middle limestone unit—the beta member; and a lower sandstone and evaporite interval, which he referred to as the gamma member. Sorauf (1962) applied geographic names to McKee's informal members (Fig. 11.3). He named the alpha member the Woods Ranch Member; the beta

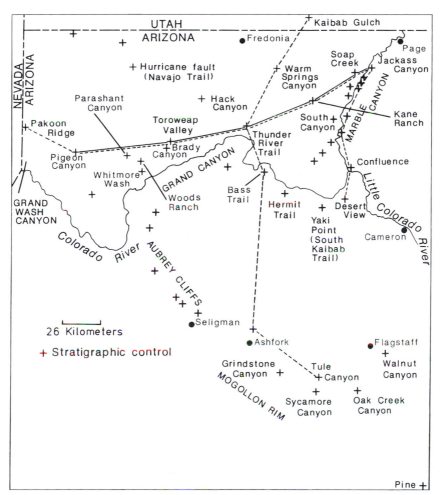

FIGURE 11.1. Index map showing location of measured sections used in this study. Dashed lines indicate lines of section used to construct panel diagram shown in Fig. 11.3. Solid lines indicate line of section used to construct the correlation diagram shown in Fig. 11.3. Solid lines indicate line of section used to construct the correlation diagram shown in Fig. 11.12.

the Brady Canyon Member; and the gamma the Seligman Member. Sorauf and Billingsley (1991) have proposed formally that these member names be accepted.

In addition to dividing the Toroweap Formation into members, McKee (1938) divided the formation into three lateral phases: the western phase, where the three members are recognizable; the transition phase, which consists of irregular-bedded sandstone; and the eastern phase, which consists of cross-bedded sandstone (Fig. 11.2). The limits of the western phase are determined by the extent of the Brady Canyon Member. To the east and southeast of the pinchout of this member, the two other members (the Seligman and Woods Ranch members) cannot be distinguished. The western phase is well-developed in the Grand

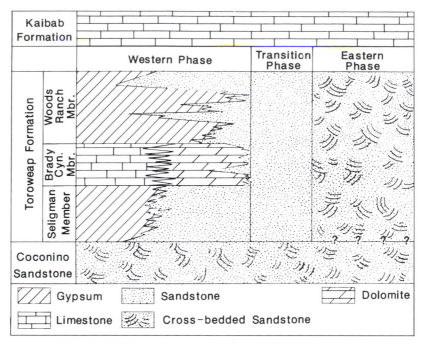

FIGURE 11.2. Schematic diagram showing members and lateral phases of the Toroweap Formation. (From Rawson and Turner-Peterson 1979.)

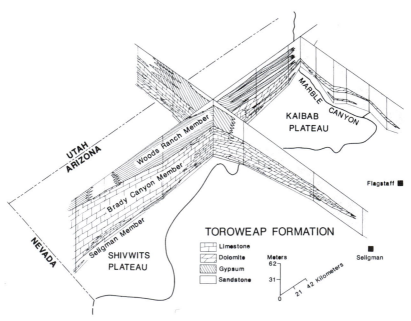

FIGURE 11.3. Panel diagram constructed chiefly in the western phase of the Toroweap Formation, along lines of sections shown in Figure 11.1. Note thinning of Brady Canyon Member to the east and southeast, as well as an increase in dolomite in that member. Evaporites in Woods Ranch and Seligman members are confined largely to the area north of the Grand Canyon. (From Rawson and Turner-Peterson 1974.)

FIGURE 11.4. Photograph of Toroweap Formation in South Canyon, near Marble Canyon (see Fig. 11.1 for location). Presence of all three members is characteristic of the western phase. Pk, Kaibab Formation. Members of the Toroweap are: Ptwr, Woods Ranch Member (slope-forming unit in this region because of gypsum beds); Ptbc, Brady Canyon Member (cliff-forming carbonate units); Pts, Seligman Member (thin sandstone interval). Pc, Coconino Sandstone.

Canyon region (Fig. 11.4). The transition phase of the Toroweap is particularly well-developed in the area of Sycamore Canyon. To the south and east of this area, the sandstones are more extensively cross-bedded. The Toroweap Formation in Oak Creek and Walnut canyons typifies McKee's eastern phase and is characterized by cross-bedded sandstone (Fig. 11.5).

LITHOLOGY AND STRATIGRAPHY

Stratigraphy

Geologists can identify all three members in the Grand Canyon region (Figs. 11.2 and 11.3). The lower boundary of the Toroweap Formation is at the base of the Seligman Member. It is the first non-cross-bedded or evaporite-bearing unit above the cross-bedded Coconino Sandstone. The boundary is conformable in most locations but is not distinct in the area of Marble Canyon, where cross-bedded units of the Coconino Sandstone intertongue with the lowermost beds of the Seligman Member. The first appearance of a thick carbonate unit above the non-

FIGURE 11.5. Photograph of Oak Creek Canyon showing cross-bedded sandstone typical of the eastern phase of the Toroweap Formation. Line of vegetation, indicated by arrow, marks the boundary between the Toroweap Formation and the underlying Coconino Sandstone, a unit that also contains cross-bedded sandstone. Pk, Kaibab Formation; Pt, Toroweap Formation; Pc, Coconino Sandstone.

cross-bedded sandstone defines the upper contact of the Seligman Member. It is less than 45 feet (15 m) thick in the Grand Canyon, but may be as much as 450 feet (152 m) thick in the North Muddy Mountains in Nevada (Bissell 1969).

The Brady Canyon Member forms the massive cliff of carbonate above the relatively thin Seligman Member in the Grand Canyon region (Fig. 11.4.) Limestone predominates to the west, whereas dolomite is abundant to the east (Figs. 11.2 and 11.3). The Brady Canyon Member has its greatest development in the western Grand Canyon, where it is up to 280 feet (93 m) thick. It thins uniformly in an easterly direction to a depositional edge just east of Marble Canyon. An aphanitic dolomite unit with prominent desiccation cracks marks the upper contact here.

The Woods Ranch Member typically extends from the top of the aphanitic dolomite that forms the uppermost unit of the Brady Canyon Member to the base of the cliff-forming limestone of the overlying Kaibab Formation. Repetitive intervals of evaporate, limestone, and sandstone form a distinctive slope in most of the Grand Canyon region. Where the Woods Ranch Member lacks evaporate, it usually forms cliffs. The Woods Ranch Member shows no consistent thickening or thinning trends throughout most of the study area. Geologists have measured a maximum thickness of about 180 feet (60 m).

Along the Mogollon Rim of central Arizona, the three members of the Toroweap disappear because of lateral facies changes. That fact, plus the presence of an additional member at the base of the Kaibab Formation, requires that different criteria be used to define the lower and upper contacts of the Toroweap Formation.

In the Sycamore Canyon area, the upper contact of the Toroweap is difficult to identify because the sandstone beds of the Fossil Mountain Member (the basal unit of the Kaibab Formation in this area) resemble the sandstone beds of the Toroweap Formation.

In the Oak Creek Canyon area, cross-bedded sandstone characterizes the entire Toroweap Formation, making it difficult to differentiate this formation from the underlying Coconino Sandstone. However, a truncation surface marked by vegetation in surface exposures separates beds contemporaneous with the Toroweap Formation from the underlying Coconino Sandstone (Fig. 11.5).

Lithofacies

Evaporite Lithofacies Interbeds of evaporate, thin-bedded carbonate, and fine-grained sandstone characterize a significant part of the Woods Ranch Member (and to a lesser degree, the Seligman Member). The lithofacies is restricted in both members to the area north of the Grand Canyon (Figs. 11.6 and 11.7). The evaporate beds, which most often are gypsum, frequently contain laminae of limestone and dolomite (Figs. 11.8a and b) several millimeters thick. The evaporate in these laminated intervals usually is about one centimeter thick. Evaporite beds occur in sequences characterized by a basal carbonate unit up to 1.5 feet (0.5 m) thick, gypsum beds up to 3 feet (1 m) thick, and sandstone beds that generally are 1.5 feet (0.5 m) thick. The carbonate beds, which typically are limestone rather than dolomite, are thinly laminated and vuggy (Fig. 11.8c). Gypsum or anhydrite nodules fill the vugs in places. This suggests that all of the vugs originally may have been filled with evaporate minerals. Frequently, the sandstone beds are poorly cemented, and it is difficult to discern sedimentary

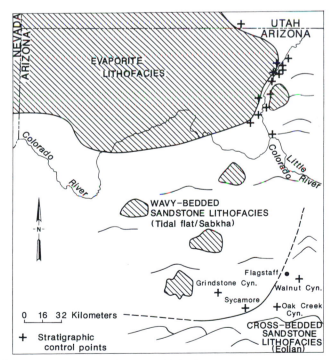

FIGURE 11.6. Map showing distribution of lithofacies in Seligman Member and equivalent strata in the Toroweap Formation. Map is based on the most abundant lithofacies present at each locality. (Modified from Rawson and Turner-Peterson 1974.)

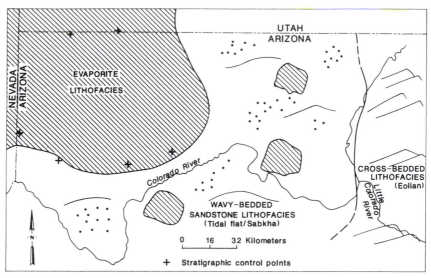

FIGURE 11.7. Map showing distribution of lithofacies in Seligman Member and equivalent strata in the Toroweap Formation. Map is based on the most abundant lithofacies present at each locality. (Modified from Rawson and Turner-Peterson 1974.)

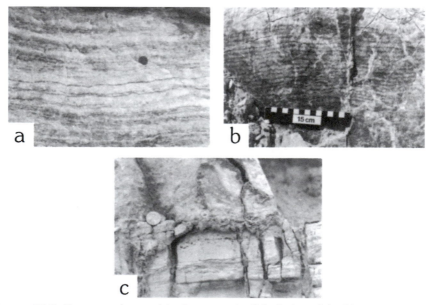

FIGURE 11.8. Features observed in the evaporite lithofacies of the Toroweap Formation. Photographs are from the Woods Ranch Member along the Thunder River Trail, Grand Canyon. (a) Gypsum bed (white bands) with thin laminae of limestone (gray), with penny for scale; (b) Laminated bed that contains alternations, on the scale of a few millimeters, of gypsum (white) and limestone (gray); (c) Thin-bedded limestone bed with vugs near the top (below and to the right of the quarter), overlain by bed of gypsum. Locally, similar vugs are filled with evaporite minerals, suggesting that these vugs originally contained evaporites.

structures. Locally, however, geologists can identify cross-bedding. With the exception of one thin limestone bed near the top of the Woods Ranch Member (a unit that contains abundant pelecypods of the genus *Schizodus*), this lithofacies is unfossiliferous. The *Schizodus*-bearing limestone bed, which contains ooids locally, has been named informally the "Hurricane Cliffs tongue" of the Woods Ranch Member (Altany 1979).

The evaporite lithofacies probably represents deposition in a shallow, subaqueous, shelf environment in which warm restricted-marine waters were evaporated, promoting the precipitation of evaporite minerals. Originally, geologists thought that this lithofacies represented sabkha deposition (Turner 1974; Rawson and Turner-Peterson 1974, 1979, 1980); we now know that laminated evaporites indicate precipitation in standing water rather than formation by displacive growth within sabkha sediments (Schreiber 1986). The thinly laminated evaporate and carbonate (Fig. 11.8b) is similar to the anhydrite–carbonate couplets in the laminated sulfate facies in subaqueous evaporates of the Castile Formation of the Delaware Basin of southeastern New Mexico and western Texas. Some carbonate units in the evaporate lithofacies of the Toroweap Formation contain desiccation cracks, which indicate local subaerial exposure and thus shallow, rather than deep, water deposition. Most likely, this lithofacies was part of a shallow shelf sequence. The carbonate units also lack fossils, which suggests that conditions in the restricted-marine waters were not conducive to marine life. Vertical repetition of certain lithologies—a basal carbonate bed, middle evaporate bed, and upper sandstone unit—suggests that cyclic sedimentation characterized deposition of this lithofacies. Thin carbonate laminae within the evaporate beds reflect periodic freshening of the waters that were depositing the evaporite. We do not know if this freshening occurred on a seasonal basis.

Siliciclastic Lithofacies

Irregular-Bedded Sandstone Fine- to medium-grained sandstone and siltstone characterize the intervals of the Woods Ranch and Seligman members that are laterally adjacent and equivalent to the evaporite lithofacies. This lithofacies dominates in areas just to the east and southeast of the evaporate lithofacies of the Woods Ranch and Seligman members (Fig. 11.6 and 11.7). In addition, this lithofacies predominates in the entire Toroweap Formation and beyond the depositional edge of the Brady Canyon Member (Fig. 11.2), typifying the transition facies of McKee (1938). Carbonate and evaporate beds are rare in this lithofacies, in contrast to their abundance in the evaporate lithofacies. Typical sedimentary structures include wavy lamination (Fig. 11.9a), flaser lamination (a discontinuous form of wavy lamination), lenticular lamination, and fluid-escape structures. Minor cross-bedded, fine-grained sandstone units with sets of inversely graded laminae occur in this lithofacies. Locally, these units exhibit high-index ripples whose axes trend directly down the cross-bedding laminae. Geologists also have noted wave and adhesion ripples (Fig. 11.9b) as well as brecciated (and rare) thin carbonate beds (Fig. 11.9c).

Intraformational brecciation is a common feature in this lithofacies, particularly in Marble and Sycamore canyons, within the transition facies defined by McKee (1938). The breccia most often consists of blocks of cross-bedded eolian sandstone within a matrix of flaser (or lenticular) laminated sandstone. The cross-bedded blocks typically show evidence of having collapsed downward (Fig. 11.9d). In the Sycamore Canyon area, laterally extensive beds of brecciated sandstone are interbedded with cross-bedded sandstone.

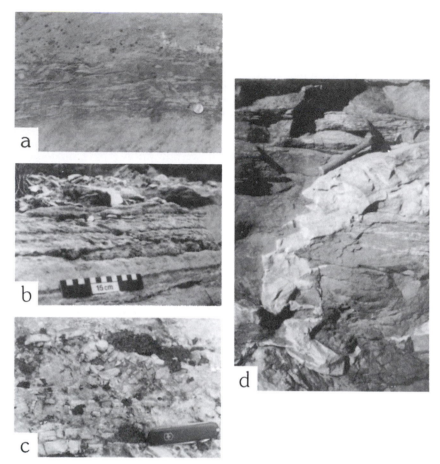

Figure 11.9. Features observed in the wavy-bedded sandstone lithofacies of the Toroweap Formation. Photographs a–c are from the Woods Ranch Member along the South Kaibab Trail, near Yaki Point in the Grand Canyon. (a) Wavy and lenticular bedding, with penny for scale. (b) Adhesion ripples at the top of a cross-bedded sandstone. (c) Brecciated carbonate beds, with knife for scale. (d) Cross-bedded, light-colored sandstone bed (immediately below hammer) collapsed into underlying units. This may have been caused by removal of evaporites by dissolution. From the Woods Ranch Member in Jackass Canyon, near Marble Canyon, Arizona.

Most geologists interpret the irregular-bedded sandstone lithofacies as tidal-flat deposits. Flaser bedding, wavy bedding, and lenticular bedding are characteristic of intertidal environments (Reineck and Wunderlich 1968), whereas brecciated carbonate units similar to those in the Toroweap Formation generally occur in supratidal zones of tidal-flat complexes (Shinn 1983). The brecciation found in carbonate beds probably is related to the dolomitization of limestone beds in a supratidal setting. This dolomitization results in shrinkage that is caused by a loss of volume. The abundance of fluid-escape structures is consistent with a tidal flat environment. Cross-bedded units that exhibit inverse, graded bedding and high-index ripples on the slipface almost certainly are eolian dune deposits.

Although dune deposits are not abundant in this lithofacies, they do indicate the occasional migration of eolian sediments across the tidal-flat surface. Limited paleocurrent data from the eolian sandstones in this lithofacies show a southerly transport direction, which is consistent with the paleowind directions for the Permian in the Colorado Plateau region (Peterson 1988).

One possible interpretation for areas in the transition phase that contain abundant intraformational sandstone breccias is a supratidal or sabkha environment. This interpretation relies in part on the regional facies interpretation that Rawson and Turner-Peterson (1979) have proposed for the Toroweap Formation. Most of the brecciation occurs in the southeastern part of the study area, in the part of the transition facies that contains the largest number of eolian sandstone units. This is particularly true in the Sycamore Canyon area, where laterally extensive, brecciated sandstone units interbed with cross-bedded eolian sandstone. In the overall facies distribution, brecciated units occur landward of the intertidal deposits and are intercalated with eolian sandstones. Fryberger et al. (1983) have suggested a supratidal or siliciclastic-sabkha environment to explain similar laterally extensive, brecciated units. It is possible that the evaporite formed by displacive growth within a siliciclastic sabkha. Subsequent removal of this evaporate by the movement of relatively fresh groundwater through the sandstones would have caused collapse and brecciation within the sabkha units.

Cross-Bedded Sandstone The irregular-bedded and brecciated sandstone units of the transition phase of the Toroweap Formation grade laterally, in a southeasterly direction, into cross-bedded sandstone of the eastern phase of the Toroweap (Fig. 11.2). The eastern phase is characterized by fine- to medium-grained, cross-bedded sandstone that is indistinguishable from sandstone in the underlying Coconino Sandstone (Fig. 11.5). Most frequently, the cross-beds are wedge and tabular planar (Fig. 11.10a), with some trough-shaped sets. Average set thickness is about 6 feet (2 m), and the average dip direction is S11°W, with a consistency ratio of 0.90 (Rawson and Turner-Peterson 1979). The sets of cross-bedded sandstone contain thick avalanche or sand flow toes as much as 0.5 inches (5 cm) thick (Fig. 11.10b) and inversely graded laminae (Fig. 11.10c). Slumping and deformational features are not uncommon in this lithofacies (Fig. 11.10d).

We base an eolian interpretation for this lithofacies on the presence of diagnostic eolian characteristics, such as large-scale cross-bedding, avalanche (sand flow toe) deposits, and inversely graded ripple laminae. The deformational structures shown in Fig. 11.10d are similar to those studied in modern environments and seem to develop exclusively in dunes that are wet or that have been wetted (McKee and Bigarella 1972). The eastern-phase Toroweap Formation may have been deposited in a coastal dune environment. An eolian interpretation also is consistent with the southeasterly transport direction for the cross-bedded sandstone lithofacies; this direction is the same as that determined for the eolian rocks of the underlying Coconino Sandstone. It also is consistent with paleowind directions during the Permian in the Colorado Plateau area (Peterson 1988).

Carbonate Lithofacies Geologists have identified several carbonate lithofacies within the Brady Canyon Member of the Toroweap Formation. Rawson and Turner-Peterson first described these lithofacies in 1974.

Skeletal Packstone Skeletal material in the grain-supported carbonate in this lithofacies consists of disarticulated crinoid columnals; bryozoans; brachiopods; ostracods; gastropods; endothyrids; echinoid and brachiopod spines; and trilo-

FIGURE 11.10. Features observed in the cross-bedded sandstone (eastern phase) of the Toroweap Formation. (a) Section composed almost entirely of a tabular- and wedge-shaped, cross-bedded sandstone, Oak Creek Canyon, Arizona. (b) Avalanche (sand flow) toes observed in cross-bedded sandstone, Oak Creek Canyon, Arizona. (c) Inversely graded laminae in cross-bedded sandstone, Sycamore Canyon, Arizona. Note preservation of foreset laminations in middle part of photograph. 10X hand lens for scale. (d) Deformational structures in cross-bedded sandstone, Oak Creek Canyon, Arizona. These structures seem to develop exclusively in eolian dunes that are wet.

bite fragments. The matrix is micrite, which is common to all of the carbonate lithofacies. Although only a few grainstones are present, we can see rare quartz grains in thin section. The quartz, which is fine-silt-size, increases in abundance from west to east. This lithofacies is common in the western part of the Grand Canyon region.

Pelletal Wackestone This lithofacies consists of carbonate units that contain elliptical-to-spherical pellets ranging from 0.004 to 0.008 inches (0.1 to 0.2 mm) in diameter. No true ooids occur, except in the "Hurricane Cliffs tongue" of the Woods Ranch Member. The pelletal wackestone facies commonly is altered to a dolomicrite, which geologists include in this facies when they can recognize the original texture. Dolomitized pelletal wackestone is prevalent in the eastern part of the study area and also occurs in some western sections, particularly near the top of the Brady Canyon Member.

Sandy Dolomite A sandy dolomitic lithofacies is abundant in the Marble Canyon area. It also occurs near both the base and the top of the Brady Canyon Member in the western part of the study area. The dolomite is silt-size, with abundant fine-to-coarse, sand-size quartz grains.

Aphanitic Lime Mudstone and Dolomite Aphanitic lime mudstone and dolomite commonly occur at the base and top of the westernmost exposures of

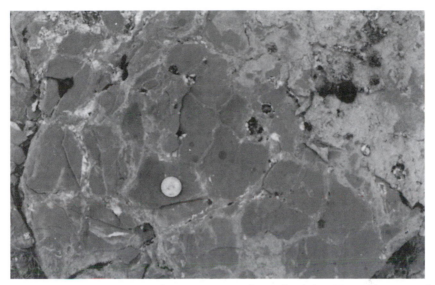

FIGURE 11.11. Desiccation cracks on upper surface of a dolomitic mudstone at the top of the Brady Canyon Member. Desiccation cracks such as these are abundant at this same interval across northern Arizona and probably indicate widespread withdrawal of the sea at the end of deposition of the Brady Canyon member. Quarter for scale.

the Brady Canyon Member in the study area. This lithofacies probably is present to the east as well, but quartz grains mask the texture. The carbonate is aphanitic (0.004 mm in diameter), and the dense rock that it forms breaks with a conchoidal fracture. To date, geologists have not found fossils or fossil fragments in this unit. Desiccation cracks, common in this lithofacies, are readily observed at the top of the Brady Canyon Member in most localities (Fig. 11.11).

Distribution and Interpretation of Carbonate Lithofacies

During the formation of the Brady Canyon Member, carbonates with an open-marine fauna were deposited to the west, whereas carbonates with a restricted-marine fauna were deposited to the east (Fig. 11.12). This reflects an incursion of the sea from the west. The mud-supported texture that characterizes all lithofacies of the Brady Canyon Member indicates that these sediments were laid down in quiet-water conditions and that there was no significant reworking. The fauna and textural types are consistent with shallow, quiet-water conditions on a carbonate shelf. Distribution of lithofacies in the Brady Canyon Member changed through time. Figure 11.13 shows a representative distribution of lithofacies at one particular time (T-2 in Figure 11.12). Gradation from open-marine fauna in limestone beds in the west to restricted-marine fauna in dolomitic mudstone in the east, as shown in Figure 11.13, is characteristic of the Brady Canyon Member. Of particular interest is the persistence of a dolomitic mudstone at the top of the member (R-3 in Fig. 11.12). This represents a westward progradation of a lithofacies deposited in the most restricted of marine conditions. Also noteworthy is the frequent occurrence of desiccation cracks on the upper surface of this

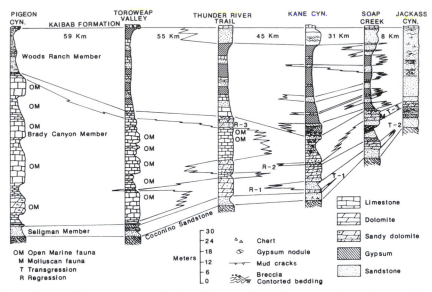

FIGURE 11.12. West-to-east correlation diagram of the Toroweap Formation, northern Arizona (see Fig. 11.1 for locations). Evaporites in the Seligman Member represent the first incursion of the sea following deposition of the underlying eolian Coconino Sandstone. In the Brady Canyon Member, open-marine fauna predominate to the west, and restricted-marine fauna predominate to the east. Several transgressions and regressions are apparent within the Brady Canyon Member. Regression R-3 may represent progradation westward in response to a slowing down of a relative sea level rise, and the top of the Brady Canyon Member (marked with desiccation cracks) may represent a relative drop in sea level. The evaporite lithofacies of the Woods Ranch Member is thought to reflect another relative sea level rise. (Modified slightly from Rawson and Turner-Peterson 1974.)

dolomite unit, indicating widespread subaerial exposure at the end of deposition of the Brady Canyon Member.

Current models of dolomite formation do not explain the greater abundance of dolomite in the eastern part of the Brady Canyon Member. The idea of reflux dolomitization implies a syndepositional process, where dolomitization occurs in response to the movement of hypersaline brines from a lagoonal environment into adjacent or underlying carbonate sediments (Adams and Rhodes 1960). Although evaporites are present in the Woods Ranch Member of the Toroweap Formation, the distribution of these evaporites does not coincide entirely with the distribution of dolomite in the underlying Brady Canyon Member (Fig. 11.3). An alternative model, the "Dorag" dolomitization model (Badiozamani 1973), implies the replacement of earlier formed limestone intervals in a "mixing zone" of seawater and fresh water.

Although some of the carbonates in the Brady Canyon may have been dolomitized in this way (e.g., some pelletal wackestone beds), replacement textures are not abundant in the carbonate beds. As Hardie (1987) points out, we do not understand the process of dolomitization very well. Distribution of dolomite in the Toroweap Formation, as in many ancient examples, clearly is

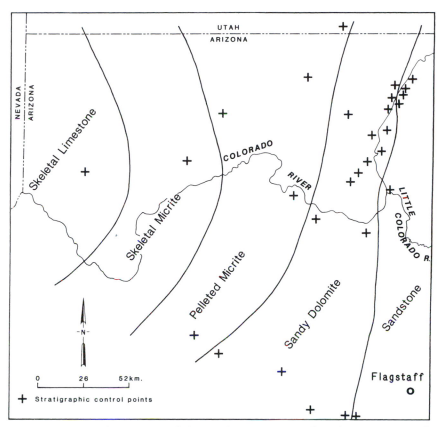

FIGURE 11.13. Lithofacies map of the Brady Canyon Member during time T-2 (see Fig. 11.12). (From Rawson and Turner-Peterson, 1974.)

greater in the more restricted-marine lithofacies. This association suggests that restricted circulation and great evaporation promote dolomite precipitation, but we do not know the mechanism by which it occurs.

PALEONTOLOGY

McKee (1938) summarized the paleontology of the Toroweap Formation in his original work on the Kaibab and Toroweap Formations. Except for the "Hurricane Cliffs tongue" of the Woods Ranch Member (which contains the marine bivalve *Schizodus*), the Brady Canyon Member is the only fossiliferous member of the Toroweap Formation. McKee recognized two major faunal facies in the Brady Canyon Member: an open-marine fauna to the west and a molluscan fauna to the east. The open-marine fauna includes brachiopods, bryozoans, crinoids, and horn corals. The molluscan fauna includes bivalves and gastropods, with a few scattered scaphopods and cephalopods. Kirkland (1962) compiled a faunal listing of subsequent finds in the Toroweap Formation. Belden (1954), Mullens (1967), Miller and Breed (1964), Beus and Breed (1968), Turner-Peterson (1974), and Rawson and Turner (1974) reported additional species in the Toroweap Formation.

AGE AND CORRELATION

Fossils contained in the Brady Canyon Member in northern Arizona suggest a late Leonardian age for the Toroweap Formation. Recent studies of bryozoans in Nevada (Gilmour and Vogel 1978) verify this timeframe. The late Leonardian is equivalent to the Kungurian (263 to 258 million years ago) and, possibly, late Artinskian (268 to 263 million years ago) ages on the radiornetric timescale (Harland et al. 1982).

As the Toroweap Formation grades southeastward into the cross-bedded sandstone lithofacies that characterizes the eastern phase, it becomes indistinguishable from eolian rocks of the underlying Coconino Sandstone. Similarly, geologists believe that the Toroweap Formation is equivalent to the White Rim Sandstone Member of the Cutler Formation in southeastern Utah. The Toroweap Formation also is equivalent to the San Andres Limestone in northwestern New Mexico.

DEPOSITIONAL HISTORY

We usually interpret the overall depositional history of the Toroweap Formation in terms of relative sea-level changes with time. Subaqueous evaporate and tidal-flat sediments in the Seligman Member, associated with a relative rise in sea level, represented the first incursion of the sea from the west. Eolian sandstone within the member indicates times when the sea withdrew from the region. The development of a thick carbonate sequence in the Brady Canyon Member indicates an incursion of the sea as far east as the Marble Canyon area.

Near the end of the Brady Canyon deposition, a progradation of carbonate lithofacies occurred. This slowing down of the relative sea-level rise is reflected in the progradation of restricted-marine lithofacies at the top of the Brady Canyon Member to the west (R-3 in Fig. 11.12). The period of subaerial exposure indicated by abundant desiccation cracks at the top of the dolomitic mudstone marks a relative lowering of sea-level and subaerial exposure of the entire shelf.

With another incursion of the sea, probably in response to a relative rise in sea level, the shelf flooded again. Cyclic sedimentation of carbonate, evaporate, and sandstone in the Woods Ranch Member reflects conditions similar to those that persisted during the formation of the Seligman Member. Here, the evaporates and carbonates indicate periods of subaqueous deposition, whereas the eolian sandstones suggest times of subaerial exposure. Because no evidence for a barrier exists in the Woods Ranch Member of the Toroweap Formation, it is difficult to interpret the evaporate in terms of the traditional barred-basin model (Schreiber 1986). The evaporate lithofacies of the Woods Ranch Member extends as far as the Permian outcrops in Nevada—and without significant change. It thus appears that a shallowing of seawater across a broad shelf caused the restricted circulation that is required to generate a hypersaline brine for evaporate precipitation.

Away from the area of dominantly marine carbonate–evaporite sedimentation, tidal-flat, sabkha, and eolian depositional environments persisted throughout deposition of the Toroweap Formation. In the transition phase of the Toroweap, sediments were deposited chiefly in a tidal-flat environment. Siliciclastic sabkhas developed along the eastern and southeastern margins of the tidal flats. Farther to the east and southeast, eolian deposition that had begun during the formation of the underlying Coconino Sandstone persisted (Figs. 11.2 and 11.5). Evidence of moisture during eolian deposition in the eastern phase of the Toroweap, as suggested by the deformational features, contrasts with the scarcity of such features in the underlying Coconino Sandstone. We attribute this difference to the sea's proximity to the dune fields during the formation of the eastern phase of the Toroweap. This is in contrast to the drier, inland conditions that persisted during deposition of the Coconino Sandstone. Southwesterly transport directions in both the Coconino Sandstone and the Toroweap Formation are consistent with paleowind directions for the Permian.

SUMMARY

The Toroweap Formation, in its variety of lithofacies, reflects transgressions and regressions of an eastward-advancing and westward-retreating sea. The formation was deposited in open and restricted-marine environments, tidal flats, sabkhas, and eolian dune fields. During times of transgression, the stable shelf was flooded by shallow-marine waters that were conducive to marine life. The climate probably was semiarid to arid, and large dune fields developed in the inland regions. The shoreline commonly was in the vicinity of the Grand Canyon, which explains the striking lateral and vertical changes in lithofacies. It is this very variety of lithofacies in a relatively small area that makes the Toroweap Formation such a fascinating unit for geologists.

• 12 •

KAIBAB FORMATION

Ralph Lee Hopkins and Kelcy L. Thompson

INTRODUCTION

The Kaibab Formation comprises the caprock of the Grand Canyon and forms the surface of the Kaibab and Coconino plateaus, which is the area through which the deepest part of the canyon has been carved. When visitors view the Kaibab Formation from scenic points and trailheads within the national park, they can recognize it easily as the gray, stepped cliff directly above the vegetated slope of the Toroweap Formation (Fig. 12.1). The Kaibab, the youngest Paleozoic rock unit on the southern Colorado Plateau, is composed of a variety of lithologic types that were deposited within a complex shallow-marine setting during the Permian. It represents the final chapter in the geologic story recorded by the sedimentary layers of the Grand Canyon.

Because it is at the top of the Grand Canyon stratigraphic section (and, therefore, is easy to see), geologists have studied the Kaibab Formation extensively. This chapter summarizes the cumulative knowledge gained through decades of observations by numerous geologists and is dedicated to Edwin D. McKee, in whose footsteps we all have followed.

NOMENCLATURE

The detailed description and naming of strata included within the Kaibab Formation has a long and interesting history. The earliest recorded observations of Permian rocks in northern Arizona were made by Jules Marcou (1856). Originally, geologists considered the units represented by the present-day Kaibab and Toroweap Formations a single formation, the Aubrey limestone (G.K. Gilbert 1875). Walcott (1880) placed these strata within the Upper Aubrey Group. The name "Kaibab" first was applied by Darton (1910) for exposures on the Kaibab Plateau north of the Grand Canyon. Noble (1914, 1922), along with Longwell (1921), provided the initial description and correlation of Kaibab strata across the Grand Canyon region. Geologists then subdivided these strata into five basic topographic and lithologic units. Reeside and Bassler (1922) named the uppermost beds the Harrisburg Gypsiferous Member for exposures at Harrisburg Dome in Utah. Noble (1928) proposed a type section for the Kaibab Limestone in Kaibab Gulch (in Utah, along the East Kaibab Monocline).

It was not until the classic work by McKee (1938) that the Kaibab Limestone was split into two formations. McKee proposed that the term Kaibab be applied only to the massive upper limestone and the unit directly above it, suggesting

Figure 12.1. View of precipitous Kaibab cliff along the south rim near the South Kaibab Trail (KT) showing Fossil Mountain (K_f) and Harrisburg (K_h) members. Note pinch out of sandstone until laterally within the Fossil Mountain member (*arrow*). Slope-forming Woods Ranch Member of the Toroweap Formation (P_t) everywhere underlies the Kaibab Formation in the walls of the Grand Canyon. Photograph looking east from Mather Point.

that these rocks be called the "Kaibab Formation" because they are composed of a variety of lithologic types. The lower limestone unit and adjacent slope-forming units became the Toroweap Formation (Turner, Chapter 11, this volume). McKee's scheme recognized three members within each formation and named them "alpha," "beta," and "gamma" in descending order.

In an attempt to establish more formal rock units, Sorauf (1962) proposed a change in terminology for both the Kaibab and the Toroweap Formations. He suggested that the rocks included with the alpha (or upper) member be called the Harrisburg Member. The beta (or middle) member of the Kaibab became the Fossil Mountain Member, named for Fossil Mountain along the south rim near the Bass Trail. Rocks included within the gamma (or lower) member, confined by McKee to the Mogollon Plateau south of the Grand Canyon, were not present in Sorauf's field area. Geologists interpret the gamma member as a facies within the Fossil Mountain Member (Lapinski 1976; Cheevers and Rawson 1979). Most subsequent workers have utilized the revisions in nomenclature suggested by Sorauf (1962), and the designation was formalized by Sorauf and Billingsley (1991).

DISTRIBUTION

Rocks of the Kaibab Formation form a continuous layer across the Grand Canyon and the surrounding region. The best exposures in cross section occur along the cliffs of the canyon and its tributaries. Although recent exposure and erosion of Kaibab rocks beneath the Permo-Triassic unconformity obscure thickness trends,

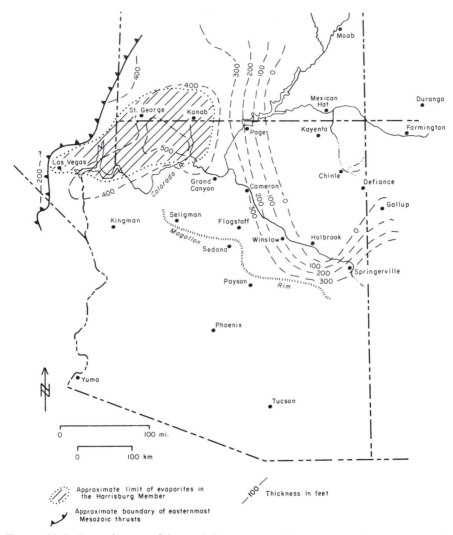

Figure 12.2. Isopach map of the Kaibab Formation illustrating total thickness trends across the Grand Canyon and surrounding regions. Maximum thickness of the Kaibab occurs in northwestern Arizona, where the Harrisburg Member contains significant evaporites. Data are compiled from many sources.

we know that the formation gradually thickens to the west (Fig. 12.2). Along the rim of the canyon, it generally ranges between 300 and 400 feet (90–120 m) in total thickness. Geologists have measured the greatest thickness in northwestern Arizona, in an area west of Kanab Creek, and northwest of the Colorado River. Here it exceeds 500 feet (150 m). East of the Grand Canyon, the Kaibab thins dramatically and is absent along the Defiance and Monument upwarps. Outcrops southeast of Winslow and south of Holbrook clearly show stratigraphic thinning (McKee 1938; Mather 1970; Cheevers 1980).

To the south, the Mogollon Rim, or escarpment, defines the limit of Kaibab exposure. This area represents the southern edge of the Colorado Plateau. Northward, the Kaibab Formation extends into southern and central Utah. It is well-exposed along the Hurricane Cliffs, Virgin River Gorge, and Beaver Dam Mountains in the southwestern part of the state (Nielson 1981). The formation continues on into the Circle Cliffs–Waterpocket Fold and the San Raphael Swell regions in central Utah (Davidson 1967) and into the subsurface across southern Utah (Irwin 1971). Some of this formation contains oil (e.g., the Upper Valley Field near Escalante, Utah). Geologists have found Kaibab outcrops as far north as the Deep Creek Mountains and Confusion Range in Utah, as well as in the isolated mountains of northeastern Nevada.

In southern Nevada, westernmost outcrops of the Kaibab Formation occur in a number of scattered mountain ranges in the Las Vegas area (Longwell 1921; Bissell 1969). The formation appears to thin westward (Fig. 12.2).

STRATIGRAPHY

Facies changes within the Kaibab Formation and in adjacent units complicate internal and regional stratigraphic relations.

Lower Contact

At the Grand Canyon, the Kaibab Formation is underlain everywhere by gypsum and/or contorted sandstones of the Woods Ranch Member of the Toroweap Formation (Fig. 12.1). Originally, geologists believed that the Kaibab–Toroweap contact was unconformable. They based this impression primarily on the presence of local intraformational breccias and erosional surfaces (McKee 1938). Further study has shown that these features are related to collapse following the dissolution of evaporitic facies within the underlying Woods Ranch Member, indicating that the contact is conformable or only locally disconformable. The basal Kaibab is the first cherty carbonate or sandstone unit located above the *Schizodus* bed (Hurricane Cliffs tongue) of the Woods Ranch Member.

To the south and east of the Grand Canyon, the evaporites and contorted sandstones (sabkha complex) of the underlying Woods Ranch Member are transitional and interstratified with cross-bedded sandstone facies (eolian dune complex). These rocks ultimately become indistinguishable from the Coconino Sandstone (Turner, Chapter 11 this volume). As a result, the Kaibab Formation in the Mogollon Rim region directly overlies the Coconino Sandstone. In northeastern Arizona and southeastern Utah, the White Rim Sandstone underlies the Kaibab.

Fossil Mountain–Harrisburg Contact

The contact between the Fossil Mountain and Harrisburg members of the Kaibab Formation is conformable, though different workers have placed it at slightly different stratigraphic levels. The similarity in the facies of both members at eastern localities has made it difficult to establish the contact. At the Grand Canyon, however, there are distinct textural, mineralogical, and faunal changes within the upper portion of the Kaibab sequence. Typically, a distinctive white, butterscotch, or red nodular-to-bedded chert horizon marks the base of the Harrisburg Member. This "marker-chert" most often is coincident with the disappearance of the normal-marine fauna and limestone mineralogy that is most characteristic of the cliff-forming Fossil Mountain Member (Sorauf 1962; Clark 1980; Hopkins 1986). In areas of southwestern Utah where the marker-chert is absent, geologists place the contact at the obvious change to slope-forming gypsiferous beds in the basal part of the Harrisburg Member.

East and south of the Grand Canyon, the contact between members is less obvious. It is easy to distinguish the Fossil Mountain Member in this region because it contains fauna of normal-marine affinity—particularly the productid brachiopod *Peniculauris bassi* (McKee 1938). At its eastern and southeasternmost extent, the lithology and fauna within the Fossil Mountain and Harrisburg members are similar. In this area, geologists place the contact at a distinct change from the thick-bedded sandstones and sandy carbonates of the Fossil Mountain Member to thinner-bedded units of the Harrisburg. The latter contain a more abundant molluscan fauna. Along its depositional edge, for example, southeast of Winslow and south of Holbrook, the Kaibab sequence is difficult to subdivide into members (Cheevers and Rawson 1979). Because the Harrisburg Member has disappeared from this area, only the shoreward facies of the Fossil Mountain Member have been preserved.

Upper Contact

In northern Arizona and southern Utah, the Triassic Moenkopi Formation occurs above the Kaibab Formation. Because of their less resistant nature, however, Moenkopi redbeds at the Grand Canyon have been removed almost entirely by erosion. One consequence of this erosion is that the Kaibab Formation caps many of the vast plateaus that border the Grand Canyon. Only rarely are the uppermost beds of the Kaibab preserved.

In northwestern Arizona, southeastern Nevada, and southwestern Utah, discontinuous conglomerate-filled channels and breccia deposits occur between the Kaibab Formation and Timpoweap Member of the Moenkopi Formation. Reeside and Bassler (1922) termed these deposits the Rock Canyon conglomerate for a channel 250 feet (75 m) deep and 700 feet (210 m) wide in Rock Canyon, which is north of Antelope Spring, Arizona. At several localities (e.g., in the Beaver Dam Mountains), channels of the Rock Canyon conglomerate have scoured completely through the Harrisburg Member and into the underlying Fossil Mountain Member (Nielson 1981). The tectonic significance of the Rock Canyon conglomerate is unresolved, though associated features may represent paleokarst depressions.

In areas of southwestern Utah and southern Nevada where carbonates of the Timpoweap Member of the Moenkopi Formation overlie uppermost carbonates of the Harrisburg Member, the formational contact can be difficult to determine (Bissell 1969; Nielson 1981). Geologists also have trouble distinguishing the contact when gypsiferous beds of the Lower Red Member of the Moenkopi Formation occur directly above gypsiferous beds of the Harrisburg Member (Bissell 1969; Cheevers 1980). To the north, in the mountains of western Utah and eastern Nevada, the Kaibab Formation is overlain conformably by the Permian Plympton Formation and Gerster Limestone of the Park City Group.

LITHOLOGY AND COMPOSITION

The Kaibab Formation is a complex sedimentary package composed of a variety of rock types. Due in part to mixing between carbonate siliciclastic sediment and intense post-depositional (diagenetic) changes in composition [in particular, silicification (chert formation) and dolomitization], rock ledges of the Kaibab appear similar at first glance. A number of detailed studies have delineated internal facies characteristics and distribution, providing resolution with respect to major changes in lithology, mineral composition, and faunal constituents.

Fossil Mountain Member

The Fossil Mountain Member of the Kaibab Formation is a prominent cliff that weathers to form distinctive pinnacles, or "hoodoos," below the rim of the canyon. The member thickens gradually westward and typically ranges between 250 and 300 feet (75–105 m). At Fossil Mountain along the south rim, where it is over 200 feet (60 m) thick, cherty limestones that contain abundant whole fossils characterize the member (McKee 1938; Cheevers 1980; Hopkins 1986).

As observed in outcrop along both rims of the canyon, the Fossil Mountain Member exhibits a pronounced change in lithology, mineralogy, and faunal constituents from west to east (Fig. 12.3). In western Grand Canyon, for example (in an area west of Fossil Mountain on the south rim and North Bass Trail on the north rim), the member contains a characteristic cherty, fossiliferous limestone with an abundant and diverse normal-marine fauna. This fauna includes brachiopods, bryozoans, crinoids, sponges, and solitary corals. Carbonate textures are dominated by skeletal wackestone (matrix-supported texture), with only minor amounts of admixed detrital quartz. Packstone intervals (grain-supported texture) are common locally, but typically form an insignificant percentage of the sequence. Sandstone comprises less than 10 percent of total lithofacies and commonly occurs near the base. Dolomite occurs only as scattered rhombs replacing a micrite matrix (carbonate mud).

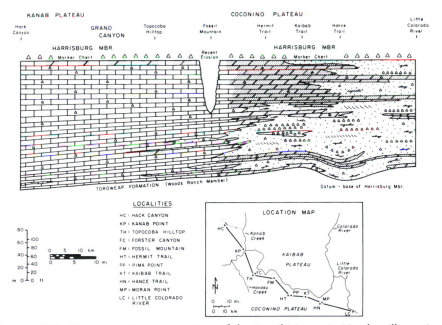

FIGURE 12.3. Diagrammatic cross section of the Fossil Mountain Member illustrating the west-to-east change in lithology and mineral composition at the Grand Canyon. See Fig. 12.4 for list of symbols. (Modified from Hopkins 1986.)

In contrast, the Fossil Mountain Member becomes increasingly siliciclastic to the east. Here, dolomite is the predominant mineralogy, and a restricted-marine fauna is most characteristic (Fig. 12.3). Within the zone of most pronounced lithologic transition, at Hermit Trail, for example, sandstone and sandy carbonate comprise nearly 50 percent of the member.

Carbonate lithofacies consist of skeletal wackestone and mudstone textures that have been altered to dolostone. We can see a similar change on the north rim near Point Sublime. In the basal portion of the sequence, sandstone beds occur in close association with relatively siliciclastic-free, skeletal carbonate units, though mixing between lithologies near unit contacts is common. Scouring along basal contacts occurs locally, and in some cases, siliciclastic units pinch out laterally (Fig. 12.1). Preserved sedimentary structures are scarce because units typically are bioturbated intensely. We can recognize horizontal, ripple, and low-angle laminations within certain sandstone intervals. To the east along both rims, the percentage of sandstone increases significantly. Ultimately, the Fossil Mountain Member consists of approximately 75 percent sandstone or sandy dolostone (e.g., at Desert View on the south rim and Cape Royal on the north rim).

An obvious lithologic characteristic of the Fossil Mountain Member is the amount and variety of chert (McKee 1938; Hopkins 1986). Chert is common as spherical nodules associated with siliceous sponges, which are most abundant in normal-marine carbonate facies in the western portions of the Grand Canyon. Irregular and branching chert forms also are typical within these facies. Many form by the selective replacement of burrow structures. Closely spaced nodular-to-bedded chert occurs as thin, laterally continuous intervals within sandstone facies. Petrographic study reveals that these chert horizons contain abundant relict sponge spicules and apparently formed by the recrystallization of biogenic silica during shallow burial. These horizons are most common in the eastern portions of the Grand Canyon, where they weather to form distinct recesses along cliff faces.

Another important chert type is a small, white nodule of cauliflower shape that occurs in both carbonate and sandstone lithologies, primarily where dolomitization has been pervasive within the member. These nodules represent silicified evaporites and suggest that a major portion of dolomitization in the Fossil Mountain Member was associated with the migration of hypersaline pore fluids in the shallow subsurface. The selective silicification of skeletal material represents an additional chert type. This silicification results in excellent preservation of many fossils. Chert also is common as lenses and irregular nodules within sandstone units. The fundamental control on the origin of chert in the Fossil Mountain Member, and in the Kaibab Formation as a whole, is attributed to the primary distribution and abundance of siliceous sponges and spicules within the depositional environment (Hopkins 1986).

Harrisburg Member

The Harrisburg Member of the Kaibab Formation forms the uppermost cliffs and receding ledges along both rims of the canyon. This member consists of an assemblage of gypsum, dolostone, sandstone, redbeds, chert, and minor limestone. It generally is thinner than the underlying Fossil Mountain Member. It is difficult to determine its true thickness and extent, however, because of removal associated with the Permo-Triassic unconformity, evaporite dissolution, and recent erosion. Complete sections range from 80 feet (25 m) at eastern exposures along the Little Colorado River (Blakey and Middleton, unpublished data) to 300 feet (90 m) near Whitmore Wash (Sorauf 1962). The member thickens dramatically west and northwest of Kanab Creek and is thickest in northwestern Arizona, southwestern Utah, and southern Nevada, where gypsum comprises a considerable portion of the sequence (Fig. 12.2). In fact, gypsum currently is mined from the Harrisburg Member at Blue Diamond Hill, west of Las Vegas, Nevada. At its type section at Harrisburg Dome east of St. George, Utah, it is about 280 feet (85 m) thick (Reeside and Bassler 1922).

Despite considerable change in lithology and thickness, we can recognize at least six informal stratigraphic units within the Harrisburg Member (Figs. 12.4 and 12.5; Clark 1980; Blakey and Middleton, unpublished data). It is possible that a seventh unit is present is some areas. If this is the case, erosion has removed most of the evidence of this unit (Clark 1980).

The basal unit (unit 1) appears gradational with fossiliferous beds of the underlying Fossil Mountain Member. It is marked by a ragged cliff or site recess 20–40 feet (6–12 m) thick. The unit consists of bedded, nodular, and lenticular chert (marker-chert) that is gradationally overlain by sandstone and sandy dolostone. Locally, silicified evaporite nodules are abundant. Horizontal, ripple, low-angle, and hummocky cross-stratification, preserved in places within the chert, occurs on a local basis. Petrographic study reveals the chert to contain abundant sponge spicules, varying amounts of detrital quartz, and scattered peloids, but only sparse skeletal fragments.

A carbonate ledge 5–12 feet (1.5–3.5 m) thick constitutes a second unit. At western localities, this unit has a limestone mineralogy and is characterized by packstone and wackestone textures containing a variety of fossil fragments that include pelecypods, gastropods, crinoids, bryozoans, foraminifera (forams), and ostracods. At sections along Kanab Canyon and to the northwest, this unit has an oncolite-bearing cap and is locally brecciated. Sections to the east become increasingly dolomitic. They are dominated by mudstone textures that show cryptalgal laminations and contain intraclasts and calcite-filled vugs.

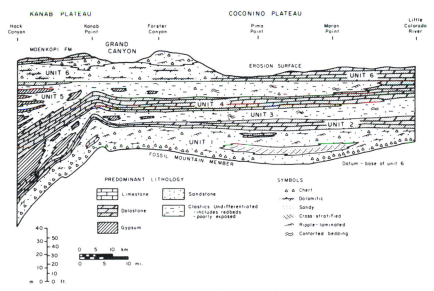

FIGURE 12.4. Diagrammatic cross section of the Harrisburg Member illustrating change in thickness and lithology laterally across the Grand Canyon. See Figure 12.3 for line of section. (Modified from Blakey and Middleton, unpublished data).

FIGURE 12.5. View of receding ledges of the Harrisburg Member above cliff-forming Fossil Mountain Member. Note warping of beds (*arrow*) related to evaporite dissolution. Photograph by R.C. Blakey looks northwest from Kanab Point.

Unit 3 is a poorly exposed, slope-forming sequence that is lithologically variable and undergoes extreme changes in thickness. At eastern localities, this unit is about 20 feet (6 m) thick. Sandstone, sandy dolomitic mudstone, and local cryptalgal-laminated, dolomitic mudstone characterize unit 3. To the west and northwest of Kanab Creek, however, portions of the sequence that contain considerable bedded gypsum and contorted red sandstone can exceed 100 feet (30 m) in thickness. In this region, this unit may show dramatic local changes in thickness and some warping of adjacent beds (Fig. 12.5).

A persistent medial ledge corresponds to unit 4, which averages 20 feet (6 m) in thickness and ranges from 10 to 40 feet (3–12 m). This unit consists predominantly of sandy dolomitic mudstone that locally contains thin lenses of fossil hash, intraclastics, and cryptalgal-laminated horizons. Fossil fragments include pelecypods, gastropods, and ostracods—with crinoids, bryozoans, and brachiopods occurring less commonly. The unit generally contains less sand to the west and toward the top. Local brecciation and warping is related to the dissolution of gypsum within unit 3 below.

Unit 5, which forms the upper slope, exhibits lithologic and thickness trends that are similar to those of unit 3. At eastern localities, this unit is poorly exposed, generally less than 20 feet (6 m) thick, and consists of ripple-laminated sandstone that contains chert locally. West and northwest of Kanab Creek, however, this unit approaches 80 feet (24 m) in thickness and consists of gypsum and interstratified siliciclastic redbeds. Unlike unit 3, this unit contains a number of thin, laterally persistent cryptalgal-laminated dolostone beds. Apparently, it pinches out east of the Grand Canyon (Fig. 12.4).

Uppermost beds (unit 6) form the chert-rubble erosion surface across much of the Grand Canyon region. Only locally is the unconformable contact with the Moenkopi Formation preserved. Where nearly complete, as in Robinson Wash south of Hacks Canyon (Blakey and Middleton, unpublished data), this interval approaches 50 feet (15 m) in thickness and consists of a lower ledge-forming, sandy, fossiliferous dolostone; middle slope and ledge-forming, cherty, ripple-laminated sandstone; and upper slope and ledge-forming cryptalgal-laminated and peloidal dolostone. West and northwest of Kanab Creek, the member contains a prominent molluscan fauna, which includes whole *Bellerophon* gastropods that are spectacularly jasperized locally. To the east, this unit is composed predominantly of dolostone containing increasing numbers of pelecypods, gastropods, and peloids that form packstone textures.

PALEONTOLOGY

Marine invertebrate and vertebrate fossils have been studied by many authors. The paleontology of the Fossil Mountain Member includes work done on the following organisms: brachiopods (McKee 1938; Beus 1964, 1990); bryozoans (Keppel 1932; Condra and Elias 1945a,b; McKinney 1983); conodonts (Thompson 1995); ctenacanthoid sharks (Thompson 1995); nautiloids (Miller and Unklesbay 1942); paleoniscid fish (Thompson 1995); petalodont sharks (David 1944; Ossian 1976; Hansen 1978); sponges (Griffen 1966); and trilobites (Cisne 1971, 1977; Brezinski 1991); the work of DeCourten (1976) on trace fossils should also be noted. Studies by Batten (1964), Beus (1965), Brady (1955, 1959, 1962), Chronic (1953), Mather (1970), Nicol (1944), and Snow (1945) were limited to fossils found within the Harrisburg Member. Many of the fossil forms in the Kaibab Formation (e.g., trilobites) became extinct toward the end of the Permian.

Fossil Mountain Member

In western sections of the Grand Canyon, skeletal limestones in the Fossil Mountain Member contain a diverse macrofaunal assemblage that includes a variety of brachiopods, fenestrate and ramose bryozoans, crinoids, siliceous sponges, and solitary corals (Fig. 12.6). Fossils often are whole and unabraded—indicating little or no transport.

Two types of macrofossils are most characteristic of the Fossil Mountain Member at western localities: (1) large productid brachiopods (*Peniculauris bassi*), which are often silicified and found along bedding plane exposures in life position (concave up)—sometimes with delicate spines still attached; and (2) siliceous sponges (*Actinocoelia maeandrina*; Finks 1960), which commonly occur in the center of spherical chert nodules. Typically, brachiopods and sponges decrease in abundance toward the top of the member, whereas the number of fenestrate and ramose bryozoans and associated crinoid debris significantly increases.

Recent work reveals several different kinds of microfossils (Fig. 12.7), which include: (1) gnathodid, hindeodid, and sweetinid conodonts; (2) seven groups of chondrichthyan dermal denticles (placoid scales); (3) seven types of chondrichthyan teeth; (4) two types of osteichthyan teeth; (5) two orders of ostracods; and (6) two kinds of sponge spicules (Thompson 1995). All are helpful in determining the paleoenvironment at the time the Fossil Mountain Member was deposited; this is especially the case for the conodont fossils, because the morphologic characteristics of all three conodont genera indicate a shallow, warm water environment. Additionally, the association of these genera, along with the marked absence of neogondolellid conodonts, is consistent with nearshore to intermediate marine environments.

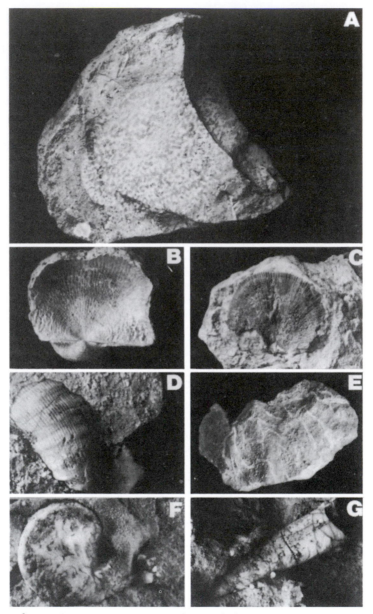

FIGURE 12.6. Fossil types representative of the Kaibab Formation at the Grand Canyon. Examples from the Fossil Mountain Member: (a) Siliceous sponge (*Actincoelia maendrina*) within chert nodule; (b) propductid brachiopod (*Penciculauris bassi*); (c) strophomenid brachiopod (*Derbyia*); (d) solitary or rugose coral; (e) limestone composed of branching or ramose bryozoans and crinoid debris. Examples from Harrisburg Member: (f) coiled gastropod (Bellerophontid); (g) scaphopod (*Prodentalium*). (Photographs by Ted Melis.)

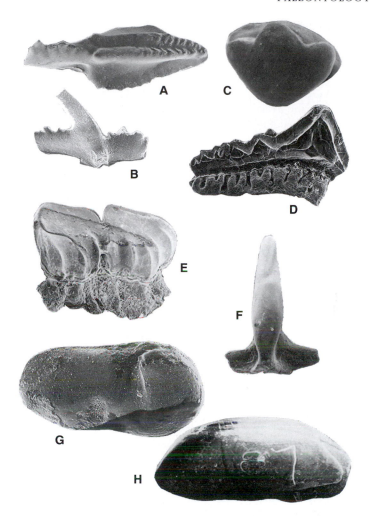

Figure 12.7. SEM microphotographs of selected microfossils from the Kaibab Formation, 50× magnification (all specimen numbers refer to the Museum of Northern Arizona, Flagstaff, Geology Collection). Conodont elements: (a) *Neostreptognathodus newelli* (MNA · X 137); (b) *Sweetina festiva* (MNA-X 138). Chondrichthyan teeth: (c) *Cooleyella peculiaris* (MNA · X 135); (d) Hybodont tooth (MNA · X 136). Chondrichthyan dermal denticles: (e) *Cooperella striatulata* (MNA · X 133); (f) *Moreyella typicalis* (MNA · X 134). Ostracod steinkerns: (g) Cavellinid ostracod (MNA · X 139); (h) Bairdeacean ostracod (MNA · X 140). (Microphotographs by Kelcy Thompson.)

Macrofaunal and microfaunal assemblages in the western portion suggest unrestricted, open-marine conditions characterized by seawater of normal-marine salinity, with deposition below a fair-weather wavebase. Numerous branched, horizontal, and irregularly inclined burrow traces characteristic of low-energy, offshore environments indicate a homogenization of sediment by biogenic reworking. Hopkins (1986) proposed that the apparent vertical transition in macrofossils, from abundant brachiopods and sponges to an assemblage dominated by bryozoans and crinoids, may have reflected an environment of increasing water depth from a few tens of meters to nearly 100 meters. However, paleoecological interpretations based on the conodont assemblage suggest that shallow marine conditions prevailed.

In the eastern sections of the Grand Canyon, the Fossil Mountain Member is dominated by a molluscan fauna characterized by pelecypods (*Schizodus* most common) and poorly preserved gastropods. These fossils typically occur as fossil molds. This systematic decrease in skeletal types of normal-marine affinity within a carbonate facies is associated with increasing siliciclastics. The most pronounced faunal transition found to date takes place in the upper portion of the member between the South Kaibab Trail and Hance Trail on the south rim, and between the North Kaibab Trail and Cape Royal on the north rim.

Lateral change in macrofaunal assemblages within the Fossil Mountain Member reflects a shoreward transition to increasingly restricted-marine environments. Shallow-water, low-energy levels, limited water circulation, elevated temperature, and salinity characterize these environments. Whole productid brachiopods, particularly *P. bassi*, are present in some sandy dolomitic units to the east, suggesting that they may have been more tolerant to changing conditions or that conditions locally were transitional to normal-marine. Brachiopods that show indications of transport may have moved shoreward during storms.

Harrisburg Member

Molluscan faunal assemblages characterized by a variety of pelecypods and gastropods dominate fossils in the Harrisburg Member. In addition, scaphopods are very abundant locally. Nautiloid cephalopods and trilobites are present but vary in their distribution. Geologists can identify ostracods and foraminifera in thin section. These faunal types represent hardy individuals tolerant of a greater range in environmental conditions. Along with gypsum deposits and silicified evaporite nodules, they indicate a partially to highly restricted, shallow-marine environment.

Brachiopods, bryozoans, crinoids, and other normal-marine organisms typically are rare, occurring as small fragments. However, they are present in increasing numbers within the carbonate beds found to the west. This suggests a possibility that seawater of normal or near normal salinity may have returned from time to time to the western Grand Canyon region during Harrisburg deposition.

AGE AND CORRELATION

Geologists have debated the age of the Kaibab Formation until relatively recently. While most agree that the Kaibab was Leonardian in age, based on comparisons of brachiopod faunas (McKee 1938; Fisher 1961; McKee and Breed 1969; Welsh et al. 1979; Kues and Lucas 1989) and the distribution of the siliceous sponge *A. maeandrina* Finks (Finks et al. 1961; Griffen 1966), the question remained as to whether part of the formation was slightly younger or in part Wordian (Guadalupian) in age (Cooper and Newell 1948; Newell 1948; Sorauf 1962).

The conodont assemblage obtained from the western portion is extremely useful in precisely dating the age of the Fossil Mountain Member; the presence of the gnathodid species *Neostreptognathodus newelli*, along with the brachiopod *P. bassi*, correlates well with Wardlaw and Collinson's Zone 3 in the Park City Group in eastern Nevada and western Utah (Wardlaw and Collinson 1978).

The absence of other indicator fossils from lower zones (*P. ivesi*, Zone 1; *N. sulcoplicatus*, Zone 2) and higher zones (*Thamnosia depressa*, Zone 4; *Neogondolella bitteri*, Zone 5) helps to further constrain the time range to be Roadian (latest Early Permian, latest Leonardian) age (Thompson 1995). This brachiopod and conodont fauna help to establish correlation with other Permian sequences in regions beyond the Grand Canyon, including the uppermost part of the Grandeur Member of the Park City Formation in Wyoming, the Meade Peak Phosphatic Shale Member of the Phosphoria Formation in eastern Idaho and western Wyoming, the Garden Valley Formation in central Nevada, the Plympton Formation (Park City Group) in Utah, and the Road Canyon Formation in west Texas (Thompson 1995).

In general, the Kaibab Formation records widespread marine deposition about 260 million years ago, though in western Utah the Kaibab is, at least in part, older than the Kaibab in southern Utah, southern Nevada, and the Grand Canyon (Wardlaw 1986).

DEPOSITIONAL HISTORY

The close association of carbonate and siliciclastic sediments in the Kaibab Formation reflects a complex depositional history marked by major shifts of subtidal, shallow-marine environments. The overall depositional setting represents a mixed carbonate-siliciclastic ramp that existed along the southeastern margin of the Cordilleran miogeocline during the early Roadian (Fig. 12.8). The Kaibab ramp extended across northern Arizona and into southern Nevada, at times exceeding 200 miles (125 km) in width. In this setting, minor fluctuations in the relative position of sea level resulted in abrupt changes in depositional environments. Considering the quiescent tectonic setting of the Grand Canyon region during the Permian, it is most likely that these cycles were caused by glacial–eustatic sea-level oscillations (Kendall and Schlager 1981).

Fossil Mountain Member

The Fossil Mountain Member represents an overall transgressive phase of sedimentation punctuated by repeated regressive events of varying regional extent. Significant eastward shifts in carbonate facies across the Grand Canyon region record a relative rise in sea level. The westward distribution of siliciclastic facies, on the other hand, reflects seaward progradation of nearshore environments during a relative fall in sea level. The abrupt lateral and vertical lithofacies changes within the sequence (Fig. 12.4) reflect these lateral facies migrations.

Beyond the westward limit of siliciclastic progradation, carbonate sedimentation in the Fossil Mountain Member generally was continuous. These outer-ramp carbonate environments consisted of a diverse normal-marine faunal association that effectively baffled, trapped, and stabilized sediment, resulting in widespread limestone units dominated by wackestone textures. Discrete organic buildups, such as sponge patch reefs, have not been documented. Local packstone textures reflect higher-energy conditions that may represent localized shoals, transgressive lag deposits, or storm events. Dissipation of wave, tidal, and current energy across the broad, gently dipping seafloor was sufficient to restrict circulation in nearshores siliciclastic-dominated environments without the development of distinct physical barriers (Irwin 1965). Because there was no exchange with seawater of normal-marine salinity, and the rate of evaporation was high, the salt content of water within inner-ramp environments increased. We see this

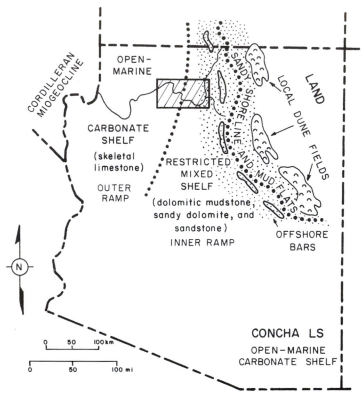

FIGURE 12.8. Hypothetical paleogeographic map of northern Arizona illustrating generalized environments and facies relationships during Fossil Mountain deposition. (After Blakey and Knepp 1988).

from an eastward decrease in skeletal types of normal-marine affinity and the presence of molluscan faunas. The occurrence of silicified evaporite nodules within dolomitized facies also supports this thesis.

The sedimentary characteristics of sandstone units and their relationship with carbonate facies suggest deposition within inner-ramp, nearshore environments. Deposition may have occurred as relatively featureless sand sheets, lower shoreface deposits, or low-relief bars and sandwaves. A variety of processes may have created the currents necessary for sediment transport—including longshore processes, local wind regimes, and episodic storm events. Much of the sediment, however, experienced an intense biogenic reworking that resulted in structureless facies. The most probable source of siliciclastic detritus is coastal eolian complexes thought to border the Kaibab sea to the east (Fig. 12.8). Unlike the underlying Toroweap Formation, however, geologists have not documented any intertonguing between marine and eolian deposits in the Kaibab Formation.

Thin, laterally continuous chert horizons, most common as interbeds within siliciclastic facies at eastern localities, could have formed by suspension settling of sponge spicules and carbonate mud within slight topographic depressions. These spicules may have moved shoreward as a suspension cloud during storm events, or they may have been winnowed by gentle fair weather and/or tidal currents.

Harrisburg Member

Sedimentary strata of the Harrisburg Member represent a distinct change from the diversely fossiliferous, skeletal carbonates characteristic of the Fossil Mountain Member. The six informal, sedimentary units that comprise the Harrisburg sequence reflect deposition within predominantly restricted-marine environments during cyclic westward retreat of the Kaibab sea. Repeated shifts in depositional environments are recorded by the alternation between carbonate, siliciclastic, and evaporite deposits across the Grand Canyon region (Fig. 12.4).

The Harrisburg sequence developed as a result of repeated transgressive–regressive cycles, with lower-order oscillations indicated by minor lithologic and textural variations within individual units. Carbonate deposition generally occurred within shallow, subtidal environments. Molluscan faunas, mudstone textures, and cryptalgal laminations suggest a low-energy, restricted setting, though local fossil hash, intraclastic, and oncolite horizons indicate periodic higher-energy conditions. The occurrence of brachiopods, bryozoans, and crinoids within carbonate units at western localities documents brief pulses of normal-marine conditions.

Sandstone intervals correspond to relative falls in sea level and result in a net transport of sediment derived from nearshore and coastal dune environments. Stratification is a product of the migration of small-scale bedforms, such as ripples and perhaps sand waves, generated by a combination of wave, tidal, and storm currents.

Thick accumulations of evaporites in the Harrisburg document at least two periods of extreme restriction associated with major regressive phases. The intense evaporation associated with an arid climate favored chemical sedimentation. Deposition of massive and bedded gypsum deposits probably occurred within localized hypersaline basins or lagoons bordered by a coastal mudflat–sabkha complex. The considerable thickness of the evaporite deposits suggests that seawater of normal-marine salinity replenished these restricted areas periodically.

Rapid change in the lithology and thickness west of Kanab Creek reflects periods of increased differential subsidence in this portion of the Kaibab ramp. The relationship, if any, of Harrisburg evaporite basins to recurrent movement along basement faults is uncertain. Other, more local thickness variations, along with warping and brecciation of adjacent carbonate units, are more likely related to post-depositional dissolution. Contorted beds in the upper part of the Harrisburg to the east suggest that the evaporites originally were more widely distributed and have been removed from much of this area.

SUMMARY

Rocks of the Kaibab Formation are testimony to the ancient seaway that covered most of the Grand Canyon region approximately 260 million years ago. The cyclic interbedding of carbonate and siliciclastic sediments documents a complex depositional history characterized by repeated shifts of subtidal, shallow-marine environments. Rocks of the Fossil Mountain Member record a west-to-east transition from fossiliferous open-marine limestone to restricted-marine sandy dolostone. The overlying Harrisburg Member reflects deposition during the cyclic retreat of the Kaibab sea. A short walk down any of the canyon rim trails allows visitors to easily examine the great variety of rocks and fossils that comprise the Kaibab Formation.

13

MESOZOIC AND CENOZOIC STRATA OF THE COLORADO PLATEAU NEAR THE GRAND CANYON

Michael Morales

INTRODUCTION

If you stand on either the north or south rim of the Grand Canyon, the soles of your shoes will rest on the cracked and weathered limestone of the Kaibab Formation. This topmost rock unit of the canyon was deposited near the end of the Paleozoic Era. As you peer into the deep chasm below, you will see a mile (1.6-km)-thick section of strata that accumulated during the Proterozoic Eon and Paleozoic Era. Now turn your gaze skyward and imagine a section of rocks extending *above* your feet for approximately another mile, about the same distance above the rim as the bottom of the canyon is below the rim. This exercise will give you an idea of the great thickness of marine and terrestrial rock layers that were deposited on the top of the Kaibab Formation in several intervals during the Mesozoic Era (Hintze 1988). These sediments once covered the entire southwestern portion of the Colorado Plateau Physiographic Province (Fig. 13.1), an area that includes the Grand Canyon (Billingsley 1989).

In the vicinity of the canyon, denudation stripped away the Mesozoic strata during the Late Cretaceous in a major episode of erosion associated with the uplift of the southwestern Colorado Plateau (Lucchitta, Chapter 15, this volume). During the early Cenozoic Era (Paleocene and Eocene epochs), one-half mile (0.8 km) or more of terrestrial sediments and volcanics accumulated, only to be removed almost entirely in another episode of erosion during the Oligocene (Elston and Young 1989). Widely dispersed remnants of these deposits, primarily unnamed gravels and interbedded freshwater limestones, still crop out in the southwestern part of the Grand Canyon area (Elston and Young 1989; Young 1989). Thus, nearly all Cenozoic and Mesozoic sedimentary rocks have been removed from the Grand Canyon area, leaving mainly strata of middle and late Paleozoic age (Kaibab, Toroweap, Coconino, Redwall, Supai) as the topmost bedrock units surrounding the canyon. Scattered patches of Pleistocene volcanic rocks cover sedimentary units in and near the western Grand Canyon (Hamblin, Chapter 17, this volume). South of the canyon, Tertiary and Quaternary volcanic rocks of the Mount Floyd, San Francisco, and Mormon Mountain volcanic fields (Fig. 13.2) cover the Paleozoic strata (Chronic 1983). These fields include a multitude of cinder cones (e.g., Sunset Crater), volcanic domes (e.g., Bill Williams

FIGURE 13.1. The Colorado Plateau Physiographic Province.

Mountain), composite volcanoes (e.g., San Francisco Mountain or Peaks), and many lava flows (Holm 1987; Holm and Moore 1987).

North and east of the Grand Canyon, where less erosion has occurred, Mesozoic and Cenozoic strata of the southwestern Colorado Plateau are preserved in vast badland outcrops of cliffs, canyons, and plateaus. These rocks are especially visible in two regions: north of the canyon, in the area of northern Arizona and southern Utah that Clarence E. Dutton named the Grand Staircase, and east of the canyon to Black Mesa, Arizona. This chapter summarizes the main aspects of rock units, excluding surficial Quaternary deposits, that rest on top of the Kaibab Formation in areas of the Colorado Plateau near the Grand Canyon (Table 13.1).

THE GRAND STAIRCASE

The topography north of the Grand Canyon is made up of alternating cliffs and flatlands that form a series of erosional steps of increasing elevation through Mesozoic and Cenozoic deposits. The Grand Staircase section of this region (Stokes 1986) includes the Uinkaret, Kanab, Kaibab, Markagunt, and Paunsaugunt

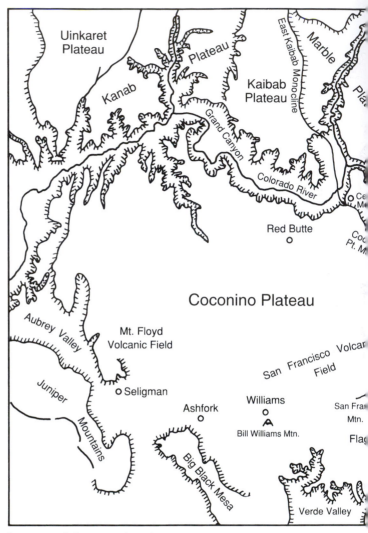

Figure 13.2. Physiographic map of the central and eastern Grand Canyon and surrounding areas. (After Gregory 1950, Cooley et al. 1969, King 1977, and Billingsley and Hendricks 1989.)

plateaus; Antelope Valley; Telegraph Flat; the Little Creek, Wygaret, Kolob, and Skutumpah terraces and their equivalents to the east; the Block Mesas region; and the Moccasin Terrace (Fig. 13.3). The plateaus and terraces are bordered on their west and east sides by the major north/south-trending Hurricane, Toroweap–Sevier, and West Kaibab–Paunsaugunt faults (and associated monoclines and cliffs) and by the East Kaibab monocline (and related faults). In three areas, Zion and Bryce national parks and Cedar Breaks National Monument, deep canyons and extensive badlands have been carved into the rocks. Figure 13.4 illustrates the slightly north-dipping strata and stepped topography of the Grand Staircase.

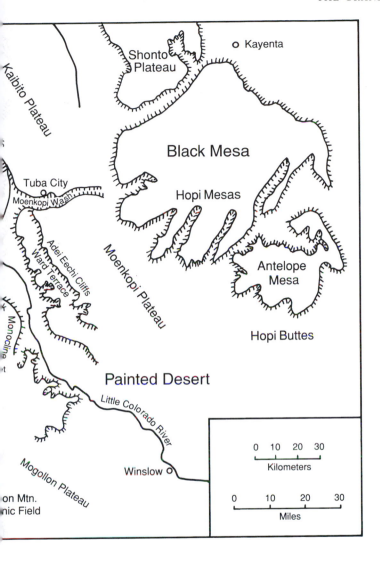

The first line of cliffs north of the Grand Canyon is the Chocolate Cliffs. The name is derived from the red-brown mudstone of the Lower and Middle Triassic Moenkopi Formation that forms the majority of the cliff profile. This escarpment sometimes is called the Shinarump Cliffs after sandstones and conglomerates of the Shinarump Member of the Upper Triassic Chinle Formation that cap the cliffs. Although the Chocolate Cliffs officially terminate at the northern end of Kaibab Plateau, the same rock units form a cliff and slope line along the southern margin of Paria Plateau. There, however, the Moenkopi–Shinarump cliffs are not as well developed as the true Chocolate Cliffs. Erosional unconformities, which represent gaps in the rock/time record, separate the Moenkopi from the underlying Kaibab and overlying Chinle formations. The Moenkopi contains both marine and nonmarine sediments, whereas the Shinarump member of the Chinle is composed primarily of fluvial channel deposits. The flatlands and slopes above

TABLE 13.1. Alphabetical List of Formations of the Southwestern Colorado Plateau North and East of the Grand Canyon[a]

Formation Name	Age	Major Lithologies	Main Depositional Environment	Fossil Groups
Bidahochi	Mio-Plioc.	Clay, volcanics	Lacustrine, volcanic	Terres. and fresh. verts; inverts; & plants
Carmel	M. Jur.	Limestone, mudstone	Marine	Marine inverts., verts., & algae
Chinle Claron	Lt. Tri.	Mudstone, sandstone	Fluvial, lacustrine	Terres. plants; fresh. inverts.
Cow Springs Sandstone	M. Jur.	Sandstone	Eolian	None?
Dakota inverts;	Lt. Cret.	Sandstone, mudstone, coal	Marginal marine, fluvial	Terres. plants, verts., & marine inverts.
Entrada Sandstone	M. Jur.	Sandstone, mudstone	Fluvial, eolian	None?
Kaiparowits	Lt. Cret.	Mudstone, sandstone	Fluvial	Terres. & fresh. verts., inverts., & plants
Kayenta	E. Jur.	Siltstone, sandstone	Fluvial, eolian	Terres. plants, verts.; dinosaur tracks
Mancos inverts. Shale	Lt. Cret.	Shale	Marine	Marine plants, verts., &
Moenave	E. Jur.	Sandstone, mudstone	Eolian, fluvial	Fresh. fish, crocodiles, dinosaurs, & reptile tracks
Moenkopi	E-M Tri.	Mudstone	Marine, fluvial, tidal flat	Marine inverts.; terres. & fresh. verts., inverts., & plants; vert. & invert. trace fossils
Morrison	Lt. Jur.	Mudstone, sandstone	Fluvial	Terres. & fresh. plants, verts. (esp. dinosaurs), inverts., & trace fossils
Navajo Sandstone	E. Jur.	Sandstone, limestone	Eolian lacustrine	Terres. reptiles, plants, & invert. trace fossils; dinosaur tracks
Summerville	M. Jur.	Sandstone	Eolian	None?
Straight Cliffs	Lt. Cret.	Sandstone, mudstone	Fluvial, marginal	Marine & fresh. inverts. fresh. marine & terres. verts.
Temple Cap	E-M. Jur.	Mudstone, sandstone	Fluvial	None?
Toreva	Lt. Cret.	Mudstone, sandstone,	Littoral, fluvial, coal	Vert. fragments?
Tropic Shale	Lr. Cret.	Shale	Marine	Marine plants, verts., & inverts.

(*continued*)

TABLE 13.1. Alphabetical List of Formations of the Southwestern Colorado Plateau North and East of the Grand Canyon[a] (*Continued*)

Formation Name	Age	Major Lithologies	Main Depositional Environment	Fossil Groups
Wahweap	Lt. Cret.	Sandstone, mudstone	Fluvial	Fresh. & terres. verts. & inverts.
Wepo	Lt. Cret.	Sandstone, mudstone,	Fluvial, paludal, coal	None?
Wingate Sandstone	E. Jur.	Sandstone	Eolian	Dinosaur tracks
Yale Point Sandstone	Lt. Cret.	Sandstone	Littoral	None?

[a]Data compiled from many sources. Mio-Plioc., miocence-Pliocene; Paleoc.?, Paleocene; Cret., Cretaceous; Tri., Triassic; Jur., Jurassic; E., Early; M., Middle; Lt., Late; inverts., invertebrates; fresh., freshwater; verts., vertebrates; terres., terrestrial.

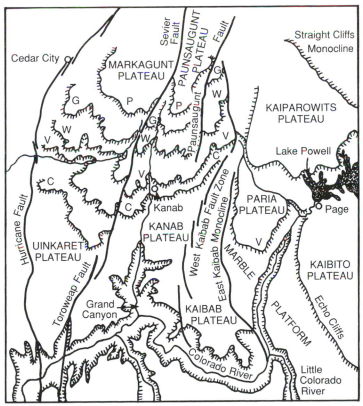

FIGURE 13.3. Physiologic map of the Grand Staircase section of the Colorado Plateau. C, Chocolate Cliffs; V, Vermilion Cliffs; W, White Cliffs; G, Gray Cliffs; P, Pink Cliffs. (After King 1977, Stokes 1986, and Billingsley and Hendricks 1989.)

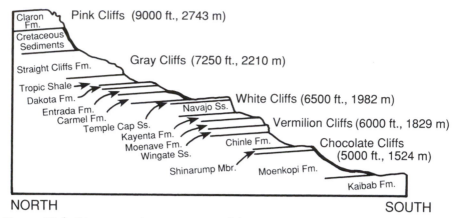

FIGURE 13.4. Diagrammatic cross section of the Grand Staircase—vertical scale greatly exaggerated. (After King 1977, Stokes 1986, Hintze 1988, and Clemmensen et al. 1989.)

the Chocolate Cliffs, including Little Creek Terrace, Telegraph Flat, and related areas to the east, are formed by different fluvial and lacustrine members of the Chinle Formation.

The next step up the Grand Staircase is the very prominent Vermilion Cliffs. Starting at the base, red and purple deposits of the Chinle Formation, Wingate Sandstone, and the Moenave and Kayenta formations make up these cliffs, with the lower part of the Navajo Sandstone forming the caprock. The latter four formations comprise the Lower Jurassic Glen Canyon Group and are of primarily eolian and fluvial origin. An unconformity separates the Wingate Sandstone from the underlying Chinle Formation; however, contacts between the other units are gradational to intertonguing. Above the Vermilion Cliffs, the Navajo Sandstone forms the bench that includes Mocassin Terrace, Wygaret Terrace, and their equivalents to the east.

The next escarpment northward is called the White Cliffs; it is composed primarily of the prominent buff-to-white colored Navajo Sandstone. Fluvial deposits of the Lower and Middle Jurassic Temple Cap Formation cap the cliffs, and there is an erosional unconformity between the two formations. The White Cliffs terminate on the west side of the Kaiparowits Plateau. The flatlands and slopes above the White Cliffs include the Kolob and Skutumpah terraces and their equivalents. Zion Canyon is carved into southern Kolob Terrace along the White Cliffs. The southern portions of the two terraces are made up primarily of the Dakota Formation overlying the Carmel Formation, which rests on the Navajo Sandstone. To the east, the Entrada Formation is present between the Carmel and the Dakota. The contact between each of the four formations is unconformable. Both the Carmel and Entrada are Middle Jurassic formations of the San Rafael Group. The former, however, is of marine origin, whereas the latter is predominantly tidal flat and eolian. Depositional environments of the Dakota Formation range from terrestrial to marginal marine.

The Gray Cliffs form a relatively low line that runs across the Kolob and Skutumpah terraces and extends to the eastern side of the Kaiparowits Plateau. They are formed by the gray Tropic Shale capped by the Straight Cliffs Formation. The Tropic Shale disconformably overlies and interfingers with the Dakota Formation. The Tropic Shale and the Straights Cliffs Formation have an intertonguing contact. Deposits of the Tropic Shale are marine, while those of the Straight Cliffs Formation are terrestrial to marine.

The northern portions of Kolob and Skutumpah terraces, formed by the Straight Cliffs Formation, continue northward from the Gray Cliffs to the final step up the Grand Staircase, the Pink Cliffs. These are made of Upper Cretaceous rocks of uncertain affinities (called Wahweap and Kaiparowits formations in early literature) that are capped by light pink and orange lacustrine and fluvial deposits of the lower Cenozoic (Paleocene?) Claron Formation (Hintze 1988). An unconformity separates the Claron from underlying strata. The Pink Cliffs form the southern escarpment of the Markagunt Plateau on the west and the Paunsaugunt Plateau on the east (Fig. 13.3). The badlands of Cedar Breaks and Bryce Canyon are carved into the Pink Cliffs on the southwestern Markagunt and southeastern Paunsaugunt plateaus, respectively.

SOUTH RIM TO BLACK MESA

The topography east of the Grand Canyon includes cliffs, terraces, plateaus, and generally north/south-trending fault zones and monoclines that cross a transect from the south rim eastward to Black Mesa. Unlike the Grand Staircase, this region of the southern Colorado Plateau has no general name. It includes the Coconino, Kaibito, and Moenkopi plateaus; Marble Platform; Ward Terrace; Black Mesa (including the Hopi Mesas); and the Little Colorado River Valley, which contains the western part of the Painted Desert (Fig. 13.2). The northeastern edge of the Coconino Plateau is marked by the East Kaibab, Coconino Point, and Black Point monoclines (and their associated faults). The Echo Cliffs, a result of erosion along the Echo Cliffs monocline, form the western border of Kaibito Plateau. The related Adei Eechii Cliffs comprise the western boundary of Moenkopi Plateau. Between the Kaibito and Moenkopi plateaus runs Moenkopi Wash, which extends eastward to Black Mesa. Figure 13.5 illustrates the stratigraphy and topography along the transect from the eastern part of the canyon's south rim to Black Mesa.

In the central and eastern portion of the Grand Canyon, the south rim represents the northern border of the Coconino Plateau. The Kaibab Formation caps the plateau here, except in rare places where remnants of Triassic rocks are preserved. At Cedar Mesa, slightly east of the south rim's Desert View overlook, the Moenkopi Formation overlies the Kaibab Formation; it, in turn, is capped by a resistant layer of Cenozoic volcanic rock. At Red Butte, south of Grand Canyon Village, the Moenkopi is covered by Shinarump deposits, which also are overlain by a younger lava flow.

On its northeastern border, the Coconino Plateau folds downward along the east to northeast-dipping East Kaibab, Coconino Point, and Black Point monoclines. This folding is especially noticeable at Gray Mountain, southeast of Desert View (Barnes 1987). Tidal-flat, fluvial, and eolian deposits of the Lower and Middle Triassic Moenkopi Formation border the edge of the plateau and extend to the middle of the Little Colorado River Valley and Marble Platform. Farther eastward, overlying red and purple deposits of the fluvial lower members of the Upper Triassic Chinle Formation make up the Painted Desert. A step up and to the east of the Painted Desert in the Little Colorado River Valley is Ward Terrace. This flat area is composed of the mainly gray lacustrine sediments of the Chinle's upper members.

East of Marble Platform and Ward Terrace, the topography includes two erosional steps upward (Fig. 13.5). The first is formed by the Echo and Adei Eechii

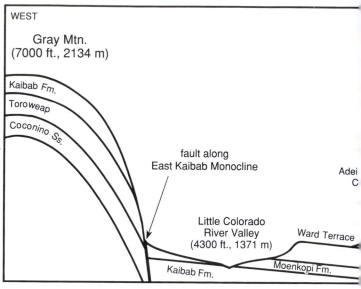

FIGURE 13.5. Diagrammatic cross section from eastern Grand Canyon to Black Mesa—vertical scale greatly exaggerated. (After Brown and Lauth 1958, Cooley et al. 1969, and Chronic 1983.)

cliffs, which border the Kaibito and Moenkopi plateaus, respectively. The second step includes the cliffs and top of Black Mesa. Units of the Upper Jurassic Glen Canyon Group form the Echo and Adei Eechii cliffs, which are capped by the Navajo Sandstone. The Wingate Sandstone is present only in the southern part of the Adei Eechii Cliffs; it thins northward to an erosional pinchout south of Moenkopi Wash. Strata of the Echo Cliffs dip steeply to the east because of the Echo Cliffs monocline. However, the beds of the Adei Eechii Cliffs dip only slightly in the same direction because the monocline does not extend that far south. Most of the surface of the Kaibito and Moenkopi plateaus is formed by the Navajo Sandstone, except in places where less intense erosion has left younger strata, such as the unconformably overlying Carmel Formation.

The lower flatlands and slopes of Black Mesa are composed of, in ascending order, the Carmel Formation, Entrada Sandstone, Cow Springs Sandstone (west side) or Summerville Formation (east side), and Morrison Formation—all part of the Middle and Upper Jurassic San Rafael Group (Cooley et al. 1969). The Entrada has conformable contacts with units below and above it, whereas the Cow Springs and Summerville, which laterally intergrade and interfinger, have an uncomfortable contact with the overlying Morrison. The upper slopes and cliffs of Black Mesa are made of Upper Cretaceous deposits of the Dakota Formation, Mancos Shale (correlative with the Tropic Shale), Toreva and Wepo formations, and Yale Point Sandstone. The latter three units comprise the Mesaverde Group. An unconformity separates the Dakota from the underlying Morrison. The Mancos has gradational boundaries with the Dakota below and Toreva above. The lower and upper contacts of the Wepo are intertonguing. Deposits of the San Rafael Group, except the marine Carmel Formation, are terrestrial in origin, as are parts of the Dakota, Toreva, and Wepo formations. Other portions of the

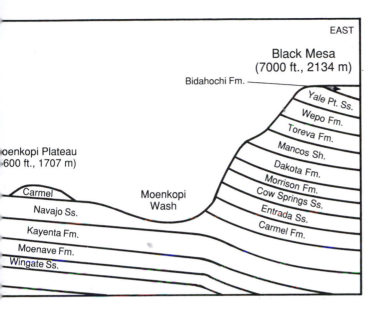

latter three formations represent nearshore environments (Harshbarger et al. 1957; Wilson 1974). Sediments of the Mancos Shale are of marine origin.

Most of Black Mesa's top surface is formed by the Wepo Formation, but the Yale Point Sandstone caps it to the northeast. In the extreme southern portion of Black Mesa, in the area of the Hopi Mesas, the Miocene–Pliocene Bidahochi Formation unconformably overlies the Mesaverde Group (Cooley et al. 1969). Lacustrine sediments dominate the Bidahochi, but fluvial and volcanic deposits are also present.

SUMMARY

Although the Proterozoic and Paleozoic rocks of the Grand Canyon record ancient environments, prehistoric life, and geologic events over hundreds of millions of years, they do not provide a complete picture of the area's geologic history. For the rest of the story, we must look to strata resting on top of the Kaibab Formation in and near the canyon.

14

POST-PRECAMBRIAN TECTONISM IN THE GRAND CANYON REGION

Peter W. Huntoon

INTRODUCTION

This chapter will summarize in chronological order the post-Precambrian structural events that are recorded in the rocks of the Grand Canyon. Laramide compression and Late Cenozoic extension will receive particular attention because these events caused most of the deformation found here. A reading of this chapter will be greatly facilitated by having the geologic maps by Huntoon et al. (1981, 1982, 1996) and Billingsley and Huntoon (1983) at hand.

The typical section in this chapter will place structural events occurring in the Grand Canyon region into the larger contexts of cordilleran tectonics and plate tectonics. The evidence used to time the events, and the types and geometries of the structures associated with them, will be described at the local and regional levels. Causative stresses will be deduced. The influence of the events on paleogeography and on the movement of sediments into and out of the region will be outlined.

This is a skeletal history. Many pages and even chapters have been torn from the book by erosion during periods of uplift. One of the largest gaps is the 130 million years missing from the Paleozoic record inclusive of Ordovician through Middle Devonian time. Similarly, most of the Mesozoic record has eroded from the plateaus surrounding the canyon, and too much Cenozoic history has been carried away by the Colorado River thanks to Laramide and Tertiary. To draw some inferences, we will step away from the canyon to adjacent regions where fragments of the record still remain.

The terms "Colorado Plateau region" and "Grand Canyon region" are used for orientation purposes. However, be aware that the Colorado Plateau did not become fully defined until Miocene time. "Mogollon Highlands" refers to the Mesozoic and Cenozoic uplifts occurring south of the Colorado Plateau.

THE COLORADO PLATEAU

The Grand Canyon is eroded into the southwestern corner of the Colorado Plateau, a geologic province that is underlain by a thick continental crust that became slightly separated on the east from the continental craton during the Cenozoic Era. As shown in Fig. 14.1, the plateau is ringed by zones of intense deformation, including the extended Rio Grande rift on the east, compressional Uinta uplift to the north, and drastically extended Basin–Range to the south and

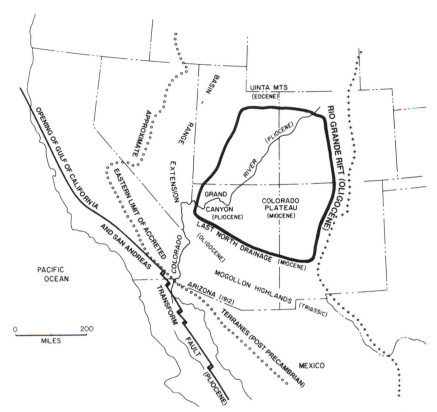

FIGURE 14.1. Selected tectonic elements in the western North American cordillera and the timing of their inception.

west. The Paleozoic rocks in the Grand Canyon are about a mile (1.6 km) thick. Nevertheless, they are but a thin, sensitive veneer resting on a 19- to 25-mile (30- to 40-km)-thick crust comprised of an ancient basement complex.

The record of Phanerozoic deformation reveals that the basement underlying the Colorado Plateau has enjoyed unusual stability since the close of Precambrian time. The Paleozoic rocks in the Grand Canyon region were deposited in the equatorial regions of the earth on the southwestern part of a growing continent. The plateau was an integral part of the larger land mass at that time. Subsequently, the continent was rafted through plate tectonic processes some 2000 miles (3200 km) northward across the face of the earth, and it rotated a few tens of degrees in a counterclockwise direction (Elston and Bressler 1977).

Between the beginning of Cambrian and the end of Cretaceous time, a period of roughly 480 million years, net subsidence of the Precambrian surface in the Grand Canyon region amounted to between 1.5 and 2 miles (2.4 to 3.2 km). In the last 70 million years, a period dominated by uplift, the Precambrian surface has risen in two pulses a total of approximately 2 miles (3.2 km). Clearly, the cratonic block upon which the Grand Canyon occurs has moved great lateral distances and has both lost and gained considerable elevation. However, it has not undergone serious internal fragmentation since the close of Precambrian time.

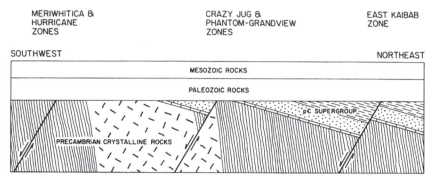

FIGURE 14.2. Schematic diagram showing the configuration of the Precambrian basement in the Grand Canyon region, Arizona, prior to Laramide contraction. Note that the Paleozoic rocks rest on successively older Precambrian rocks toward the southwest. Minor Paleozoic and Mesozoic(?) movements along the basement faults are ignored. All dips are exaggerated.

INHERITED PRECAMBRIAN BASEMENT

As shown on Figure 14.2, a late Precambrian erosion surface exposed progressively older rocks in the Precambrian complex toward the west. The oldest rocks comprise the approximately 1.7-billion-year-old crystalline complex (Pasteels and Silver 1965). The younger Precambrian sedimentary and volcanic Grand Canyon Supergroup is preserved to the east on a series of fault-bounded, northeast-tilted basement blocks. These layered rocks generally dip between 6 and 8 degrees toward the east, and they are beveled above by the erosion surface into northeast-thickening wedges.

Supergroup wedges to the east contain increasingly more complete sections. Consequently, the youngest Precambrian sediments are found in the easternmost part of the Grand Canyon. Maximum thicknesses occur along the west side of the Butte fault. Huntoon et al. (1996) showed a thickness of approximately 9000 feet (2700 m) in the vicinity of Malgosa Canyon. Elston and Scott (1976) have reported thicknesses of between 12,000 and 14,000 feet (3600 and 4200 m), based on measured sections elsewhere.

The Precambrian faults defining the boundaries of the tilted basement blocks in the eastern Grand Canyon are north- and northwest-trending normal faults with extensive records of recurrent Precambrian displacement (Walcott 1889). The bounding faults include the Butte, Phantom-Cremation, Crystal, and Muav faults. Synchronous Precambrian deformation consisting of parallel, parasitic folds and grabens developed in bands as wide as 2.5 miles (4 km) along the eastern margins of the blocks. The resulting zones commonly are spaced about 10 miles (16 km) apart. Elston (1979) estimated the age of the latest Precambrian movements along these faults at 845 to 810 million years using K–Ar dates.

The north- and northwest-trending faults offset an older set of northeast-trending reverse faults. The Bright Angel fault is the most studied of these (Huntoon and Sears 1975). Shoemaker et al. (1975) conclude that there is a good possibility that considerable right-lateral, strike-slip displacement occurred along the northeast-trending faults prior to deposition of the Precambrian Grand Canyon Supergroup.

The best evidence for recurrent Precambrian faulting in the western Grand Canyon occurs along the Hurricane fault. The metamorphic grades of deformed rocks found between the surfaces of the fault in exposures between Granite Spring and 224-Mile canyons represent pressures and temperatures unavailable in the region since the end of Precambrian time. These metamorphic products developed before deposition of the younger Grand Canyon Supergroup, revealing a long prehistory of activity along the fault. In addition, juxtaposition of Precambrian crystalline terranes with differing fracture-foliation fabrics and lithologies is common along the fault. Rocks with a marked discordance in foliation trends are also juxtaposed across the Meriwhitica fault, indicating a substantial but unknown magnitude of Precambrian offset.

PALEOZOIC AND MESOZOIC TECTONISM AND SEDIMENTATION

North America grew in a westerly and southerly direction by lateral continental accretion during Paleozoic and Mesozoic time. The North American cordillera increased by an additional 30 percent through accretion during the Mesozoic Era alone (Coney 1981). The area that was to become the Colorado Plateau was ensconced just to the east within the North American craton. Although buffeted by continental-scale tectonic events, the Grand Canyon region remained insulated by distance to the point that little happened in the form of discrete offsets along faults in the underlying basement or local volcanism. The forces operating in the Grand Canyon region were attenuated sufficiently that deformation mostly took the form of broad-scale but gentle warping of the crust and variable rates of subsidence or uplift. The lack of angular unconformities of regional extent in the Paleozoic section in the canyon reveals that, at least during Paleozoic time, the uplifts associated with the disconformities are best categorized as epeirogenic.

Cambrian to Late Triassic Tectonics

The southwestern corner of the Colorado Plateau occupied a distant inboard position on the west-facing continental shelf through Devonian time. The edge of the continent is characterized as a passive margin undergoing a progression of marine transgressions and regressions (Woodward-Clyde Consultants 1982). The Cambrian sea transgressed eastward over the Precambrian foundation, burying a generally flat-lying surface broken locally by scattered, essentially isolated, small but rugged hills. The highest hills were 1200 feet (360 m), and they finally were covered by the Muav Limestone. Examples of buried Precambrian hills occur under Isis Temple, in Modred Abyss and Monadnock Amphitheater, and along the Colorado River immediately east of Deer Creek.

The longest hiatus in the Paleozoic stratigraphic record occurs between the undivided Cambrian carbonates and Late Devonian Temple Butte Formation, a gap of approximately 130 million years. Missing are rocks inclusive of Ordovician, Silurian, and Early and Middle Devonian age, yet the contact is a disconformity throughout the Grand Canyon. Similarly, the contact between the Temple Butte Formation and the Mississippian Redwall Limestone is a disconformity, even though this hiatus was caused by orogenic uplift associated with the Antler orogeny. The Grand Canyon region was being subjected to uplift and subsidence; however, the crust was not being deformed internally. The conclusion is that the region was distant from active orogenic zones.

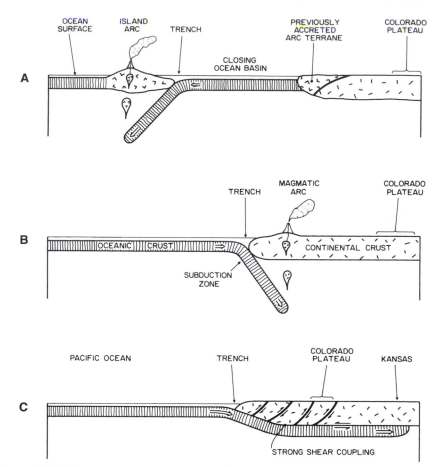

Figure 14.3. Convergent margin orogens along western North America. (a) Intraoceanic arc-trench orogen active periodically in post-Precambrian through Late Triassic time. Notice that the ocean basin closes, allowing island arc to accrete to continent, then another subduction zone and its island arc can form offshore and likewise eventually accrete to continent. (b) Slow landward subduction causing development of magmatic arc inboard on continent above steeply descending slab active from Late Triassic to Late Cretaceous time. (c) Rapid subduction resulting in shallow slab descent and slab underplating of continent to produce buoyant uplift and strong shear-coupling with eastward telescoping of continental crust during Laramide time. Vertical scales are greatly exaggerated.

Convergence of the oceanic and North American plates during the Antler orogeny in latest Devonian–Mississippian time resulted in the accretion of huge tracts of land to western North America and downwarping of the Cordilleran miogeocline along a north–south axis through eastern Nevada and southeastern California (Dickinson 1981). Accretion appears to have been accomplished as shown in Fig. 14.3a through plate convergence in an offshore arc-trench system. This mechanism involved the oceanward subduction of the oceanic lithosphere lying between an offshore trench and the continent. As the crust was consumed in the trench, both the trench and island arc migrated toward the edge of the continent. Once the trench and continent met, the island arc became sutured to the continent, and a new arc-trench system developed offshore to repeat the process. The continent built progressively oceanward from central Nevada and eastern California in this fashion.

Orogenic belts associated with accretionary growth lay to the west of the Grand Canyon region and were subjected to large-scale eastward thrusting, uplift caused by crustal thickening, and deep-seated metamorphism. The north-trending depositional basin through east-central Nevada collected the detritus shed eastward off these uplifts. Marine limestones, such as the Devonian Temple Butte and the Mississippian Redwall limestones, were deposited in the Grand Canyon region on the eastern margin of the basin and thickened toward it.

One of the best-preserved erosional events in the Paleozoic section involves the Late Mississippian emergence of the Redwall Limestone. A series of westward-draining valleys are found in the western part of the canyon that were incised as much as 400 feet (120 m) into the limestone. The slightly uplifted Redwall surface was pervasively karstified so that the landscape took on the appearance of the modern Yucatan Peninsula. The Mississippian (Chesterian) Surprise Canyon Formation fills the valleys and caves; yet it and the Supai Group lie disconformably on virtually all the older rocks.

The only offsets along faults in the Grand Canyon that have been identified as dating from the Paleozoic Era are inferred from highly localized angular unconformities at the top of the Mississippian Redwall Limestone. These disturbances occurred in the interval between deposition of the Redwall Limestone and Surprise Canyon Formation. At least 150 feet of the Redwall section, including the entire Horseshoe Mesa Member, were truncated by erosion across the crest of a minor anticline at a site along the Tanner Trail (McKee and Gutschick 1969). This erosion surface is overlain unconformably by the Pennsylvanian Watahomigi Formation. The fold is attributed to displacement along the underlying Precambrian Butte fault. Huntoon and Sears (1975) observed a similar truncation of the upper 30 feet (9 m) of the Horseshoe Mesa Member in a band one-quarter-mile wide over the Bright Angel fault. The truncation and an anomalous 10-degree west dip of the Redwall Limestone were caused by reverse motion along the underlying Proterozoic fault, resulting from reactivation of that structure. These deformations are attributed to horizontal, generally east–west compression in the context of present compass directions that were caused by contraction associated with the Antler orogeny.

The Pennsylvanian and earliest Permian Ouchita orogeny developed when a terrane that now comprises the Yucatan Peninsula and South America was sutured to the North American continent along a collisional belt extending through northern Mexico, central Texas, and southeastern Oklahoma (Dickinson 1981). The resulting grand-scale internal stresses within the North American plate were sufficient to cause the intracratonic block uplifts and associated basins of the ancestral Rocky Mountains. The Uncompahgre uplift and Paradox basin developed northeast of the Grand Canyon region in response to this collision, and the Defiance uplift to the east gained additional elevation. Clastic sediments, such as those in the Supai Group, were shed from these highlands and transported south and west into the Grand Canyon region, which remained low. Some graded into carbonate facies toward the west across the Grand Canyon region.

There was possible Pennsylvanian rifting in the Antler belt in the western Basin-Range region, followed by local Early Permian emergence. The Grand Canyon region first collected the Coconino Sandstone, a coastal dune deposit facing the ocean to the west. The Coconino Sandstone, which is the shoreward facies of the Seligman Member of the Toroweap Formation, was buried by the southeastward transgressing marine members of the Toroweap Formation, and later the Kaibab Formation, as the land subsided. The marine shelf environment persisted, except for a brief period of uplift to the west and south at the end of Permian time, until after deposition of the marine Moenkopi Formation in Early and Middle Triassic time. The Late Permian–Early Triassic Sonoma accretionary event was occurring far to the west.

The huge interval of Paleozoic time spanning 325 million years was characterized by net subsidence and net accumulation of sediments that now thicken westward from 3500 to 5000 feet (1100 to 1500 m) through the Grand Canyon. The top of the accumulating sedimentary pile had fluctuated within several hundred feet of sea level, with most of the time spent below the water.

Late Triassic to Late Cretaceous Uplift

A worldwide reorganization of plate motions between Late Triassic and Late Jurassic time resulted in the opening of the Gulf of Mexico and the Atlantic Ocean through sea floor spreading. Concurrently, the Pacific Ocean crust began to subduct under North America in contrast to earlier oceanward subduction (Fig. 14.3B). The result was the inboard development of a Mesozoic magmatic arc south and west of the Grand Canyon region that was characterized by batholithic intrusions, arc volcanism, and thermotectonic uplift. These events first caused general emergence of the Colorado Plateau region and strong uplift of the contiguous Mogollon Highlands to the southwest, and much later they caused uplift and eastward thrusting in the Sevier Highlands to the northwest during Cretaceous time (Woodward-Clyde Consultants 1982).

Most of Mesozoic time in the Grand Canyon region was characterized by low but emergent conditions in which continental sedimentation within intracratonic basins prevailed. The broad, gently sloping Mogollon Highlands began to rise during deposition of the marine Moenkopi Formation, and ultimately it caused regression of the Moenkopi sea toward the northwest. The uplift established a sediment dispersal pattern characterized by north and northeast flowing streams heading in the highlands that persisted until Cenozoic time in the Grand Canyon region. Thus the highlands became a major source for detritus in the Moenkopi Formation and succeeding Mesozoic units in the region. Large inland seas lay mostly to the northwest. The first of the Late Triassic units was the voluminous Chinle Formation, which, in the southwestern Colorado Plateau region, was comprised of fluvial detritus originating from the highlands south of the plateau region and air fall ash that probably came from the northwest. Debris eroded from the highlands included arc volcanics, sedimentary cover, and unroofed Precambrian basement. Erosion in the highlands appears to have exposed Paleozoic and Precambrian terranes as early as Late Triassic time based on clasts found in the Chinle Formation (Stewart et al. 1972). Continued continental sedimentation or erosion prevailed in the Grand Canyon region almost to the end of Mesozoic time, with the exception of the southward encroachment of the marine Middle Jurassic Carmel Formation. Although the Mesozoic section is almost completely eroded, upwards of 4000 feet (1200 m) of predominantly fluvial and eolian Mesozoic sediments were deposited in the region based on thicknesses and paleotrends in outcrops to the east and north.

The last marine transgression during the Mesozoic Era took place in Late Cretaceous time. The sea came from the northeast, and it resulted in the deposition of the Dakota Formation and Mancos Shale. Substantial volumes of older rocks had been eroded from the Grand Canyon region by this time, especially from the elevated area to the southwest. The surface buried by the marine deposits was very smooth and sloped gradually toward the northeast. Harshbarger et al. (1957) reported that the Dakota Formation lies unconformably on progressively older rocks toward the south in eastern Arizona. The unit rests directly on the Paleozoic section at McNary. However, it is doubtful if the sea extended to the southwest across the entire Grand Canyon region.

Neither discrete offsets along reactivated Grand Canyon basement faults nor development of local structural basins and uplifts during pre-Laramide Mesozoic time have been documented within the confines of the Grand Canyon. The primary reason is the paucity of Mesozoic strata in the area. Minor reactivation of basement faults undoubtedly occurred, particularly in response to east–southeast-directed compression associated with the Sevier orogeny along the northwestern margin of the Colorado Plateau during Cretaceous time. However, it cannot be proven here.

LARAMIDE OROGENY

Laramide orogenesis in western North America, herein used to embrace latest Cretaceous (Maastrichtian) through Eocene events, involved widespread uplift and a significant eastward expansion of the belt of cordilleran deformation beyond the previous limits of accretion, orogenesis, and arc magmatism. Laramide deformation more than doubled the surface area encompassed within the pre-Laramide cordillera and defined the cordillera as we know it in the United States (Fig. 14.1). Crustal contraction and eastward transport in a zone extending from the trenches along the west coast to the eastern limits of the Rocky Mountains characterized Laramide deformation. Types of deformation included east-verging thrust faulting and reverse displacements along reactivated Precambrian basement faults. The faulting was accompanied by the development of monoclines and anticlines in the covering sedimentary rocks. Arc magmatism swept eastward across the cordillera, then waned in intensity (Snyder et al. 1976).

The plate tectonic explanation currently favored for Laramide orogenesis was a flattening of the angle of subduction of the Pacific oceanic plate under North America caused by rapid rates of subduction (Fig. 14.3c). Dickinson (1981, p. 125) summarized the attributes of this concept as follows. (1) The belt of magmatism moved inland as the locus of melting near the top of the subducted slab shifted away from the subduction zone. (2) Magma generation waned as slab descent became subhorizontal because the slab no longer penetrated as deeply into the asthenosphere. (3) Shallower descent of the slab increased the degree of shear and the area of interaction between the descending slab and the overriding cratonic crust. As rapid subduction took place, the subducted hot, buoyant, oceanic plate appears to have underplated North America as far east as the Great Plains, thereby contributing to the uplift of the west. The area that was to become the Colorado Plateau was caught in the eastward compressing Laramide cordillera.

The Laramide orogeny profoundly impacted the Grand Canyon region. It caused: (1) widespread uplift, (2) east–northeast crustal shortening, (3) compartmentalization of the Colorado Plateau region into subsidiary uplifts and basins, and (4) widespread erosion. The result, shown on Figure 14.4, was the development of generally north-striking, east-dipping monoclines as the underlying basement failed in response to east–northeast contraction. Laramide monoclinal

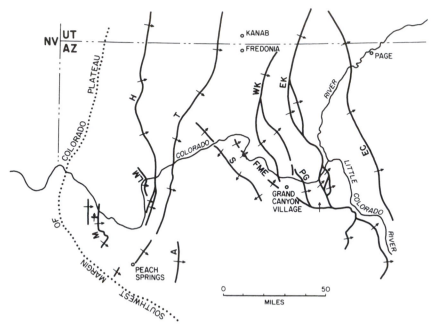

Figure 14.4. Locations of the Laramide monoclines in the Grand Canyon region, Arizona. From west to east: M, Meriwhitica; LM, Lone Mountain; H, Hurricane; T, Toroweap; A, Aubrey; S, Supai; FME, Fossil-Monument-Eremita; WK, West Kaibab; PG, Phantom-Grandview; EK, East Kaibab; EC, Echo Cliffs.

folding in the Grand Canyon region was accompanied by mild regional warping of the intervening structural blocks, a process that produced uplifts such as the Kaibab Plateau and downwarps such as Cataract basin.

Laramide erosion uncovered progressively older rocks to the south and west, including the Precambrian basement along the southwestern edge of the Colorado Plateau region. The enormous volume of detritus eroded from the Grand Canyon region and areas to the south was transported northeastward into the intracontinental basins of Utah and beyond. Areas in the Grand Canyon region that were monotonous lowlands at the close of Cretaceous deposition became an uplifted, dissected landscape characterized by north- and east-flowing, sediment-choked streams. Developing topography included the elevated Laramide plateaus and high parts of folds, as well as step-bench topography.

Chapin and Cather (1983) presented evidence for an Early Eocene northeastward reorientation of Laramide compressive stresses within the Colorado Plateau region. This caused 60 miles (100 km) of north–northeast translation of the Colorado Plateau along right-lateral, strike-slip faults that partially decoupled the Colorado Plateau from the North American continent along the future Rio Grande rift. The Early Eocene reorganization of stresses appears to have resulted in minor development, or reactivation, of northwest-trending monoclines in the Grand Canyon region. Northward crowding of the Colorado Plateau into the Wyoming Foreland Province was accommodated by crustal shortening manifested as thrust faulting and regional folding in Wyoming (Chapin 1983) . The result was the east- and southeast-trending basins and ranges in Wyoming. The southernmost of these was the east-trending Uinta uplift bounded both to the north and south by thrust faults dipping under the range. Bernaski (1985) summarized data which reveals that the Uinta uplift began to rise in Early Eocene time. Thus, Laramide structures outlined the eastern and northern boundaries of the Colorado Plateau. The Rio Grande rift along the eastern boundary became better defined and more strikingly decoupled from the North American craton as a result of extension beginning in Late Oligocene time.

Grand Canyon Monoclines

Most Laramide monoclines in the Grand Canyon region formed in the Paleozoic and Mesozoic sedimentary cover in response to reverse movements along favorably oriented faults in the Precambrian basement (Fig. 14.5). Three lines of evidence demonstrate that most faults under the monoclines were inherited from Precambrian time: (1) juxtaposition of crystalline rocks having different lithologies and fracture-foliation fabrics that cannot be restored by removal of Laramide offsets; (2) juxtaposition of Precambrian Supergroup strata that cannot be restored to prefault conditions by removal of Laramide offsets; and (3) presence of the Precambrian synorogenic Sixtymile Formation along the west side of the Butte fault in the eastern Grand Canyon.

The maximum offset across a Grand Canyon monocline is at least 2500 feet (750 m) along the East Kaibab monocline. The regional trends of the monoclines are generally north–south, but they are characterized by great sinuosity. The longest is the East Kaibab monocline, which is ~190 miles (~300 km). East–west spacings between the monoclines in the Grand Canyon region vary from 7 to 30 miles (11 to 50 km).

Total crustal shortening resulting from deformation within the monoclines was less than 1 percent across the region (Davis 1978). The reasons for this low percentage are as follows: Spacings between the monoclines are large in comparison to local shortening across them, and the dips of the underlying faults are steep.

Monocline Trends

The trends of the Laramide monoclines are sinuous, and they tend to branch (Fig. 14.4). Branching is well-developed along the East Kaibab monocline. Here, prominent northwest-trending branches such as the Phantom–Grandview segment splay from the main fold. The Hurricane monocline bifurcates southward into two parallel branches. Branching also yields en echelon patterns such as those observed between the various segments of the East and West Kaibab monoclines. Some monoclines are segmented with the intervening gaps exhibiting no discernible deformation. An example is the Fossil, Monument, and Eremita segments, which together constitute a detached western extension of a weakly developed branch of the East Kaibab monocline. Changes in trend and complicated branching are linked directly in outcrops on the floor of the Grand Canyon to Precambrian fault patterns which have been reactivated (Huntoon 1993).

Monocline Profile Geometry

Most segments of the Grand Canyon monoclines are developed in the Phanerozoic section over a single, high-angle reverse fault in the Precambrian basement

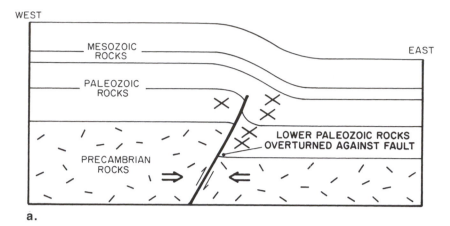

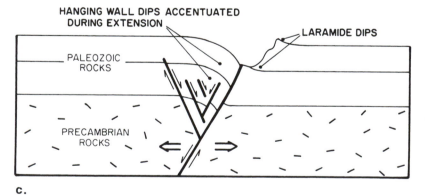

FIGURE 14.5. Stages in the development of a typical monocline-fault zone, Grand Canyon region, Arizona. (a) Laramide folding over reactivated Precambrian fault; Precambrian fault was normal. (b) Late Cenozoic normal faulting. (c) Late Cenozoic configuration after continued extension.

FIGURE 14.6. Precambrian Cardenas Lava (Yc) displaced down in normal fashion against older Dox Formation (Yd) along the west-dipping basement fault underlying the East Kaibab monocline as viewed northward from Tanner Rapid, eastern Grand Canyon, Arizona. The fault was reactivated in Laramide time with approximately 600 feet of reverse (left up) offset to cause monoclinal folding of overlying Paleozoic and younger rocks at this location. Yn, Nankoweap Formation, Yg, Galeros Formation.

(Fig. 14.6). Laramide displacements along the fault generally produced an abrupt offset at the top of the Precambrian basement. The dip of the fault typically is between 60 and 70 degrees. In profile, the anticlinal and synclinal axial surfaces in the monocline converge downward on, and terminate against, the underlying fault at or below the Precambrian–Cambrian contact. Consequently, the dips of the strata increase and the width of the fold decreases with depth in a monocline. The height to which the fault propagated into the overlying Paleozoic strata is proportional to the offset at the Precambrian–Cambrian contact. The dis-

FIGURE 14.7. Mechanical crowding within the Cambrian Tapeats Sandstone at the synclinal hinge of the East Kaibab monocline, south wall of Chuar Canyon, eastern Grand Canyon, Arizona. Notice almost vertical dips at this level in the fold. Arrow points to man for scale.

placement on the fault gradually attenuated with elevation largely through ductile deformation of the Paleozoic rocks so that it rarely extends above the top of the Pennsylvanian–Permian Supai Group. Deformation in close proximity to the coring fault includes: (1) minor horizontal shortening folds and kink bands in the footwall block, (2) highly localized drag folding adjacent to the fault surface, and (3) numerous conjugate sets of minor thrust faults.

Shortening across a monocline at all levels is equal to the heave of the Precambrian–Cambrian contact across the underlying reverse fault. Severe crowding developed in the basal Paleozoic rocks in the vicinity of the synclinal hinge (Fig. 14.7). These rocks commonly are riven with conjugate thrusts that mechanically thickened the strata in the syncline and operated to move material out of the syncline away from the fold. The basal Cambrian beds are commonly overturned and highly attenuated adjacent to the coring fault.

The thicknesses of the Cambrian through Pennsylvanian rocks between the anticlinal and synclinal axial surfaces in the East Kaibab monocline are attenuated between 30 and 60 percent (Fig. 14.8). This contrasts with comparatively gentle dips of less than 15 degrees with virtually no attenuation at the level of the Permian strata (Fig. 14.9). The Permian rocks occupying the anticlinal hinge are rarely thinned by brittle failure in the form of downward propagating grabens because of space-compensating horizontal shortening across the monocline.

The Precambrian–Paleozoic contact in the footwall block to the east of the East Kaibab monocline is broadly flexed for 3 to 5 miles (5 to 8 km) in the area immediately north of the Grand Canyon. This flexing adds 1000 feet (300 m) of structural relief to the fold where it is best developed.

FIGURE 14.8. East Kaibab monocline at Chuar Butte, eastern Grand Canyon, Arizona. Ductile thinning of Supai Group (IPPs) and Redwall Limestone (Mr) in center is obvious. Note increasing steepness of dips with depth. View is toward the north, Kwagunt Butte to left.

FIGURE 14.9. East Kaibab monocline north of the Grand Canyon, Arizona. Dips in the upper Paleozoic section are moderate at this stratigraphic level. Plateau surfaces are comprised of the Permian Kaibab Formation. Ptk, Kaibab and Toroweap Formations, Pc, Coconino Sandstone, Ph, Hermit Shale, IPPs, Supai Group. View is toward the north, Cock's Comb in the center.

Influence of Basement Strength on Monocline Profiles

The composition of the rocks below the Precambrian–Paleozoic contact is highly variable from west to east across the Grand Canyon (Fig. 14.2). The Paleozoic rocks were deposited directly on crystalline rocks to the west, whereas they are deposited on increasingly thicker sections of progressively younger Precambrian Supergroup rocks to the east. The strength of the unfaulted crystalline rocks tends to be reasonably isotropic. (Isotropy implies that the strength of the rocks was the same regardless of direction prior to failure.) However, the Supergroup sediments are highly anisotropic as a result of bedding, especially the thick Galeros and Kwagunt formations of the Chuar Group. These rocks are ductile, particularly when saturated. Consequently, the presence of the Supergroup sediments causes variations in monocline profiles.

Profile variations are most easily observed by the degree of folding of the Precambrian–Cambrian contact in the hanging wall block, as well as by the level within the fold where the anticlinal axial surface converges on the reactivated basement fault. The ideal monocline in the Grand Canyon region used here as a standard for comparison is one developed over a single reactivated fault that dips 60 degrees, and it is contained wholly within isotropic, rigid, crystalline rocks. Examples include the Hurricane, northern part of the Meriwhitica, and southern part of the Crazy Jug monoclines. Reactivation of the fault under the monocline produced a step-like offset at the Precambrian–Cambrian contact (Fig. 14.10a). Both the anticlinal and synclinal axial surfaces in the overlying fold converge downward on the intersection between the Precambrian–Cambrian contact and the fault surface on the respective sides of the structure. Thus, the Precambrian–Cambrian contact remains planar and the fold does not extend down into the Precambrian crystalline basement.

In contrast, the Precambrian–Cambrian contact in the hanging wall is folded down toward the reactivated fault in locations where Grand Canyon Supergroup strata are preserved in the hanging wall block (Fig. 14.10b). This occurs along the East Kaibab, Grandview–Phantom, and Crazy Jug monoclines. Dips of the contact in the hanging-wall block adjacent to the fault range up to 20 degrees. The degree of flexing and setback of the anticlinal hinge from the fault increase in proportion to the thickness of the underlying Supergroup section. This variant is a function of the considerably greater ductility of the Grand Canyon Supergroup sediments in contrast to the rigidity of the crystalline rocks.

The Precambrian–Cambrian contact in the footwall block remains essentially planar until it very closely abuts the reactivated fault regardless of whether or not sections of the Grand Canyon Supergroup are present in the footwall block. Consequently, the synclinal axial surface always converges on the intersection between the contact and the fault surface in the footwall block in the Grand Canyon monoclines.

Causative Stresses

Hubbert (1951) used basic mechanical principals and experimental results to demonstrate that faults tend to develop in intersecting conjugate sets (pairs of faults with parallel trends but opposing dips). His analysis assumes that the rock was brittle and reasonably isotropic with respect to structural strength prior to failure. The relationship between the orientations of the causative principal stresses and the angles of fracture is summarized as follows. The axis of the maximum principal stress bisects the acute angle formed by the intersecting shear planes; the axis of the minimum principal stress bisects the obtuse angle. In en-

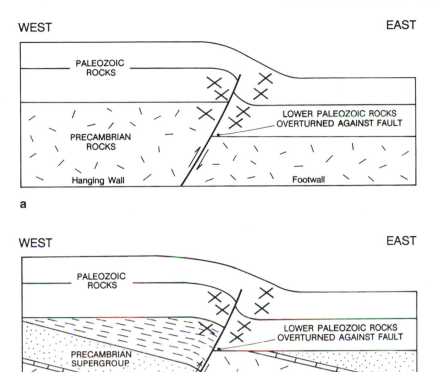

Figure 14.10. Idealized composite profiles of Grand Canyon monoclines contrasting those with and without the ductile Precambrian Grand Canyon Supergroup in the hanging wall of the underlying reactivated fault. Small crosses represent small scale conjugate thrust faults. (a) Precambrian crystalline rocks in hanging wall. (b) Precambrian sedimentary rocks in hanging wall.

vironments where vertical stresses dominate, the maximum principal stress is vertical, and the faults are predicted to dip between 60 and 62.5 degrees. In contrast, horizontal compression results in horizontal maximum principal stresses and the development of thrust faults that theoretically dip between 27.5 and 30 degrees.

The stress regime responsible for the development of the monoclines involved east–northeast-oriented horizontal maximum principal stresses and vertical minimum principal stresses. Orientations of the maximum principal stresses have been deduced from conjugate shear fractures in both Precambrian and lower Paleozoic rocks at numerous locations along the monoclines (Huntoon 1993). The conjugate shears occur at all scales from microscopic to mesoscopic, and they appear as intersecting second-order thrust faults. A larger-scale example is shown on Figure 14.11. These observations corroborate those of Reches (1978b), who used a variety of stress indicators to determine that the average orientation of the maximum principal stress was N67°E along the Palisades segment of the

FIGURE 14.11. Conjugate thrusts in the lower Paleozoic limestones east of the Lone Mountain monocline at Granite Park, western Grand Canyon, Arizona. View is toward the northeast.

East Kaibab monocline. His analysis utilized stress orientations deduced from the Paleozoic rocks from calcite twinning, minor faults, kink bands, and minor folds.

In contrast to the second-order thrusts, the basement failed along steeply dipping first-order Precambrian normal faults that were already in place to accommodate Laramide strain. What became the upthrown block after Laramide displacement had been the downdropped block in Precambrian time. The presence of these weaknesses rendered the rocks anisotropic, which destroys the relationship between the principal stress and fracture orientations predicated by Hubbert. Consequently, the dips of the reactivated faults do not reveal information about Laramide stress regime.

We had to find a monocline that did not develop over a preexisting basement fault if we were to use the dips of basement faults to deduce the causative stresses. Huntoon (1981) found a segment of the Meriwhitica monocline which satisfied this criterion. It developed over a six-mile-wide block of unfaulted crystalline basement lying between two reactivating Precambrian faults. The unbroken basement block had to fail along a new fault for the monocline to develop through the area. The block did fail, and the basement exposures lying west of Milkweed Canyon shown on Figure 14.12 reveal that the fault dips between 17 and 27 degrees as anticipated. East–northeast horizontal compression was confirmed as the cause.

A similar thrust fault in an embryonic stage of development underlies an incipient monocline in the south wall of Diamond Canyon 1.5 miles (2.4 km) east of the Hurricane fault. In this case, the southwest-dipping fault crosscuts well-developed north–northeast striking vertical foliation showing no regard for that preexisting fabric. Once again, horizontal compression is implied.

FIGURE 14.12. Laramide thrust fault (*arrows*) underlying the Meriwhitica monocline in a small canyon to west of the junction of Milkweed and Spencer canyons, western Grand Canyon, Arizona. Thrust fault developed in previously unfaulted Precambrian crystalline basement in response to horizontally oriented Laramide maximum principal stresses.

Davis and Tindall (1996) deduced that there is a component of right-lateral strike-slip motion along the basement fault underlying the northern part of the East Kaibab monocline. Their findings are based on the orientations and motions along minor faults in the Cretaceous rocks within the fold. They estimated that lateral slip was as much as three times the vertical offset at that location. This is consistent with the motion expected along a reactivated basement fault that was not oriented perpendicular to the maximum principal stresses.

Age of Monoclines

It is difficult to establish the timing for the inception of monoclinal folding in the Grand Canyon using stratigraphic evidence because the Late Cretaceous section has been eroded from the region. However, Late Cretaceous rocks containing unconformities are present in the southern high plateaus of Utah and elsewhere in the Rocky Mountain region, and these establish a Maastrichtian initiation for Laramide deformation (Anderson et al. 1975; Dickinson et al. 1987).

The presence of angular unconformities between the Late Cretaceous Kaiparowits, Campanian–Paleocene(?) Canaan Peak, Paleocene(?) Pine Hollow, and Paleocene–Eocene Wasatch formations described by Bowers (1972) reveals that folding was occurring in the southern Utah basins during Laramide time. The Grand Canyon region and areas to the south were undergoing concurrent uplift. Laramide erosion in the Grand Canyon area removed most of the remaining Mesozoic strata from the Permian surface east of the Aubrey monocline

(Fig. 14.1) and Paleozoic strata as low as the upper part of the Cambrian Muav Limestone to the west. The Meriwhitica, Hurricane, and southern segment of the Toroweap monoclines were beveled during this erosional event prior to deposition of Paleocene–Eocene rocks on the Laramide erosion surface. The beveling of the monoclines indicates that the monoclines were developing concurrently with the regional warping that produced the Late Cretaceous and Paleocene unconformities in Utah.

Young (1979) made a case for late-stage recurrent folding along the Meriwhitica monocline. He found Eocene(?) lacustrian limestones that are restricted to Laramide paleocanyons on the hanging-wall block upstream from the monocline. He concluded that renewed folding caused the axis to rise sufficiently to pond water in the channel. The ponded sediments represent the youngest episode of Laramide folding documented in the Grand Canyon region to date.

An analysis of fission track data collected from Grand Canyon rocks led Naeser et al. (1989) to conclude that Laramide uplift and monoclinal folding commenced about 60 million years ago followed by a second pulse of uplift beginning in Late Eocene time between 40 and 35 million years ago. These findings are consistent with the timing of tectonism deduced from the incomplete stratigraphic record.

EOCENE TO MIOCENE STABILITY

Rapid Laramide rates of oceanic crustal subduction in trenches along the west coast slowed about 45 million years ago, causing the descent angles for the oceanic slab to steepen under the North American plate (Dickinson 1981). The result was a return of arc magmatism to the southern cordillera and a significant relaxation of east–west compressive stresses across the region which had deformed during Laramide contraction. Deformation ceased in the Grand Canyon region as a consequence.

Tectonic quiescence at shallow crustal levels prevailed in the Grand Canyon region from about Late Eocene through Early Miocene time, longer than in the Basin-Range to the west and south. The southwestern part of the Colorado Plateau remained undifferentiated from the Mogollon Highlands to the south. Drainage was toward the northeast in valleys that originated in the highlands and terminated in basins on the plateau or possibly exited from the plateau to the north. The rapid, early Tertiary erosion that accompanied Laramide uplift waned. This left a regional northeast-sloping erosion surface that was beveled across successively older rocks toward the highlands.

Inherited Laramide Erosion Surface

The major structural uplifts and basins that currently characterize the Colorado Plateau around the Grand Canyon were in place at the close of Laramide time, including the Shivwits and Kaibab plateaus and the Cataract basin. Appreciable relief existed across these features, and erosion by streams had already deeply dissected canyons into the elevated areas and removed large quantities of Mesozoic and Paleozoic sediments from the surfaces. Step-bench topography was present, with the largest step being the southwest-facing Mogollon escarpment in the Permian strata. It trended northwest–southeast across the western Grand Canyon at a location not far south of its present slightly retreated position along the Shivwits Plateau and Aubrey Cliffs.

The Laramide erosion surface is still preserved over much of the Grand Canyon region behind the canyon rims. It occupies a position on successively younger strata from west to east because the Hurricane, Toroweap, and Aubrey monoclines progressively stepped the Paleozoic strata down to the east, so the Laramide surface was beveled across successively younger strata in that direction. Cambrian rocks formed the surface to the west, whereas Permian and Triassic rocks formed the surface to the east.

At Long Point, at the southern edge of Cataract basin, Young (1999) documented Paleocene–Eocene lacustrian deposits resting on the Kaibab Formation. These lake deposits indicate that the Laramide erosion surface was virtually flat at that location. He attributed the ponding to the development of Cataract basin adjacent to the rising Kaibab uplift during the last stages of Laramide deformation.

Laramide Drainage System

Remnants of the Laramide drainage system are preserved as hanging valleys west of the Toroweap monocline and south of the Colorado River (Young 1999). The meandering courses of the oldest paleovalleys in the vicinity of what is now the Hurricane fault reveal that gradients were gentle at the time they formed. Remnants of the Laramide erosion surface preserved under Miocene volcanics in the vicinity of the Meriwhitica monocline on the Hualapai Plateau indicate that the monocline was beveled to very low relief. Strata as low as the upper part of the Cambrian Muav Limestone were stripped from the monocline. Hindu Canyon, one of the paleovalleys, crossed the fold with little regard for its presence despite structural offset of about 1000 feet (300 m). The floors of the paleovalleys, as well as the oldest sediments preserved in them, probably date from Late Cretaceous(?) and Paleocene(?) time. Northward flow is revealed by channel slope, imbricated pebbles, and increasing Precambrian clast concentrations toward the south within the stream deposits.

The Music Mountain Formation, an early Tertiary arkose, filled the oldest Laramide valleys in the vicinity of Peach Springs canyon. It then spread over the adjacent plateaus. The unit is characterized by abundant Precambrian clasts, rarity of volcanic clasts, deeply weathered profiles, and a deep, red appearance in many outcrops (Koons 1948). The aggregation indicates that the climate was wet and that paleoslopes in the channels were extremely gentle to flat. The weathering reveals that the surface was long-lived.

The channel filling, followed by uplift during late Laramide time, first caused meander cutoffs and then deep incision of the Peach Springs paleovalley in the vicinity of the Hurricane fault zone (Fig. 14.13). The resulting canyon was cut over 1600 feet (500 m) below the Laramide surface. Unresolved is where the stream in it exited the region. One plausible scenario is that it turned eastward or merged with an eastward flowing river that occupied a Laramide canyon superimposed on the Uinkaret and Kaibab uplifts which discharged northeastward into the intercontinental basins of Utah and beyond. In this view, the inferred canyon was a precursor to the Grand Canyon that late in Cenozoic time was reoccupied and deepened into the modern Grand Canyon by the oppositely flowing Colorado River. Another popular scenario proposed by Young and Brennan (1974) is that the Peach Springs drainage continued northward between the Shivwits and Kaibab plateaus over what is now the north rim of the Grand Canyon. Their model requires that the northward tilt of the southwestern part of the Colorado Plateau had to be between 0.5 and 1 degree greater than at present. The additional tilt is needed for the steam to have cleared the present north rim dur-

FIGURE 14.13. Deeply incised Laramide paleovalleys in the Peach Springs Canyon area, western Grand Canyon, Arizona. The two meander loops (*small arrows*) are older and less deeply incised than the Peach Springs paleocanyon (*large arrows*), which is colinear with the Hurricane fault. Offset Early Tertiary rocks along the floor of Peach Springs Canyon reveal that Late Cenozoic normal faulting postdated incision of both channels. Arrows show paleocurrent directions. (U.S. Geological Survey photograph.)

ing the period of maximum canyon incision near the mouth of Peach Springs Canyon.

The onset of Oligocene–Miocene arc volcanism to the southwest of the Colorado Plateau was heralded by increased amounts of volcanic and pyroclastic fragments deposited in the upper part of the Eocene–Miocene Buck and Doe Conglomerate. The unit was depositing in the then-aggrading canyons on the Hualapai Plateau. It is overlain by basalt flows and the ~18.5-million-year-old Miocene Peach Springs Tuff (Young and Brennan 1974). The volcanic activity was a harbinger of extensional tectonism that would impact the region strongly.

LATE CENOZOIC EXTENSIONAL TECTONISM

Initiation of normal faulting in the Grand Canyon region in Middle(?) to Late Miocene time was the outgrowth of complex plate tectonic interactions along

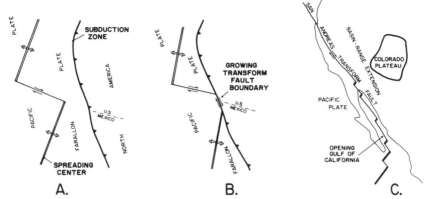

FIGURE 14.14. Stages in the Cenozoic collision of the East Pacific spreading center and Farallon subduction zone, western North America. (a) Spreading center and subduction zone converge during Early Tertiary time. (b) Spreading center begins to subduct near U.S.–Mexican border in Late Oligocene time, and interplate motion is accommodated along transform fault that grows north and south at continental margin as Farallon plate continues to subduct. (c) Remaining part of Farallon plate subducts between Late Oligocene and Late Miocene time, resulting in annihilation of both the spreading center and subduction zone; interplate motion accommodated by San Andreas transform fault beginning in Pliocene time and continuing to present.

the western continental margin that began in Late Oligocene time. Figure 14.14 illustrates the sequence of events. The description that follows is summarized from Dickinson (1981, p. 129).

The Pacific–Farallon ridge and the Farallon–American trench obliquely converged upon each other near the border of the United States and Mexico during Late Oligocene time. The result was annihilation of the ridge and trench at the point of contact, along with the mechanical substitution of right-lateral transform faulting along the area of contact. The transform fault progressively grew northward and southward as the trench-ridge system continued to converge. Transform faulting successively stepped inboard onto the continent during this period. The San Andreas fault emerged as the principal transform fault along the plate boundary near the beginning of Pliocene time. These events caused the cessation of subduction, eventual termination of arc magmatism, and widespread extensional tectonism in the region bordering the San Andreas fault. The result was wholesale extension within the Basin–Range during Late Oligocene through Middle Miocene time, along with east–west opening of the Rio Grande rift beginning in Late Oligocene time (Fig. 14.1).

The major phase of Basin–Range extension along the lower Colorado River area began in Late Oligocene time, considerably earlier than the first surface manifestations of normal faulting on the Colorado Plateau. Detachment faulting in the Whipple Mountains was characterized by east–northeast-dipping, low-angle faults in which the upper plate glided down slope to the northeast on an unextended lower plate (Davis et al. 1980). Similar extension also may have begun at middle and lower crustal levels under the southwestern part of the Colorado Plateau during this period.

Speculations regarding Late Oligocene–Late Miocene extension at deep crustal levels under the plateau and at shallower levels in the Basin–Range ap-

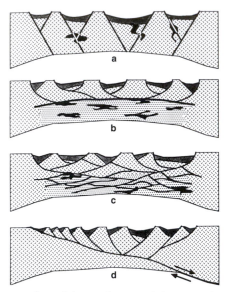

FIGURE 14.15. Summary of models used to explain extension in the Basin–Range Province west and southwest of the Colorado Plateau, Arizona (a) Classic horst-graben model. (b) Subhorizonal decoupling zone model. (c) Shear zone bounded lens model. (d) Crustal penetrating shear model. Either models c or d could have produced crustal thinning and subsidence along the southwestern Colorado Plateau in Late Oligocene–Early Miocene time. (From Allmendinger et al. 1987, Fig. 7.)

pear to be served best by invoking either a gently northeasterly dipping, crustal-penetrating normal fault or slip between shear-bounded lenses, respectively illustrated in Fig. 14.15c and 14.15d. Either allows for: (1) tectonic thinning of the crust under the western part of the plateau from about 25 to 19 miles (40 to 30 km), (2) slight down-to-the-southwest subsidence of the southwestern corner of the plateau, (3) structural differentiation of the plateau from the Basin–Range in Miocene time, and (4) progressive tectonic erosion of the southwestern and western edges of the plateau. Motion of the Colorado Plateau had to include a slight clockwise rotation of 2 to 4 degrees owing to the southward widening of the Rio Grande rift (Fig. 14.16). Extrusion of lower plate rocks from under the western part of the plateau possibly accompanied rotation. Thus the Colorado Plateau was rotating into extensional space created within the Basin–Range Province at all crustal levels.

Figure 14.17 illustrates that northwest-striking blocks calving off the thin, trailing edge of the plateau could account for tectonic erosion of its western and southwestern margins. The upper plate blocks now comprise the Hualapai, Cerbat, Juniper, Aquarius, and Peacock ranges. These northeasterly tilted mountain blocks probably rest in tectonic contact on lower plate rocks, such as observed in the Whipple Mountains.

By Miocene time, the upper part of the crust along the western margin of the Colorado Plateau was undergoing the first significant crustal extension to affect the surface since late Precambrian time. By about mid-Pliocene time, the Grand Canyon region began to experience east–west extension that first caused down-to-the-west normal faulting along many of the Laramide monoclines; and,

FIGURE 14.16. Clockwise rotation of the Colorado Plateau into extensional space created in the Basin–Range during Late Oligocene–Early Miocene time. Arrows in the Basin–Range schematically illustrate that all rocks within the province moved away from the Colorado Plateau at rates which increased to the west as the crust within the Basin–Range simultaneously extended.

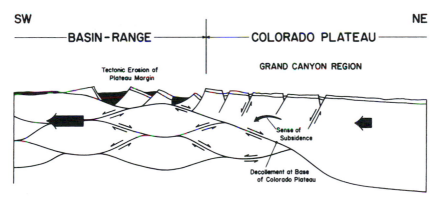

FIGURE 14.17. Cartoon illustrating tectonic erosion and subsidence along the southwestern edge of the Colorado Plateau over extending shear-bounded lenses in Late Oligocene–Early Miocene time. Lens concept from Hamilton (1982). Heavy arrows show absolute motions within the Basin–Range and Colorado Plateau provinces. Fine arrows show relative motions between shear surfaces. Notice that the motion between the lenses causes the crust to both thin and lengthen. Vertical scale greatly exaggerated, particularly at top.

Figure 14.18. View toward the northeast along the main strand of the Late Cenozoic Hurricane fault from the top of Diamond Peak, western Grand Canyon, Arizona. Arrows point to the Cambrian Tapeats Sandstone on the respective sides of the fault. The east (right)-dipping Laramide Hurricane monocline is not developed in this reach.

more recently, the onset of internal fragmentation of the intervening blocks by normal faulting coupled with the development of extensional basins.

Age of Onset of Faulting

Definitive timing for the inception of late Cenozoic normal faulting in the eastern Grand Canyon has eluded researchers because Tertiary sedimentary and volcanic rocks that could be used to bracket the onset of faulting are missing. More information is available from the western Grand Canyon, where Tertiary rocks such as those shown in Fig. 14.18 provide insights. Figure 14.19 summarizes what is now known.

The earliest Cenozoic normal faulting occurred along the listric west-dipping Grand Wash fault, which is the western boundary of the Colorado Plateau in Arizona. The maximum offset along the fault in the vicinity of the mouth of the Grand Canyon has been estimated by Lucchitta (1979, Fig. 13) to be down to the west 10,000 ft (3000 m). Downdropping, accompanied by down-to-the-east rotation of the western block, created the Grand Wash trough, which filled with the Miocene and younger clastic and evaporite rocks of the informally named Rocks of the Grand Wash Trough (Billingsley et al. in press). It appears that the initial faulting was synchronous with the latter part of the major phase of Basin-Range extension to the southwest, which Hamilton (1982) assigned to Late Oligocene through Late Miocene time. Young and Brennan (1974) observed that eruption of the ~18.5-million-year-old Peach Springs Tuff predated most of the offset along the southern part of the fault. They based their findings on slight pre-tuff, southwestern-facing scarp development along the fault, and the fact that

FIGURE 14.19. Two down-to-the-west, Late Cenozoic normal faults sever the east (left)-dipping Lone Mountain monocline south of Parashant Canyon (foreground), western Grand Canyon, Arizona. Faulted surface is the top of the Permian Esplanade Sandstone.

the tuff is tilted on ranges comprising the hanging wall blocks to the west that were later partially buried by the Rocks of the Grand Wash Trough. Therefore, major movement occurred along the fault after eruption of the tuff, but prior to deposition of the upper part of the Rocks of the Grand Wash Trough. These relationships bracket the major offset within Middle Miocene-Late Miocene time, findings consistent with those of Faulds et al. (1997) for the similar Red Lake basin, which is the next structural basin along the Grand Wash fault to the south.

Young minor displacements along the fault appear to account for some of the folding of the Miocene–Pliocene Hualapai Limestone, as well as for minor faulting of a gravel deposit located 3.5 miles (5.5 km) south of the Diamond Bar Ranch (Lucchitta 1967) and scarps in the alluvium north of Lake Mead in the vicinity of Grand Gulch and Squaw canyons (G.H. Billingsley, personal communication, 2000).

The Hurricane fault is the southern, waning extension of the Wasatch fault, which in Utah comprises the physiographic boundary between the Colorado Plateau and Basin-Range Province. The maximum offset across the Hurricane fault zone in the Grand Canyon is in excess of 2600 feet (800 m) at Three Springs Canyon.

Billingsley (2000) found that basalts dated at about 3.5 million years in the Mt. Trumbull–Bundyville area north of the Grand Canyon rest on the same erosion surface developed on Chinle strata on either side of the fault, allowing him to conclude that the offsets along the Hurricane and Toroweap faults are younger than 3.5 million years in the Grand Canyon region. The implications of this are profound because it reveals that the major Cenozoic faults observed in the western Grand Canyon largely, if not totally, postdate erosion of the canyon.

Based on the scanty data available in the Grand Canyon region, it appears that normal faulting is migrating eastward into the Colorado Plateau with time. The record for this is well-documented in south central Utah (Rowley et al. 1981). Jackson (1990) supports this view by making the case that activity along the younger faults, such as the Toroweap fault, is waxing whereas activity along the faults to the west is waning.

Figure 14.20. Two prominent Late Cenozoic normal faults offset the Permian Esplanade Sandstone down to the northeast. These faults comprise the southwestern boundary of a graben and occur four miles southeast of the mouth of Whitmore Canyon, western Grand Canyon, Arizona.

FIGURE 14.21. Quaternary basalt (right of cone) and alluvium displaced down to the west along the Toroweap fault in Prospect Valley south of the Colorado River, western Grand Canyon, Arizona. Notice that the offset of the basalt is greater than that of the alluvium. View is toward the southeast.

Normal Faulting Along Preexisting Monoclines

The strain resulting from Pliocene to recent east–west extension in the Grand Canyon region was first accommodated along reactivated Precambrian faults such as the Hurricane fault (Fig. 14.18). Normal faults severed the Laramide monoclines, displacing the west block down opposite to Laramide downfolding to the east (Fig. 14.19). This produced continuous faulting along most of the monoclines lying west of the East Kaibab monocline. Prominent local northwest, north, and northeast fault trends developed, reflecting local trends along the reactivated basement faults. The faults in the western Grand Canyon propagated laterally beyond the ends of the Laramide monoclines as extension continued, a situation that is well-expressed along the southern part of the Hurricane fault and various segments of the Toroweap fault.

New faults developed in the rocks adjacent to the reactivated basement faults as extension progressed (Huntoon et al. 1981). In some locations, the faults are arranged in two, or even three, intersecting sets and break the surface into a mosaic of fault-bounded blocks. Each of the sets contains conjugate faults that converge downward, thus producing numerous grabens (Fig. 14.20). The net effect is east–west lengthening as well as vertical thinning of the Paleozoic section, culminating in the development of extensional sag basins in the most densely faulted areas. The presence of numerous faults with different orientations allowed the Paleozoic rocks to deform without significant tilting of many of the fault-bounded blocks. This style of deformation has been treated theoretically by Reches (1978a) and is well-exposed along the Hurricane fault zone where a predominance of dip-slip displacement is revealed by near vertical slickenlines (Hamblin 1965).

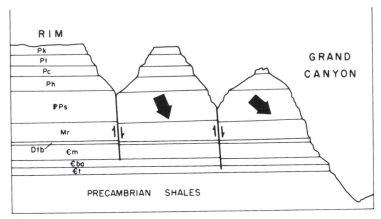

FIGURE 14.22. High-angle gravity faults develop across a ridge between deep canyons as the buttes settle into the ductile Bright Angel Shale, Grand Canyon, Arizona. Pk, Kaibab Formation; Pt, Toroweap Formation; Pc, Coconino Sandstone; Ph, Hermit Shale; IPPs, Supai Group; Mr, Redwall Limestone; Dtb, Temple Butte Formation; Cm, Muav Limestone; Cba, Bright Angel Shale; Ct, Tapeats Sandstone.

Subsidiary Faults and Extensional Basins

Continued extension has resulted in the emplacement of several laterally extensive faults internal to the blocks lying between the Laramide monoclines. Some of these, such as the Bright Angel fault in the eastern Grand Canyon, were produced by reactivation of Precambrian faults that had not been reactivated during Laramide compression (Huntoon and Sears 1975). Others undoubtedly are the result of new faulting within the extending basement rocks.

Basins caused by densely spaced intersecting faults have developed in areas where extension has been particularly great. Some occur along the major reactivated fault zones such as the area between the Toroweap and Hurricane faults along the Colorado River, as well as along the Hurricane fault zone in the area centered on Parashant Canyon (Huntoon et al. 1981). Closely spaced faulting in the latter area has progressed sufficiently to create a 10-mile (16-km)-diameter structural basin with approximately 600 feet (180 m) of structural closure. Evidence that development of these basins is geologically young is the fact that incision of the canyon has not kept pace with subsidence so the Colorado River does not flow on bedrock across them, but rather on late Tertiary and younger sediments.

Two notable extensional basins occur interior to the blocks between the reactivate Laramide fault zones: (1) the 320 square miles or 320 miles^2 (820 square kilometers or 820 km^2) north-trending Blue Springs basin with about 1000 feet (300 m) of closure centered around Blue Springs in the Little Colorado Canyon (Cooley et al. 1969, Plate 1) and (2) the more than 600 square miles or 600 miles^2 (1500 square kilometers or 1500 km^2) Markham fault zone in Cataract basin. The high-angle normal faults allowing for subsidence in the Blue Springs area are exceptionally well-exposed in three dimensions in the Little Colorado gorge.

The predominately northwest trending faults in the Markam fault zone are intersected by a few northeast-trending grabens. In addition, a cluster of northwest-trending normal faults, most down to the west, also occurs at Rose Well Camp, 10 miles (16 km) to the west of the Markham fault zone. The youthfulness of both the Markham and Rose Well faulting is revealed by closed topography and active sedimentation along the downthrown sides of many of the faults. The Kaibab bedrock along one of the faults crossing the northeastward flowing Rogers Wash in the Rose Well zone is offset up to the north, thereby damming the channel and producing somewhat more than 30 feet (10 m) of topographic closure at Sink Tank. The total offset across the fault could not be measured owing to burial of the downthrown southern block by modern sediments. However, neither recent sedimentation south of the fault nor incision of the resistant Kaibab Formation north of the fault has kept pace with subsidence of the downdropped block.

Increased Fault Densities with Depth

Fault densities in the Paleozoic section increase with depth within the major fault zones. Excellent examples of this can be observed in the Hurricane, Toroweap, and Eminence zones (Huntoon et al. 1981, 1996). This geometry implies that the subsidiary faults propagated upward from shallow levels within the upper part of the Precambrian basement or, in some cases, from the brittle lower Paleozoic carbonate section. The displacements along individual faults attenuate with elevation in the Paleozoic section through ductile deformation within the Pennsylvanian and Permian sediments.

Extensional Sag

Infolding toward faults of the Paleozoic strata in the hanging wall is a common feature along many large-displacement normal faults in the region. This phenomenon has been called reverse drag. It occurs along faults in which the dip of the fault surface decreases with depth. The cause is creation of space as the hanging-wall block pulls away from the footwall block. The space fills by infolding of the strata from the hanging wall. The phenomenon is well-developed along the Bright Angel fault at Grand Canyon Village. In this case, the dip of the fault surface is 70 degrees in the Paleozoic section, which contrasts with 60 degrees on the reactivated part of the fault in the Precambrian basement.

Similarly, the dips of the folded strata in the hanging wall of a faulted monocline become progressively accentuated as extension continues. Here also, the cause is a fault surface along which the dip diminishes with depth (Fig. 14.5C). Hamblin (1965) demonstrated that successive Cenozoic lava flows that cross the Hurricane monocline are each less steeply infolded than those below. It is clear, therefore, that the degree of infolding is proportional to the amount of displacement across the fault.

The distinction between a faulted monocline and a fault exhibiting reverse drag is that in the case of the faulted monoclines, the strata in the footwall are upturned toward the fault. If the dips in the hanging wall block are appreciably greater than those in the upturned footwall block, extensional sag has taken place along the faulted monocline.

Recurrent Faulting

The most spectacularly exposed and complete records of late Cenozoic faulting in the Grand Canyon region occur along the Hurricane and Toroweap faults. The resolution of the number and timing of Pliocene and Quaternary displacements is constrained only by the number of such young rock units deposited across the faults. Highlights from outcrops along the Hurricane fault near Whit-

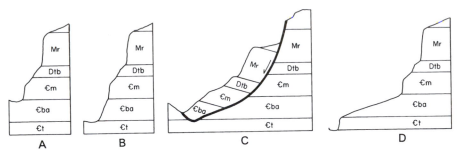

FIGURE 14.23. Stabilization of a Redwall cliff profile through detachment of a rotational slide, Grand Canyon, Arizona. (a) Channel floor has not exposed the ductile Bright Angel Shale so cliff profile is stable. (b) The channel has begun to cut into the Bright Angel Shale; and as it deepens, the shale slope is too steep to buttress the load of the cliff. (c) The rocks comprising the Redwall cliff fail. (d) Erosion of debris results in stable cliff profile. Mr, Redwall Limestone; Dtb, Temple Butte Formation; Cm, Muav Limestone; Cba, Bright Angel Shale; Ct, Tapeats Sandstone.

more Wash serve to illustrate the quality of the record. The Paleozoic rocks are offset approximately 1000 feet (300 m), whereas older Pleistocene basalt flows that fill an old channel of Whitmore Wash to a depth of about 600 feet (180 m) are displaced only 75 feet (23 m). A slightly eroded scarp in Whitmore Wash extends 8 miles (13 km) north from the Colorado River and offsets by 50 feet (15 m) a stage IV Pleistocene basalt flow that cascaded into the canyon from the east. The same scarp also displaces younger alluvium by only 15 feet (5 m). With only these four units, we are able to bracket a minimum of four faulting events. Fenton (1998), using cosmogenic dating, has expanded these findings and constrained the timing of specific increments of displacement along both the Hurricane and Toroweap faults.

Scarps in Pleistocene(?) alluvium along the Aubrey, Toroweap (Fig. 14.21), and Hurricane faults demonstrate Quaternary activity in the western Grand Canyon region. Some of this activity is Holocene (Jackson, 1990). Holm (1987) presents evidence for Quaternary faulting at the northeastern edge of the San Francisco volcanic field immediately to the southeast of the Grand Canyon. Although Holocene offsets have not been proven in that area, Akers et al. (1962) found a fresh scarp in a Quaternary gravel that they suspected might be Holocene in age.

Disruption of Drainage

The western margin of the Colorado Plateau is demarcated by the west-facing Grand Wash fault scarp from the area north of the Utah state line southward around the entire Hualapai Plateau to a terminus south of the Cottonwood Mountains. Displacement was sufficient to sever the northward flow of streams across that province boundary by late Early Miocene time. Young and Brennan (1974) place the timing of final drainage disruption at about the time the Peach Springs Tuff erupted. Their interpretation is based on the fact that sediments preserved above the tuff on the Hualapai Plateau do not contain clasts derived from the region south of the plateau. Severing of the northward flowing streams across the Mogollon escarpment took place as early as Oligocene time in the region south and east of the eastern Grand Canyon (Peirce et al. 1979). Two other processes operated to further disrupt these streams from crossing onto the Colorado Plateau: (1) extensional subsidence and fragmentation of the headlands in the

Basin and Range Province and (2) partial burial of the southern plateau margins by Miocene volcanics.

Concurrent subsidence of the southwestern margin of the Colorado Plateau allowed for eventual establishment of the west-flowing Colorado River across the region in earliest Pliocene time following deposition of the Miocene–Early Pliocene Hualapai Limestone in the Grand Wash trough. One popular model for establishment of the modern Colorado River through the Grand Canyon is headword erosion of a gully from the Grand Wash cliffs across the Kaibab Plateau where it captured the ancestral Colorado River (Hunt 1969; Lucchitta, Chapter 15, this volume). This mid-Tertiary river, for which no trace exists, was postulated to exit the plateau to the southeast of the Grand Canyon region or to have ponded in Miocene Lake Bidahochi playa in northeastern Arizona, prior to capture.

An emerging, alternative model is that as the late Cenozoic climate got wetter, the Colorado River prograded southwestward from its source in the Rocky Mountains by filling the basins before it with sediments and overtopping their lowest margin. A modern analogue is the recent downstream progradation of the Mohave River which has become integrated across several formerly isolated structural basins in eastern California, and which now discharges into Death Valley. Upon reaching the Grand Canyon region, the Colorado River reoccupied a Laramide canyon already superimposed across the Kaibab and Uinkaret plateaus, and it prograded westward through structural and topographic lows on the old Laramide erosion surface lying just south of the Laramide Mogollon escarpment in the western part of the canyon. Regardless of how the Colorado River got here, the first sediments from it did not reach the Grand Wash Trough until Pliocene time. It excavated the Grand Canyon to within 50 feet (15 m) of its present depth by early Pleistocene time (McKee et al. 1968).

Near-Surface Structural Localization of Volcanism

Basalts ranging in age from Late Miocene through Holocene erupted through vents on the plateaus immediately adjacent to the Grand Canyon. Best and Brimhall (1970) note the following relationships between volcanism and structure on the Uinkaret Plateau. (1) Normal faulting and volcanism have operated simultaneously throughout most of Late Cenozoic time, although the inception of faulting predates basaltic volcanism. (2) A shift in fault activity from the Grand Wash to the Hurricane–Toroweap zones has been paralleled in time by an eastward shift in volcanism. (3) Vents throughout the region lie between fault lines and are independent of them with but a few exceptions. (4) Many of the basalts carry mantle-derived peridotite inclusions, indicating that the magma originated from the upper mantle. Dutton (1882) and Koons (1945) observed the tendency for cones on the Uinkaret Plateau to align parallel to faults but to occur in the areas between them. For example, the Hancock Knolls are comprised of 11 centers in a 9-mile (14-km) alignment 6 miles (10 km) west of, and parallel to, the Supai monocline and fault. Three Pleistocene basalt plugs form a second parallel alignment in exposures on the Esplanade bench 2 miles west of the Supai monocline.

Deeply eroded outcrops of dike swarms on or below the Esplanade surface in the central and western Grand Canyon indicate that the dikes are localized along fractures and minor faults that parallel nearby normal faults. For example, a small dike swarm exposed in the inner gorge between two plugs in the alignment 2 miles west of the Supai monocline reveals that extensional fissures deep in the Paleozoic section served as feeders to the discontinuously spaced plugs higher in the section. Other examples include (a) dike swarms a mile west of the Hurricane fault and south of the mouth of Whitmore Canyon and (b) another swarm in the walls of Tincanebitts and Dry canyons. The fact that vents do not preferentially occur along the principal late Cenozoic faults indicates that, in general, upward movement of magma was independent of the faults at great depths. Rather, the dikes and vents tended to localize on extended fractures in the Paleozoic section in close proximity to the surface. The parallelism between the intruded fractures and nearby faults implies that late Cenozoic extension either created or opened the fractures.

MODERN GEOPHYSICAL SETTING

Although the Grand Canyon is situated on the western edge of the Colorado Plateau physiographic province, the modern geophysical properties of the crust under it appear to be transitional in character between those of the stable interior of the plateau and the extending Basin–Range Province to the west. For example, the thickness of the crust is 25 miles (40 km) or more under most of the plateau, but the crust tapers to about 19 miles (30 km) thick from the easternmost Grand Canyon to the Grand Wash fault (Smith 1978). The crust is only 15 miles (24 km) thick in southern Nevada and western Utah.

Heat flow through the plateau interior typically is 1.5 heat flow units (1 heat flow unit = 10^{-6} calorie/centimeter-second), but from east to west across the Grand Canyon it rises to two units (Blackwell 1978). Increased heat flow in the western Grand Canyon appears to be substantiated by a warming of groundwater discharged from springs west of Kanab Canyon. Temperatures of waters from springs in the lower Paleozoic section in the eastern Grand Canyon range from 50°F to 73°F, whereas those sampled from the same rocks to the west range from 64°F to 86°F (Loughlin and Huntoon 1983).

The intermountain seismic belt coincides with the crustal transition zone between the Basin–Range and Colorado Plateau provinces. The seismic belt is characterized by high seismicity and tectonic extension. Seismicity associated with it extends as far east as the West Kaibab fault zone (Smith and Sbar 1974). Fault plane solutions for earthquakes within the belt in southern Utah reveal that east–west extension is occurring and that focal depths are shallow, most at depths of less than 10 miles (16 km). The faults have steep dips, and motion on them tends to be near vertical. Smith and Sbar (1974) report that an earthquake on November 11, 1971, near the Hurricane fault north of Cedar City, Utah, produced three north-trending fractures in alluvium. The longest exhibited horizontal east–west extension and could be traced for half a mile. Focal depths during the quake were from near surface to slightly over a mile (1.6 km) deep.

Citing a combination of seismic refraction, reflection, low resistivity, and pressure wave velocity data, Keller et al. (1975) postulated and the presence of a mantle upwarp that occupies a band at least 50 miles (80 km) wide under the transition zone. The upwarp extends approximately 30 miles (50 km) eastward under the Colorado Plateau. Its eastern limit appears to correlate with a lateral change in crustal magnetization reported by Shuey et al. (1973).

The coincidence between the postulated mantle upwarp and the intermountain seismic belt led Keller et al. (1975) to speculate:

> The presence of the upper crustal low-velocity layer may be related to the mechanism of Cenozoic faulting and seismicity. Thus the presence of (a low-velocity layer) east of the Wasatch front could provide an explanation for the presence

of Cenozoic block faulting east of the province boundary. The shallow seismicity characteristic of the intermountain seismic belt also extends east of the Wasatch front and is roughly coincident with the easternmost zone of Cenozoic normal faulting.

Late Cenozoic basaltic volcanism also characterizes much of this same region, a finding that is consistent with extensional tectonics.

GRAVITY TECTONIC STRUCTURES

The extreme topographic relief of the Grand Canyon produces huge stress gradients within the canyon walls. The associated failure of the rocks yields valley anticlines, high-angle gravity faults, and rotational landslides that are unrelated to deep-seated processes. The rates of motion involved in the development of these structures is probably almost imperceptible in human terms. All classes of structures discussed here are actively forming someplace in the canyon now.

Valley Anticlines

A valley anticline whose axis parallels the Colorado River is present between Fishtail and Parashant canyons. Identical valley anticlines also occur in the principal tributary canyons to this sixty-mile reach including Kanab and Tuckup canyons. The coincident trends of the sinuous anticlines and the canyons reveal a genetic link between the two. Huntoon and Elston (1980) deduced that the anticlines formed by lateral flowage of the saturated shaly parts of the Cambrian Muav Limestone and underlying Bright Angel Shale from under the canyon walls. The floor of the canyon arches up in response to the compression across it. The huge lithostatic load under the 2100-foot (640-m) canyon walls drives the flowage (Sturgul and Grinshpan 1975).

The limbs of the typical valley anticline dip away from the canyon at angles of up to 60 degrees, and the folding extends over 800 feet (270 m) into the canyon walls. Numerous sets of minor, low-angle, conjugate thrust faults parallel the axis of the anticline in the Muav Limestone. The intersecting thrusts dip both toward and away from the river, and they are particularly numerous and have larger offsets where the Muav Limestone is steeply folded. However, they occur in beds dipping as little as two degrees. The thrusts serve to thicken and shorten the beds in the floor of the canyon perpendicular to the trend of the river.

Gravity Faults

High-angle gravity faults occur across narrow ridges between deep canyons in which the Bright Angel Shale is exposed. They represent brittle failure of the rocks overlying the shale as those rocks founder above the ductile shale (Fig. 14.22). The faults are steeply dipping to near vertical, and displacements die out with depth. They do not penetrate below the Bright Angel Shale.

This class of faults is important in the morphologic evolution of canyon scenery. Once formed, erosion proceeds along them, segmenting what was formerly a ridge between two canyons into a chain of buttes. As erosion continues, the buttes segregate and, ironically, the faults vanish.

FIGURE 14.24. Former channel of the Colorado River that was blocked by a Pliocene(?) rotational slide that fell from the south side of Cogswell Butte (right), Grand Canyon, Arizona. View is downstream.

Rotational Slides

Rotational slides are a most important factor in canyon widening. As shown in Fig. 14.23, they are massive blocks or rows of blocks that detach from the canyon wall and glide into the canyon. The detachment surface is an upward-facing concave normal fault. As the block glides toward the canyon, it rotates backward against the curved fault surface.

There are two common settings for rotational slides in the Grand Canyon. The largest failures involve the Redwall cliff inclusive of the Cambrian Bright Angel Shale upward through the Permian Esplanade Sandstone. This section is commonly 1600 feet (490 m) or more thick. Smaller rotational slides involve the 250- to 400-foot (75- to 120-m)-thick Cambrian Tapeats Sandstone that detaches above the ductile Precambrian Galeros Formation in the eastern Grand Canyon.

The cliffs fail because oversteepened slopes on the shales do not provide sufficient buttressing to support the lithostatic loads under the cliff when the Colorado River or its tributaries first incise into the shale. Huge tiers of blocks calve off the canyon wall until the slope on the shale is wide enough to support the rocks above it. Huntoon (1975) proposed that a stable canyon profile is attained through a series of catastrophic rotational slides shortly after the shale is exhumed. The unusually large setback of the north rim along the east side of the Walhalla Plateau in the eastern Grand Canyon was caused by progressive calv-

FIGURE 14.25. Row of Pliocene(?) rotational slide blocks that detached from the Red-wall cliff, Surprise Valley, Grand Canyon, Arizona. A second, closer row is buried by alluvium in the center of the valley.

ing of the rocks above the Galeros Formation. Those cliffs have retreated almost to the western margin of the Galeros substrate.

Spectacular examples of geologically young rotational slides involving failure of the Redwall cliff line the Colorado River between Deer and Fishtail canyons. A tier of huge blocks calved off the north wall and completely filled a one-mile reach of the Colorado channel. The blockage displaced the river 0.3 miles (0.5 km) to the south where it cut a parallel channel. The elevation of the blocked channel is the same as the modern channel attesting to the youth of the slide. Sediments ponded in the temporary lake behind the slide are preserved in thick deposits on the south side of the river two miles upstream from Deer Canyon.

A much older rotational slide off the south side of Cogswell Butte blocked the Colorado River halfway between Deer and Tapeats canyons. The buried channel there lies 210 feet (70 m) above the modern river (Fig. 14.24). The river carved the narrows to the south as it bypassed this blockage. Similarly, an even older slide from the south wall preserves a remnant of the Colorado channel lying 540 feet (175 m) above current river level upstream from Tapeats creek.

The courses of tributaries were also buried and displaced by equally large rotational slides. Paleochannels at the mouth of Deer Canyon record that the

FIGURE 14.26. Carbon Butte, the small left-dipping butte in the center, detached from the ridge to the left and glided one mile down a valley tailing in its wake blocks of Tapeats Sandstone torn from its base, eastern Grand Canyon, Arizona. View is toward the northeast, with Chuar Butte in the background.

stream was first displaced west by a slide off the west side of Cogswell Butte, then east by a slide from the west side of Deer Canyon. The scenic narrow slot behind the falls at the mouth of the Deer Canyon which lies between the older paleochannels is the most recent displaced position of the creek.

The oldest slides occur in Surprise Valley, which contains two or more rows of huge rotational blocks (Fig. 14.25), some which displaced the mouth of Tapeats Canyon to the east. The buried paleochannel of Tapeats Canyon lies about 900 feet (290 m) above the Colorado River. The fact that this channel remnant lies one-fifth of the depth of the Grand Canyon above the modern canyon floor hints at the great antiquity of that slide. The paleochannel could date from Tertiary time.

Large, young rotational slides are common in the western Grand Canyon downstream from Whitmore Wash. The most notable is a 1.2-mile-long detachment from the west that blocked the Colorado River near 205-Mile Canyon. It displaced the river eastward along its entire length.

Carbon Butte is the farthest traveled rotational slide block in the Grand Canyon (Ford et al. 1970). The butte (Fig. 14.26) includes rocks from the Bright Angel Shale up through the Redwall Limestone, and it occupies a position between the east and west forks of Carbon Canyon in the eastern Grand Canyon. The butte detached from the Redwall cliff on a listric fault bottoming between the Cambrian Tapeats Sandstone and the underlying Kwagunt Formation. The mass slid southward, trailing behind it large chucks of Tapeats Sandstone torn from its base. The block came to rest one mile (1.6 km) and 1800 feet (550 m) below its starting point. The track it followed was a valley eroded on resistant Precambrian strata along the south-plunging axis of a Precambrian syncline.

POSTSCRIPT

Late Precambrian time in the Grand Canyon region was characterized by erosional beveling across an uplifted, block-faulted mountainous terrane produced by extensional tectonism. That same stress regime has been reimposed on the region. The outcome is fairly well assured. The plateaus surrounding the Grand Canyon will continue to fragment by extensional faulting. Erosion will continue to wash the elevated rocks to the sea. The canyon will gradually disappear. Someday the seas will return and deposit new rocks here. Perhaps Ecclesiastes (1:9) said it best: "The thing that hath been, it is that which shall be; and that which is done is that which shall be done: and there is no new thing under the sun."

• 15 •

HISTORY OF THE GRAND CANYON AND OF THE COLORADO RIVER IN ARIZONA

Ivo Lucchitta

INTRODUCTION

Many desert rivers, including the Nile River of Africa, the Tigris and Euphrates rivers of Asia, and the Colorado River of the western United States, obtain their waters from mountain ranges far removed from the warm and parched lands that border most of their courses. This combination of abundant water and warm climate has made parts of these rivers lifelines, allowing the cultivation of large areas that otherwise are desert.

The Tigris, Euphrates, and Nile led to the establishment of ancient civilizations such as those of Mesopotamia and Egypt—and probably to the development of settled urban living as we know it today. It is no coincidence that sites such as Ur, Babylon, Thebes, and Memphis were sited on the banks of these rivers.

The Colorado River cannot claim to be the cradle of civilization; its virtue, instead, is to support modern cities—such as Los Angeles, Las Vegas, and Phoenix—that could not flourish in its absence. Many of these cities are far removed from the Colorado, whose waters are brought to them by a system of artificial impoundments and aqueducts. The Colorado, therefore, has enabled us to carry out a great experiment, based on contemporary technology, in settling areas that are in themselves inhospitable.

Great demands are made on the waters of the Colorado River, yet the amount of available water is finite. This has resulted in a host of social, political, and engineering problems that have centered chiefly on how many reservoirs should be built and who should get the water. Many of these problems are fertile areas of endeavor for geologists concerned with the practical application of their science.

For other geologists, however, the river has been the source of quite different interests and controversies. These interests are theoretical in nature and have to do with the general problem of how rivers are born and develop. When and how did the Colorado River come into being? How do rivers like the Colorado evolve? When did canyon cutting and correlative uplift occur? How and why has the Colorado cut across the many belts of high ground astride its course? How quickly was the Grand Canyon cut? Might the answers to these questions give us an insight into how rivers in general establish their courses?

This chapter considers the history of the Colorado River and its Grand Canyon (Fig. 15.1). To understand what follows, it is important to remember two con-

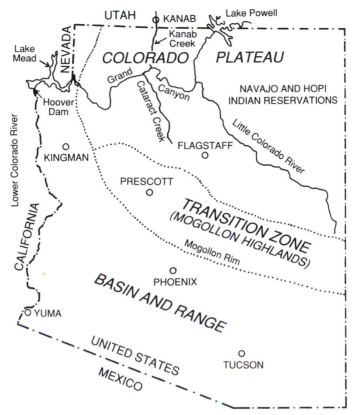

FIGURE 15.1. Location map, Arizona.

trasting views regarding the history of the Colorado River. The first is that the river from birth has been part of an integrated drainage system with a course approximating the present one.

According to the first view, the river evolved pretty much as it is today and at some well-defined time, such as the Eocene. A statement that is made about any part of the river applies to the river as a whole; the entire river is young or old, as the case may be.

The second view is that most rivers are continually changing entities that have evolved from various ancestors and will continue evolving into progeny whose configuration depends on factors such as tectonism and climate. According to this view, the answer to the question "When was the Colorado River born?" can only be another question: "How much departure from the present configuration is one willing to tolerate and still speak of the Colorado?"

It is important to remember that the Colorado traverses two contrasting terrains in Arizona. The first is the canyon country, typified by the Grand Canyon. This is highly dissected terrain, commonly with substantial topographic relief. The second is the plateau country, which is typified by most of the Navajo and Hopi reservations (Fig. 15.1). This landscape is characterized by low relief, wide mature valleys, and scarps that develop on beds of contrasting resistance and retreat down structural slopes. The plateau country is older and more widespread than the canyon county, which is encroaching on it.

For the first 60 years or so after John Wesley Powell's 1875 journey of discovery, geologists subscribed to the idea of a river with a simple history; it was born with the same course that it has now. The questions of paramount importance were: When was it born and when did the uplift of the region (which as considered responsible for the cutting of the canyons) occur? Because erosion of the plateau country is pervasive, these early geologists inferred that the erosion was also deep—the "Great Denudation" of Dutton (1882). Consequently the denudation, the canyon cutting, and the uplift ultimately responsible for both must have occurred a long time ago, presumably shortly after retreat of the great inland seas at the beginning of the Tertiary Period. According to this view, the Colorado River, the uplift, the canyon cutting, and the Great Denudation all began in Eocene time—and perhaps even earlier in the Tertiary.

The origin of the river and the Grand Canyon seemed safely established. Attention, therefore, was focused on geomorphic problems highlighted by the textbook-like character of the Grand Canyon region, where sparse vegetation and simple structure make it possible to see landforms clearly and to trace them for great distances. These characteristics led to the development of several concepts of fundamental importance in geomorphology, among which are the principles of antecedence, superposition, consequence, and anteposition, all having to do with the relations between drainage systems, structure, and topography (Davis 1901, 1903; Babenroth and Strahler 1945; Strahler 1948).

Storm warnings signaling danger for the view that the Colorado River is old were hoisted in the 1930s and 1940s by geologists studying the Basin and Range country (Fig. 15.1). These geologists found that interior-basin deposits of late Miocene and Pliocene age are common along the course of the Colorado River. They also could find no evidence for an older drainage system that could be called the Colorado. In conformity with the concept of a monophase history for the river, these geologists concluded that the entire river, and thus the Grand Canyon as well, was no older than late Tertiary (Blackwelder 1934; Longwell 1936, 1946).

The next development occurred in the plateau country of Arizona, Utah, and Colorado. Here, widespread evidence, eventually summarized by Hunt (1969), showed that drainage systems, locally departing from the present course of the Colorado River but arguably ancestral to it, existed certainly in the Miocene and very probably as early as the Oligocene. They might have existed even earlier, but if so, the evidence is gone. There was now a major paradox: The same river seemed to be at least as old as Miocene–Oligocene in its upper reaches, but no older than latest Miocene or Pliocene in its lower ones.

In an attempt to shed light on the paradox, E. D. McKee and the Museum of Northern Arizona sponsored studies on critical areas at and near the mouth of the Grand Canyon. Results by Lucchitta (1966) and Young (1966) showed no stratigraphic or morphologic evidence of a through-flowing drainage system during the deposition of Miocene interior-basin materials related to Basin–Range deformation. Nor could the lack of evidence for through-flowing drainage be bypassed by looking elsewhere along the course of the lower Colorado River or the southwest margin of the Colorado Plateau in Arizona (Fig. 15.1). In this area, interior-basin deposits are ubiquitous, and the deposits older than Basin–Range rifting indicate drainage northeastward, from what now is the Basin and Range Province onto what now is the Colorado Plateau.

The northeast drainage existed as recently as the emplacement of the Peach Springs Tuff, as 18-million-year-old ignimbrite that flowed onto the Colorado Plateau. Before rifting of the Basin–Range, therefore, drainage was not to the west or southwest (as would be required for a river with a course similar to that of the present Colorado) but in the opposite direction—to the northeast.

The next step in the conceptual journey was the idea of a polyphase history for the river. Hunt (1969) contributed to it by proposing drainage systems initially departing markedly from the present Colorado River, but gradually evolving into this configuration. However, Hunt, as well as Lovejoy (1980), still postulated an ancestral Colorado river flowing westward from the Colorado Plateau even before Basin–Range deformation, a concept not supported by the evidence.

The concept of a polyphase history for the Colorado river was developed fully for the first time by McKee et al. (1967). These authors accepted the antiquity of the upper part of the drainage system, as documented by Hunt, but could not accept a continuation of this drainage westward through the Grand Canyon into the Basin and Range Province. Instead, they proposed that the ancestral Colorado followed its present course as far as the eastern end of the Grand Canyon, but then continued southeastward (not westward) along the course of the present Little Colorado (Fig. 15.1) and Rio Grande rivers into the Gulf of Mexico. In Pliocene time, a youthful stream, emptying into the newly formed Gulf of California and invigorated by a steep gradient, eroded headward and captured the sluggish ancestral river somewhere in the eastern Grand Canyon area. It was then that the river became established in its present course, and the carving of the Grand Canyon began.

The concept is pivotal because it introduces the idea (even though the point is not made explicitly) that drainage systems evolve continually—and do so chiefly through headward erosion and capture and in response to tectonic movements. During this process, the configuration and course of a drainage system may change so much that it becomes difficult and rather arbitrary to continue calling the ancestral drainage system by its present name.

Evidence accumulated since 1967 argues against drainage southeastward along the Little Colorado and Rio Grand rivers, as proposed by McKee et al. (1967). On the other hand, evidence has continued to grow that an ancient river could not have flowed through the western Grand Canyon region (Fig. 15.2) into the nearby Basin and Range Province (Lucchitta 1972, 1975; Young 1970; Young and Brennan 1974).

Analysis of deposits along the course of the lower Colorado River in the Basin and Range Province has confirmed that this part of the river is no older than latest Miocene. It also has shown that the capture of the ancestral Colorado River is documented by the appearance within river deposits (Imperial Formation of Miocene and Pliocene age) in California's Salton trough of coccoliths otherwise found only in the Cretaceous Mancos Shale of the Colorado Plateau (Lucchitta 1972).

An attempt to synthesize current information led Lucchitta (1975, 1984) to postulate that the ancestral Colorado did not flow to the southeast along the valley of the Little Colorado River, as proposed by McKee et al. (1967). Instead, the river cross the Kaibab Plateau along the present course of the Grand Canyon, then continued northwestward along a strike valley in the area of the Kanab, Uinkaret, or Shivwits plateaus (Fig. 15.2) to an as yet unknown destination. After the opening of the Gulf of California, this ancestral drainage was captured west of the Kaibab Plateau by the lower Colorado drainage. According to this concept, the upper part of the Grand Canyon in the Kaibab Plateau part of the canyon in this area and in all of the western Grand Canyon postdates the capture and was carved in a few million years, a process aided by nearly 0.6 miles (0.9 km) of regional uplift since the inception of the lower river (Lucchitta 1979).

This hypothesis is based on (1) the occurrence of gravels of probable river origin in the area of the Kanab, Uinkaret, and Shivwits plateaus and (2) the observation that northwest-trending drainages along strike valleys were common and persistent before canyon cutting, as evidenced by fossil valleys preserved

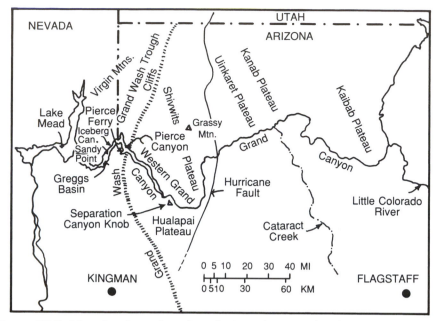

FIGURE 15.2. Map showing selected geologic and geographic features in northern Arizona.

under Miocene lavas in many places in the southwestern Colorado Plateau and by ancient valleys in the plateau country. Examples of such valleys are those of Cataract Creek (Fig. 15.2) and the Little Colorado River (Fig. 15.1), which predate canyon cutting and have not yet been affected appreciably by it.

The old problem of how the Colorado could have crossed the Kaibab Plateau can be analyzed by going backwards in time and restoring rocks removed from this upwarp in the past few million years (Fig. 15.3). This shows that a river such as the ancestral Colorado could have flowed readily across the Kaibab (then lower topographically than its surroundings) in an arcuate racetrack corresponding to the present configuration of the eastern Grand Canyon. The racetrack was localized by north-facing monoclinal flexures that cross the Kaibab and interrupt the general southward plunge of this dome.

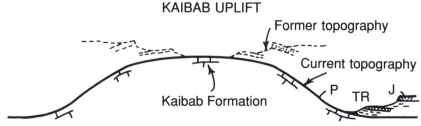

FIGURE 15.3. Cross section showing uplift of the Kaibab Plateau both a few million years ago and currently. P, Permian; TR, Triassic; J, Jurassic.

THE EVIDENCE

The Colorado River and its tributaries have been cutting down vigorously during the last several million years of their history. Such erosion—or at least non-deposition—may have been typical of this river system through most of its life. This characteristic creates great difficulties for the geologist intent on reconstructing the history of the river because such a history can only be pieced together from evidence left behind by the river.

The evidence is of two kinds: (1) river deposits such as gravel, sand, and silt and (2) landforms such as river valleys and canyons. Of the two, the deposits are by far the more useful because most can be attributed unequivocally to a specific river, on whose provenance, direction of flow, and age they provide valuable information. In contrast, a landform such as a valley can result from a river other than the one of interest or even from the action of an entirely different agent, such as a glacier.

Any river (even one that is cutting down) leaves behind deposits. For rivers that are cutting down, such as the Colorado, most of the deposits are removed soon after deposition. This means that deposits with which the geologist can work are few and scattered, especially in the Grand Canyon. Furthermore, the deposits preserved tend to reflect only the most recent part of the river's history.

Because of these factors, study of the Colorado River consists largely of a detective-like piecing together of circumstantial evidence, most of which is negative. In other words, the evidence is more likely to show that the Colorado River did not go through some area at a specified time than to document its existence in some specific place at a specific time. When circumstantial evidence is not negative, it typically attributes to an inferred Colorado River the properties known to be widespread at the time in question. For example, if at some time in the past most drainages followed northwest-trending strike valleys, one can reasonably infer that the Colorado also followed such a valley.

These points may seem too obvious to be worth repeating, but they need to be made once more because many people are taken aback by the lack of hard data pertaining to the history of the Colorado River. It is true that we do not have much direct evidence, given the problems mentioned above, but the circumstantial information is of many different kinds and from different places. Collectively taken, it enables us to construct a solid history that has a good chance of being correct, at least in its major aspects.

The history of the Colorado River and the Grand Canyon is best subdivided into three periods of tectonism that profoundly affected the drainage patterns of their time. These periods are:

1. *Pre-rifting*. The interval between the beginning and the middle of the Tertiary, at which time basin–range rifting got underway along the present course of the Colorado River, and the Colorado Plateau became distinct structurally and morphologically from the adjacent Basin and Range Province.
2. *Rifting*. The time of basin–range extension, with intensity tapering off toward the end of the interval—five to eight million years ago, at most. This was a time of widespread interior drainage in the Basin and Range Province.
3. *Post-rifting*. During this time—between five to eight million years ago and the present—rifting ceased, the Gulf of California opened, and through-flowing drainage became established.

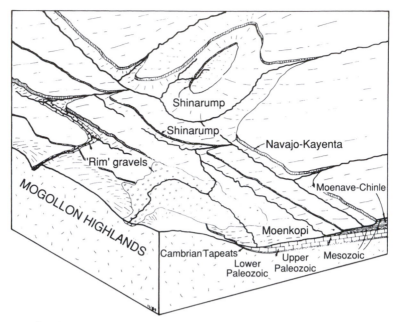

FIGURE 15.4. Block diagram showing schematically the relation of Mogollon High-lands to the Colorado Plateau before Miocene rifting. Diagram shows scarps com-posed of hard-over-soft couplets retreating down structural slope, the inferred topog-raphy on the Kaibab uplift, and the trellis drainage network. Looking about northwest, Hualapai Plateau would be at left side of diagram, the area of the Little Colorado River at the right side.

Pre-rifting

Before the onset of rifting in late Oligocene to Miocene time, the terrain south and southwest of the Colorado Plateau (Fig. 15.1) was higher than the nearby plateau, both topographically and structurally (Fig. 15.4). It remains high struc-turally today, even after the rifting. This belt of uplift, often referred to as the Mogollon Highlands, presumably was formed during the orogenic events at the end of the Mesozoic Era and the beginning of the Tertiary period. It caused the gentle northeastward tilting of strata near the southwest margin of the plateau, as witnessed by the widespread "rim gravels" (Finnell 1962, 1966; McKee and McKee 1972; Peirce 1984; Peirce et al. 1979; Peirce and Nations 1986). Along much of the rim, these gravels were deposited on rocks high in the Paleozoic section. They occur locally on Mesozoic rocks as well, yet they contain lasts of Precambrian igneous and metamorphic rocks that could have come only from the belt of uplift south of the rim, where such rocks were exposed.

Along the southwestern margin of the Colorado Plateau, in the area of the Hualapai Plateau, the rim gravels are present within ancient canyons trending northeastward, down the structural slope (Fig. 15.4). Clast provenance, imbrica-tion, and gradient of the canyon floors all indicate derivation from the south-west (Young 1966, 1970). A similar drainage direction is indicated by the 18-mil-lion-year-old Peach Springs Tuff, an ignimbrite that flowed northeastward onto what now is the Colorado Plateau from a source area in the Basin and Range Province (Young 1966; Young and Brennan 1974; Glazner et al. 1986). In the area southwest of the present plateau rim, erosion already had cut down to the Precambrian basement by Miocene time, and Phanerozoic strata were retreating northward from a structural and topographic high near Kingman (Lucchitta 1967, 1972).

At the southern margin of the Colorado Plateau, "rim gravels" are widely distributed (Fig. 15.4) along the Mogollon Rim (Fig. 15.1), which is the physiographic southern edge of the Colorado Plateau. Even though gravel-filled channels are not as prominent there as they are on the Hualapai Plateau, gravel-filled channels do occur, notably along the Little Colorado River. These gravels delineate the probable drainage pattern near the southern margin of the Colorado Plateau for much of Tertiary time (Lucchitta 1984).

For most of its course, the Little Colorado River flows in a mature and subdued valley that trends northwestward—parallel to the regional strike of Mesozoic and Paleozoic units (Figs. 15.1 and 15.5). The valley is at the erosional feather-edge of the Triassic Moenkopi Formation on the Permian Kaibab Formation. The northeast side of the valley is in the Moenkopi, capped by the resistant Shinarump member of the Triassic Chinle formation, which dips gently to the northeast and is very resistant to erosion. Because of this resistance, the topographic surface is near the top of the Kaibab over wide areas of the Colorado Plateau.

The hard-over-soft couplet represented by the Shinarump over the Moenkopi is the lowest, stratigraphically, of several such couplets within the Mesozoic section (Fig. 15.6). The scarps formed by the couplets face southwest and with time

FIGURE 15.5. High-altitude (U-2) aerial photograph looking north to northeast. The Little Colorado River is in the foreground. Dark mass in middle distance is Black Mesa. Wide valley of the Little Colorado River and scarps forming its northeast side are clearly visible.

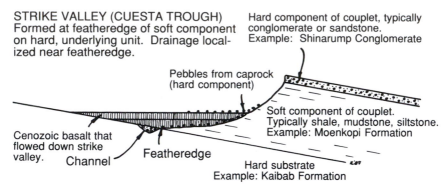

STRIKE VALLEY (CUESTA TROUGH)
Formed at featheredge of soft component
on hard, underlying unit. Drainage localized near featheredge.

Hard component of couplet, typically
conglomerate or sandstone.
Example: Shinarump Conglomerate

Pebbles from caprock
(hard component)

Soft component of couplet.
Typically shale, mudstone, siltstone.
Example: Moenkopi Formation

Cenozoic basalt that
flowed down strike
valley.   Channel   Featheredge

Hard substrate
Example: Kaibab Formation

FIGURE 15.6. Diagrammatic representation of typical strike valley formed at foot of
hard-over-soft scarp retreating (northeast) down structural slope. Also shown are the
channel located near featheredge of slope-forming unit, basalt that in many places
flowed down strike valleys, and debris derived from caprock. Looking northwest.
Width of valley typically is a few kilometers.

migrate northeastward down the structural slope. The couplets that are stratigraphically highest and youngest have migrated the farthest to the northeast.
Those that are lowest and oldest, on the other hand, are found to the south or
southwest. These couplets are closest to the belt of uplift (Mogollon Highlands)
from which they started. The Grand Staircase described by Dutton (1882) is composed of these couplets.

A northeastward migration can be documented for the valley of the Little
Colorado River, where the ancient gravel-filled channels are cut into the top of
the Kaibab Formation southwest of the present channel of the river. Both the
northwest-trending valleys at the foot of the Shinarump–Moenkopi couplet and
the northeastward migration of such valleys with time can be documented on
the Shivwits Plateau (Figs. 15.2 and 15.6), where various basaltic lavas of late
Miocene age have flowed down successive stands of the valleys (Lucchitta 1975).

Most tributaries of the Grand Canyon are short, steep, and immature (Fig.
15.7). The Little Colorado River, Cataract Creek, and Kanab Creek (Figs. 15.1, 7)
have the length and appearance of conventional, mature river valleys. The Little Colorado and the Cataract Creek flow northwestward—parallel to the strike
of beds; Kanab Creek also shows the influence of structure on its course by flowing south around the western flank of the Kaibab uplift, which has overprinted
and modified the regional northwestern strike.

These three streams illustrate features characteristic of the pre-Grand Canyon
drainage. One feature is the control of drainage by structure, which is represented by the gentle dip of beds modified locally by folds and faults. another is
the marked effect of lithology, as represented by couplets within the Mesozoic
section that differ in resistance to erosion and form strike valleys and scarps. Together, these features have given the region a trellis drainage pattern that consists of northwesterly segments parallel to strike and northeasterly segments trending down the structural slope.

The streams bringing rim gravels onto the Colorado Plateau and ancient
drainages such as the Little Colorado must have been functional for a substantial period of time because there is little evidence on the southwestern Colorado
Plateau of widespread and long-lasting Tertiary ponding in most pre-rifting time.
The conclusion is that the drainages must have been tributary to a master stream
that presumably also reflected the effects of structure and stratigraphy (Fig. 15.4).

FIGURE 15.7. High-altitude (U-2) photograph looking west, down much of Grand Canyon. Shivwits Plateau is visible in the distance on the right. The short, steep, and immature tributaries to the Grand Canyon are evident, in contrast to the well-developed Cataract Creek in the middle distance on the left. The confluence of Kanab Creek with the Colorado River is in the middle right of the picture.

The inference is that this was the ancestral Colorado River and that it flowed chiefly in northwest-trending strike valleys. The Colorado, however, did not flow through the western Grand Canyon region. Here, the presence of gravels and the Peach Springs Tuff shows instead that drainage near the edge of the Colorado Plateau was to the northeast.

Additional evidence that the river did not flow through the western Grand Canyon itself is provided by a knob on the south rim of the Grand Canyon near Separation Canyon (Fig. 15.2). The knob is composed of a mid-Miocene basalt flow that caps gravel and colluvium containing upper Paleozoic rocks (Young 1966; Lucchitta and Young 1986). In this area, such rocks crop out only on the Shivwits Plateau to the north. Today, the knob and the Shivwits Plateau are separated by the Grand Canyon, which could not have existed when the colluvium and basalts were emplaced (Young 1966).

Rifting

Mid-to-late Miocene basin–range rifting affected most of the western Cordillera and produced ranges separated by deep structural basins filled with thick sequences of deposits. The basins sank so rapidly that through-flowing drainage was not able to maintain itself, resulting in widespread interior drainage. In western Arizona, rifting began in mid-Miocene time and was essentially over by the end of that epoch.

The southwestern Colorado Plateau was much less affected by rifting than was the Basin and Range terrain to the west and south. Normal displacement occurred on many small faults and several larger ones, such as the Hurricane, but the intensity of faulting did not approach that in the Basin and Range. The most noteworthy effect of the rifting was the foundering of the formerly high areas south and southwest of the plateau. The foundering formed structural basins beyond the edge of the plateau and interrupted the old drainages that had deposited the rim gravels (Peirce et al. 1979; Young 1966; Young and Brennan 1974; Lucchitta 1967, 1979). Ponding and deposition of lake beds occurred locally on the plateau, presumably as a result of minor warping. The most conspicuous such lake beds are the Bidahochi Formation of latest Miocene and Pliocene age and the Willow Springs Formation of Pliocene to Pleistocene(?) age. Such deposits, however, are minor in comparison with the ubiquitous interior-basin deposits of the nearby Basin and Range Province.

Particularly interesting with respect to the history of the Colorado River is the Miocene Muddy Creek Formation, as exposed by the Grand Wash trough and the Pierce Ferry area (Figs. 15.2 and 15.8) at the mouth of the Grand Canyon. No evidence of a river emptying into the Grand Wash trough is present in the lithologies and facies distribution of the Muddy Creek, which, instead, records interior drainage with derivation of clastic material from nearby highlands. Analysis of directions of transport and topographic closure leads to the same conclusion.

West-draining canyons are present in the Grand Wash Cliffs, a prominent, west-facing fault scarp formed during basin-range deformation. The scarp marks the mouth of the Grand Canyon and defines the western edge of the Colorado Plateau (Figs. 15.2 and 15.8). The west-draining canyons are short and steep. Where they debouch into the Grand Wash trough, the washes that carved the canyons have deposited fans of locally derived material. One such fan emerges from Pierce Canyon, whose mouth is only 1.5 miles (2.5 km) north of the Grand Canyon (Figs. 15.2 and 15.8). The fan was deposited across the mouth of the present Grand Canyon. This could not have happened if the Grand Canyon and the Colorado River existed in their present location at the time.

The highest unit of the Muddy Creek Formation in the Grand Wash trough is the Hualapai Limestone, which interfingers downward with clastic rocks of the Muddy Creek. The Hualapai was deposited in a shallow lake, which extended about 25 miles (40 km) west from the Grand Wash Cliffs. As with other lithologies of the Muddy Creek, the Hualapai contains no evidence for a major river emptying into the lake in which the limestone precipitated. Instead, the lake waters were highly charged with calcium carbonate and other salts. Deposits younger than the Hualapai record through-flowing drainage.

The Hualapai was the youngest unit to be deposited in the area near the mouth of the Grand Canyon before the lower Colorado River was established. However, geologists have not yet dated it directly. The Muddy Creek Formation near Hoover Dam contains basalts that are five to six million years old (Anderson 1978; Damon et al. 1978). A tuff about 1600 feet (500 m) below the Hualapai Limestone in the Pierce Ferry area has been dated by fission-track methods at eight million years (Bohannon 1984). Another tuff low within the limestone in an area about 25 miles (40 km) west of the mouth of the Grand Canyon also has yielded an age of about eight million years (Blair 1978). South of Hoover

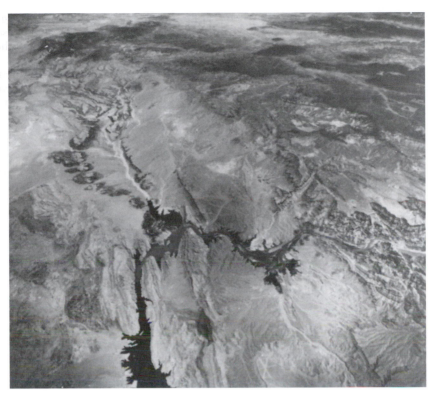

FIGURE 15.8. High-altitude (U-2) photograph looking about north from near mouth of the Grand Canyon (right foreground). Lake Mead is in foreground. North-trending valley in center of picture and west of the prominent Grand Wash cliffs is the Grand Wash Trough. Iceberg Canyon is the narrow, north-trending arm of the lake in the lower left. Pierce Ferry is at wide area of the lake in the lower center of the picture.

Dam (Fig. 15.1), the Bouse Formation (an estuarine deposit associated with the opening of the Gulf of California) records the presence of the Colorado River in its present lower course. The Bouse has been dated by the K–Ar method at 5.47 ± 0.2 million years (Damon et al. 1978). In the Pierce Ferry area, basalts intercalated with Colorado River gravels flowed down the valley of the Colorado when it was within 330 feet (100 m) of present grade (Lucchitta 1967, 1972). The flows have been dated by K–Ar methods at 3.8 million years (Damon et al. 1978). This evidence indicates that the end of interior-basin deposition and the establishment of through-flowing drainage along the lower Colorado River in its Basin and Range course occurred between four and six million years. No lower Colorado River existed before that date.

Evidence of various kinds from the western Grand Canyon region leads to similar conclusions. According to Lucchitta (1975), the Shivwits Plateau is capped for the most part by upper Miocene lavas that overlie uppermost Paleozoic and lowermost Mesozoic rocks. A long and narrow finger of the Plateau juts southward for about 20 miles (30 km) into the Grand Canyon (Fig. 15.7). This finger is surrounded on three sides by the Grand Canyon. Relief of nearly 1 mile (1.5 km) within a horizontal distance of a few miles between the top of the plateau and the Colorado River results in an extremely rugged topography of canyons, cliffs, and buttes. In contrast, the basalt capping the plateau, dated by the K–Ar method at 7.5 and 6.0 million years (Lucchitta and McKee 1975; Lucchitta 1975), overlies a remarkably flat and smooth surface with only a few meters of relief. In several places, cone-shaped vent area for the basalt are truncated by the cliff-forming edge of the Shivwits Plateau, yet there are no remnants of Shivwits lava within the Grand Canyon.

Lavas of the Shivwits Plateau locally overlie the erosional featheredge of the Triassic Moenkopi formation on the resistant Permian Kaibab Formation (Fig. 15.6). The featheredge formed northwest-trending strike valleys bounded on the northeast by a scarp capped by the Triassic Shinarump Member. This is a situation common on the Colorado Plateau. The Shinarump shed its pebbles into the valley in a way that preserved the pebbles both beneath and on top of the basalt (Lucchitta 1975). Today, the scarps are gone, and their place is taken by canyons tributary to the Grand Canyon (Figs. 15.6 and 15.9).

On Grassy Mountain, which is on the Shivwits Plateau 15 miles (25 km) northeast of the Grand Canyon (Fig. 15.2), a six-million-year-old basalt overlies remnants of gravels that contain pebbles of igneous and metamorphic rocks. The gravels also contain metamorphosed volcanic rocks similar to rocks of Protero-zoic age cropping out south of the present Colorado Plateau margin. The gravels rest on the Moenkopi Formation. The only reasonable source for the pebbles is to the south because in other directions, erosion has cut down only to Paleozoic and Mesozoic rocks. Neither the pebbles nor the arkosic matrix of the gravels shows much sign of weathering; this suggests that the gravels are not re-worked from older deposits. The lack of weathering within the gravels, together with the lack of a weathered zone between the gravels and the basalt, suggest that the two are of approximately the same age. The conclusion is that streams flowed northward across the present course of the western Grand Canyon as re-cently as six million years ago (Lucchitta 1975). These streams presumably were tributary to a larger stream that may well have been the ancestral Colorado River, but the course of this river would have been north or northwest of the western Grand Canyon. This interpretation is supported by other remnants of gravels that

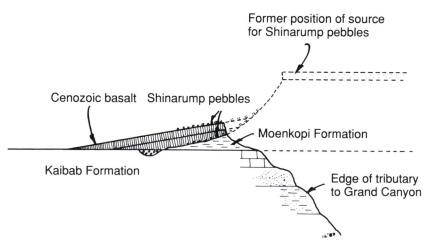

Figure 15.9. Diagram showing how scarp previously forming northeast side of Shivwits strike valleys occupied by basaltic lavas now is the site of a deep canyon tributary to the Grand Canyon. Looking approximately northwest.

are composed of exotic lithologies and are found in several places in the Arizona Strip country north of the Grand Canyon.

Collectively, the various features of the southwestern Colorado Plateau indicate that no rugged canyon topography and no Grand Canyon existed in the western Grand Canyon region as recently as six million years ago, the age of the young lavas in the southern Shivwits Plateau (Lucchitta 1975). The Colorado River must have become established in the western Grand Canyon after that date. It may, however, have flowed northwestward through the Strip country long before.

Post-Rifting

The evidence summarized above indicates that the transition from interior drainage to the through-flowing drainage of the lower Colorado River probably occurred between five and six million years ago, or at the Miocene–Pliocene boundary. Miocene, Pliocene, and Quaternary deposits of the Colorado River are distributed widely along its lower course. Included are estuarine deposits (upper Miocene and Pliocene Bouse Formation); probable deltaic deposits (Miocene and Pliocene Imperial chiefly reflecting downcutting; and fine-grained deposits (Pleistocene Chemehuevi Formation) produced by temporary aggradation. Information on these deposits has been summarized by Lucchitta (1972), who interpreted the data in light of the history of the lower Colorado River.

The Bouse Formation is particularly informative. In the Yuma area (Fig. 15.1), the Bouse is distributed widely in the subsurface even well away from the Colorado. Upstream along the river, however, the formation is restricted largely to the river valley. Fossils in the Bouse indicate a brackish estuarine environment that became progressively less salty northward along the river valley (Smith 1970). The Bouse rests on an erosional unconformity with deposits that reflect interior drainage and that locally interfinger upward with gravels of the Colorado River (Metzger 1968; oral communication, 1969).

The evidence indicates that the lower Colorado River became established early in Bouse time (latest Miocene) along its present course and that its prograding deltaic deposits progressively filled the estuary in the southward direction. Eventually, the delta reached the Salton trough of California, where it is represented by the thick Imperial Formation. This formation contains a well-defined horizon above which Cretaceous coccoliths found elsewhere only in the Mancos Shale of the Colorado Plateau make their appearance; the horizon signals the capture of the old, upper Colorado River by the developing and headward-eroding lower Colorado.

A 3.8-million-year-old basalt flow associated with indurated gravels of the Colorado River occurs about 330 feet (100 m) above present river grade at Sandy Point (Fig. 15.2) in the upper Lake Mead area (Longwell 1936; Lucchitta 1967). This basalt is part of the extensive basalts of the Grand Wash (Figs. 15.2 and 15.9) and flowed for several miles along the valley of the Colorado River, which must have been nearly as low then as it is now.

In the western Grand Canyon basaltic flows occur at river level for a long distance. These flows, which have their source in the area of the Uinkaret Plateau (Fig. 15.2), are the intracanyon lavas of McKee et al. (1967) and have been dated at about 1.2 million years. The lavas shows that at the time of their emplacement, the western Grand Canyon was a deep as it is today.

The upper Cenozoic deposits along the lower course of the Colorado thus show that the river came into being five to six million years ago when the Gulf of California opened up. The river extended itself southward, or downstream, by filling it estuary with deltaic deposits. It also extended itself by headward erosion and the integration of former interior drainages. By the time the delta reached the Salton trough area, the head of the river had captured the old upper Colorado River. By 3.8 million years ago, the river was essentially at its present grade in the upper Lake Mead area, and by one million years ago (at the latest), it was at its present grade in the western Grand Canyon. According to these figures, the western Grand Canyon was excavated during the interval between six million years ago (younger Shivwits lavas) and one million years ago (intra-canyon basalts). More likely, it began to be cut shortly before five million years ago as it is today. The rapid downcutting almost certainly was aided by the nearly 0.6 mile (0.9 km) of uplift experienced by the western edge of the Colorado Plateau and adjacent Basin and Range Terrain since the time when the Lower Colorado River was established in its present course (Lucchitta 1979).

Much of the course of the lower Colorado River in the western Grand Canyon and upper Lake Mead areas can be explained through structural or topographic features. Thus, the Virgin Canyon section (Fig. 15.2) probably is at the margin of an ancient fan spilling southward from the south Virgin Mountains; the Greggs Basin–Iceberg Canyon section (Figs. 15.2, 8) is along Wheeler fault; the Pierce Ferry section is through the lowest spot in the old Muddy Creek basin; the Hualapai Plateau section is along a strike valley developed at the foot of the upper Grand Wash cliffs (southwest-facing rim of the Shivwits Plateau); and the section along the east side of the Shivwits Plateau is developed near the Hurricane fault.

SUMMARY

A striking conclusion that emerges from the study of the Colorado River is that even a canyon as intricate and immense as the Grand can be carved in a surprisingly short time—five million years, probably substantially less, in spite of the tough and remarkably undeformed rocks that make up its walls. Perhaps the key to understanding this phenomenon is the fact that in canyons the volume of material removed per unit of downcutting is small compared with that for more open valleys. In other words, the rate at which material is carved from a canyon is small in relation to the rate at which the floor of the canyon is lowered (Lucchitta 1967).

Another and perhaps even more striking conclusion is that one can draw a parallel between the development of physical systems, such as drainage networks, and the Darwinian concept of biological evolution based on survival of the fittest and natural selection. In river systems, the external agent that triggers change is tectonism rather than the random mutations of biology. Competition between drainages occurs through changes in gradient, whereby rivers whose gradient is increased are favored, and those whose gradient is reduced are handicapped.

The battles for survival are fought with headward erosion and stream piracy or capture. The ultimate result is a succession of drainage configurations that change with time in response to external forces—chiefly, the deformation of earth's surface. Because any particular configuration is merely a still frame within the movie of evolution, ancestors may bear little resemblance to their descendants.

• 16 •

HYDRAULICS AND GEOMORPHOLOGY OF THE COLORADO RIVER IN THE GRAND CANYON

Susan Werner Kieffer

INTRODUCTION

The Colorado River and its tributaries drain much of the southwestern United States, ultimately emptying into the Gulf of California in Mexico. Only the Mississippi River exceeds the Colorado in length within the United States. The potential use of the water of the Colorado River for irrigation, hydroelectric power, and domestic purposes was recognized more than a century ago. In 1905, severe floods on the Colorado River caused extensive damage in the Imperial Valley of California, and political pressure arose for construction of flood control/storage dams on the river. The development and regulation of the river expanded rapidly until, at the present time, the Colorado River is sometimes referred to as "the world's most regulated river" (National Research Council 1987, p. 18). Additional information about the history of the river's development is summarized in Kieffer et al. (1989), and so further discussion in this chapter focuses on the portion of the Colorado River that lies mainly within Marble and Grand Canyons—that is, between Glen Canyon Dam and Lake Mead.

It is convenient to divide the history of the Colorado River within the Grand Canyon into three (very unequal) time spans: (1) the time of unregulated flow, prior to the finishing of Glen Canyon Dam in 1963; (2) the time of reservoir filling from 1963 to 1983; and (3) the current operational period, when Lake Powell is filled to a capacity that optimizes requirements for water storage and power generation. The hydraulics and sediment transport capacity of the river have been very different in each of these times.

The downcutting of the Grand Canyon was accomplished by a river capable of carrying large sediment loads during the time of unregulated flow. Monitoring at the U.S. Geological Survey's gage station near Phantom Ranch from 1921 to 1962 showed that typical average daily discharges were near 17,000 cubic feet per second (cfs). The mean annual flood was 77,500 cfs, but larger floods were not uncommon. For example, a flood of 300,000 cfs occurred in 1884. Sediment loads were large during this time; the average daily sediment load past the Phantom Ranch gage exceeded 300 tons per day.

During the reservoir-filling period, the average daily discharge was reduced to 11,000 cfs. Sediment that originates upstream of Lake Powell has been trapped in the lake since 1963, and so the average sediment load has been reduced to about 50 tons per day. Much of the sediment now available downstream from

Glen Canyon Dam is contributed by the two large tributary streams, the Paria (near Lees Ferry) and the Little Colorado River. Thus, the Colorado River (whose name in Spanish means "colored red" in reference to its former heavy sediment load) now often runs clear.

Between Lees Ferry (near the Arizona–Utah border) and Diamond Creek (about 220 miles downstream), the water surface of the Colorado River drops from 3116 feet in elevation (944 m) down to 1336 feet in elevation (405 m). The water surface does not maintain a constant gradient between these two elevations, but consists, instead, of a series of pools that are relatively flat and tranquil and rapids that are steep, fast, and turbulent. The bed of the river likewise is not uniform in gradient, but consists of relatively flat sections under the tranquil pools, debris dams (natural wiers) where tributaries enter, and scour holes. These characteristic configurations of the water surface and channel bottom have been created over thousands of years by hydraulic, hydrologic, and geomorphic interactions superimposed on the tectonic forces that drive the uplift of the Colorado Plateau.

The purpose of this chapter is to show how an understanding of hydraulic dynamics in individual stretches of the river near rapids can help us interpret the shape of the river channel and the geomorphic and hydrologic history of floods and erosion over the past thousands to perhaps hundreds of thousands of years.

A DESCRIPTION OF THE MAJOR FEATURES OF RAPIDS

The pool and rapid sequences of the Colorado River are very obvious in high-altitude photographs (Fig. 16.1a) because the rapids occur where the river is constricted by debris from side canyons, and these constrictions, as well as the white water of the rapids, are obvious in the photographs. Photographs taken at lower altitude (Fig. 16.1b) reveal that at the constrictions the water surface changes from a featureless pool into a white, wave-filled chute. At the bottom of the chute, the water squirts out of the constriction into the next quiet region like a jet from a fire hose. Adjacent to the jet are eddies and beaches, sites of some of the most critical ecologic zones and recreational beaches along the river.

These stretches of the river between one relatively tranquil pool (or mainstem segment) and the next are the famous rapids of the Colorado River. They

---

FIGURE 16.1.(a) (on opposite page.) High-altitude photograph (U.S. Geological Survey Series GS-VFDC-C, #6-134, taken June 24, 1982) of the Middle Granite Gorge of the Grand Canyon. River flows from bottom of photo toward top. Note that each rapid is associated with a constriction in the river at the base of a side canyon (or two side canyons in the case of Deubendorff Rapids; see text). The alternation of quiet pools with high-velocity rapids, visible in four places in this photographs, has been called the "pool and rapid" sequence of the Colorado River (Leopold 1969). (b) A lower altitude aerial photograph centered on Deubendorff Rapids, showing the transition of the river from a quite pool on the left into a foaming jet with standing waves in the constricted region, and the jet emerging like fire hose (labeled "tailwaves") into the next quiet region of the river. (Photograph by U.S. Bureau of Reclamation, 1984.)

(*continued*)

FIGURE 16.1. (*Continued*)

reveal many fundamental fluid mechanical phenomena on a grand scale: waves that stand still while water flows through them; zones of smooth, tranquil flow where large boulders protruding through the water cause hardly a ripple; zones of turbulent, aerated flow where the large boulders cause major hydraulic features; and even zones where the water flows backwards.

Common features of rapids are illustrated in Figures 16.2 and 16.3. In describing the rapids, we use a mixture of terms from hydraulics and from the vocabulary of those who have explored and described the river.

Rapids in the Grand Canyon typically form where a debris fan from a tributary canyon constricts the river. Above each rapid, the river is wide, relatively deep, and tranquil. In this chapter, the term "pool" is reserved specifically for these tranquil sections of the river immediately upstream of a rapid (Fig. 16.2a). At most discharges, a pool is a hydraulic backwater (a concept discussed later in more detail). Conceptually, a pool can be thought of as a pond formed by the debris-fan dam. At discharges between about 7000 and 30,000 cfs, water velocity is low in the pools, typically less than 1 foot per second (0.3 m/s). Water in the pools is deep, typically more than 15 feet (5 m) at the low end of this discharge range and more than 30 feet (9 m) at the higher end of this discharge range (Kieffer 1987).

At the downstream end of a pool, water accelerates gradually in the constriction (Figs. 16.2b, c; 16.3a, b, c), reaching velocities more than an order of magnitude higher than the velocities in the pool, even at discharges as low as 7000 cfs (Fig. 16.3d). Velocities of this magnitude have been found at most of the major rapids where velocity measurements were performed (they were not performed at Cremation–Bright Angel and 24.5-Mile Rapids). The highest veloc-

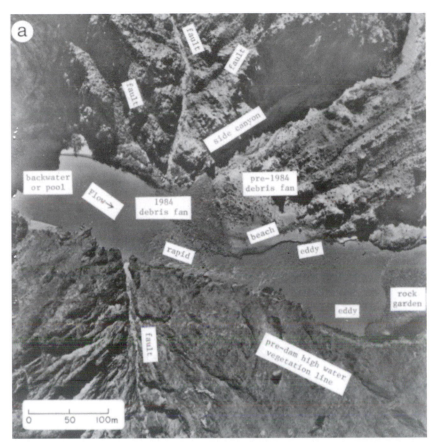

FIGURE 16.2. (a) Aerial photograph of Granite Rapids at a discharge of 5000 cfs, showing the geomorphic features common to many rapids. Terms defined in text. [Photograph by U.S. Bureau of Reclamation, 1984. Regional faults from Dolan et al. (1978).]

ities measured were approximately 33 feet per second (10 m/s) at Hermit Rapids (Kieffer 1988, Map 1897 F). In the converging portion of the channel (where rapids are found), standing waves (or laterals) bound a tongue of smooth, accelerating water, upon which may stand smooth, undulating, nonbreaking waves called rollers. In the constriction and the diverging portion of the river channel, crisscrossing lateral waves typically intersect to form high-amplitude breaking waves or haystacks.

Below each rapid is a zone in which the depth is intermediate between that of the shallow constriction and the deep pool. In this zone, the so-called "runout" of the rapid, water velocity still is relatively fast, typically 10 to 15 feet per second (3–5 m/s), but it decelerates toward the ambient conditions in the next pool, the so-called "tailwater" conditions in hydraulics. Strong vertical motions occur in the water of this region. The shear that results can give rise to turbulent boils with as much as 1 foot (0.3 m) of superelevation, indicating a vertical velocity of at least 8 feet per second (2.4 m/s) (Leopold 1969).

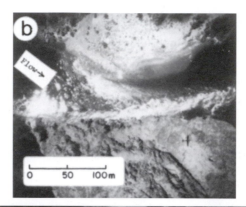

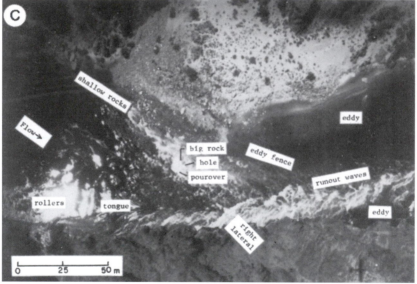

FIGURE 16.2. (*Continued*) (b and c) Aerial photographs of Granite Rapids at 30,000 cfs showing typical wave structures in rapids. (Photographs by U.S. National Park Service.)

Obstacles in the bed of the channel (such as rocks or bedrock protrusions) also cause a variety of wave patterns—including holes, curlers, rooster combs, and sculpted waves. A definition of these less commonly known terms follows (refer also to Fig. 16.2):

1. *Tongue:* Smooth water between the first two strong lateral waves (right and left) at the top of a rapid.
2. *Roller:* A wave that stands oblique, often perpendicular, to the current and breaks back onto the current; the term "nonbreaking roller" is used to indicate the smooth, rolling waves often found on the tongue.
3. *Lateral:* A wave standing oblique to the current near the top of a rapid, usually emanating from shore.
4. *V-wave:* The composite wave formed when opposing laterals intersect.

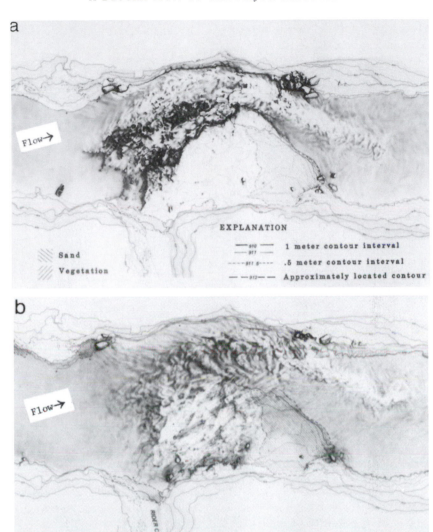

FIGURE 16.3. (a–d) Parts of preliminary hydraulic map I-1897-A, House Rock Rapids (Kieffer 1988). (a–d) Map views; contour intervals indicate with solid lines are 1 meter; dashed lines on beach in (a) are 0.5-m contours. (a) Topographic contours and standing waves at 5000 cfs; (b) the same at 30,000 cfs; (c) water-surface profile at 5000 cfs; (d) water-surface velocities and streamlines of floats at 5000 cfs. In (a)–(d), flow direction is from left. In (d), numerals indicate velocities (in m/s) along the streamlines between the adjacent dots. Trajectories of the floats were determined from analysis of movies taken from the camera station indicated. The boat, shown only for scale, is a standard commercial motor raft that is 10 m (33′) long. These maps are for schematic illustration of hydraulic features only and are not intended for navigation purposes.

(continued)

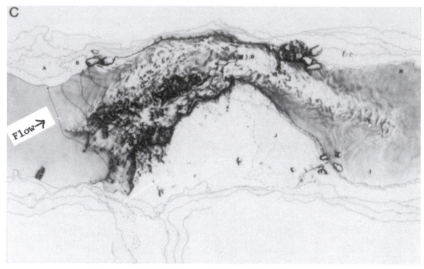

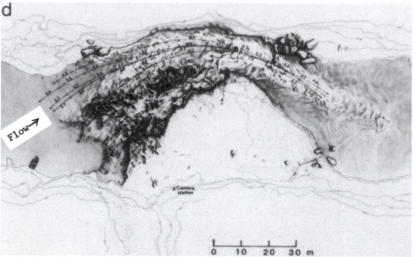

FIGURE 16.3. (*Continued*)

5. *Eddy fence:* The shear zone between two currents with different velocity magnitudes or directions.
6. *Pourover:* A zone where water "pours over" an obstacle, obtaining a large, downward component of velocity.
7. *Hole:* A trough in a standing wave, usually deep.
8. *Haystack:* A pyramidal wave (shaped like a haystack), usually breaking on top and sending spray in all directions.
9. *Rooster comb:* A haystack elongated in the downstream direction.
10. *Runout:* A zone of standing, generally nonbreaking (or weakly breaking) waves at the bottom of a rapid; more or less synonymous with "tailwaves."
11. *Rock garden:* An area of rocks in the channel within or downstream of the diverging section of a rapid.

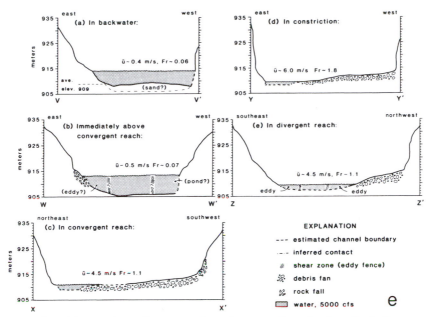

FIGURE 16.3. (e) Location of cross sections is approximately described above each cross section and view through cross section is downstream. Average water velocity is given by *u*, and Froude number is given by Fr.

To visualize some of these features and to understand the hydraulics of the rapids described and interpreted in the rest of this chapter, a reader may rely on, or create, intuitive mental background pictures of the rapids by several methods. An excellent description of the river, its rapids, and the eddies and whirlpools below then can be found in Beer (1988), an account of two men who swam the whole length of the river. Alternately, many features of the rapids can be seen on video (Kieffer 1986). If the reader has been to (or can hike to) Phantom Ranch via the Bright Angel or Kaibab trails, careful study of Bright Angel Rapids located at the bottom of these trails will be useful, as will study of the much smaller, but analogous, Bright Angel Creek. Other places where features of the flow that are discussed here can be seen are in small steep-gradient streams (particularly in the Pacific Northwest), in storm gutters along city streets when water is flowing, and in amusement parks that have water rides!

For the reader who has not been around any rapids, it also may be helpful to spend a few minutes at the kitchen sink—using the faucet to simulate the river and a cookie sheet, some curved dishes or pie pans, and a few utensils to create miniature rapids. Place a cookie sheet at an angle under the faucet, and turn on the water, watching the change in flow patterns as the discharge through the faucet is increased. Try confining the flow into a channel with a few straight-edged implements (yardsticks, knives, etc.), and then try making a converging–diverging flow by placing curved implements on the cookie sheet. As the flow changes characteristics, try placing the tip of a spoon in various flow regions to simulate an obstacle (like a rock) in a rapid, and note the different wave behaviors that can be obtained. This experiment has the potential to be messy, as well as instructive!

The interpretations of the hydraulics of the rapids in this chapter are based on a comparison of the rapids with slightly fancier and better-controlled versions of the kitchen-sink experiment—laboratory flumes. The comparison is only semi-quantitative because the Colorado River channel is much more complex than laboratory flumes, and we have relatively little data about the details of the channel shapes. In the river channel, wave patterns are irregular on a microscale because the complex and detailed channel geometry disturbs regular wave patterns (for example, individual large boulders change the local energy of the flow and create locally complex hydraulic patterns). The wave patterns also are irregular in time because discharge changes with time.

The use of laboratory flume data involves scaling problems of several orders of magnitude because laboratory flumes typically are of meters in dimension, whereas the Colorado River dimensions are one or two orders of magnitude larger. This is descriptively conveyed by Larry Stevens (1985, p. 25): "Let's put it this way; at 32,000 cfs, 1000 tons of water are moving through the river channel every second. If an average elephant weighs about 5 tons, this means that the flow of the river is equal to 200 elephants skipping by every second. A hole in the river may take up about a third of the channel, so the hydraulic dynamics in that hole are about the same as 67 elephants jumping up and down on your boat." In spite of the difference in scale between the Colorado River and a laboratory flume, we can learn a great deal about the hydraulics of the river by applying basic open-channel hydraulic principles. In turn, increased knowledge of the properties of the river and its interaction with the channel margins can help us better manage the river corridor (as in the current U.S. Bureau of Reclamation Glen Canyon Environmental Studies project), understand potential hazards for recreational boating, and interpret the geomorphology of the river channel.

LOCATION OF THE RAPIDS AND DESCRIPTION OF CHANNEL MORPHOLOGY

Many geologists have speculated on the origin of the pool–rapid–runout sequence (Leopold 1969; Dolan et al. 1978; Howard and Dolan 1981). Important factors in determining the origin and location of the rapids are gradient and variations of gradient along the river, spacing of the rapids, and the relation of the rapid spacing to the location of structural controls. We can divide the channel of the river into stretches that have different geomorphic and hydraulic characteristics (Howard and Dolan 1981). The frequency and, to some extent, the magnitude of the rapids depends on their location in these stretches. Typical stretches are as follows:

1. A wide valley with a freely meandering channel (for example, miles 67–70, near Tanner Rapids)
2. Valleys of intermediate width with tributary fan deposits (in these valleys, the river usually has cut into soft sandstones or limestones; for example, the few miles just downstream of the Little Colorado River near mile 61.5)
3. Narrow valleys in fractured igneous and metamorphic rocks (for example, "Granite Narrows" through miles 77–112; Fig. 16.1)
4. Narrow valleys of roughly uniform width and few constrictions in massive Muav Limestone (for example, miles 140–165)

5. Nearly flat stretches where the channel bottom is sandy (for example, miles 1–10 and parts of Marble Canyon)

The rapids occur almost exclusively where floods in tributary canyons, controlled by local or regional faulting or jointing, have delivered large boulders into the river channel (Dolan and Trimble 1978) (see the example of Fig. 16.2a, showing the faults and debris fan at Monument Creek where Granite Rapids is formed). Because the tributary canyons are much steeper than the Colorado River channel, floods in the tributaries can deliver boulders into the main channel that may be too large for even the large natural floods of the Colorado River to move [the discussions of Graf (1979, 1980) are relevant to this problem, though not based specifically on data from the Grand Canyon].

Debris fans from the tributary canyons can form on one or both banks of the Colorado River at the tributary junctions because the controlling faults cross the canyon (Howard and Dolan 1981). If meteorologic and drainage conditions are conducive, floods can occur in opposing tributaries, thus forming two debris fans. The relative size of the fans depends on the availability of movable material in the contributing drainages, on the frequency and magnitude of floods in each drainage, and on the tributary gradient (for a recent study, see Webb et al. 1987). It is, however, more common to have one of the debris fans be significantly larger than the other (see Figs. 16.1 and 16.2). The river usually erodes through the weaker parts of the debris fan, but the erosion may extend into the bedrock wall at the distal end of the debris fan if this material is easily eroded. This process can result in the formation of a pronounced change in the course of the river, and many rapids occur on local curves of the river (Fig. 16.1b).

In spite of the variations in the nature of bedrock along the course of the river discussed above and the structural control exerted by faults and joints on the location of the rapids, the river channel at the rapids is remarkably uniform in shape where it cuts through the tributary debris fans (Fig. 16.4). The unconstricted channel is about 300 feet (90 m) wide at 10,000 cfs (the width varies with discharge but this effect is not significant at the level of overall generality

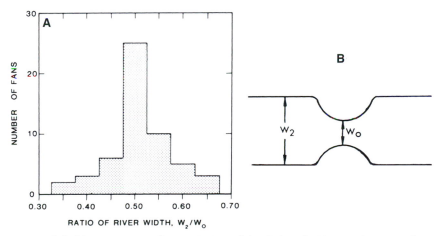

FIGURE 16.4. (a) Histogram of constriction of the Colorado River as it passes through 59 of the largest debris fans in the 248-mile (400-km) stretch below Lees Ferry (Kieffer 1985). (b) Idealized sketch of converging–diverging geometry of a rapid, showing widths W_0 and W_2 used to define the "constriction" of the rapid in (a).

considered in this chapter). At each rapid, the channel narrows appreciably, typically to about one-half of the unconstricted width. Figure 16.4 shows the ratio of surface width of the river immediately upstream of a rapid to the width in the narrowest part at 54 major debris fans; this ratio is termed the "constriction" in this chapter. The remarkable feature of the histogram of Figure 16.4 is the uniformity of this shape parameter—sharply peaked with nearly half of the rapids at the value 0.5. Why is the channel so uniform in shape at so many rapids, and why is this value specifically 0.5? The discussion of hydraulic–geomorphic interactions presented in this chapter will suggest an answer to this question.

The values of constriction plotted in Figure 16.4 were measured on 1973 air photographs (Fig. 16.2a is taken from this 1973 series of photographs). Because Glen Canyon Dam was operating under daily fluctuating flow conditions when the 1973 air photograph series was taken, the discharge at different rapids along the photograph series varied from about 7000 to 30,000 cfs. In the analysis that follows in this chapter, I need to use an average width for the river at places above, in, and below the rapids. This average width must be one that, together with an average depth and an average flow velocity, satisfies the requirement for conservation of discharge. A problem is that the surface width of the water measured from the photographs and shown in Figure 16.4 is not identical to the average width of an idealized channel. For example, in the histogram of Fig. 16.4, Crystal Rapids is the point at the furthest left, at a constriction of 0.33. However, idealization of the channel shape to a rectangular cross section suggests that the average channel constriction at Crystal was about 0.25 in 1973. This average value, rather than the value based on surface widths, is used in the hydraulic discussions in this chapter.

At this time, we cannot make a histogram like Figure 16.4 based on actual channel cross sections or on average widths because we lack detailed surveys of the river channel. However, the author has prepared a series of maps of the ten largest rapids along the river (House Rock, 24.5-Mile, Hance, Cremation–Bright Angel, Horn Creek, Granite, Hermit, Crystal, Deubendorff, and Lava Falls; Kieffer 1988). These maps have a topographic base sufficiently accurate to allow the value of constriction and its dependence on river stage to be determined more accurately. They also have additional data that allow visualization of hydraulic features in the rapids. Figure 16.3 is an example of parts of the map of House Rock Rapids.

Each hydraulic map contains the following: (a) topographic contours of the channel (Fig. 16.3a–d); (b) hydraulic information at two or more discharges (compare Figs. 16.3b and 16.3c); (c) water-surface elevations at different discharges, that is, rating curves and water surface profiles, shown implicitly by comparison of the shorelines in Figures 16.3b and 16.3c, and explicitly in figure 16.3b; (d) velocity and streamline data at one or two discharges (Fig. 16.3d); (e) channel cross sections, showing both the shape of the channel and the hydraulic conditions of the water (Fig. 16.3e); and (f) a detailed discussion of the characteristics of the rapid and interpretation of the data (not shown in Fig. 16.3).

Careful study of the channel shape in three dimensions reveals that at rapids the channel is constricted both laterally and vertically. In map view (for example, as in the air photographs of Figs. 16.2a and 16.2b, or on the topographic maps of the channel in Figs. 16.3a and 16.3b), the lateral constriction can be seen easily if the discharge is low enough that the river does not cover the debris fans. In vertical cross sections that are perpendicular to the flow direction (such as those of Fig. 16.3e) or those that are parallel to the flow directions (such as the fathometer tracings of Fig. 16.5), we can see that the channel of the Colorado River is constricted also by a vertical bulge in the bed, caused by the debris fan.

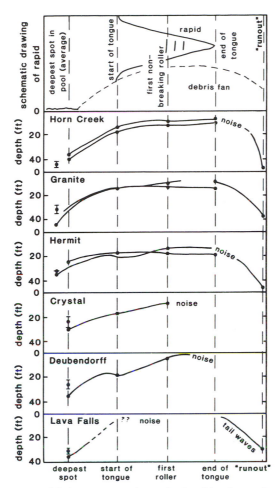

FIGURE 16.5. Summary of fathometer tracings obtained by the author, Julie Craf (U.S. Geological Survey, personal communication, 1986), and Owen Bayham (river guide, personal communication, 1986). No horizontal dimensions are implied. The top drawing is a schematic map view of a typical rapid showing the maximum depth measured in the upstream pool, the start of the tongue, the location of the first non-breaking rollers, the end of the tongue, and the runout of the rapid. Where known, discharges are indicated. Similar profiles occur in every rapid for which fathometer data were obtained.

Detailed studies of the rapids show that the constrictions typically begin underwater well upstream of the location of the first waves in the rapid. The water gets shallower and faster as the constriction gets tighter, and the most complex and largest waves in a rapid occur in the shallowest and narrowest part of the channel. The channel deepens again under the runout or tailwaves. Although the water surface drops several meters in most rapids, the elevation of the channel bottom is often the same upstream and downstream of the rapid to within experimental accuracy (~1 m). Thus, the rapids are not formed by sudden drops in channel bed elevation, but rather are a result of vertical and lateral constriction.

WHY ARE THERE WAVES IN RAPIDS?
TWO IMPORTANT HYDRAULIC
PARAMETERS AND THEIR
INTERPRETATION

In the previous section, I defined the nomenclature used to describe the rapids. In this and the next sections, I use hydraulic theory of open-channel flow to relate the descriptive terminology to hydraulic conditions in the rapids.

We can calculate two important hydraulic parameters in the pools, rapids, and runouts below the rapids from velocity and depth data presented above: the Reynolds number (Re $= uD/\mu$, where u is the average flow velocity, D is the average depth, and μ is the viscosity), and the Froude number (Fr $= u/(gD)^{1/2}$, where g is the acceleration of gravity). These dimensionless numbers indicate the state of the flow. The Reynolds number indicates whether the flow is laminar (Re $< 10^4$) or turbulent (Re $> 10^4$), and the Froude number indicates whether the flow is subcritical (Fr < 1), critical (Fr $= 1$), or supercritical (Fr > 1).

Under most flow conditions, the Reynolds number is greater than 10^5, indicating turbulent flow. For example, consider the Reynolds number of a backwater—where the flow looks tranquil and nonturbulent. If the discharge is greater than a few thousand cfs, the water velocity typically is greater than 0.3 feet per second (10 cm/s). The depth typically is greater than 3 feet (100 cm). Water viscosity is on the order of 0.01 cm²/s. Even for these apparently tranquil backwater conditions, the Reynolds number is greater than 10^5, and the flow thus is turbulent. Appearances are deceptive! In other parts of the river, u increases by up to an order of magnitude, and D can decrease by the same amount, but we generally find Reynolds numbers well in excess of 10^4. Such high Reynolds numbers imply that turbulent conditions exist nearly everywhere in the river. The turbulence has important implications for the mixing of sediment and nutrients. However, because the river is turbulent nearly everywhere, variations in the Reynolds number cannot explain the dramatic differences we find in the water structure of the pools, rapids, and runouts.

The Froude number is the dimensionless number that we need to look at to investigate differences in stability of waves and hydraulic regimes. It characterizes two states of flow that are dramatically different in energy balance and wave stability. The numerator of the Froude number depends on velocity—that is, on momentum or the square root of the kinetic energy. The denominator depends on potential energy—that is, on depth. The Froude number, therefore, is a measure of the relative importance of the kinetic and potential energies. The denominator also defines a characteristic velocity, called the critical velocity [$c = (gd)^{1/2}$], that depends only on water depth. The stability of standing waves depends on the ratio of the fluid velocity (numerator) to this characteristic velocity (denominator).

Dramatic changes in flow regime occur as the Froude number changes from less than one (subcritical) to greater than one (supercritical). These changes are similar to those that occur in transitions from subsonic to supersonic flow in gas dynamics, and this analogy may be useful for the reader in thinking about the standing waves in the rapids.

Changes in Froude number near a rapid are dramatic. In a typical backwater, $u \sim 1$ m/s and $D \sim 10$ m, so Fr ~ 0.1 or even less (a subcritical condition). In a rapid, on the other hand, $u > 5$ m/s, $D < 3$ m, so Fr ~ 1 or Fr > 1 (critical and supercritical conditions). Measurements show that Froude numbers exceeding 2 can occur in the rapids. Figure 16.3e shows Froude numbers for different parts of House Rock Rapids. Measurement of the Froude numbers in different parts of the rapids therefore suggests that the dramatic change in flow regime from backwater to rapid is caused by differences in the balance of kinetic and potential energy. These differences change the stability of waves in the channel. The general principles that apply to analysis of shallow-water flow in these different flow regimes comprise the classic theory of open-channel hydraulics (Bakhmeteff 1932; Ippen 1951; Ippen and Dawson 1951; Rouse et al. 1951; or Chow 1959).

At this point, the reader can develop an intuitive feeling for the significance of Froude numbers for different flow regimes and about the concept of standing waves by returning to the kitchen sink and putting a flat plate on the bottom of the sink under the faucet (this experiment is described in Thompson 1972, p. 525). As the faucet is turned on and the downgoing jet of water strikes the plate, water spreads laterally toward the edges. Depending on the rate of discharge from the faucet and the position of the plate, an inner ring of water that is moving rapidly outward in a radial direction surrounds the impact point where the jet hits the plate. If the reader has produced the correct experimental conditions, a ridge of water surrounds this inner ring a few inches radially outward from the jet. Beyond this ridge, the water is deeper and moves more slowly toward the edge of the plate. The ridge or ring where the water depth and velocity change is a standing wave that separates an inner region of supercritical flow from an outer region of subcritical flow. This standing wave, which is called a "hydraulic jump," is circularly symmetric about the descending jet of water. In a linear channel, however, hydraulic jumps commonly are linear, though they may stand perpendicular or oblique to the flow direction.

If you insert a probe, such as a finger, spoon, or small pebble, into the two different regions of flow on the plate, very different wave phenomena occur. In the inner, supercritical region, standing waves will be formed as "wakes" to the object, whereas, in the outer, subcritical region any waves formed by the object migrate upstream or downstream through the fluid. Thus, they are traveling, not standing, waves. Wave behavior in the critical region is highly unstable (by analogy, it is well known that wave instability makes maneuvering aircraft in the transonic regimen difficult, but that maneuvering becomes more stable under supersonic conditions). In all of these regimes, eddies caused by shear in the fluid can form downstream of an object immersed in the flow. These eddies must be distinguished from standing waves.

Hydraulic jumps are the basic standing waves in the Colorado River, and they occur in a variety of geometries. Large, oblique hydraulic jumps emanate from shore and are oriented downstream at an angle to the current (see Fig. 16.1b, or 16.2b and 16.2c). Normal hydraulic jumps stand perpendicular to the current. Finally, miscellaneous hydraulic jumps of various geometries stand around rocks and obstacles on the bed of the channel in regions of supercritical flow in the rapids.

The same change of conditions that gave rise to standing waves in the inner ring under the kitchen faucet and yet produced no standing waves in the outer, deeper water occur in the rapids of the Colorado River. The contrast in stability of standing waves between supercritical and subcritical flow explains why there are strong waves in rapids where the flow is supercritical, but there are no standing waves in backwaters where the flow is subcritical. Just as you can manipulate the strength, stability, and position of the circular hydraulic jump under the faucet by increasing or decreasing the flow rate through the faucet or by changing the position or angle of the plate, so the behavior of waves in the rapids depends on the flow rate of the Colorado River, the shape of the channel, and the gradient of the bed. Using these concepts, I now show how the different parts of a rapid and the waves can be semiquantitatively explained in terms of flume hydraulics.

THE PIECES AND PARTS OF A RAPID:
BACKWATERS
(POOLS)–RAPIDS–RUNOUTS

Backwaters (Pools)

Pools and backwaters form upstream from a rapid if changes in channel shape produce local conditions in which the given discharge cannot be accommodated in the channel cross section without a transition from subcritical to supercritical flow. To clarify this, consider a specific example of flow at 30,000 cfs. If the main channel is 100 feet wide (30 m) and 30 feet (9 m) deep, the flow per unit area is 10 ft/s (3 m/s); that is, the flow velocity required to accommodate the discharge is 10 ft/s. The flow is subcritical because the critical velocity is 17.1 ft/s (5.2 m/s) and the Froude number is 0.6. If the channel narrows to 30 feet (9 m) in width and maintains the same depth, the flow per unit width or flow velocity would have to increase to 33.3 ft/s (10 m/s) to accommodate the discharge. The flow would be supercritical with a Froude number of 1.9. Conservation of energy would not allow the fluid to accelerate in this simple way because of the change in flow regime. Instead, water would pond upstream of the rapid in a backwater to increase the depth and, therefore, potential energy of the flow, and the river would adjust itself so that the Froude number would just equal unity in the constriction.

In effect, the cross-sectional area of the whole backwater–rapid system is increased by ponding in the backwater or "pool" above a rapid (see the cross sections in Fig. 16.3e). The potential energy of the deepened pool is available for conversion to velocity in the constriction (compare depth and velocity in cross sections (a) and (b) with those in (c) and (d) in Fig. 16.3e). Within the backwater itself, the increased depth caused by the constriction reduces the velocity compared to that in an unconstricted channel of the same diameter (that is, compared to a normal main-channel flow that enters the backwater). As passengers on a raft float from the main-channel current into a backwater above the rapid, they often lose the sense of "floating" and may feel the need to row across the tranquil pond—especially if there is any upstream wind to halt all progress! For example, the backwater above Crystal Rapids, which extends approximately a mile back to Boucher Rapids, is known affectionately by river runners as "Lake Crystal." Velocities of only a few tenths of a foot per second and Froude numbers as low as 0.01 can occur in the backwaters, indicating conditions dominated by potential energy.

The Converging Section of Rapids: Tongues,
Nonbreaking Rollers, and Oblique Lateral Waves

In this section, I show how features in laboratory flumes of relatively simple geometry and well-controlled discharges (Figs. 16.6, 16.7, and 16.8) relate to the more complex and variable patterns of waves in the rapids of the Colorado River (Figs. 16.9 and 16.10).

A simplified illustration of hydraulic features in flumes with converging-diverging geometries is given in Figure 16.6. The top part of Figure 16.6a is a

(a) Subcritical conditions:

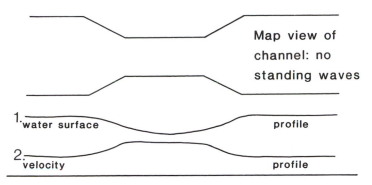

Map view of channel: no standing waves

1. water surface   profile

2. velocity   profile

(b) Supercritical conditions:

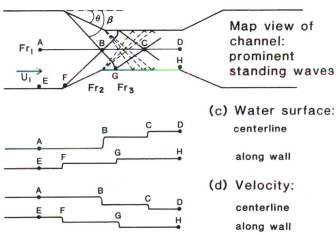

Map view of channel: prominent standing waves

(c) Water surface:

centerline

along wall

(d) Velocity:

centerline

along wall

FIGURE 16.6. Comparison of the flow fields in subcritical and supercritical flow [compiled from Chow (1959) and other basis hydraulics text]. (a) Schematic map view of subcritical flow through a constriction. Schematic profiles (1) and (2) show how the water surface elevation and velocity change along the channel. (b) Schematic map view of supercritical flow through the same constriction. (c) Changes in water surface elevation and (d) local velocity through the channel, respectively. The profiles of these quantities are along the paths A-B-C-D and E-F-G-H.

map view of subcritical flow in an idealized converging-diverging laboratory flume, and it is singularly uninteresting because there are no standing waves. The two lines below the map view of the flume are (top) water-surface elevation and (bottom) velocity profile. The increase in velocity and decrease of water-surface elevation in the constriction in subcritical flow conditions is exactly analogous to the well-known venturi effect that gas flow shows in converging–diverging nozzles. Figure 16.6b is a map view of the same flume with supercritical flow conditions. This case is much more interesting than the subcritical case because patterns of oblique, criss-crossing waves occur in the converging

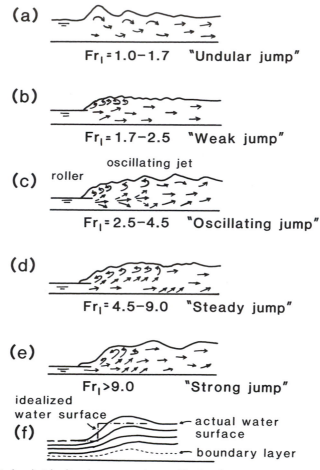

FIGURE 16.7. (a–e) Idealized cross sections of hydraulic jumps with the entering flow at different Froude numbers as shown. (From Chow 1959, p. 395.) (f) Schematic cross section of idealized and actual hydraulic jumps. (After Ippen 1951, p. 339.)

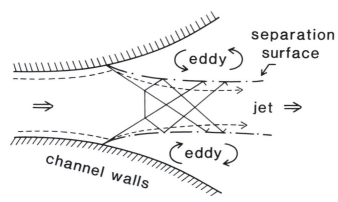

FIGURE 16.8. Illustration of the structure of a supercritical jet emerging from a constriction. (From Chow 1959, p. 471; originally from Homma and Shima 1952.)

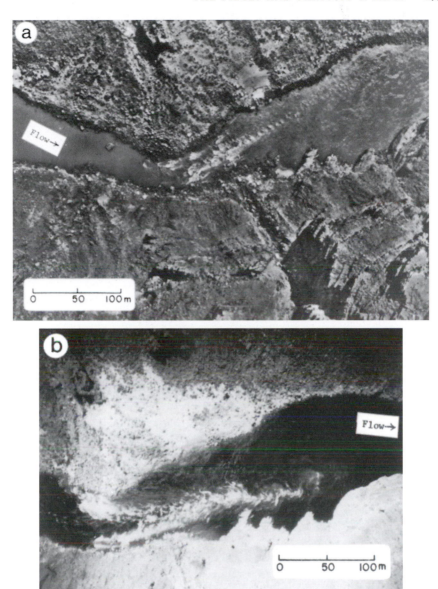

FIGURE 16.9. 24.5-Mile Rapids at discharges of (a) 5000 cfs and (b) 30,000 cfs. Note the dramatic change in the orientation of the tailwaves and the sizes of the eddies on each side of the tailwaves with discharge. [Photograph (a) by U.S. Bureau of Reclamation, 1984; photograph (b) by National Park Service, 1986.]

and constricted sections. Such waves also can occur in the diverging section but for simplicity are not shown here. The wave patterns (specified by the angle β) depend on the channel shape (specified by the angle ϕ). The height of the waves can vary across the channel—particularly where waves intersect each other. At the points of intersection, the wave heights can either be added to or subtracted from each other, depending on the type of wave.

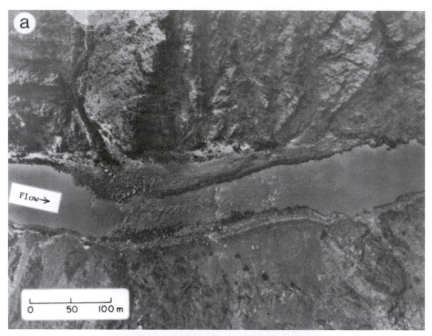

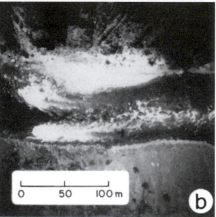

FIGURE 16.10. (a) Hermit Rapids, showing the tongue and lateral waves at 5000 cfs. (b) View of the same rapid at the same scale showing these features at 30,000 cfs. Note difference in tongue lengths, angle of lateral waves to shore, and nonbreaking rollers on the tongue. [Photograph (a) by U.S. Bureau of Reclamation, 1984; photograph (b) by National Park Service, 1986.]

Smooth water from the backwater extends farthest downstream into a rapid along the tongue in the converging part of the rapid (compare the tongue of Granite Rapids in Fig. 16.2 with the centerline A-B in Fig. 16.6b). A characteristic of most tongues is the presence of smooth, nonbreaking rollers (Fig. 16.1c). The length of the tongue and the angle of the oblique waves change with discharge and channel shape.

The converging part of a rapid is a region in which the flow changes in a complex way from subcritical conditions in the backwater to supercritical conditions in the constriction downstream; this region is often referred to as the "top" or "head" of the rapid. The measurements of water surface velocities and depths in the rapids suggest that the Froude number of the flow reaches a value of unity near the beginning of the tongue. For this reason, I interpret the non-breaking rollers seen on most tongues as undular hydraulic jumps typically found when the Froude number is in the range of 1 to 1.7 (Fig. 16.7a).

In the supercritical region of the rapid (to the sides of and downstream of the tongue), complex waves appear. The stronger breaking waves (the oblique lateral waves and the haystacks) suggest higher Froude numbers than do the undular waves on the tongue. Values greater than 2 have been measured (Fig. 16.3e). The measured values of Froude number are probably lower limits for several reasons. The Froude number depends on water velocity, and, properly defined, it must be an average velocity for the whole water depth. When floats are used to measure velocities, several effects cause the measured velocities to be lower than the average velocity: drag between the water surface and air, drag between the float and air, and "surfing" and bouncing of the floats on the waves. It seems probable that the floats we used to measure surface velocities traveled appreciably slower (10–20%?) than the mean water velocity. Nevertheless, the measurements are internally consistent, and they show that the velocities and Froude numbers increase continuously from the backwater into the constriction and, sometimes, on into the diverging region of the rapid (Fig. 16.3d).

The water's energy is dissipated in the converging and constricted part of a rapid by wave action, bottom friction, and air entrainment. The noise of the rapids is part of the mechanism by which potential energy that was stored in the water in the backwater (and that was converted to kinetic energy in the converging section of the rapid) is dissipated as the water begins to decelerate back toward tailwater conditions.

The Diverging Section: Haystacks, Tailwaves, Eddies, and Beaches

Some very nonintuitive phenomena occur in the diverging part of a constricted rapid because the flow is supercritical. Our intuition, typically based on subcritical flow, would lead us to believe that the water would decelerate as soon as the river channel widens (Fig. 16.6a). Flow velocity measurements show, however, that the velocity can increase not only into the constriction, but well into the diverging part of the channel (this is not dramatic at House Rock Rapids shown in Fig. 16.3, but can be seen in the data from the other rapids in the maps of Kieffer 1988). The flow velocity and the Froude number both increase until a hydraulic jump is encountered—the hydraulic purpose of the jump is to return the flow toward ambient main-channel conditions appropriate to the downstream reach. This process is nonlinear; the flow does not adjust smoothly to the convergences and divergences of the channel. Instead, changes occur discontinuously through standing waves (hydraulic jumps). When sufficient energy has been dissipated by the mechanisms mentioned above, the water returns to subcritical conditions and must adjust its depth and velocity to the tailwater conditions downstream.

Another nonintuitive aspect of the flow in the rapids is the influence—or, rather, lack of influence—of the downstream tailwater conditions on the upstream flow conditions. In a subcritical constriction, the velocity and depth of water in the constriction and even upstream of the constriction are influenced by the depth in the downstream tailwater—that is, the flow adjusts continuously and smoothly to changes in channel geometry. In contrast, in supercritical flow, the tailwater conditions can only influence flow conditions in the lower part of the divergence below the hydraulic jump where flow returns to subcritical conditions. Flow in the converging and constricted parts of the rapid is uninfluenced by conditions downstream of the major waves.

Channel expansion typically is very abrupt in the diverging section (Fig. 16.2). When changes in channel shape are abrupt, flow streamlines generally cannot follow channel boundary curvatures (a condition known as rapidly varying flow in hydraulics). Because the streamlines cannot follow the channel boundary, the narrow high-speed jet formed in the constriction squirts into the divergent section with nearly the same diameter that it acquired in the constriction. The jet in the divergent reach is marked by a train of gently breaking to nonbreaking tailwaves (Fig. 16.2c). Measurements show that tailwaves generally are regions where the Froude number is very close to 1 (see Fig. 16.3d). Thus, we can interpret the tailwaves as nonbreaking, undular hydraulic jumps, and we note that they have a structure similar to, but not identical with, the nonbreaking rollers on the tongue in the converging part of the rapid where the Froude number was also estimated to be 1. A major difference between the tongue and the tailwaves is the width of the main current: in the converging part of the rapid, nearly the whole discharge is gradually funneled into the tongue, and the flow accelerates until Fr ~ 1 conditions are obtained. In contrast, in the diverging part of the rapid, the flow does not expand back to the full width of the channel.

The region between the jet and the channel boundary fills with recirculating flow of relatively low velocity—an eddy (Fig. 16.8). Eddies always are found on at least one side of a strongly diverging river channel below a rapid, and they often are found on both sides (Figs. 16.1b and 16.2c). These recirculating zones, in turn, allow sand deposition. For this reason, many important ecologic sites and pleasant camping sites are downstream of the rapids. Details of the expansion of the constricted jet back to ambient conditions in the tailwater are important in determining sediment transport to these beach sites (Fig. 16.2c).

The shape of the main jet of water emerging from the constriction can be defined quite well because separation surfaces between the jet and the recirculating zones are manifested as strong shear zones (eddy fences) and "boils" (small whirlpools). The shape of the jet in the tailwater cannot be predicted accurately from theory. However, laboratory data suggest that the distance downstream that the jet will maintain constant diameter is proportional to the diameter of the constriction in which it was formed. A jet typically will maintain constant diameter for a distance downstream of several constriction diameters. In detail, this relation depends also on the Froude number of the jet in the constriction and divergence. For example, with a Froude number of two, an ideal laboratory jet would maintain constant diameter for roughly three constriction diameters downstream. As an example, at House Rock Rapids, the constriction diameter is about 30 m when the discharge is 5000 cfs (Fig. 16.3a), and the jet maintains a strong identity (as evidenced by tailwaves) for at least 100 m downstream. Within this distance, the jet velocity appears to remain constant at 10 to 15 ft/s (3–5 m/s), and the Froude number appears to be near unity. The length and orientation of the jet and tailwaves can change dramatically with discharge (Fig. 16.9), a fact that has important implications for the formation and stability of beaches under different discharge conditions.

Much of the energy that must be dissipated in the rapid (the excess backwater head and the vertical elevation drop) is expended in the waves and in bottom friction. Nevertheless, the fact that the flow often has a high velocity at the bottom of the rapid indicates that not all of the water's excess energy has been dissipated. Additional dissipation occurs through mixing with the relatively stagnant water of the eddies that bound the jet in the tailwater. The motion of the jet induces circulation in the eddies, and the two flows (jet and eddy) mix in a zone that expands around the separation line (Landau and Lifschitz 1959, p. 131).

In the photographs and airbrush illustrations of jets in this chapter (Figs. 16.1, 16.2, 16.3, 16.9, 16.10), you can see the narrowing of the tailwaves with distance downstream. The converging lines that bound the tailwaves can be taken as evidence of the extent to which the mixing zone has extended into the main current (the jet). To a first approximation, one can say that the jet has decelerated to and been mixed back into main channel (or tailwater) conditions by the end of the tailwaves. The mixing not only decelerates the jet, but it allows suspended sediment carried in the high-velocity main channel current to be transported laterally toward the channel boundaries, into the recirculating eddy, and, ultimately, onto the beaches. Flow in the recirculating zone typically is very slow, about 1 f/s (0.3 m/s). Thus, sediment, if available, can be deposited within these zones (the rate of sedimentation will depend on particle size and density, fluid velocity and density, and other factors). Detailed studies of this zone are in Schmidt and Graf (1987).

Below the eddy-beach system, another pool and rapid sequence begins, and the hydraulics described here are repeated again, each time like a theme and variations, along the length of the Grand Canyon. Each rapid is unique, yet all are similar.

With this background, we can ask how the characteristic configuration of backwaters, rapids, and fast-water runouts has been created and evolved over thousands of years of hydraulic, hydrologic, and geomorphic interactions superimposed on the tectonic forces that drive the uplift of the Colorado Plateau and the downcutting of the Grand Canyon. A glimpse into these processes was offered by a series of unique events in 1983.

THE 1983 FLOODS: A UNIQUE WINDOW TO HYDRAULIC–GEOMORPHIC INTERACTIONS

In 1983, unusually high releases from Glen Canyon Dam, which controls the flow through the Grand Canyon, provided the opportunity to observe how channel geometry at a rapid changes as the discharge history of the river changes. Most of the interesting events were at Crystal Rapids (Fig. 16.11). This rapid was relatively insignificant before 1966 (for example, it was not mentioned in Powell's 1875 report). Figure 16.11a shows the configuration of the rapid in 1963, the year that Glen Canyon Dam was closed. Three years later, a large storm over the north rim of the Grand Canyon caused flash flooding in a number of tributaries (Cooley et al. 1977). In particular, debris poured down Crystal Creek and Bright Angel Creek, which is 10 miles (16 km) upstream from Crystal Creek.

The debris flow down Crystal Creek made Crystal Rapids one of the most dangerous rapids on the river for recreational boating. Large boulders carried in the debris flow tightly constricted the river channel at Crystal Rapids (Fig. 16.11b). Discharges at Glen Canyon Dam between 1966 and 1983 varied between 3000 and 35,000 cfs. These discharges were sufficient to cut a narrow channel through the distal (south) end of the debris flow and to cause occasional shifts of boul-

FIGURE 16.11. Crystal Rapids at three different times. (a) March 1963, discharge approximately 5000 to 6000 cfs. (Photographs by A.E. Turnber, Bureau of Reclamation.) (b) Same view, March 1967, approximately three months after a debris flow in Crystal Creek. Discharge is 16,000 cfs. (Photograph by Mel Davis, Bureau of Reclamation.) Note that even though the stage at 16,000 cfs is higher than at 6000 cfs, the debris fan is larger in (b). (c and d) Pair of aerial photos, showing the configuration of the debris fan and the rapid after the 1983 flood. Discharge approximately 5000 cfs in both photographs. The top photograph is from (a). About 10–15 m of lateral erosion has taken place along the shore adjacent to the rock indicated by the arrow. (Bottom photograph by Bureau of Reclamation, October 1984.)

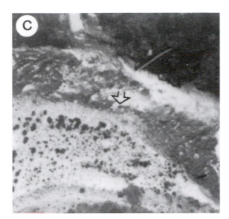 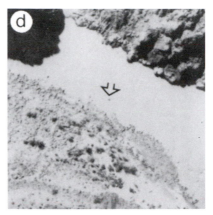

FIGURE 16.11. (*Continued*)

der positions within the channel. Overall, however, the channel at Crystal remained tightly constricted, and it retained the general shape documented by the 1967 photograph shown in Figure 16.11b. The shape parameter or "constriction" defined in Figure 16.3 (but modified as discussed in the text) was about 0.25 in 1973.

During the spring of 1983, rapid snowmelt at the headwaters of the Colorado River forced operators of Glen Canyon Dam to increase discharges to 92,000 cfs to keep Lake Powell from overtopping the spillways of the dam. This discharge was nearly three times larger than any discharge through Crystal Rapids since the 1966 debris flow, and it was comparable to the annual Colorado River floods prior to Glen Canyon Dam. Thus, a relatively young debris fan was subjected to its first "flood." The changes in Crystal Rapids (compare Fig. 16.11c with 16.11d) during these high flows provided an opportunity to observe some of the dynamic processes that contour the river channel over geologic time.

During the 1983 flood, the channel at Crystal Rapids was widened, and the shape parameter increased significantly from 0.25 to 0.4. This corresponds to an increase of 35 to 50 feet (10–15 m) in width, and this increase dramatically altered the hydraulic characteristics of the rapid. Nevertheless, the rapid is still quite different from the other rapids that have constrictions of 0.5 (remember from Fig. 16.3 that most rapids have constrictions of 0.5). Local gradients within Crystal Rapids are steeper than those generally found at other rapids, waves are larger, and the dependence of wave structure on discharge is more variable. By watching the evolution of the rapid toward the configuration of the more mature rapids, I have worked out the following ideas on the hydraulic–geomorphic interactions between the Colorado River and the debris dams that episodically block its course.

EROSION OF THE DEBRIS FANS
BY THE RIVER

The boulder deposits that constrict the river to form rapids are emplaced by debris flows from steep tributary canyons. The steep gradient of these canyons allows these streams to carry large boulders into the main channel (which has a much smaller gradient than the tributaries). These boulders cannot be moved by typical main-stem flows in unconstricted reaches of the river. Although the ini-

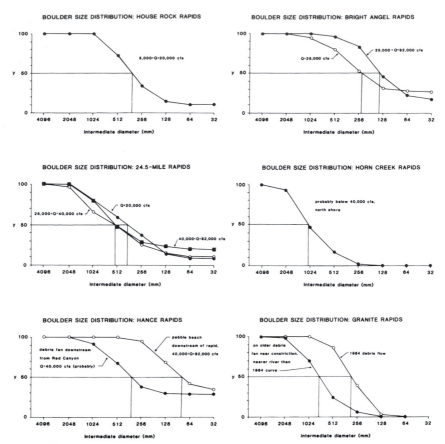

FIGURE 16.12. (Part I opposite page, Part II above) Particle size distributions measured at the places indicated at rapids (Kieffer 1987). The twelve graphs are arranged in downstream order. The ordinate, y, is the percent of particles smaller than the given (intermediate) diameter. The horizontal and vertical lines in each graph are to guide the reader's eye to the median diameter of the particles at the rapid.

tial size distribution of material in the debris flows is quite broad (Webb et al. 1987), the smaller material, up to large cobble size, is washed away by the river at relatively low rates of discharge. Excellent descriptions of sediment transport through the canyon can be found in Howard and Dolan (1976, 1981), and the U.S. Bureau of Reclamation Glen Canyon Environmental Studies (1987) contains recent detailed studies of sand and silt transport in the main channel. Boulders 3 to 10 feet (1–3 m) in diameter and common on the debris fans (Fig. 16.12). These boulders resist erosion (either by chemical or mechanical abrasion processes or by movement at low discharges), and so they stabilize the debris fans with the geometry shown in Figure 16.3. Because at least half of the channel generally is cleared of all but the very largest boulders at mature rapids, it is obvious that the boulders can be moved under some conditions.

Although hydrologists have documented the transport of sand and silt-sized sediment past the Lees Ferry and Grand Canyon (Bright Angel) gage stations, we know very little about the mobility of large particles in the vicinity of rapids.

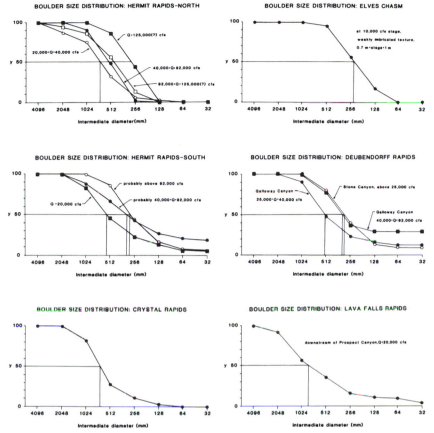

FIGURE 16.12. (*Continued*)

Specifically, quantitative modeling of the movement of boulders in the Grand Canyon has not been done. Laboratory experiments on much smaller particles and theoretical studies have shown that the size and amount of material transported are related to water velocity, depth, and, therefore, rate of discharge. Graf (1979, 1980) analyzed the stability of boulders in the Green River in Utah and concluded that the largest boulders were stable, even against motion during the largest floods. Because the Colorado has many features in common with the Green River, it generally has been assumed that large boulders, once emplaced, also are relatively stable in the rapids in the Grand Canyon.

The ability of the river to clear out debris emplaced in the channel (its "competence") is proportional to variations in velocity and water depth. The discussion in the first half of this chapter demonstrated that changes in these parameters of more than an order of magnitude occur within a rapid. There are, therefore, substantial differences in transport capacity from one section of a rapid to another. The photographic documentation of changes in Crystal Rapids shown in Figure 16.11 and the statistical analysis of rapid shapes shown in Figure 16.4 show that the water can move large boulders in the channel—at least until the river has cleared itself to about one-half of the characteristic unconstricted width. This capacity of the river to clear out large debris from at least half of the chan-

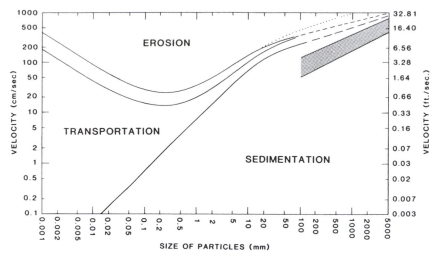

FIGURE 16.13. Summary of the relations between stream velocity and size of movable boulders. The lines on top represent the Hjülstrom criterion is for particles in a uniform bed. Bed roughness and particle shape appear to cause particles to move at lower velocities. The lower stippled area to the right represents criteria developed from field work by Helley (1969).

nel width produces the characteristic "nozzle-like" geometry seen at the rapids. The processes that accomplish this erosion involve a complex feedback between the hydraulics of the river (discharge as a function of time; depth and velocity as a function of position in the river) and the nature of the material in the bed (grain size and position in a bed of variable cross section).

One criterion widely used for predicting the transport of smaller material in a river is the Hjülstrom relation (Hjülstrom 1935; Strand 1986), which relates water velocity to the size of the largest particles that can be transported (Fig. 16.13). This Hjülstrom relation, extrapolated to large boulder sizes in Figure 16.12, suggests that a velocity of 20 f/s (6 m/s) would be capable of moving a boulder 1 to 2 ft (0.5 m) in diameter and that it would be possible to move material up to 3 to 7 ft (1–2 m) in diameter. These values depend on how the Hjülstrom curves are extrapolated.

As discussed in the previous section, water in the highest-velocity part of the rapids can have velocities exceeding 30 ft/s (10 m/s) even at the lower end of the range of discharges from Glen Canyon Dam generators (about 3000 cfs). By the Hjülstrom criterion, flows at these discharges should be capable, therefore, of clearing the channel of boulders in the 3- to 7-ft (1- to 2-m) size range.

A second criterion for boulder transport is the concept of unit stream power, originally introduced by Bagnold (1966, 1980) and recently applied to paleohydrogeologic problems by O'Conner et al. (1986). Unit stream power is the stream power (rate of expenditure) per unit area, Ω. Bagnold (1966) defined it as

$$\Omega = xQS_f/w = \beta u$$

where x is the specific weight of the fluid (assumed to be 9800 N/m³ for clear water), Q is the discharge (specifically, that component of the discharge carried by the main current), S_f is the friction slope, w is the channel width, β is the total channel shear, and u is the main channel velocity. A more convenient form of this equation is

$$\Omega = x n^2 u^3 / R^{1/3}$$

where n is the Manning coefficient of roughness (assumed for these calculations to be 0.035) and R is the hydraulic radius of the channel, taken here to be the average depth.

For example, the reader can use Figure 16.3 to calculate the unit stream power in House Rock Rapids at 5000 cfs discharge. Although the maximum water surface velocity reaches 25 ft/s (7.5 m/s), we will assume an average velocity of 21 ft/s (6.5 m/s) for a conservative estimate. The average depth is about 3 feet (1 m) in the narrowest part of the rapid. The unit stream power, therefore, is on the order of 3300 Newtons per meter per second. Unit stream power and sediment transport relations (summarized in O'Conner et al. 1986) suggest that a river with this unit stream power could transport boulders up to about 7 feet (2 m) in diameter. This conclusion agrees with the inferences from an extrapolation of the Hjülstrom diagram. Both criteria suggest that the Colorado River within the Grand Canyon is capable of moving boulders comparable in size to those moved during the largest floods known from paleohydraulic reconstruction techniques (for example, the Missoula flood, Baker 1973, 1984; and Katherine Gorge floods in Australia, Baker and Pickup 1987).

These two criteria also lead to the conclusion that the river can move large boulders even at low discharges in which the flow occupies only a relatively small part of the total river channel (note how little of the river channel is occupied by water at the 5000 cfs discharge conditions shown in the cross sections of Fig. 16.3). Another way to state this is to say that the relatively shallow, high-velocity flow characteristic of high-gradient rivers can move rather large boulders. Thus, that part of the channel of Crystal Rapids that was exposed to discharges up to 35,000 cfs was depleted of small to relatively large material prior to the 1983 flood. During larger floods, the deepest part of the main channel is cleared of even larger boulders. However, even the parts of the channel exposed to relatively shallow overflow during the floods have boulders removed, as shown by the size distributions of debris fan material that has been exposed to main-stem floods (Fig. 16.12). Preliminary measurements by the author of boulder size distributions above the river shoreline that correspond to the 92,000 cfs discharge of 1983 showed depletion of small boulders, suggesting larger floods in the past. These observations suggest that a detailed study of boulder size distributions at different places on the debris fans and talus slopes along the Colorado could be used to infer flood histories.

There are, however, serious complications in interpreting boulder size distributions in terms of simple erosion by the main stem because each debris fan has a complicated erosional and depositional record. If a debris fan had a wide variety of particle sizes when it was emplaced, and if it was not graded in size laterally, then erosion of this fan by a large flood would be expected to remove larger particles low on the fan (where the flow is the deepest and fastest) and to remove progressively smaller particles higher on the fan. The size distributions measured on the north and south banks of Galloway Canyon at Deubendorff Rapids (Fig. 16.12) are consistent with such a simple emplacement and erosion model. However, material removed from the upstream parts of the fan may be deposited on the downstream sides in the recirculating zones discussed in the first part of this chapter; thus, the size distribution depends not only on the vertical elevation on the fan, but on the relative upstream–downstream location (for example, Hance Rapids, Fig. 16.12).

Boulder size distributions on other debris fans are rarely this simple to interpret. In addition to the erosional and depositional processes mentioned above, complexities arise because (a) initial particle size distributions in the debris flow are not known; (b) winnowing or piping of fine particles may be important; and (c) the initial history of the damming and breaching of the fan generally is unknown. For example, photographs of the Crystal Creek debris fan taken in 1967 (Fig. 16.11b) show that it had a surface veneer of boulders nearly all of the way up to the mouth of Crystal Creek—far above any stage reached by Glen Canyon Dam discharges between 1966 and 1967. Because of this absence of known floods prior to the time the 1967 photograph was taken, I speculate that the 1966 debris dam caused ponding of water to this level and that the overflow of this original pond removed a substantial number of boulders all the way to the top of the debris fan. In Kieffer (1987) I discuss in detail the implications of each size distribution shown in Figure 16.12.

These many erosional and depositional processes are important in the details of the channel shape and particle size distribution where the river cuts through a debris fan to form a rapids. The geometry of the channel is the result of a balance between the local increase in the river's erosive power within the constricted zone and its decreased transport capacity in slower parts of the channel. Water accelerates from velocities on the order of a fraction of a foot per second (a few tenths of a meter per second) in the backwater above a rapid to 15 to 25 ft/s (4 to 6 m/s) on the tongue of the rapid, to values greater than 30 ft/s (10 m/s) in the constriction and part of the diverging section. Velocities of about 15 ft/s (4 m/s) then are maintained through the tailwaves. Thus, in the constriction and the diverging section of the rapid immediately below the constriction, velocities are more than adequate to clear the channel of boulders.

The position of the constriction in a rapid can change as the rate of discharge changes because of variations in topography. The position and shape of the constriction also changes with time as the river channel evolves in response to floods. An excellent example of how the position of a constriction depends on discharge can be seen at Horn Creek Rapids. At low discharges (less than 30,000 cfs), the constriction is near the top of Horn Creek Rapids. At 92,000 cfs, this constriction is nearly drowned out by the backwater from a constriction about 1000 feet (several hundred meters) downstream (Kieffer, 1988, I-map 1897-E).

The position and shape of a constriction in a rapid also changes as a result of tributary flash floods, as at Crystal and Bright Angel Rapids in 1966. In 1983, the position of constriction at Crystal Rapids again changed, moving upstream in response to high discharges. These changes are examples of why observations of rapids over a large range of discharges and, by implication, a long period of time are necessary for compiling data on channel processes and their rates.

Because there is an increase in erosive power in the constriction and in the diverging section of a rapid, boulders move downstream until the channel widens and deepens sufficiently for the water to decelerate (typically by a transition from supercritical to subcritical flow). Large boulders are moved out of the constrictions of rapids, transported through all or part of the divergent sections, and then deposited hundreds of feet (in some instances, up to 1 km) downstream to form the "rock gardens" or cobble bars found below rapids (Fig. 16.2a). A rapid, therefore, evolves into two parts: the original debris deposit (reworked, at least on the surface, by overflow) and the rock garden (or cobble bar), usually found downstream of the initial deposits.

Crystal Rapids, 1966–1987

In 1983, we were able to document changes in the hydraulic behavior and channel shape at Crystal Rapids; the following summary is taken from Kieffer (1985). The rapids in the Grand Canyon were exposed to discharge levels three times that which had occurred since 1963, and this was a particularly significant event for Crystal Rapids because of the large debris fan emplaced in 1966. In addition to their geomorphic significance, the hydraulic events during 1983 had a significant effect on commercial and private rafting in the Grand Canyon. About 10,000 people each year navigate the 250-mile (400-km) stretch through the Grand Canyon. The debris flow in 1966 and the flood of 1983 both emplaced boulders and caused waves and eddies in Crystal Rapids that have made this area difficult to navigate.

Although the geometry of the river channel prior to the studies of Kieffer (1988) is largely unknown, studies of air photographs taken in 1973 and calculations of plausible cross-section shapes (Fig. 16.14) suggest that the constriction of the channel was about 0.25—that is, the surface width of the rapid in its narrowest part was only 1/4 of the width of the main channel upstream of the rapid. The water-surface elevation dropped about eight feet (2.5 m) between the head of the rapid and a large obstacle known as "The Crystal Hole" that was several hundred meters downstream from the entrance to the rapid (locales are shown on Fig. 16.14). The water surface dropped about another 8 feet (2.5 m) through the rock garden below the rapid (Leopold, personal communication, 1984). Because the Crystal Creek flood of 1966 appears to have been large enough to have strewn boulders across the entire river channel, it appears that either the initial breaching event (when the river cut through this debris dam) or power plant releases up to 35,000 cfs from Glen Canyon Dam between 1966 and 1973 were sufficient to move rocks out of the distal end of the debris flow into the rock garden. The result was that by the time the 1973 air photographs on which the data of Figure 16.3 were based, the river had carved a channel with the value of constriction equal to 0.25.

A significant clue to the channel hydraulics in 1983 was the comparison of the waves in the mature rapids with the behavior of the wave that occupied the region of the Old Crystal Hole when discharges in late June and early July 1983 increased above earlier maximum discharge levels. Waves in the more mature, less constricted rapids than Crystal disappeared at discharges on the order of 90,000 cfs, unless the topography covered by the water at higher stages formed a different constriction in three dimensions. This phenomenon of the disappearance of local waves within rapids is the so-called "drowning out" of a rapid at high water. Instead of drowning out, the wave associated with the Old Crystal Hole in Crystal Rapids increased in height as discharges increased. At discharges in the range 50,000 to 70,000 cfs, this wave formed a formidable barrier across the river: it was more than 15 to 20 feet in height (5–6 m) and spanned nearly the entire navigable channel (Fig. 16.15).

The size, location, and even the sound of this wave changed with discharge. In the years prior to the 1983 flood, the trough-to-crest height had been about 10 feet (3 m) at 20,000 cfs and about 3 feet (1 m) higher at 30,000 cfs. Between 1966 and 1983, the wave was associated with a large rock in this location rather than with its critical position in the neck of the constriction. As discharges rose to around 50,000 to 60,000 cfs in June 1983, boatmen and passengers reported that the wave surged to a height between 15 and 30 feet (5 and 9 m); it was verified photographically at about 15 to 20 feet (5–6 m) (Fig. 15). Perhaps most interestingly, as discharges reached 92,000 cfs in early July, river observers noted

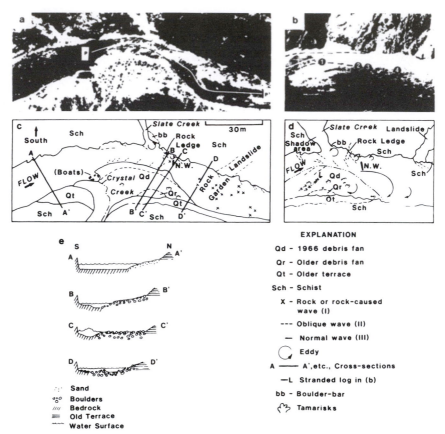

FIGURE 16.14. (a) Crystal Rapids on June 16, 1973. (U.S. Geological Survey Water Resources Division air photograph.) (b) View of part of the rapid at approximately the same scale during the high flow of 1983. (c) and (d) Keys to features on (a) and (b). (e) Schematic cross sections along lines A-A', B-B', C-C', and D-D'. Relative widths correct; vertical scale exaggerated. Rise of debris fan from the river level in (a) to the old alluvial terrace indicated by Qt is about 5.5 m. The stage at 92,000 cfs is just at the base of this terrace. Assumed boundaries for deep channel are shown by a light dashed line in (a) and (c). P-P' was the preferred navigation route through the rapids prior to 1983. The Crystal Hole, formed by a normal wave (hydraulic jump), is indicated by N.W. This wave is not easily visible in (a) because of photo scale. The white line in (b) indicates the path of kayaks whose velocities were measured at approximately 9 m/s.

that the wave height decreased to 10 to 15 feet (3–4.5 m). At discharges over 50,000 cfs, the wave appeared to be located about 100 feet (30 m) downstream from its pre-1983 position at 30,000 cfs (compare the position of the wave, labelled N.W. in Figs. 16.14c and 16.14d). Observers reported that at 50,000 to 60,000 cfs the wave emitted a low roar like a jet engine, but it did not generate the same loud roar at 92,000 cfs, though loud booms were clearly audible every few seconds.

After the 92,000 cfs discharges of 1983, surface wave patterns within the rapid altered dramatically, and the local gradient within the rapid changed (Fig. 16.1c).

FIGURE 16.15. Photograph of the wave, interpreted as a normal hydraulic jump, that formed across much of the main channel in late June 1983, when discharges were raised to about 60,000 cfs. (Photograph taken June 25, 1983; copyrighted by Richard Kocim, reprinted with permission.) Pontoons on the raft are each 1 m in diameter; midsection is about 3 m in diameter. More than 30 passengers were on board; one head is visible on the lower side of the raft. From the scale of the raft, the trough-to-crest height of this wave can be estimated at more than 5–6 m.

This new hydraulic situation has persisted. The drop in elevation of 5 to 10 feet (2–3 m) that was spread between the top of the rapid and the old Crystal Hole is concentrated now in a narrow zone of only a few tens of meters near the top of the rapid. As a result, the oblique waves on the right side of the tongue at the entrance to the rapid have increased dramatically in height. The change in bed slope and water-surface gradient at the head of the rapid suggests that about 30 feet (100 m) of headward erosion occurred during the high discharges.

In addition, the channel widened by 30 to 50 feet (10–15 m) at its narrowest point during this flood (compare Figs. 16.11c and 16.11d). This means that the constriction value changed to 0.40—approaching the 0.5 value typical of the older, more mature debris fans. Widening occurred in and downstream of the constriction in the zone of supercritical flow. Widening did not occur solely during peak flows, but it appears to have begun as soon as the discharges exceeded the controlled flows of the prior two decades. This conclusion is unsupported by direct measurement. However, a very similar series of events—including a debris flow in 1966—occurred at Bright Angel Creek, 10 miles (16 km) upstream from Crystal Rapids and in a similar geologic setting. There, as the discharge rose through the range of 60,000 to 70,000 cfs, the bed at Bright Angel gage station was scoured by about 8 feet (2.4 m). Presumably, a scour of similar magnitude occurred at the same discharges at Crystal because of the similarity in channel morphology and material at the two locations.

Water velocity varies throughout the length of a rapid because of geometry and gradient changes. Unfortunately, no systematic measurements of water velocity could be made at Crystal during the flood; however, on June 27, when the discharge peaked at 92,000 cfs, kayaks were filmed going through the rapid (approximately along the white line shown in Fig. 16.14a). Analysis of the films showed maximum velocities of 28 to 32 ft/s (8.5–9.8 m/s). Although there can be no rigorous correlation of these velocities with average water velocity, the kayaks appeared to be moving with the current. If the average water velocity was even close to these values, the river was capable of moving boulders several meters in diameter (Fig. 16.13). Large, moving boulders presumably were the source of the loud, booming noises heard by the author.

The velocities measured, the depths inferred from measurements of stage, and the behavior of the large wave that developed in Crystal Rapids all indicated conditions of supercritical flow during the flood. The flow was forced into supercritical conditions by the geometry of the converging–diverging channel. The large wave that stood across the channel had characteristics generally associated with a "normal hydraulic jump." Shallow-water flow theory allowed me to analyze the relation between discharge and backwater energy—conventionally expressed as depth (Fig. 16.16a), wave height (Fig. 16.16b), velocity in the constriction (Fig. 16.16d), velocity in the diverging supercritical section of the rapid and in the diverging subcritical section below the hydraulic jump (Fig. 16.16c), and the change in water velocity through the hydraulic jump (Fig. 16.16e).

Figure 16.16b shows that the measured wave height increased with discharge until the discharge reached about 60,000 cfs; the calculations indicate that this behavior would be expected for a channel with a pre-flood constriction of 0.25. At higher discharges, the model calculations suggest that the wave height should have continued to increase, but, instead, it decreased. The curves in Figure 16.16b suggest that changes in the shape parameter caused the decrease in wave height. In particular, it appears that by the time discharges reached about 60,000 cfs, the channel had begun widening and that by the time of peak discharges, the constriction had changed to about 0.40.

Although the parts of Figure 16.16 look complicated, a careful study of these figures reveals the complex interactions that were occurring. The meteorological situation forced the engineers to increase the discharges constantly, causing water velocities to increase. Channel widening occurred as a result of erosion at the high velocities. The channel widening alone would have allowed a decrease in velocity, but the increasing water level of Lake Powell necessitated further increases in discharge.

During this flood event, the Colorado River continued to contour the channel at Crystal Rapids into a shape in which the velocities in the most highly constricted portion of the channel were equal to the threshold velocity for the transport of the major boulders. Material removed from the constriction was transported several hundred meters downstream into the area of the rock garden. This section of the river was modified substantially by the 1983 flood.

For nearly a year following the peak flows in June and July of 1983, high discharges prevented direct observations of the effects of the flood. By the time that Crystal could be examined again at low water (October 1984), the record of the 1983 events was partially overprinted by sustained discharges at 60,000 and 25,000 cfs. Nevertheless, field evidence indicated that the erosion postulated on the basis of the open-channel hydraulics theory did indeed occur (Figs. 16.11c and 16.11d).

Observers found that the eroded section of the channel at Crystal Rapids had a fresh cutbank in the boulders. Similar cutbanks have been observed after the 1984 debris flows down Monument Creek at Granite Rapids and at Elves Chasm in 1984. These cutbanks are evidence of the action of the river contouring its own channel, and if they can be related to specific discharges, they provide valuable clues about the geomorphic evolution of the debris fans.

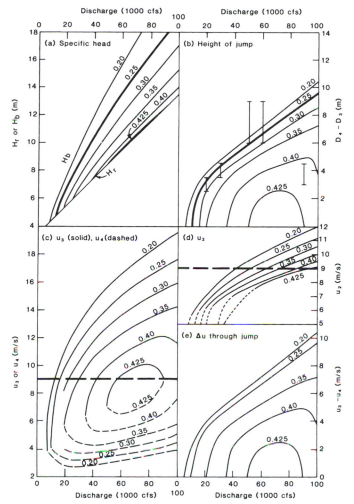

FIGURE 16.16. Model calculations for flow through Crystal Rapids (from Kieffer 1985). (a) Specific head (Hr) measured at Bright Angel Creek vs. discharge, with backwater heads (Hb) calculated for Crystal Rapids for the constrictions, w2/w0 (0.20, 0.25, . . .) indicated. The specific head can be thought of as the normal depth of the river, and the backwater head as the depth cause by the constriction. (b) Calculated height of the hydraulic jump for constrictions indicated. Bars denote observed values. (c) Calculated values of flow in the supercritical region of the diverging section of the rapid (u3) and in the subcritical region of the diverging section (u4). These two sections are separated by the hydraulic jump. Dashed line at 9 m/s indicates the velocity at which larger boulders at Crystal Rapids were assumed to be transportable by the current. (d) Calculated values of velocity in region 2 (the constriction) for constrictions indicated. Flow subcritical where dashed. (e) Velocity change through hydraulic jump that separated regions 3 and 4.

A MODEL FOR THE GEOMORPHIC-HYDRAULIC EVOLUTION OF THE RIVER CHANNEL AT DEBRIS FANS

The two decades of observations at Crystal Rapids are but a glimpse into the history of much larger debris flows (e.g., from Prospect Canyon at Lava Falls) and much larger flood events that occurred throughout the history of Grand Canyon downcutting. Figure 16.17 shows a generalization and extrapolation of the ideas developed at Crystal Rapids. This figure should be interpreted to represent but one cycle of recurring episodes of deposition and modification.

In this model, I have arbitrarily chosen the beginning of the sequence as a time when the main channel is relatively unconstricted—that is, major floods are assumed to have occurred on the Colorado River since the last time this tribu-

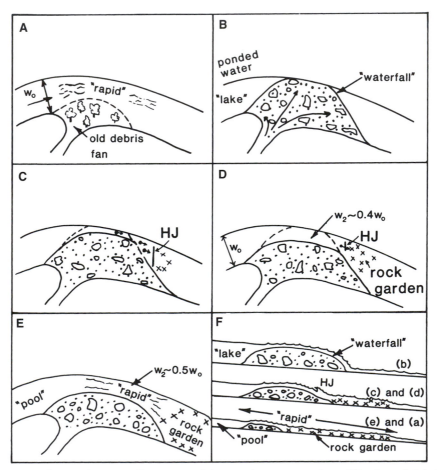

FIGURE 16.17. Schematic illustration of the emplacement and modification of debris fans, the formation and evolution of rapids, and the formation of rock gardens. (a) Initial channel geometry. (b) Side-canyon flood. (c) Erosion: small floods. (d) Erosion: moderate flood, supercritical flow. (e) Erosion: large flood, subcritical flow. (f) Longitudinal cross sections. See text for further explanation.

tary canyon had a major debris flow (see Fig. 16.17a). Between Figures 16.17 and 16.17b, it is assumed that unusual weather, climate, or an accumulation of debris within the catchment basin caused a large debris flow to emerge suddenly from the tributary and to dam the Colorado River channel. The flow in the river is disrupted and ponded by the emplacement of a debris dam (Fig. 16.17b). A "lake" forms behind the debris dam, and at some time, depending on the height of the dam and the discharge in the main stem, a waterfall forms as this dam is overtopped. Schuster and Costa (1986) suggest that the dam would be breached (typically at the distal end of the debris fan) nearly instantaneously (hours to days)—and perhaps catastrophically. The evolution of a rapid from a lake and waterfall then would begin.

Unless there is a major breach in the debris dam with the first breakthrough of the ponded water (for example, more than 50 percent of the material is removed), the constriction of the main stem initially is severe (Fig. 16.17b). Floods of differing size and frequency then erode the channel to progressively greater widths (as shown in Fig. 16.17c–e). Small floods (Fig. 16.17c) enlarge the channel slightly (again, a nonquantitative term that depends on the prior history of the fan during breaching and on the material composing the fan). At first, the depletion of fine material through the debris fan may undermine the positions of large boulders. This undermining can cause rather dramatic changes in the scale of individual boulders and waves within the rapid. Indeed, it may be the reason that changes in wave behavior in Crystal Rapids seemed rather frequent in the late 1960s and early 1970s. As discussed above, the movement of a single, large boulder can cause dramatic changes in local wave structure in a rapid within regions of supercritical flow.

Moderate floods (Fig. 16.17d) enlarge and widen the channel further. The channel may become wide enough that at low discharges the flow is weakly supercritical or even subcritical. For example, the 1983 discharges of 92,000 cfs at Canyon Rapids (considered high by standards of Glen Canyon Dam discharges) widened the constriction at Crystal sufficiently that the strength of the Old Crystal Hole was diminished during subsequent lower flows at which it previously had been a substantial obstacle. On the other hand, the lateral waves became stronger because of headward erosion, which created additional potential energy at the top of the rapid. These waves, rather than the Old Crystal Hole, now form the main rafting hazard, though the hole is still a strong wave at discharges above about 30,000 cfs. Lateral widening, vertical scouring, and headwall erosion of the underwater debris dam occur simultaneously (Fig. 16.17f). Large floods further widen the channel at the debris fans and erode upstream through the fans (Fig. 16.17e). This cycle (a–e) can be repeated over and over through geologic time as floods in tributaries and on the main stream occur.

In summary, the local geometry of the river channel is subject to change if a substantially larger discharge is put through a rapid. This discharge can occur from flooding caused by meteorological events, from flooding caused by local emplacement and the breaching of natural dams in the river (debris of laval flows), or, now that Glen Canyon Dam is in place, from operational procedures at the dam. When the local gradient in the channel changes from such flood events, new waves can arise and old waves can become small or disappear.

The Rapids: Past, Present, and Future

The shape of the river channel at a debris fan at any instant of geologic time reflects contouring by different flood events—including any flood event that may have accompanied the emplacement and breaching of the debris dam. Even discharges as low as several thousand cfs appear to have sufficient velocity to clear the channel of large boulders, though such discharges are typically only a few percent of the total channel cross-sectional area. Fine-grained, transient sediment may partially mask the larger scale erosion (e.g., Howard and Dolan 1981, Fig. 7).

We now return to the questions raised earlier in this chapter. "Why is the shape of the channel, i.e., nozzle, eroded through the debris fans so uniform, and why is the value of constriction specifically 0.5?" The answer lies in Figure 16.16d (and in extrapolations of this figure to higher discharges found as Fig. 13 in Kieffer's 1985 publication).

The fact that the Colorado River is less constricted at most of the tributary debris fans than it is at Crystal Rapids suggests that discharges higher than 92,000 cfs have occurred in their history. We know this to be true: A flood of 220,000 cfs occurred in 1921, and a flood estimated at 300,000 cfs occurred in 1884. We can reasonably assume that even larger floods have occurred since many of these debris fans formed, a time that may exceed 10^4 years (Hereford 1984).

Because of the higher velocities associated with higher discharges, larger floods will make the channel at a debris fan wider. Extrapolation of the calculations done for Crystal Rapids, based on a threshold transport velocity of 30 ft/s (9 m/s), suggest that floods as large as 400,000 cfs are required to open the channel up to the constriction of 0.5. The accuracy of this estimate cannot be stated because we know too little about the threshold velocity for erosion (or other similar criterion). There are other variables to consider as well: (1) the reliability of our extrapolations using standard power-law functions of the dependence of depth, velocity, and head-on discharge; (2) our lack of knowledge on the rate at which vertical cutting and headward erosion occurred (we are assuming that the channel comes to an equilibrium shape during each flood); and (3) our inability to consider the true geometry of the river channel. As further data become available, we will be able to construct more accurate models.

Despite these uncertainties, we know that discharges an order of magnitude greater than discharges from the power plant at Glen Canyon Dam (and approximately a factor of five greater than the 1983 flood levels) contoured much of the river channel at the debris fans and gave the rapids their characteristic configuration. Without floods of this magnitude in the future, the character of the rapids will change as tributaries flood. The change will be toward more highly supercritical conditions as the constrictions become narrower, both laterally and vertically.

17

Late Cenozoic Lava Dams in the Western Grand Canyon

W. K. Hamblin

INTRODUCTION

"We have no difficulty as we float along, and I am able to observe the wonderful phenomena connected with this flood of lava. The canyon was doubtless filled to a height of 1,200 to 1,500 feet, perhaps by more than one flood. This would dam the water back; and in cutting through this great lava bed, a new channel has been formed, sometimes on one side, sometimes on the other. . . .

What a conflict of water and fire there must have been here! Just imagine a river of molten rock running down a river of melted snow. What a seething and boiling of waters; what clouds of steam rolled into the heavens!"

J. W. Powell, Aug. 25, 1869

From the time Powell first viewed the remnants of basalt adhering to the walls of the inner gorge in the western Grand Canyon over 100 years ago, relatively few people have had the opportunity to see and study this isolated area. But more and more visitors are discovering the viewpoint at Toroweap, where they are privileged to see one of the most spectacular displays of volcanism in North America (Fig. 17.1).

The volcanic features of this area are much more complex than one might first imagine. What Powell observed during his epic trip down the Colorado River was only a small fraction of the region's volcanic phenomena. Over 150 lava flows have poured into the canyon during the last 1.5 million years, and they have left an incredible record of volcanic events and their influence on the Grand Canyon. Some flows were extruded on the Uinkaret Plateau and cascaded over the north rim of the canyon into Toroweap Valley and Whitmore Wash. Others were extruded within the canyon itself and spread out over the Esplanade Platform before forming spectacular frozen lava falls that plunged over the rim of the inner gorge into the Colorado River 3000 feet (900 m) below. In several places, volcanoes are perched precariously on the very rim of the canyon, and remnants of others cling to the steep walls of the inner gorge. In addition, the dikes, sills, and volcanic necks exposed in the canyon are all associated with the complex sequence of recent volcanic events in the Uinkaret Plateau.

The spectacular lava falls that spill over the Esplanade into the inner gorge cap remnants of an older sequence of flows that formed huge lava dams, some of which were over 2500 feet (600 m+) high. One was more than 84 miles (135 km) long. The lava dams backed up the water of the Colorado River to form

FIGURE 17.1. Volcanic features in the western Grand Canyon. View looking northeast at the cascades and remnants of lava dams near the mouth of Toroweap Canyon. Vulcan's Throne is perched on the rim of the inner gorge. Recent extrusions of basalt flowed across the Esplanade platform and cascaded into the inner gorge, where they cap large remnants of major complex lava dams. Smaller remnants of other dams can be seen high on the north wall of the canyon.

temporary lakes upstream. Several of these lakes extended upstream through the Grand Canyon into Utah, slightly beyond the present extent of Lake Powell. As the lake behind the barrier overflowed, a new gorge was eroded through the lava dam, leaving only remnants of the basalt clinging to the walls of the canyon. Later eruptions formed new dams, which subsequently were breached and largely destroyed by the overflow of the Colorado River.

At least 13 major lava dams were formed in the Grand Canyon during the last million years. These dams, together with remnants of the sediment deposited in the lakes behind the dams, provide a fascinating record of this unusual and most recent series of events in the history of the Grand Canyon.

METHODS

The volcanic phenomena in the western Grand Canyon region are exceptionally well exposed, but they present some unusual problems to the field geologist because of the extremely rugged and largely inaccessible nature of this part of the canyon. Most of the basalt remnants exist as thin slivers clinging to the vertical cliffs of the inner gorge. This means that they are inaccessible for the most part and cannot be reached by trails from the canyon rim.

To study the exposures of lava throughout their 84-mile (135-km) extent along the river, we had to float equipment and supplies 180 miles (288 km) downstream from Lees Ferry at the beginning of each field season and establish a series of temporary base camps along the river. This allowed us to enter and leave the canyon and work slowly downstream so that our river operations could last the entire season.

However, working from the river in this manner presented its own problems. Although the basalts are almost 100 percent exposed and stand out in stark contrast to the tan and reddish Paleozoic strata, many of the exposures are largely inaccessible because they exist as vertical cliffs hundreds to thousands of feet above the river. In an attempt to solve this problem, we photographed the entire canyon wall from view points on the opposite side of the river, using a hand-held aerial camera. Enlarged prints were made and fitted together to produce a photo mosaic (cross section) on which we could plot all of our measurements and geologic observations. Elevations of the contacts between flow units were measured from the river using a theodolite.

We plotted our original data on enlarged vertical aerial photographs, and later we transferred our mapping to 7.5-minute topographic maps as the maps became available.

We also established several base camps on both the north and south rims of the Esplanade to study the more complicated areas between Toroweap and Whitmore Wash. We obtained additional data by using light aircraft and a helicopter.

Although some tantalizing questions about the details of certain relationships among the lava flows and the significance of various flow units remain, the study we made has established a significantly large database. And from this base, reasonably safe interpretations can be made about the late Cenozoic history in the western Grand Canyon.

Methods of Determining Relative Ages

The relative age of most of the flow remnants in the canyon is expressed clearly by superposition or by juxtaposition. The process by which juxtaposition of the flow remnants is produced is shown in the series of diagrams in Figure 17.2. The idea is simple. The first intracanyon flow entered the canyon from cascades or from centers of extrusion in the canyon itself. The lava partially filled the canyon, causing a lake to form upstream. Eventually, the backwater overflowed the barrier and eroded most of the basalt, leaving only thin vestiges of the lava flow adhering to the canyon walls in place of the once continuous flow. A subsequent flow entering the canyon would be juxtaposed against the older.

The remnants of lava flows in the Grand Canyon may be recognized and correlated throughout the canyon on the basis of several criteria. Although the petrography of some flows are similar, others are unique. Therefore, some flows can be recognized without difficulty on the basis of petrographic characteristics

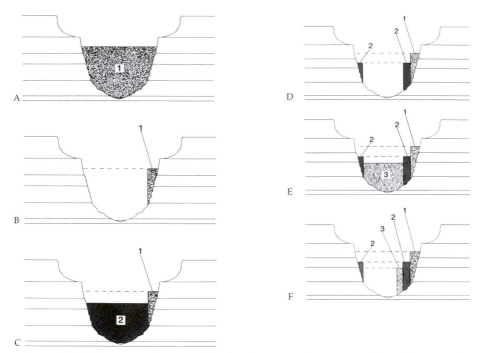

FIGURE 17.2. Diagrams showing the development of juxtaposed lava flows in the Grand Canyon. (A) A lava flow partly fills the canyon. (B) Erosion leaves small remnants of flow 1 adhering to the canyon wall. (C) Flow 2 refills the canyon. (D) Erosion removes most of flow 2, leaving remnants juxtaposed against flow 1 and against the canyon wall. (E) Flow 3 fills the canyon. (F) Erosion of flow 3 leaves remnants of flows 1, 2, and 3 stacked side-by-side according to relative age.

alone—even in a small outcrop. Diabase flows, for example, are readily distinguished from the dense, black, aphanitic basalt that characterizes other flow units.

In addition to petrographic characteristics, the internal structure of a number of flows is a distinguishing characteristic. One flow can be recognized by its abnormally thick joint columns in the basal colonnade. These range from 7 to 18 feet (2 to 5 m) in diameter. Others have unique jointing in the entablature, not only in size but also in the geometry of the columns. Stratigraphic sequences also are useful in correlation because river gravels of a specific lithology and thickness may occur within a specific sequence of flows. In addition, elevation and gradients of the top of all flow units were measured carefully with a theodolite. This provided an important means of correlating flow units on the basis of geomorphic relations.

Using these methods, we were able to map remnants of 13 major lava dams that were formed and subsequently destroyed in the Grand Canyon during the last one to two million years.

CHARACTERISTICS OF LAVA DAMS

It is apparent from the sequences of basalt preserved in the inner gorge of the Grand Canyon that four different types of dams were constructed during the

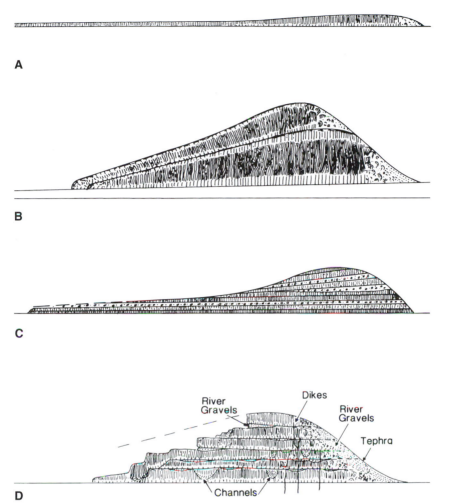

FIGURE 17.3. Types of lava dams in the Grand Canyon. (A) A simple dam formed by a single flow 150 feet to 600 feet thick. (B) High dam formed by several massive flows over 800 feet thick. (C) Compound dam composed of numerous flows 10 to 30 feet thick (3 to 9 m). (D) Complex dam built from multiple flow units 50 to 200 feet thick.

period of Late Cenozoic volcanic activity in the area. The nature of these structures is shown in Figure 17.3. Lava dams are simply intracanyon lava flows that attained a thickness sufficient to form a significant barrier to the flow of the Colorado River. They were thick, at least in part, because of the restriction of narrow canyon walls. These lava dams were asymmetrical structures, steep on the upstream margin and gently inclined downstream. The length of the dams ranged from 20 to more than 86 miles (32 to 137 km), and the height ranged from 150 to more than 2000 feet (45 to 600 m). The rate, volume, and viscosity of the lava extruded are primarily responsible for differences in the natures of the dams.

Thin Single-Flow Dams

The simplest type of barrier in the Grand Canyon was a dam constructed of a single lava flow, 150 to 600 feet thick (45 to 180 m). This type of barrier was exceptionally long, extending downstream for several tens of miles [in one instance, more than 86 miles (137 km)]. The internal structure of this type of flow is characterized by the classic columnar jointing that consists of a basal colonnade, an entablature, an upper clinker surface, and/or an upper colonnade. Most of these dams show little or no evidence of erosion on their upper surfaces.

The type of eruption that produced such a barrier probably consisted of a rapid extrusion of a moderate amount of fluid lava. Such a dam could be constructed by lava extruded within the canyon, or from lava that was extruded on the canyon rim and subsequently cascaded into the river. The volume of lava forming these barriers ranged from 0.03 cubic mile to 0.5 cubic miles. This is comparable to the average volume of flows in the volcanic fields on the Uinkaret Plateau. Judging from historic eruptions of Hawaii, Iceland, and Mexico, these dams were constructed within a period of several weeks.

Massive Dams

A distinctively different type of dam was formed in the canyon by abnormally thick, massive basaltic flows. These flows were over 800 feet (240 m) in depth. Because these are the oldest dams to form in the canyon, only a few remnants are visible, but the elevation of the upper surfaces of the remnants indicate that these barriers were not more than 20 miles (32 km) long. It also appears that the lava was extruded faster than the backwater of the Colorado River could overflow. The best estimates indicate that the volume of these dams was 0.5 to 1.2 cubic miles. This is only twice the volume of the average single-flow dam.

Compound Dams

Compound dams were constructed by a sequence of numerous flow units 10 to 30 feet thick (3 to 9 m), all of which were deposited in relatively rapid succession. These dams ranged from 300 to 900 feet (90 to 270 m) in height. The overall geometry of the compound dams was much like the simple single-flow dams, consisting of a steeply inclined front (upstream slope) and a long, gently inclined downstream slope. The volume of these dams ranged from 0.06 to 0.23 cubic miles.

Complex Dams

Complex dams were built from a series of multiple flow units 50 to 200 feet (15 to 60 m) thick. The general characteristics of each individual flow unit are similar to those of the single-flow dams, but the upper colonnade or clinker surface on flows within the complex dam often is eroded. Locally, deep channels are cut in the flow. These channels are filled with younger flows, ash, and sediment. Lenses of river gravel, sand, and, in some cases, ash are found separating the major units of basalt. The upstream slope of the complex dams consists of flow units that are inclined in a steep angle upstream. These pass rapidly into rubble, agglomerate, and tephra.

The erosional surfaces on the flows, together with interstratified river gravels, clearly indicate that the lake which formed behind the dam overflowed during the period of extrusion and most flows were eroded, to some extent, between periods of eruption.

The complex dams were quite high, ranging from 600 to more than 1400 feet (180 to 420 m) above the present gradient of the Colorado River. The steep gradient on the preserved remnant suggests that the complex dams were no more than 12 miles long. In many cases, it may have been much less because the lower end of each flow would have been eroded by upstream migration of waterfalls and rapids before the subsequent flows were extruded.

CHARACTERISTICS OF LAKE SEDIMENTS

The lakes that formed in the Grand Canyon behind the lava dams were unusual from the standpoint of their geomorphic setting, origin, and history. They had little in common with natural lakes formed in glacial terrains or in low-lying coastal regions. These were lakes formed in a deep canyon and identical in nearly every respect to the present man-made reservoirs such as Lake Mead and Lake Powell. A considerable amount of hydrologic data is available for these two reservoirs and provides the best insight into the nature of the sediment deposited in the lakes behind the lava dams, their history, and ultimate destruction.

When the Colorado River was blocked by a lava dam, coarse sand and gravel were deposited as deltaic sediments at the point where the Colorado River entered the lake. Fine-grained sediments were carried by turbidity currents far into the deeper part of the reservoir, where they were deposited as graded beds of silt and mud. In the absence of large tributaries, the main filling of the lake was accomplished primarily by deposition from the Colorado River.

The lakes formed in the Grand Canyon were surrounded by steep canyon walls. Large beaches rarely developed along the shore, and without significant input from the tributaries, the major process operating along the shores of the lakes was mass movement. Slope wash, rock falls, and the general downslope migration of colluvium (which presently is the major process operating along the canyon walls) would have continued during the lifetime of the lakes. A thin mantle of coarse, subaqueous slope debris was deposited close to the canyon walls contemporaneous with the deposition of lake silts. Each tributary continued to transport and deposit this material into the lakes at the heads of tributary bays.

Beyond the zone affected by slope processes, the major types of sediment deposited were mud and silt. Gravel would be a significant facies in some major tributary channels as a result of flash floods, and some of the larger tributaries near the upstream reaches of the lake probably constructed small deltas of sand and gravel. However, the major delta of sand and gravel formed in the area where the Colorado River emptied into the lake. Deltaic deposits prograded downstream over the silts deposited by turbidity currents.

After the dam was destroyed, most of the unconsolidated, water-saturated sediment was flushed rapidly out of the main canyon, leaving only minor remnants of sediment close to the valley walls. This material would be dominantly slope wash and colluvium with silt filling the space between the larger particles. Therefore, most of the preserved sediment near the canyon walls was not composed of typical lake silts. Instead, it was composed of colluvium with intercalated laminated silt. This type of deposit is difficult to distinguish from the present accumulation of colluvium.

DEVELOPMENT AND DESTRUCTION OF LAVA DAMS

The hydrologic data from the Bureau of Land Management's Lake Mead Survey (1963 and 1964) provide the basic information from which we are able to calculate the rates at which the lakes behind the various lava dams were filled with water and, subsequently, sediment. These data also provide some indication of the time necessary for a lava dam to be eroded away completely (Table 17.1).

Rates of Formation of Dams

Although the formation of a lava dam in the Grand Canyon was a significant event that dramatically changed canyon morphology, the time needed to create a lava dam was remarkably short by any standard and certainly would be considered instantaneous in a geologic time frame. Observations of basaltic eruptions in historic times indicate that most basaltic extrusions occur in a matter of days or weeks. The major flows in the Grand Canyon, most of which were 100 to 200 feet (30 to 60 m) thick, probably moved tens of miles down the Colorado River in a matter of days.

This conclusion is supported by the fact that the upper colonnade and/or a clinkery upper surface of the flow often is preserved, essentially unmodified by erosion. This indicates that the flow was extruded in a period of time less than that required for the lake impounded behind the dam to overflow. If extrusion occurred during a longer period of time, the lake behind the dam would overflow, and erosion would modify the upper surface features of the basalts quickly.

We can calculate quite precisely the time required for the backwater behind each dam to overflow. Based on present discharge rates of the Colorado River, the lake formed by backwaters behind a lava flow 100 feet (30 m) high would overflow in 17 days. The construction of a single-flow lava dam 2000 feet (600 m) high and tens of miles long could be completed within a few weeks. There-

TABLE 17.1. Data on the Geometry, Age, and Hydrology of the Lava Dams in the Western Grand Canyon

Dam	Elevation	Height above River	Radiometric Date[a]	Volume of Lava	Lake Length	Water Fill Time	Sediment Fill Time
1 Prospect	4000	2330	0.68 ± 0.05	4.0		23 yrs	3018 yrs
2 Lava Butte	3365	1730	0.58			10 yrs	382 yrs
3 Ponderosa	2800	1130	0.61 ± 0.02	2.5		1.5 yrs	163 yrs
4 Toroweap	3093	1443	0.56 ± 0.07	3.7		2.62 yrs	345 yrs
5 Esplanade	2600	960		1.8		287 days	92 yrs
6 Buried Canyon	2480	850	0.91 ± 0.07	1.7		231 days	87 yrs
7 Whitmore	2500	900		3.0	100	240 days	88 yrs
8 "D" Flows	2295	635	0.58 ± 0.03	1.1	74	87 days	31 yrs
9 Lava Falls	2260	600		1.2		86 days	30 yrs
10 Black Ledge	2033	373		2.1	53	17 days	7 yrs
11 Gray Ledge	1813	203	0.79 ± 0.13	0.3	37	2 days	10.3 mos
12 Layered Dbs	1938	298	0.62 ± 0.05	0.3	42	8 days	3 yrs
13 Massive Dbs	1826	226	0.14 Ma	0.2		5 days	1.4 yrs

[a]Dates from Dalrymple et al. (1998).

fore, it is clear that any lava flow retaining an uneroded upper surface was extruded in a matter of a few days.

Dams built from multiple flow units required more time and involved cycles of partial erosion between periods of extrusion. The short time necessary for the complete erosion of a dam, however, puts definite time constraints on the development of any barrier to the Colorado River. Regardless of size and history of eruption, the formation of every lava dam in the Grand Canyon was instantaneous from the perspective of a geologic time frame.

Rates of Reservoir Fill

The time required for reservoirs behind these dams to become filled completely with water and sediment also was extremely rapid. The hydrologic data from the various lakes are summarized in Table 17.1. These data indicate that the lakes formed behind the smaller barriers—those 150 to 400 feet (45 to 120 m) high—would overflow in 2 to 17 days. Lakes formed behind the higher dams (200 to 1000 feet or 60 to 330 m high) overflowed in 22 years. Thus, the lava dams were subjected to erosion soon after they were formed, even before the interior of the lava was completely cool.

The volume of sediment carried by the Colorado River has been measured for many years by the Bureau of Reclamation. These data indicate that reservoirs behind the lava dams silted up in only a few hundred years at most. Many of the smaller reservoirs silted up in a few months . Thus, the sediment load of the Colorado River was soon transported over the dam, causing normal erosion by abrasion of the river channel after the dam was formed. The data in Table 17.1 indicates that a reservoir formed behind the dam 150 feet (45 m) high would be filled with sediment in 10.33 months. A dam 1150 feet (345 m) high would be full of sediment in only 345 years. The highest dam would be filled with sediment in 3000 years. Thus, every phase of the construction of the dam and the formation of the reservoir or lake behind it, and the ultimate filling in of the lake with sediments, occurred in a very short time.

Erosion and Destruction of Dams

Although we do not know the precise details in which lava dams were eroded, some boundary conditions can be established, based on downcutting of major stream systems. Normal downcutting of the stream channel by abrasion was undoubtedly a significant process of erosion. It began as soon as water overflowed the dam and reached maximum efficiency when the lake silted up and a normal sediment load was transported over the dam. Another important process was the migration of rapids and waterfalls that initially formed on the downstream end of the flow (Rogers and Pyles 1979). Two important characteristics of the intracanyon flows facilitated the migration of waterfalls (Fig. 17.4):

1. Intracanyon flows were deposited directly on the sand and gravel bed in the channel of the Colorado River. This layer of unconsolidated sediment would be eroded easily by undercutting at the plunge pool below the waterfall.
2. The vertical columnar jointing in the basalt constituted an all-pervasive structural weakness throughout the flow. The hexagonal columns produced by the jointing impart a low cohesive strength to the rock body so that the columns would readily topple into the plunge pool beneath the waterfall.

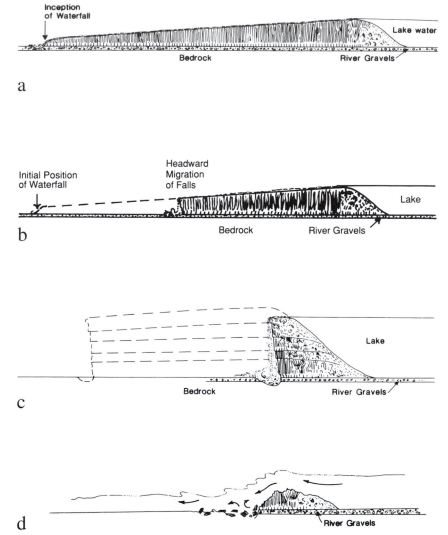

FIGURE 17.4. Diagrams showing erosion of a lava dam by headward migration of waterfalls. (a) As soon as the backwater overflowed the lava dam, a small waterfall would form at the downstream end of the lava flow. (b) Upstream migration of the falls would be accelerated by undercutting of the unconsolidated river sediments beneath the flow and by the weakness of the rock resulting from vertical columnar jointing. (c) As the waterfalls migrated headward, the stability of the dam could be jeopardized by the pore pressure in the columnar jointing. (d) Some dams may have failed catastrophically. This event could have been minimized if contemporaneous downcutting lowered the level of the overflow before the falls migrated to the head of the dam.

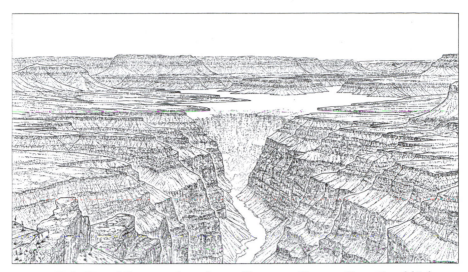

FIGURE 17.5. Waterfalls over a lava dam at Toroweap Canyon. Exceptional high waterfalls [up to 2460 feet (750 m) high] would form as headward erosion destroyed the dam. The higher dams would be 15 times as high as Niagara Falls with more than twice the discharge.

In addition, hydraulic plucking of the columns probably occurred in a river with a discharge as high as that of the Colorado. The combined effects of the unconsolidated sand and gravel substratum, and columnar jointing in the basalt, resulted in rapid upstream migration of the waterfalls. As the waterfalls migrated upstream and approached the head of the dam, they became higher and higher. The process of erosion then was accelerated by the increase in potential energy for undercutting and scouring. Erosion of the highest dams must have presented a spectacular scene. As the waterfalls advanced toward the highest section of the dam, a tremendous scour hole must have been generated at the base of the falls. Exceptionally high waterfalls—those over 2500 feet (759 m)—would form on the highest dams (Fig. 17.5). As erosion proceeded headward, the stability of the dam almost certainly was jeopardized by the enormous pore pressure in the columnar joints. At some critical point, dams may have failed as catastrophic events, with a rapid discharge of a tremendous volume of water and saturated sediment that had accumulated in the lake behind the dam.

Documentation of the rate of waterfall migration in some areas provides important insight into possible time frames of the destruction of the lava dams. Niagara Falls, for example, has migrated a distance of more than 11 miles (18 km) in 8000 years, an average rate of three feet per year. If this figure is typical for the migration of a waterfall on a large river, the lava dams in the Grand Canyon, which generally were less than 20 miles long, would take a maximum of 20,000 years to be completely destroyed by headward migration of waterfalls alone. It also seems safe to conclude that the time interval for the various phases of the buildup and destruction of the dam would be on the following orders of magnitude:

1. Single-flow dams would be formed in a matter of several days, whereas the higher compound lava dams would take up to several years for construction.

2. Water would fill the reservoir behind these dams in a matter of months. At most, only 7 years would be required for the water to fill the reservoir behind the highest dam completely.
3. Sediment would fill the small reservoirs within 1 to 7 years. In the higher reservoirs, it would take between 100 and 1000 years to fill the lake completely with sediments. Most of the dams probably would be destroyed within 10,000 to 20,000 years after they were formed.

It is apparent from these observations that the 13 lava dams occupied the canyon for a relatively brief period of time, probably no more than a total of 240,000 years.

THE LAVA DAMS

In the Grand Canyon the oldest lava dams whose remnants are still preserved are (a) two huge structures formed by thick, massive flows and (b) a multiple flow structure composed of numerous flows 50 to 150 feet thick. These oldest known flows are referred to as the Lava Butte Dam, the Prospect Dam, and the Ponderosa Dam.

FIGURE 17.6. (a) Photograph of the Prospect Dam. Although only one remnant of this dam remains, the thick flows suggest that the dam did not extend downstream for more than a few miles. The absence of an erosional surface between flows suggests that extrusion was rapid and the barrier was completely constructed before the backwater in the lake overflowed.

(*continues*)

The Prospect Dam

The only remaining remnant of the Prospect Dam is a sequence of exceptionably thick flows exposed in the large alcove just east of Prospect Canyon. Here, a vertical cliff of basalt almost 2000 feet (600 m) high extends from the talus slopes near the Colorado River up to the level of the Esplanade (Fig. 17.6). Unfortunately, the boundaries between the flow units are difficult to see from viewpoints on the river or from viewpoints on the north rim near Vulcan's Throne. Observations made from a helicopter, however, reveal that the Prospect flows are extremely thick and that there are no more than three or four major flow units, each of which is more than 800 feet (240 m) thick. The Prospect flows, therefore, are some of the thickest flows in the canyon.

The oldest units in Prospect Valley are a sequence of bulbous or elliptical basalt bodies with intersperse tephra. These flows are overlain by the thick, massive Prospect flow.

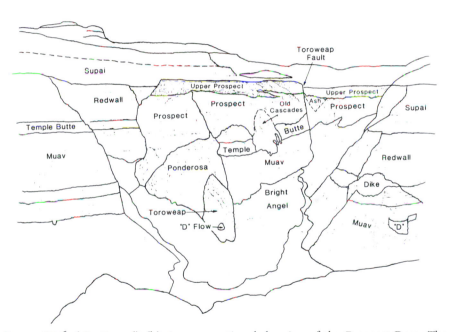

FIGURE 17.6. (*Continued*) (b) A cross-sectional drawing of the Prospect Dam. The structure of the dam consisted of several units: (1) the elliptical units formed by basalt interacting with backwater; (2) hydroexplosive tephra at the front of the dam; (3) the main flows with long, sinuous columnar jointing, (4) the basalt cap formed shortly after overflow of the lake. The Upper Prospect Flow consists of a horizontal unit nearly 400 feet (120 m) thick which once spread across the top of Prospect Canyon and across part of the Esplanade. It formed a cap rock covering the older flows and cinder cones that filled the ancient canyon. Headward erosion along the Toroweap Fault has re-excavated Prospect Canyon and formed a steep, V-shaped gorge over a mile long. Note the offset of the Upper Prospect flow near the apex of the canyon produced by recurrent movement along the Toroweap fault. Recent movement also has displaced the alluvium covering the Upper Prospect flow.

The elliptical structures probably were produced by the interaction of lava with the water of the Colorado River. Typical pillow structures would not be expected to form in abundance from a flow entering a river because most of the lava would not be covered completely with water. Only the upstream margins of the flows would interact directly with the river. The downstream segment of the flow would remain essentially dry until the backed-up river water overflowed the barrier.

Above the elliptical units are several exceptionally thick and massive flows that constitute the main body of basalt. These flows form a sheer vertical cliff over 1000 feet (300 m) high. The internal structure of these flows consists of long, slightly sinuous, columnar joints that tend to radiate out from a central point like huge shocks of wheat hundreds of feet high and only tens of feet wide. The only obvious stratigraphic break visible in this sequence of flows appears midway up the cliff at the approximate elevation of the contact between the Muav and Temple Butte limestones. A cross section showing the relationships between the flows in the Prospect alcove is shown in Figure 17.7.

On top of the sequence of thick flows exposed in the alcove east of Prospect Canyon is a single flow approximately 400 feet (120 m) thick that is characterized by massive columnar joints. This "Upper Prospect Flow" is not confined between the walls of the ancient Prospect Canyon like the underlying units. Instead, it spreads across the top of the canyon and over parts of the Esplanade (Fig. 17.6). This flow probably blocked the river where it entered the canyon, but because the flow is eroded back from the walls of the inner gorge, we cannot determine with certainty whether this unit formed a dam at an elevation of 4000 feet (1200 m) or if it cascaded into the inner gorge after the Prospect Dam was formed. The horizontal position of the flow at the very walls of the inner gorge, plus the considerable thickness of the flow and its massive columnar jointing, gives no suggestion that the flow cascaded into the Grand Canyon. Rather it supports the conclusion that a lava dam once existed at this elevation.

In the Grand Canyon the Prospect Dam is the oldest lava dam from which we have found remnants. There could be older dams associated with the ex-

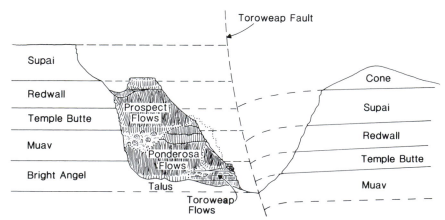

FIGURE 17.7. The flows in Prospect Canyon as seen from the west side of Vulcan's Throne. Over a mile of headward erosion has occurred in Prospect Canyon along the Toroweap fault, re-excavating a sharp, V-shaped gorge in the basalt and underlying Paleozoic rocks (Fig. 17.6). The thick, massive Prospect flow forms a sheer, vertical cliff almost 2000 feet (600 m) high near the left margin of the drawing.

trusions capping Mount Trumball and Mount Dellenbaugh, but we found no evidence of their existence in the inner gorge. The old age of the Prospect Dam is indicated by the degree to which the flows in Prospect Canyon have been eroded. The steep, V-shaped gorge of Prospect canyon, which is over a mile long, has been eroded in the Prospect flows. This face shows that the Prospect flows are much older than the basalt that fills similar major tributaries to the Grand Canyon, such as Toroweap Valley and Whitmore Wash, both of which are filled with a sequence of flows only slightly modified by erosion. The flows in Toroweap Valley and Whitmore Wash, therefore, almost certainly are much younger than those that filled Prospect Canyon.

From the elevation of the Prospect Dam remnant, we are able to reconstruct the shoreline of the lake it formed. Prospect lake was the largest and deepest lake to form in the Grand Canyon. It extended all the way up through the Grand Canyon beyond Lake Powell and upstream to beyond Moab, Utah (Fig. 17.8). The rather extensive terrace deposits of gravel, sand , and silt in the Bull Frog area of Lake Powell and at Moab occur at elevations of approximately 4000 feet (1200 m) and probably are remnants of the Colorado River delta formed in the lake. The waters in the lake were deep enough to inundate most of the tributaries in the Grand Canyon for a significant distance. This meant that the configuration of the shoreline assumed a distinct dendritic pattern. The shoreline throughout much of the Grand Canyon in the vicinity of the park headquarters was very close to the base of the vertical cliffs of the Redwall Formation. Calculations show that the lake behind the Prospect Dam required 22 years to fill with water and 3018 years to fill with sediment.

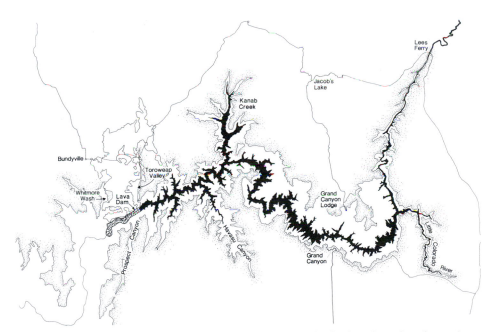

FIGURE 17.8. The Prospect Lake. The Prospect Dam was the highest lava dam formed across the Colorado River. The lake formed behind the dam that extended all the way through the Grand Canyon and up into Utah. It formed more than 1.2 million years ago.

Lava Butte Dam

Several remnants of a sequence of thin flows interbedded with river sediment are preserved high on the north wall of the inner gorge at mile 180.8 about 4 km downstream from Vulcan's Throne. One remnant caps an isolated butte that stands approximately 130 m above the surrounding area. If this remnant was formed by topographic inversion—that is, by erosion along the flow margins—it suggests that the Lava Butte Dam is one of the oldest dams preserved in the canyon. The remnants are not juxtaposed with other flows, however, so its rel-

FIGURE 17.9. (a) The Ponderosa flow at Mile 181.6, as seen from the south rim. The Ponderosa in this area is 800 feet (240 m) thick and extends from near river level to the base of the cascades. Much of the lower part of the Ponderosa is obscured by younger flows that are juxtaposed against it. The Ponderosa Dam was formed from a single, thick flow in which the classic three-part columnar jointing is well-developed. It probably was a relatively short dam and did not extend downstream much beyond Mile 190.

(continues)

ative age cannot be determined. The Lava Butte Dam reached a height of 560 m above the present river, but there are insufficient remnants to permit an accurate estimate of the original lava volume.

The Ponderosa Dam

The Ponderosa Dam was similar to, but somewhat smaller than, the Prospect Dam. It was constructed by a single flow at least 1000 feet (300 m) thick. The top of the barrier was at an elevation of 2800 feet (840 m), or 1130 feet (339 m) above river level. The best exposures are at the west end of the Esplanade Cascades on the north side of the river (Mile 181.6) and in the alcove east of Prospect Canyon (Fig. 17.9).

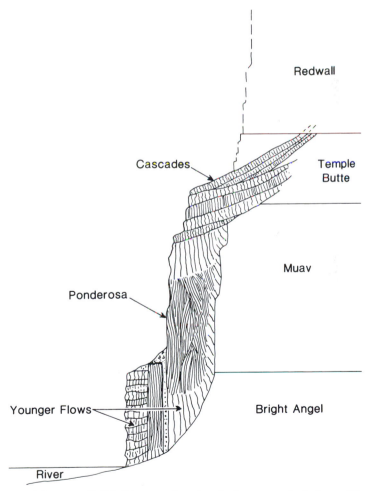

FIGURE 17.9. (*Continued*) (b) Diagram showing the juxtaposed relations of Ponderosa and younger flows. The Ponderosa is juxtaposed against the Paleozoic rocks and is capped by younger flows and cascades. Remnants of two younger dams are juxtaposed against the Ponderosa.

The lake that formed behind the Ponderosa Dam extended through the park headquarters area of the Grand Canyon upstream to somewhere near Nankoweap Rapids. It took only a little more than a year (450 days) for the Ponderosa Lake to fill and overflow. Within 162 years, the lake was full of sediment.

Complex Dams

After the destruction of the massive barriers of the Prospect and Ponderosa dams, three smaller dams were built. These were complex dams built from multiple units 50 to 200 feet (15 to 60 m) and represent a different style of extrusion. They are referred to as the Toroweap, Esplanade, and Buried Canyon dams after large exposures preserved on the north side of inner gorge at Mile 179 (mouth of Toroweap Canyon), Mile 181 (below the Esplanade Cascades), and Mile 183 (the Buried Canyon).

These complex dams were built from a series of flows interbedded with various amounts of river gravel and tephra. Most of the flows are characterized by a well-developed basal colonnade and a thick entablature consisting of slightly sinuous columns. One of the important features of complex dams is that the upper surface of each flow commonly shows evidence of erosion prior to deposition of the overlying younger unit. Locally deep channels are cut in the flows and then filled with younger basalt, gravel, and tephra. The major units of basalt, therefore, are separated frequently by lenses of gravel, sand, and, in some cases, tephra. Judging from this evidence of erosion in the upstream part of each flow during the formation of the dam, it is likely that waterfall migration destroyed a large segment of the downstream part of each flow before the extrusion of the succeeding younger flows. The downstream profile of the final complex dams, therefore, were likely steep and relatively short (Fig. 17.3d).

Each of the major flow units in the complex dams ranged from 150 to 200 feet (45 to 60 m) in thickness, and each caused an immediate barrier to form across the Colorado River. The small lakes accumulating behind these barriers rose rapidly and overflowed within a matter of a few days. (Calculations based on discharge rates of the Colorado River indicate that a barrier 100 feet high would overflow in only 2.3 days.) The overflow of the river then began to erode and modified the blocky rubble on the upper surface of the flow, together with much of the upper colonnade. Small waterfalls that formed at the end of the flow immediately began to migrate upstream. Subsequent extrusion of a younger flow, therefore, probably overlapped the eroded end of the older unit. The flows in the complex dams, however, must have been formed by a fairly rapid series of volcanic extrusions with only short time intervals between extrusions—time sufficient only for the upper segment of the overflows to be modified by erosion of the Colorado River.

Based on the present rate of discharge and the current sediment load of the Colorado River, overflow of a barrier 200 feet (60 m) high probably occurred within 3 days. It seems most likely that the small lake was completely silted up within 10 months. Without question, the small lakes formed behind each lava flow were filled completely with sediment prior to the next extrusion. Thus, by the time the complex dams were constructed, the lake behind it was completely silted up. Because a dam formed by a single flow 150 to 200 feet (45 to 60 m) would be completely eroded within 20,000 years, the construction of the complex dams must have been completed in less than 100,000 years.

The Toroweap Dam

Remnants of the Toroweap Dam occur on the north wall of the inner gorge beneath Vulcan's Throne. They are largely obscured from viewpoints at the Toroweap Campground, but from the south rim (at the mouth of the Prospect Canyon) and from the river, the thick sequence of flows can be seen adhering to the canyon wall (Fig. 17.10).

Excellent exposures of the upstream part of the Toroweap Dam are found on the north wall of the inner gorge, just upstream from the mouth of Toroweap Canyon (Fig. 17.11). Here, a detailed cross section of the internal structure of the Toroweap Dam is exposed to reveal the structure of the dam at points where the lava made contact with the waters of the Colorado River. Each of the major units shows three distinct types of structures. The front of the flow is characterized by a series of billowy, elliptical structural forms, many of which are 20 or 30 feet (6 or 9 m) in diameter. The size of the elliptical structures grade down to less than 3 feet in diameter.

We believe these features formed as a single flow entered the inner gorge and interacted with water from the Colorado River (Fig. 17.12). When the flows first entered the river, some of the material probably formed pillow structures. Because the flow was thicker than the depths of the Colorado River, however, much of the flow was not covered immediately with water. This material proceeded downstream, essentially dry. Backwater of the Colorado River probably overflowed the barrier in a matter of a few days. As a result, water flowed over the top of the hot lava, greatly influencing the nature of the cooling and the in-

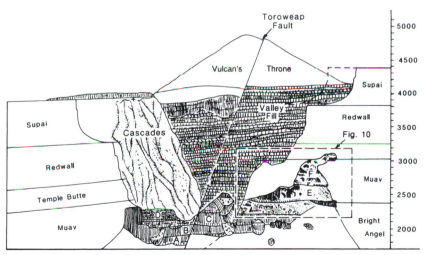

FIGURE 17.10. Diagram showing the relationships between the major flows at the mouth of Toroweap Valley. Remnants of the Toroweap Dam are exposed from river level to an elevation of approximately 3000 feet (900 m). They extend far beyond the mouth of Toroweap Valley and are juxtaposed against the Bright Angel, Muav, and Temple Butte formations. Toroweap Valley is filled with a sequence of thin flows to the level of the Esplanade. Recent cascades cover the western part of this sequence. Vulcan's Throne rests upon the flows t hat fill Toroweap Valley. The Toroweap fault displaces the entire sequence, including Vulcan's Throne.

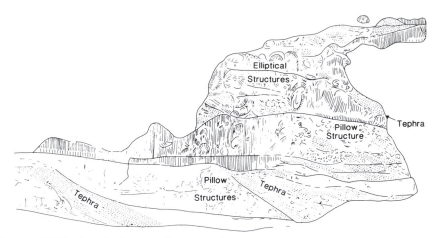

FIGURE 17.11. Sketch showing the internal structure of the Toroweap Dam, exposed in remnants high on the north wall of the inner gorge at Mile 178.4. The individual flows are 150 to 200 feet (45 to 60 m) thick and are characterized by basal colonnade and entablature jointing. Eastward, the flows develop elliptical structures that, in turn, grade into pillow basalts and stratified tephra. The tephra is inclined upstream. These structures suggest that this area is near the head of the lava dam.

ternal columnar jointing. Water from the overflow of the river would seep easily into large cracks in the lava, causing part of the flow adjacent to the fracture to cool very rapidly. Cooling in these cases would proceed inward from the upper surface and inward from the walls of the vertical cracks. Many geologists believe that this type of cooling produced the elliptical, pseudo-pillow structures found here.

At the point of contact with the river water, the lava was quenched, and water interacted with the basalt, causing it to contract and explode. The hydroexplosive activity resulting from the lava making contact with the water probably produced a small amount of tephra that accumulated in the steadily rising lake behind the barrier. This material was deposited at the angle of repose to form the upstream dipping layer of tephra. Undoubtedly, single-flow units developed lobes or offshoots that were diverted upstream a short distance over the elliptical structures and tephra. Thus, several minor flow lobes likely developed from the major flow unit.

In addition to the large remnant of the Toroweap Dam preserved below Vulcan's Throne, several small, isolated remnants are preserved high on the canyon walls on the south side of the inner gorge. These remnants, some 1400 to 1600 feet (420 to 480 m) above the present level of the river, can be observed easily from the Toroweap Campground.

The Toroweap Dam was at least 1443 feet (433 m) high, making it one of the higher dams in the Grand Canyon. Its history includes five major periods of extrusion, each followed by short periods of backwater overflow and erosion. Remnants of the Toroweap sequence are juxtaposed against the Ponderosa near the mouth of Prospect canyon, indicating that it is younger than both the Ponderosa and Prospect dams. Radiometric dates obtained from the oldest units (flow A) indicate that the Toroweap Dam is 1.16 to 1.18 million years old (McKee et al. 1967). More recent measurements indicate that the Toroweap Dam is only

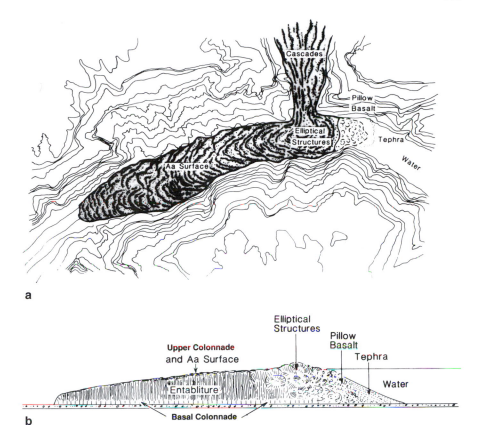

FIGURE 17.12. The structure of a hypothetical flow 200 feet (60 m) thick in the Grand Canyon. a. Top view, b. Lateral view. Only the upstream margin of the flow would interact with water from the river. This would produce hydroexplosive tephra, which would be deposited at the upstream margin of the dam. Some pillow structures would probably be produced where lava came into direct contact with the river. Most of the flow would move downstream essentially dry until the backwater overflowed the barrier. The Colorado River would overflow the barrier of a single flow within two days and would influence the cooling of the lava. The overflow may have had an effect on the development of the entablature in the downstream part of the flow.

0.56 ± 0.07 (Dalrymple et al., 1998). This indicates that the Toroweap Dam is older than the one formed in the Esplanade and Buried Canyon.

Toroweap Lake was one of the larger lakes to form in the Grand Canyon. It extended all the way through the present National Park Visitor's area and upstream into the vicinity of Lees Ferry (Fig. 17.13). Throughout much of the park region, the lake's depth was approximately 750 feet (225 m). In the region of the visitor's center, the lake essentially flooded all of Granite Gorge—with the shoreline occurring very close to the vertical wall of the Tapeats Sandstone.

The gravel, sand, and silt that formed the terrace deposits at Lees Ferry at an elevation of 3600 feet (1080 m) probably were deposited as a delta built by the Paria and Colorado rivers where they emptied into the Toroweap Lake. Like-

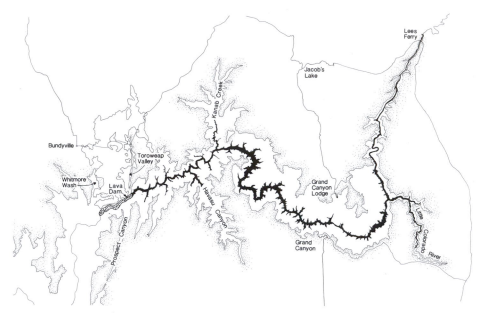

FIGURE 17.13. The Toroweap lake. The Toroweap lake extended upstream through the Grand Canyon to the area of Lees Ferry. The shoreline throughout much of the canyon was near the cliff of the Tapeats Sandstone.

wise, the silt deposits that form the main floor of Havasu Canyon probably represent a major remnant of Toroweap Lake deposits.

If the Toroweap Dam formed instantaneously, the lake behind it would overflow in 2.6 years. It would be completely full of sediment in 345 years. Because the Toroweap Dam was constructed over a period of time, the lake undoubtedly was essentially full of sediment by the time the dam was completed. After the extrusion that built the Toroweap Dam terminated, erosion by headwater migration of the waterfall and the downcutting of the stream channel probably destroyed the dam in less than 10,000 years.

The Esplanade Dam

On the north side of the river between Miles 181 and 182, a sequence of flows that formed a complex dam at an elevation of 2600 feet (780 m), or 960 feet (288 m) above river level, is preserved beneath the Esplanade cascades (Fig. 17.14). These flows can be seen from viewpoints west of the present Toroweap Campground and resemble in some respects the Toroweap sequence. Both are preserved beneath major cascades.

The Esplanade sequence also is similar in some respects to that preserved in the Buried Canyon at Mile 184. These similarities have led some geologists to consider the Toroweap, Esplanade, and Buried Canyon lavas as part of a single complex lava dam. However, the stratigraphic sequence in each dam is distinctly different.

Several small remnants of the Esplanade Dam are preserved in tributary canyons on the south wall of the inner gorge between Mile 183.5 and Mile 184.

FIGURE 17.14. Remnants of the Esplanade Dam at Mile 181–182. The lower units exposed near the lower right part of the photograph are similar to the lower flows of the Toroweap Dam. The upper units below the cascades are distinctive in that the columnar jointing consists of rough, thick columns without a basal or upper colonnade.

They provide important documentation concerning the height of the Esplanade Dam. All are preserved in "hanging valleys" at an elevation of 2600 feet (780 m).

The lake that formed behind the Esplanade Dam was similar to, but slightly smaller than, the Toroweap Lake and extended upstream only to Hance Rapids. It was exceptionally narrow throughout its entire extent and did not extend any appreciable distance up the tributaries. It probably was full of sediment by the time the youngest flow was extruded. Headward migration of waterfalls undoubtedly destroyed the dam within 10,000 years.

The Buried Canyon Dam

One of the most remarkable exposures of basalt in the Grand Canyon is on the north side of the river at Mile 183, halfway between Toroweap Valley and Whitmore Wash. Here, a segment of the Grand Canyon was filled with a sequence of older basaltic flows, which preserved a buried canyon (Fig. 17.15). The basalts here are not remnants clinging to the wall of the canyon as is the case in most areas. In this region, the basalts fill the entire lower part of the original inner gorge, just as they did when they clogged the canyon and formed a dam. The floor and walls of the original canyon are exposed in a perfect cross section. This unique exposure is preserved because the Colorado River shifted to the outside of a slight meander bend at the time it overflowed the lava dam. Subsequent downcutting formed a completely new canyon, leaving a section of the original gorge filled with basalt preserved on the inside of the meander.

The Buried Canyon contains nine major flow units totaling 650 feet (195 m) in thickness. The top of the highest flow stands at an elevation of 2480 feet (744

a

b

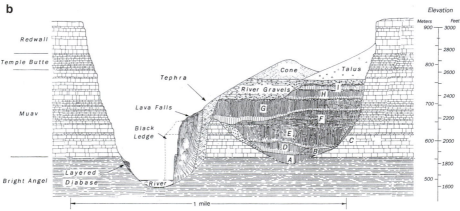

FIGURE 17.15. (a) The Buried Canyon Dam at Mile 183. (Note the boat near the sandbar for scale.) This sequence of flows filled the "original" inner gorge of the canyon. It is preserved because the Colorado River shifted southward around a meander bend and cut a new gorge in the Paleozoic strata instead of eroding the basalt. The Buried Canyon Dam was formed by a series of flows 150 to 200 feet (45 to 60 m) thick, all of which show evidence of erosion prior to the extrusion of the next younger lava. (b) Cross section showing the major flows in the Buried Canyon Dam. The nine major flow units shown in this section are separated by two major erosional surfaces. The entire sequence of basalt is overlain by 165 feet (50 m) of river gravels capped by a recent cinder cone.

FIGURE 17.16. Remnants of the Buried Canyon Dam preserved downstream at Mile 184–185.

m), or 860 feet (258 m) above the present river level. The flows are overlain by a deposit of river gravels 160 feet (48 m) thick and are capped by a younger volcanic cone.

In addition to the remnants of the dam preserved in the Buried Canyon at Mile 184, a number of small remnants are preserved as "hanging valleys" in small tributary valleys on both sides of the river between Miles 184 and 185 (Fig. 17.16). These remnants represent segments of the higher basalts that formed the dam and flowed from the inner gorge up the small tributaries.

Several flows in the Buried Canyon sequence were dated by Brent Dalrymple from the U.S. Geological Survey. They range in age from 0.89 to 1.14 million years old. Based on these figures, geologists calculate that the buildup of the dam took place over a period of approximately 250,000 years. Large erosional channels in the basalt sequence indicate that the dam was built and partly destroyed four times during this period of time.

The lake behind the Buried Canyon Dam was similar to that formed by the Toroweap Dam and Esplanade Dam (Fig. 17.13).

The Whitmore Dam

One of the most obvious expressions of the juxtaposition of the remnants of a sequence of ancient lava dams in the Grand Canyon is in the vicinity of Whitmore Wash from Mile 187 to 190. Large remnants of a sequence of relatively thin lava flows can be seen filling Whitmore Wash and adjacent canyons when

FIGURE 17.17. The Whitmore Dam as seen from the air above Mile 192. View looking northeast. Large remnants of the Whitmore Dam are preserved on both sides of the river and form wide terraces. The source of the flows was from cascades into Whitmore Wash.

the area is viewed from the air above Mile 190 (Fig. 17.17). Downstream, this sequence is preserved as prominent terraces that outline large segments of the original dam on both sides of the canyon. What makes these exposures exceptionally striking is the way they are preserved. The remnants of the dam are not preserved in protected alcoves (which is more typical) but, instead, form large terraces—some more than one-quarter mile wide and over a mile long. Thus, they are large enough to constitute a significant geomorphic feature in the canyon, recognizable in aerial photographs and on topographic maps. This is the only dam in which relatively large segments of the original upper surface are preserved.

Whitmore Dam is distinctive because it was composed of numerous thin-flow units, most of which range from 10 to 20 feet (3 to 6 m) in thickness (Fig. 17.18). More than 40 individual flow units fill the valley of Whitmore Wash. Little or no sediment separates the flow units in the exposures at the mouth of Whitmore Wash or in the adjacent valleys, indicating that the flows were extruded in rapid succession.

The source of the flows that built the Whitmore Dam is very clear. As Figure 17.17 shows, lavas were extruded from eruptive centers near the southern tip of the Uinkaret Plateau and formed eruptive centers on the Esplanade. The lavas flowed westward and cascaded into Whitmore Wash. They then moved due south where the river makes an abrupt turn to the south. There was, therefore, no real deviation in the trends of the lava flows as they moved from Whitmore Wash into the Colorado River.

Because the dam was built up by numerous thin flows, the water of the Colorado River immediately overflowed each lava barrier. At various times, it

FIGURE 17.18. Remnants of the Whitmore Dam preserved high on the north canyon wall about one-half mile downstream from Whitmore Wash. The Whitmore Dam was distinctive in that it was composed of numerous, thin flows. Remnants of younger, lower dams are juxtaposed against the canyon wall in the central part of this photograph.

also deposited river gravels. In this particular situation, there was a strong tendency for the river to impinge against the south wall of the canyon because the lavas were moving into the canyon from the north. Subsequent flows were deposited on the river gravels near the south wall, whereas lavas deposited near the north and west walls were superimposed on one another with interlayers of gravel.

Whitmore Dam was 900 feet (270 m) high and over a half-mile wide, making it the widest lava dam in the Grand Canyon. The head of the dam was right at the confluence of the Whitmore Wash and Colorado River. If it had been constructed instantaneously, only 230 days would be required for the lake behind it to fill with water and overflow. However, the interbedded gravels in the Whitmore basalts indicate that the lake was filled with water and sediment and that it was overflowing at the same time that the lavas were being extruded.

Because the lake behind Whitmore Dam was close to the elevation of the Esplanade Dam, its configuration and shoreline probably were similar to that shown in Figure 17.13. The lake, extending upstream through the Grand Canyon to Hance Rapids, was 100 miles (160 km) long, and throughout most of its extent, it was very narrow. In Havasu Creek, the lake extended up to the base of the Mooney Falls. The level surface of lake silts between Mooney Falls and Beaver Falls most likely are deposits formed in Whitmore Lake.

D Dam

A number of small, isolated remnants of a sequence of thin basalt flows are pre-served on both sides of the canyon near Lava Falls Rapids at Mile 179.5. These flows clearly are younger than the Toroweap sequence but are older than a se-ries of younger dams, most remnants of which are further downstream. They represent a barrier across the canyon 635 feet (190 m) high. The sequence of lava originally was described by McKee and Schenk (1942) and is one of the most distinctive series of flows in the canyon. It consists of a series of flow units 5 to 15 feet (2 to 5 m) thick that are separated by thin layers of clinkers and ash. Remnants of the D flow are relatively small and inconspicuous; unless one is consciously looking for them, they may be missed entirely.

The Younger Dams

Five younger dams ranging in height from 200 to 600 feet (60 to 186 m) also were formed in the Grand Canyon. Most of these dams were constructed from a single flow that extended downstream several tens of miles, but one was formed from multiple, thin diabase flows. Numerous remnants of these younger dams are preserved on both sides of the canyon in the areas of least vigorous ero-sion. The size of the remnants varies from quite small to more than a mile long. Their geomorphic and stratigraphic relationships, for the most part, are indicated by juxtaposition. The best exposures for demonstrating the age sequence of the younger intracanyon flows are in the vicinity of Whitmore Wash, where five ma-jor intracanyon flows are exposed (Fig. 17.18).

The lakes formed behind the younger dams were barely more than 1/8 mile wide, occupying little more than the present river channel (Fig. 17.19). They ex-

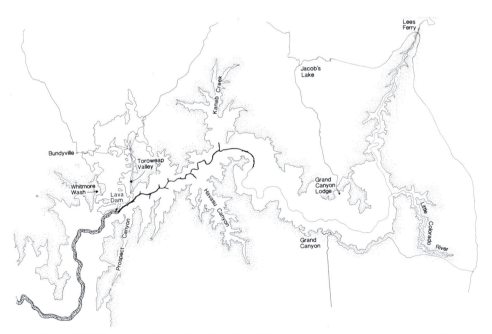

FIGURE 17.19. Lakes formed in the Grand Canyon by younger, single-flow dams.

tended upstream to the middle Granite Gorge, but did not reach the Grand Canyon Visitors Area. The lakes were long, narrow, and deep channels. The largest filled with water in 17.5 days and was silted up completely in 6.5 years.

LAVA CASCADES

Two major lava cascades, the "Toroweap Cascades" and the Esplanade Cascades" (Fig. 17.1), can be seen from the vicinity of Vulcan's Throne. In addition, a number of smaller "frozen lava falls" are found in the inner gorge east of Whitmore Wash.

The Toroweap Cascades, which spill over the rim of the inner gorge just west of Vulcan's Throne, originated from centers of intrusion high on the southern tip of the Uinkaret plateau just south of Mt. Emma. They flowed over the outer rim of the canyon, across the Esplanade, and then cascaded over the rim of the inner gorge into the Colorado River 3000 feet (900 m) below. The cascades represent some of the most recent volcanic activity within the canyon and clearly are equivalent to the younger flows on the Uinkaret Plateau. They are younger than Vulcan's Throne and likely represent volcanic eruptions within the last 10,000 to 20,000 years. The cascades are relatively thin flows, rarely more than 30 feet thick, and on the Esplanade they retain many of their original surface features such as pressure blisters. On the steep slopes, the cascades are jumbled and poorly defined. Near the western margins of the Toroweap Cascades, a relatively thick sequence of steeply inclined flows indicates that considerable volumes of lava entered the Grand Canyon from Toroweap Valley.

The Esplanade Cascades, which cap the Esplanade Dam, are similar to the Toroweap Cascades in that they represent the most recent volcanic activity within the area. The source of the cascades clearly was from centers of eruption on the Esplanade. The lava spread over the relatively flat surface of the Esplanade and spilled over the rim of the inner gorge at several places, both in this area and to the west of it.

Intrusions

The lava cascades between Toroweap Valley and Whitmore Wash are so striking when seen from the air, or from the vicinity of Vulcan's Throne, that there is a tendency to conclude that the cascades were the source of all intracanyon flows. There are, however, numerous dikes and cones within the inner gorge—indicating that much of the lava forming the lava dams may have been extruded within the inner gorge of the canyon. One of the largest, and certainly the most spectacular, evidences of volcanic events within the canyon is a huge mass of basalt referred to as Vulcan's Forge or Lava Pinnacle. It is located near the center of the Colorado River at Mile 177.9 (Fig. 17.20). Another volcanic neck similar in size to Vulcan's Forge is exposed on the south wall of the canyon at Mile 180.2. It is exceptional because the face of the canyon wall cuts across the neck and exposes a vertical cross section of the structure nearly 700 feet (210 m) high. The neck is surrounded by a white alteration zone that makes it exceptionally conspicuous from viewpoints on the Esplanade about a mile west of Vulcan's Throne.

Several systems of thin dikes are exposed on both the north and south walls of the Grand Canyon. These systems, however, are relatively obscure because they are only 3 to 5 feet (1 to 2 m) wide and follow a vertical joint system to

FIGURE 17.20. Vulcan's Forge at Mile 178. Vulcan's Forge is a volcanic neck in the middle of the Colorado River a short distance upstream from Vulcan's Throne.

the top of the Esplanade. They extend upward with little variation in width from the bottom of the canyon up to beyond the base of the Redwall Limestone. Several additional thin dikes intrude the older intracanyon flows and laminated ash in the vicinity of Lava Falls Rapids and downstream at Mile 182.5.

The most prominent dike, however, is located high on the south rim of the Grand Canyon near the mouth of Prospect Valley (Fig. 17.21). This dike is 30 to 40 feet (10 to 12 m) thick and forms a high wall projecting above the surrounding country rock. In contrast to the smaller dikes, the Prospect dike trends in an east–west direction, parallel to the Colorado river. A dike of similar size, but with a much less imposing topographic expression, is exposed along the rough trail descending the cascades from near Vulcan's Throne to the river below. It also trends in an east–west direction but is only slightly more than 10 feet (3 m) thick.

Cones

Most of the cinder cones found within the inner gorge are associated with the Toroweap fault. Vulcan's Throne is the most impressive, but a similar large cone is positioned on the canyon rim at the mouth of Prospect Canyon. In addition, remnants of five cones are found down in the canyon along the Toroweap fault zone. Two of these occur below Vulcan's Throne, and two are found along the Toroweap "trail" to the Colorado river. Also, a large cinder cone, located along the fault, is buried beneath the lavas in Prospect Canyon and now is exposed at the head of the new valley excavated along the trace of the Toroweap fault.

Several large remnants of cinder cones also adhere to the wall of the inner gorge in the vicinity of the Esplanade Cascades. The largest of these, called "Bill's Cone," is approximately the same size as Vulcan's Throne. Although much of

FIGURE 17.21. Large dike just west of the mouth of Prospect Canyon.

this cone has sloughed off into the canyon, the upper part of the cone and crater appears to be relatively unmodified by erosion. Small remnants of older cones adhere to the canyon wall in the same vicinity.

RATES OF EROSION

The series of lava flows within the Grand Canyon provide an unusually well documented record of rates of erosion, slope retreat, and the adjustment of streams to equilibrium. This is possible because the relative ages of the 12 lava dams have been determined by juxtaposition, and radiometric dates provide a series of time benchmarks for the major extrusions. In addition, we have measured the thickness, length, and volume of the lava dams so that we can calculate the rates at which erosion has cut through the dams. What we found is that the Colorado River was able to erode through the lava dams at an astounding rate. When we consider the total thickness of all the rock through which the river has cut its channel in the last million years or so, it is clear that the river has the capacity to erode through any rock type almost instantaneously and that the profile of the Colorado River essentially is at equilibrium. In addition, when equilibrium is upset, adjustment back to equilibrium is extremely rapid. The rationale is as follows.

Rates of Downcutting

All available evidence indicates that prior to the extrusions of lava into the Grand Canyon approximately a million years ago, the Colorado River had cut down to its present gradient. The size and shape of the canyon walls at the time the lava dams were formed were essentially the same as that which we see today. One of the most significant conclusions of our studies, with respect to processes of erosion and canyon cutting, is that after each lava dam was formed, the Colorado River eroded through the dam, down to its original profile *but no farther.* This process of reexcavating the canyon took place at least 13 times during the last million years or so. The cumulative thickness of basalt in these 13 lava dams was at least 11,300 feet (3390 m). Thus, the Colorado River actually has cut through a cumulative vertical thickness of nearly two miles (3 km) of rocks to remove the lava dams. In all probability, the actual downcutting through the basalts took place in less than a million years because there were large time intervals between periods of dam destruction when the Colorado River was not influenced by the presence of lava. The best estimates (based on the time necessary for the destruction of the lava dams) would be that actual downcutting through the 11,300 feet (3390 m) of basalt which formed the dams took place in approximately 200,000 years. During the intervening time, the Colorado River was flowing at its normal gradient with neither large-scale deposition nor erosion taking place.

The fundamental question of the age of the Grand Canyon (how long has it taken the Colorado River to cut through a sequence of Paleozoic strata one mile deep) is not only a question of how fast could the river erode, but of how fast the region was uplifted. Based on the rates at which the river has been able to cut through the sequence of lava dams (or the Paleozoic strata when its channel was displaced beyond the lava flows), it appears that the Colorado River has the capacity to cut down faster than tectonic processes can produce uplift.

This conclusion is supported by the fact that recurrent movement on the Toroweap fault has displaced the strata more than 150 feet in slightly more than one million years. There is no indication of rapids or waterfalls associated with the fault scarp, indicating that the escarpment produced by recurrent movement along the fault is erased by erosion immediately after it forms.

Rates of Slope Retreat

In addition to providing an insight into the capacity of a river to downcut and to maintain a profile of equilibrium, the remnants of the lava dams provide important documentation concerning rates of slope retreat and the relationship between slope profiles, stream gradient, and tectonic uplift.

Throughout the sections of the western Grand Canyon where remnants of lava dams still remain, one fact is completely clear: The remnants of lava dams are preserved only in the more protected parts of the canyon—that is, on the insides of meander bends, in protected alcoves, and as hanging valleys in the mouths of minor tributaries. Those remnants clinging to the canyon walls are preserved only as thin slivers, commonly only a few tens of feet wide. Except for the youngest flows, which may be 200 to 400 feet (60 to 120 m) wide in the broader sections of the canyon below Whitmore Wash, the remnants of the dams remain as thin sheets plastered against the canyon walls.

It is important to note that actual downcutting of the river by abrasion along the river channel would produce a vertical gorge only 200 to 250 feet (60 to 75 m) wide (the average width of the Colorado River channel). Slope retreat has been responsible for widening the rest of the inner gorge. At the present time, the inner gorge at the level of Temple Butte Limestone is roughly 2000 feet (600 m) wide. This means that approximately 900 feet (270 m) of slope retreat occurred on each side of the canyon after each lava dam was formed. At the base of the Muav Limestone, the canyon is approximately 1000 feet (300 m) wide; at that level, 400 feet (120 m) of slope retreat on each side of the river has occurred.

With the destruction of each lava dam and the rapid reestablishment of the river gradient to its original profile, there also was rapid and contemporaneous retreat of the slopes back to their original profile. The process occurred so fast that the original slope profile of the canyon was reestablished before the formation of the next lava dam.

The sequence of juxtaposed remnants near the southern end of the Esplanade sequence serves as an important example. Here, remnants of four lava dams are in juxtaposition (Ponderosa, Esplanade, and two younger dams). Remnants upstream and downstream from this site indicate that five additional units were once present in this area. After each lava dam was breached, there was an average distance of 700 feet (210 m) of slope retreat back to the original canyon wall. The cumulative distance of slope retreat that actually occurred in this area of the canyon was 4900 feet (1470 m), a rate of nearly one mile (1.6 km) per million years. If we consider all of the late Cenozoic lava dams in the canyon, the cumulative distance of slope retreat at the level of the Redwall Limestone would be 11,600 feet (3480 m). At the top of the Bright Angel, the distance would be 5500 feet (1650 m).

In each case, after a lava dam was eroded, the basalt retreated to within a few feet of the original canyon wall. Then, the processes of slope retreat essentially stopped. In many places, the processes of slope retreat completely removed the basaltic flows, *but the process of slope retreat did not enlarge the canyon and go back beyond the original canyon walls.*

Several important conclusions are obvious from these facts: (1) The energy system that causes slope retreat in the Grand Canyon has the capacity to erode canyon walls at a rate of one to two miles per million years, or more. (2) The slopes recede rapidly to a profile of quasi-equilibrium. They then recede at a very slow rate. (3) Slope retreat is balanced delicately with the stream gradient. Renewed downcutting of the stream channel is accompanied by renewed slope retreat. (4) The profile of the Grand Canyon apparently is in a state of quasi-equilibrium. (5) Periods of major rapid slope retreat are initiated by tectonic uplift.

We can conclude from these studies that erosion of the Grand Canyon did not take place at an imperceptibly slow and constant rate. Instead, it was governed by the style and degree of tectonic uplift. Inasmuch as movement on the major faults in the Colorado Plateau occurs in pulses, the actual processes of erosion also must occur in pulses; and, inasmuch as a tectonic disturbance produces only a few feet of displacement at a time, erosion back to equilibrium occurs in a series of small pulses separated by relatively long periods of quiescence.

18

EARTHQUAKES AND SEISMICITY OF THE GRAND CANYON REGION

David S. Brumbaugh

INTRODUCTION

The Grand Canyon is located in a seismically active part of the earth's crust. This region, shown in Fig. 18.1, also contains a highly complex structure. The only faults represented in Fig. 18.1 are those that have been active in the last four million years (m.y.). The seismicity of this area suggests that some of these faults may be active at present.

Physiographically, the Grand Canyon lies within the Colorado Plateau, which really is a series of plateaus rising eastward across the major boundary faults: Grand Wash, Hurricane, Toroweap, and West Kaibab (Fig. 18.1). The Grand Canyon lies at the southern end of the Intermountain Seismic Belt, which stretches north–south across central Utah. The continuation of this belt of seismicity into Arizona has been the topic of debate; nevertheless, there is no argument that the Grand Canyon region is seismically active. This seismicity may represent a tectonic boundary for the Colorado Plateau, which therefore lies inside of the accepted physiographic boundary at the Grand Wash Cliffs (Brumbaugh 1987).

Little previous work of any detail exists on the seismicity of the Grand Canyon region. Earlier workers surveyed the seismicity of the entire state of Arizona but placed little or no emphasis on the Grand Canyon region (Sturgul and Irwin 1971; DuBois et al. 1982). While studies of the northern part of the state, from the Mogollon Rim to the Utah border, do exist, the Grand Canyon region is not the focus of such works.

Seismologists know that in this century many earthquakes that range in size from Mb6.2 down to Richter Magnitude 3.0 have occurred in the Grand Canyon region (Table 18.1, Fig. 18.1). Events smaller than 3.0 have occurred, but their numbers cannot be estimated reliably because seismologists do not have instrumental coverage of sufficient density to detect these small events unless they occur close to established seismograph stations.

Five events of magnitude 5.0 or greater on the Richter scale have occurred in the Grand Canyon region since 1900. On January 25, 1906 a magnitude 6.2 tremor located 40 km northwest of Flagstaff and 55 km southwest of the rim of the Grand Canyon shook northern Arizona, Utah, and New Mexico. Minor damage occurred at Flagstaff.

On September 24, 1910, an event with a magnitude of 6.0 occurred 50 km southeast of the Grand Canyon along the Mesa Butte fault system. Minor damage was reported at Cedar Wash and at Flagstaff, Arizona.

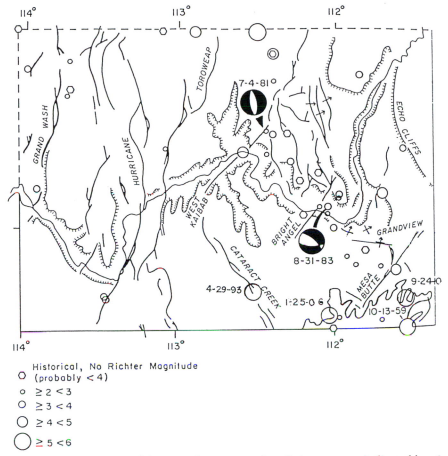

Historical, No Richter Magnitude
O (probably < 4)
o ≥ 2 < 3
O ≥ 3 < 4
O ≥ 4 < 5
O ≥ 5 < 6

FIGURE 18.1. Seismicity of the Grand Canyon region. Epicenters are indicated by circles and hexagons. Faults active in the last 4 m.y. are shown by solid lines; monoclines are shown by → . Grand Canyon escarpment = ‖‖‖‖, except where noted (Echo Cliffs). Fault plane solutions (focal mechanisms) are shown for two events: 7/4/81 and 8/31/83.

During 1959, two events of 5.0 or greater on the Richter scale occurred in the Grand Canyon region. On July 21, 1959, Fredonia, Arizona, located on the Arizona–Utah border, was shaken by the largest earthquake since 1912. This event had a calculated Richter magnitude of 5.75 and caused minor damage at Fredonia. It was felt over a 24,400-square-mile (62,500-square-kilometer) area of Arizona and Utah. This event was followed on October 13 by a smaller shock of magnitude 5.0 that was located to the southeast of the Canyon (Fig. 18.1). It was felt strongly in Flagstaff, but no damage was associated with it.

A 1993 tremor caused minor damage over a broad area. On April 29 near Valle, Arizona, a magnitude 5.4 tremor resulted in damage at Valle, Flagstaff, and the Grand Canyon. Power outages were also caused by this event at the Grand Canyon.

The rate of occurence of earthquakes in this century in the Grand Canyon region can be compared to that in the Intermountain Seismic Belt of Utah or,

TABLE 18.1. Historic Seismicity, Grand Canyon Region

Date (Mo-D-Yr)	Origin Time (UTC)	Epicenter (Lat. × Long.)	Magnitude	Intensity
1-25-06	203220/30	35.60 × 112.0	6.2 Mb	VII
9-24-10	4:05	35.65 × 111.6	6.0 Mb	VIII
8-17-38	9:08:06	36.7 × 113.7		
12-28-38	4:37:36	37.0 × 114.0		
1-31-44	4:24:58	36.9 × 112.4		
10-27-47	4:15:40	35.5 × 112.0		
7-21-59	17:39:29	37.0 × 112.5	5.75 ML	VII
10-13-59	8:15:00	35.5 × 111.5	5.0 ML	VI
2-15-62	7:12:43	36.9 × 112.4	4.5 ML	V
2-15-62	9:06:45	37.0 × 112.9	4.4 ML	V
8-28-64	6:50:46	37.0 × 113.1		
6-07-65	14:28:01	36.1 × 112.2		
9-03-66	7:53:20	36.5 × 112.3	4.4 Mb	V
10-03-66	16:03:50	35.8 × 111.6	4.4 Mb	V
12-01-66	9:20:40	36.2 × 113.9	3.7 Mb	IV
5-26-67	7:48:42	36.4 × 111.6	3.1	II
7-20-67	13:51:10	36.3 × 112.1		
8-07-67	16:24:44	36.5 × 112.4	3.9 ML	IV
8-07-67	16:40:32	36.4 × 112.6	4.0 ML	IV
9-04-67	23:27:44	36.2 × 111.7	4.6 ML	V
11-05-70	9:45:57	36.3 × 112.2		
11-24-70	16:47:56	36.4 × 112.3	3.0 ML	II
12-15-71	12:58:14	36.8 × 111.8	3.0 ML	II
8-05-79	19:10:15	36.8 × 114.0	3.7 ML	IV
1-12-81	8:59:13	35.7 × 113.5	3.5 ML	III
11-19-82	20:57:34	36.0 × 112.0	3.0 ML	II
8-31-83	8:10:09	36.1 × 112.0	3.3	III
7-18-84	14:29:33	36.2 × 112.0	3.0 ML	II
9-6-88	9:44:00	36.0 × 112.2	3.0 ML	
9-7-88	1:17:40	36.0 × 112.1	3.1 ML	
9-7-88	3:22:07	36.0 × 112.2	3.0 ML	
3-5-89	4:40:32	36.0 × 112.1	4.0 ML	
3-5-89	7:12:57	36.0 × 112.1	4.0 ML	
11-28-89	18:37:32	36.1 × 112.2	3.0 ML	
4-26-91	13:08:30	36.6 × 112.6	4.0 ML	
8-22-91	16:41:01	36.0 × 112.1	3.0 ML	
3-13-92	11:28:36	36.0 × 112.2	3.9 ML	
3-14-92	5:12:08	36.0 × 112.2	4.0 ML	
3-14-92	5:13:36	35.9 × 112.2	4.5 ML	
5-6-92	1:40:58	36.0 × 112.2	3.0 ML	
5-20-92	21:46:05	36.0 × 112.2	3.1 ML	
7-5-92	18:18:33	35.9 × 112.3	3.9 ML	
4-25-93	9:29:45	35.7 × 112.2	4.9 Mb	
4-29-93	8:21:00	35.7 × 112.2	5.4 Mb	

on a broader scale, that of the Western Mountain Region (Fig. 18.2). It should be understood that the time span for which data are available is rather small for the moderate-to-large events (MM VI–IX). Therefore, our information may not be as reflective of true rates of recurrence as we would like. Nevertheless, despite effects caused by periodicity, it is clear that recurrence rates in northern

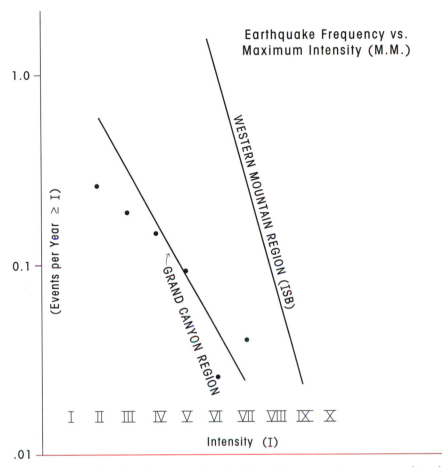

FIGURE 18.2. Earthquake frequency for the Grand Canyon region compared to the Intermountain Seismic Belt (ISB). Recurrence curves are represented by a solid line for each region. Western Mountain Region curve is from Podolny and Cooper (1974).

Arizona are lower, on the average, than those for the Intermountain Seismic Belt. The rate varies from a maximum north of the Grand Canyon, which is comparable to that of the Intermountain Seismic Belt in Utah (Kruger-Kneupfer et al. 1985), to much lower rates south of the Canyon. The slope of the recurrence curve for the Grand Canyon data (Fig. 18.2) suggests the occurrence of an MM VII (5.5–6.0) event twice every 50 years or so.

TECTONICS

The Grand Canyon region of the southern Colorado Plateau is dominated structurally by high-angle normal faults that cluster into large, northeast, north–south, and northwest-trending fracture systems. Several of these fracture systems are well-exposed where they cross the Grand Canyon (Fig. 18.1). Crossing the canyon from west to east are Grand Wash, Hurricane, and Toroweap faults, all predominantly north–south trending. The West Kaibab and Bright Angel fracture systems are prominent northeast-trending groups of normal faults. The north-west-trending Grandview–Phantom system intersects the walls of the eastern Grand Canyon. Two prominent systems, the Mesa Butte (northeast) and Cataract Creek (northwest), do not appear as canyon-crossing fault systems (Fig. 18.1).

The level of earthquake activity in the twentieth century in the Grand Canyon region, as well as the clustering of activity in the vicinity of some of the fracture systems, suggests that some of these faults are active today. Shoemaker et al. (1974) concluded that the distribution of epicenters indicated that the Bright Angel and Mesa Butte fault systems are active. Both of these, as previously noted (Fig. 18.1), are prominent, well-expressed, northeast-trending fracture systems. No surface scarps of historic age exist to indicate that any of the fracture systems are active. Nor have the 5.0 or greater events resulted in any surface rupture. Geomorphic mapping suggests the possibility that many of the northeast, north–south, and northwest-trending fracture systems may have been active in late Pliocene or Quaternary time (Scarborough et al. 1986).

If we ignore the mapped surface traces of fracture systems and concentrate only on the pattern displayed by epicenter locations, the most prominent trend of activity is a northwest trend stretching from the West Kaibab fracture system to the Mesa Butte system. Furthermore, two fault plane solutions exist (7/4/81, 8/31/83), neither one of which indicates movement on northeast-trending fault surfaces. The 8/31/83 event shows two potential fault surfaces, N40°W and N86°E. Only the N40°W trend comes close to paralleling the trends of any of the three major fracture systems. The 7/4/81 event lies at the junction of the West Kaibab and the Crazy Jug fracture systems (Fig. 18.1). The potential fault surface orientations have trends of N24°W and N30°W (Kruger-Kneupfer et al. 1985). It should be noted that the solution is a composite one; that is, it combines the 7/4/81 event with others close to it in time and space. There are those who feel that composite solutions are less reliable than single event ones. Finally, examination of the distribution of epicenters of just the largest events ($M > 4.0$), which undoubtedly are controlled by tectonic stresses, clearly shows a northwestern alignment as well.

A tectonic boundary may be defined by changes in tectonic elements. This includes parameters like structural style, crustal thickness, and seismicity. The concentrated band of seismicity in the Grand Canyon region marks the locus of a change in crustal thickness and a change in structural style. The West Kaibab fault zone marks the eastern limit of Basin- and Range-type faults on the southern Colorado Plateau. Crustal thicknesses in the western Grand Canyon region are more like those found beneath the Basin and Range terrane.

Brumbaugh (1987) has suggested that the belt of concentrated seismicity in the Grand Canyon region represents the present-day tectonic boundary of the Colorado Plateau—a boundary, furthermore, that has migrated with time, as suggested by earlier workers (Best and Hamblin 1978). This migration has expanded the area of the Basin and Range from the southwest and encroached into the Colorado Plateau.

SUMMARY

The seismicity of the Grand Canyon region of northern Arizona is contiguous with the Intermountain Seismic Belt of Utah. This, plus a similar recurrence rate north of the canyon (and a common style of faulting), suggests that the seismicity of the Grand Canyon region represents an extension of the Intermountain Seismic Belt of Utah. An analysis of the distribution of epicenters (Fig. 18.1) indicates that the majority of the events are concentrated in a rather narrow, northwest-trending band between the Mesa Butte and the West Kaibab fracture systems. This trend of activity, plus the northwestern trend of potential fault planes from the 7/4/81 event and one from the 8/31/83 event, suggest that it is the northwest-trending fracture systems, such as the Grandview–Phantom, which contain presently active faults.

It seems likely that the seismicity of the Grand Canyon region is an expression of a tectonic boundary that is marked by a change in structural style as well as in crustal thickening. This seismicity is the product of an active stretching or extension of the crust in the Grand Canyon region. This crust, moreover, is not homogeneous, but pre-fractured in northeast, north–south, and northwest directions by large fault systems. A preferential reactivation of only northwest-trending faults, such as the Grandview–Phantom, suggests that the stretching of the crust must be occurring perpendicular to this trend—that is, in a northeast–southwest direction. If this extension continues to affect the crust of the Colorado Plateau in the future, the tectonic boundary will continue to migrate into the interior of the plateau—eventually leaving the Grand Canyon region a seismically quiet area.

• 19 •

HOLOCENE TERRACES, SAND DUNES, AND DEBRIS FANS ALONG THE COLORADO RIVER IN GRAND CANYON

Kelly J. Burke, Helen C. Fairley, Richard Hereford, and Kathryn S. Thompson

INTRODUCTION

River terraces, sand dunes, and debris fans at the mouths of tributaries are the main elements of Holocene surficial geology of the Colorado River in Grand Canyon (Fig. 19.1), as shown by large-scale geologic maps. These deposits partly occupy the river corridor, the ribbon of water and land between steep bedrock walls of Grand Canyon. Because the deposits formed only during the past several thousand years, they contain little information about the origin of Grand Canyon. However, these deposits record millennia of interactions between river and tributaries in the natural regimen, before Glen Canyon Dam was built.

The pre-dam flood regimen resulted from snowmelt in the Rocky Mountains. The river flooded in late spring and early summer; flood waters, laden with large quantities of suspended sediment, moved boulders along the channel and deposited silt and sand forming high-level terraces. Over time, perhaps several millennia, the largest floods attained a relatively open channel without waterfall-like obstructions, although rapids remain as evidence that obstructions were not removed completely. Sediment deposited in the main channel by tributaries is capable of producing unnavigable obstructions (for example, see Webb et al. 1996, Fig. 35), but such debris was in time removed by the unregulated Colorado River.

With closure of Glen Canyon Dam in 1963, major floods were eliminated and the sediment load was greatly diminished (Andrews 1991). The ability of floods to remove unnavigable obstructions is limited, and sand is no longer deposited in high-level sites. The Holocene surficial geology discussed here shows how the channel system evolved under natural conditions, which in turn permits speculation about how regulated streamflow influences evolution of the system.

The high-level terraces are now eroding through incision by small ephemeral streams, revealing archeologic remains (Hereford et al. 1993). This cultural material is locally abundant along the Colorado River in association with terraces and debris fans that prehistoric people used for camping, agriculture, and construction of masonry structures (Fairley et al. 1994). Most of the remains are affiliated with the Pueblo II Anasazi, dating between 1000 A.D. and 1150 A.D., al-

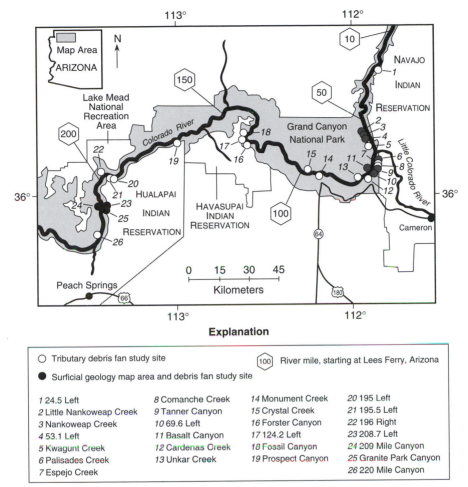

Explanation

○ Tributary debris fan study site		〔100〕 River mile, starting at Lees Ferry, Arizona
● Surficial geology map area and debris fan study site		

1 24.5 Left	*8* Comanche Creek	*14* Monument Creek	*20* 195 Left
2 Little Nankoweap Creek	*9* Tanner Canyon	*15* Crystal Creek	*21* 195.5 Left
3 Nankoweap Creek	*10* 69.6 Left	*16* Forster Canyon	*22* 196 Right
4 53.1 Left	*11* Basalt Canyon	*17* 124.2 Left	*23* 208.7 Left
5 Kwagunt Creek	*12* Cardenas Creek	*18* Fossil Canyon	*24* 209 Mile Canyon
6 Palisades Creek	*13* Unkar Creek	*19* Prospect Canyon	*25* Granite Park Canyon
7 Espejo Creek			*26* 220 Mile Canyon

FIGURE 19.1. Study sites, Colorado River, Grand Canyon National Park, Arizona.

though cultural artifacts range in age from about 800 B.C. to the early twentieth century (Jones 1986; Altschul and Fairley 1989). These same deposits are now used for recreational purposes, mainly hiking or rafting through Grand Canyon. The spectacular scenery of Grand Canyon and whitewater rapids near the eroded banks of many debris fans attract 22,000 rafters annually (Stevens 1990) who are exhilarated by the powerful waves and swift currents. In addition, most of the sand deposited by the Colorado River accumulates near debris fans (Schmidt and Graf 1990), and these deposits are popular camp sites. The sand also forms the substrate for riparian vegetation, which in turn supports the diverse ecosystem of the Colorado River (Carothers and Brown 1991).

Pre-dam terraces, debris fans, and sand dunes discussed in this chapter occupy most of the river corridor in terms of surface area and sediment volume. Post-dam alluvial deposits are substantially reduced in area and height above the river compared with pre-dam terraces. The disparity in volume between the post- and pre-dam alluvial deposits results directly from reduced streamflow. Likewise, the historic-age debris fans are substantially smaller than their prehistoric counterparts. For convenience, we consider historic time to begin about 1870–1890 A.D. with the beginning of photographic coverage of the river corridor (Turner and Karpiscak, 1980; Smith and Crampton 1987; Webb 1996). These historic-age deposits partly fill channels on the prehistoric fan surfaces whose area is about six times larger than that of the channel. For reasons that are not clear, deposition on the prehistoric surfaces has not occurred in historic time. This suggests that prehistoric debris flows were at times larger or more frequent than historic debris flows (Hereford et al. 1996a).

Recent studies of the Colorado River address the geomorphology of active processes. The hydraulics of rapids and flood-related geomorphic evolution of debris fans were studied by Kieffer (1985; Chapter 16, this volume). The process, initiation, frequency of debris flow, and deposition and reworking of debris-flow sediment in the main channel are described and analyzed in a number of papers by Robert H. Webb and colleagues. This work is discussed and summarized by Griffiths and others in this volume. Alluvial deposits of post-dam age were mapped, classified, and related to hydraulic conditions at rapids by Schmidt and Graf (1990) and Schmidt (1990). The close association between high-level river terraces and tributary debris fans was noted by Howard and Dolan (1981). Other recent studies addressed the late Quaternary surficial geology and geomorphology of the river corridor (Hereford 1996; Hereford et al. 1993, 1996a, 1997, 1997a, b, 2001, 2001a). These studies, the main topic of this chapter, identified and dated the principal Holocene surficial geologic elements of the river corridor.

GENERALIZED GEOLOGIC SETTING, CLASSIFICATION, AND AGE OF LATE QUATERNARY DEPOSITS

Holocene deposits along the Colorado River developed in a geologic framework of late Pleistocene gravel, talus, and bedrock. These deposits are best developed where the canyon at river level is wide. A minimum width of 200–400 m evidently provides the conditions necessary for deposition and preservation of sediment. For the most part, deposits are not present or are poorly developed in the upper and lower Granite Gorges where the river corridor is narrow. Classification and correlation of the late Quaternary deposits along the Colorado River in Grand Canyon are shown in Figure 19.2.

In wide reaches of the river corridor, the Holocene deposits are usually bounded by bedrock. Locally, however, the Holocene deposits are bounded at the channel margin by Pleistocene terrace-forming gravel. Two terrace levels (gvy and gvo of Fig. 19.2) form the margins of the river corridor at or near river level. The terraces are well developed near Lees Ferry, Nankoweap Rapids, Furnace Flats, and Granite Park (Fig. 19.1, river mile 0, 52, 65–74, and 208–209, respectively). The base of the younger gravel is below or slightly above river level, and the base of the older gravel is typically near river level to 20–30 m above the river. The deposits are weakly to moderately consolidated and up to 30 m thick. They consist of moderately well-rounded boulder-size clasts of Paleozoic limestone and sandstone in Grand Canyon and Mesozoic sandstone in the Lees Ferry area. Distinctive well-rounded pebbles of porphyritic rock occur sparingly in the gravels. These pebbles were derived from distant igneous sources in the laccolithic mountains of the Colorado Plateau and San Juan Mountains. Medium- to very coarse-grained sand is present in the gravel matrix or interbedded with the gravel.

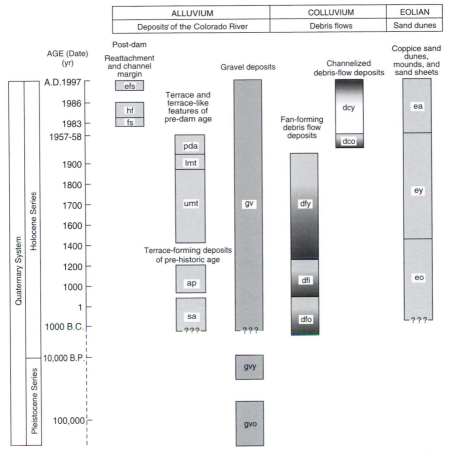

FIGURE 19.2. Correlation chart and classification of late Quaternary deposits along Colorado River. Symbols explained in text.

The gravels are two ancient channel-fill deposits of the Colorado River dating from the Pleistocene (Hereford et al. 1993; Hereford 1996). The contact between gravel and bedrock is strongly concave, and the contact slopes steeply toward the river. This contact is the margin of an ancient river channel. The Pleistocene age of the gravels is based on their elevated position above Holocene deposits and by correlation with dated late Pleistocene deposits in eastern Grand Canyon (Machette and Rosholt 1991). The younger gravel (gvy of Fig. 19.2) probably corresponds with river levels one to three, and the older gravel (gvo of Fig. 19.2) correlates with river levels four and five of Machette and Rosholt (1991). Levels one to three range from 5 ± 5 to 40 ± 24 ka; levels four and five range from 75 ± 15 to 150 ± 30 ka. In the Nankoweap Rapids area, the older gravel is substantially older than the latter age. These ages are roughly comparable to those obtained by Lucchitta et al. (1995) for gravel deposits in similar topographic positions in eastern Grand Canyon.

The younger gravel is considered late Pleistocene rather than Holocene. The Pleistocene age is inferred from the coarse grain size (gravel compared with fine-grained sand), substantial thickness (tens of meters compared with meters), and evidence of extensive bedrock erosion that resulted in widening, deepening, and realignment of the river corridor. In addition, the Pleistocene deposits are more deeply weathered at the surface and have more fully developed soil horizons.

Much of the Holocene alluvium (Fig. 19.2) overlies an unconsolidated gravel (gv of Fig. 19.2). This gravel, with a maximum exposed thickness of 4–5 m, was probably derived from tributary debris flows that deposited sediment in the main channel. The gravel is moderately rounded and better sorted than a primary debris-flow deposit, indicating reworking by the Colorado River. These gravel deposits are probably debris bars, material that accumulates downstream of active debris fans (Chapter 20, this volume).

TERRACES AND SAND DUNES

A defining moment in the late Holocene of the Colorado River in Grand Canyon was closure of Glen Canyon Dam in 1963. The profound affects on streamflow in Grand Canyon are discussed by Williams and Wolman (1984) and Andrews (1991). Briefly, the results are diminished streamflow and reduced, essentially negligible sediment load. These effects will persist for the life span of the reservoir, perhaps as long as 1200 years, assuming average sediment input remains at present levels and the dam retains its structural integrity (E.D. Andrews 1997, oral communication). Twelve hundred years is a substantial portion of Holocene-river history. For this reason, it is convenient to divide the Holocene alluvial chronology of the Colorado River into two episodes: pre- and post-Glen Canyon Dam, which may persist until 3200 A.D.

Alluvium of Pre-dam Age

Five terraces or terrace-like features and related alluvial deposits constitute the upper Holocene of the Colorado River in Grand Canyon (Hereford et al. 1993, 1996a, 1997a, 2001, 2001a; Hereford 1996). From oldest to youngest, the deposits are the striped alluvium, alluvium of Pueblo II age, alluvium of the upper and lower mesquite terraces, and the pre-dam alluvium (sa, ap, umt, lmt, and pda, respectively of Fig. 19.2). These deposits form discontinuous terraces and terrace-like features upstream, downstream, and encircling Holocene debris fans. Figure 19.3 illustrates schematically the geomorphology and geology of the terraces, the height of the terraces above the river, and the position of the terraces relative to historic streamflow levels. Each terrace occupies a distinctive topographic position with the oldest terrace highest and farthest from the river. The full suite of terraces is not present at any one location, nor are the terraces paired across the river. Terraces are unpaired because the river usually flows between bedrock or talus on one bank while the other bank is alluvium, tributary debris fan, or both. In addition, the position of the active channel relative to the older terraces varies considerably. Locally, the channel can be adjacent to any of the older terraces.

The terraces have inset geomorphic relations such that younger terraces are topographically lower than older terraces; however, the units partly overlap in the subsurface (Fig. 19.3) as indicated by the stratigraphic relations among the deposits. The area of overlap between units is an erosional hiatus that is difficult to detect (see Figs. 3 and 4 in Hereford 1996) and is easily overlooked (Lucchitta et al. 1995) without excavation, analysis, and dating.

Striped Alluvium and Alluvium of Pueblo II Age These deposits form the highest terraces of the river corridor, and they are the oldest Holocene alluvial

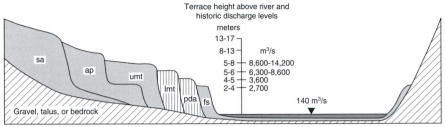

FIGURE 19.3. Schematic cross section showing stratigraphic and geomorphic relations of late Holocene alluvium, range of heights of terraces above 140 m³/s (5000 ft³/s) discharge level, and maximum discharges levels affecting historic-age terraces. sa, striped alluvium; ap, alluvium of Pueblo II age; umt and lmb, upper and lower mesquite terrace; pda, pre-dam alluvium; fs, flood sand of 1983.

deposits adjacent to the river (units sa and ap, respectively, of Fig. 19.2). Although driftwood is present on younger terraces, it is not present on these two oldest terraces. Prehistoric archeologic remains are associated for the most part with the alluvium of Pueblo II age and the striped alluvium. The remains occur on or near the surface or buried within the deposits, where they are undetectable unless exposed by erosion.

The striped alluvium consists of light-colored very fine-grained poorly sorted sand interbedded with thin beds (5–15 cm) of dark-colored sand to small-cobble gravel derived from nearby hillslopes. This interbedded relatively dark sand and gravel imparts the distinct "stripes" that are characteristic of the deposit in eastern Grand Canyon. The gravel beds increase in number and thickness in the direction of the nearby bedrock hillslopes; the gravel resulted mainly from sheetwash. The sand beds increase in number in the direction of the river. These beds are largely of fluvial origin; they contain relatively high quantities of silt and sand, and fluvial sedimentary structures are present locally, although most of the deposits lack sedimentary structures.

The alluvium of Pueblo II age derives its name from the locally abundant archeologic material of Pueblo II affinity, although diagnostic Pueblo I material is present near the base and early Pueblo III ceramic material is present on the terrace. The archeologic material consists of potsherds, flakes, rock alignments, upright slabs, walls, and other artifacts or features. This deposit consists mainly of light-colored poorly sorted very fine-grained sand of fluvial origin interbedded locally with moderately well-sorted fine-grained sand of possible eolian origin. Interbedded dark-colored sand and gravel beds forming stripes are also present, but they are not as conspicuous as those in the striped alluvium. In places, the alluvium of Pueblo II age disconformably overlies the striped alluvium. The contact between the alluviums is an eroded surface with up to one meter of relief locally, and stratification in the striped alluvium is truncated at this surface. Thus, the alluvium of Pueblo II age is stratigraphically distinguishable from the striped alluvium, although in places there is little or no topographic separation between the alluviums at the surface.

Alluvium of the Upper and Lower Mesquite Terraces The mesquite terraces (units umt and lmt of Fig. 19.2) range from narrow, discontinuous surfaces or scoured zones to well-developed readily identifiable terraces; they are topographically below the terraces of the striped alluvium and alluvium of Pueblo II age (Fig. 19.3). The name stems from the abundant western honey mesquite (*Prosopis glandulosa* var. *torreyana*; Turnet and Karpiscak 1980) present on the terraces. By and large, relatively large mesquite is present on the older, upper mesquite terrace, whereas smaller trees are present on the lower terrace. The mesquite terraces are deposits in the "old high-water zone" of previous studies (Johnson 1991; Carothers and Brown 1991; Webb 1996, Fig. 6.1). Like the striped alluvium and alluvium of Pueblo II age, alluvia of the mesquite terraces are light-colored poorly sorted very fine-grained silty sand of Colorado River origin.

Photographs of the Palisades Creek area taken in January 1890 by Robert B. Stanton (Smith and Crampton 1987, pp. 149–159; Webb 1996) show the upper and lower mesquite terraces. At that time, the upper mesquite terrace was veg-etated and appeared to not have been flooded recently. In contrast, the lower mesquite terrace was sparsely vegetated and consisted of high-albedo sand with dark elongated objects interpreted as driftwood. These differences suggest that in 1890 the lower terrace had been recently flooded whereas the upper terrace was inactive.

Driftwood in this area associated with the lower mesquite terrace contains less than 5 percent of milled and cut wood. The paucity of milled and cut wood im-plies that the driftwood is of early historic age, therefore, it was probably deposited by the flood of July 1884, the largest flood of the systematic record with estimated peak discharge of 8500 m³/s (300,000 ft³/s). In contrast, although rare, older and higher driftwood associated with the upper mesquite terrace consists mainly of beaver-cut cottonwood without milled and cut wood (Hereford et al. 1997).

The two terraces formed during the larger floods of the late prehistoric to early historic period. Judging from its relatively high topographic position above the lower mesquite terrace, sand of the upper mesquite terrace probably accu-mulated during the largest floods of the post-Anasazi era. The lower mesquite terrace, which accommodated floods with lower stage than those affecting the upper terrace, quite likely began to develop with the flood of July 1884.

Age of the Striped Alluvium, Alluvium of Pueblo II Age, and Alluvium of the Upper Mesquite Terrace

The ages of the striped alluvium, alluvium of Pueblo II age, and alluvium of the upper mesquite terrace in eastern Grand Canyon are constrained by radiocarbon dates and archeologic material (Fig. 19.4). Organic material collected from the three units was dated using the radiocarbon method. The sampling procedure, limitations of the radiocarbon dates, and stratigraphic context of the carbon are in Hereford et al. (1993, 1996a) and Hereford (1996). Deposition of the striped alluvium in eastern Grand Canyon began before 800 B.C. and could have lasted until about 300 A.D. Older dates have been obtained from the unit near Granite Park where deposition was ongoing by about 1300 B.C. (Hereford et al. 2001). In upper Marble Canyon, O'Conner et al. (1994) dated the base of a sequence that may be equivalent to the striped alluvium at 2500 B.C. This age range of 2500–1300 B.C. to 300 A.D. is consistent with the aceramic char-acter of the archeologic remains in the alluvium. This indicates that the deposit predates the Pueblo era and is most likely of late Archaic to Basketmaker III age.

Deposition of the alluvium of Pueblo II age probably began by about 700 A.D. This is supported by the presence of Pueblo I ceramics near the base of the alluvium in the Upper Unkar area (Fig. 19.1) that date from 800 to 900 A.D. Late Pueblo II to early Pueblo III ceramics are present on the surface of the alluvium. The age and stratigraphic context of these younger ceramics suggest that depo-sition of the alluvium ended between 1150 A.D. and 1200 A.D., which was prob-ably coincident with abandonment of the area by the Anasazi. The period of erosion and nondeposition (Fig. 19.4) between the two alluvia lasted about 400 years, from 300 A.D. to 700 A.D.

The alluvium of the upper mesquite terrace is younger than about 1200 A.D. and was deposited after Anasazi occupation of Grand Canyon, based on the lack of Anasazi ceramics and other material in the alluvium. Deposition of the alluvium forming the upper mesquite terrace must have begun after 1150–1200 A.D., and radiocarbon ages (Fig. 19.4) suggest that deposition could have begun as late as 1400 A.D. Thus, the erosional hiatus probably lasted 200 years, between 1200 A.D. and 1400 A.D.

The causes of this alternating fluvial deposition and erosion in Grand Canyon during the late Holocene are not well understood. Water in the Colorado River is derived from snowmelt in the headwaters of the Rocky Mountains, whereas the sediment load is mainly from tributaries of the Colorado Plateau (Andrews 1991). Thus, climate of either region could influence alluviation in Grand Canyon in ways that are not immediately apparent. Nevertheless, the alluvial chronology of the Colorado River in Grand Canyon correlates broadly with a late Holocene

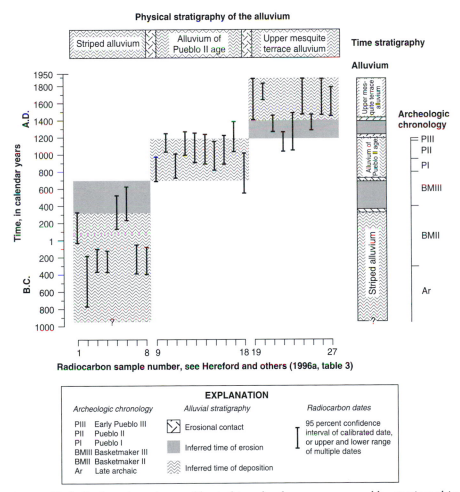

FIGURE 19.4. Radiocarbon dates calibrated to calendar years arranged by stratigraphic unit showing physical and time stratigraphy of late Holocene alluvium, along with archeologic chronology. [Modified from Hereford et al. (1996a, Fig. 19.9). Archeologic chronology of Altschul and Fairley (1989).]

chronology of the southern Colorado Plateau developed by Karlstrom (1988) and elaborated by Dean (1988, p. 129). The alluvium of Pueblo II age probably correlates with the upper Tsegi Formation, and the alluvium of the upper mesquite terrace correlates roughly with the Naha Formation, both of northeast Arizona and southern Utah (Hack 1942; Cooley 1962). Erosion in Grand Canyon around 1200–1400 A.D. coincides with widespread stream entrenchment on the southern Colorado Plateau at about this time (Hereford et al. 1996a).

Pre-dam Alluvium The pre-dam alluvium and flood debris (pda of Fig. 19.2) form a terrace, or in places a scoured surface, that is topographically beneath the lower mesquite terrace (Fig. 19.3). The dominant vegetation at this level is saltcedar (*Tamarix chinensis* Lour.; Turnet and Karpiscak 1980). These trees are normally large, mature, and partially buried in the alluvium. Tree-ring dates obtained from two trees at the Palisades Creek area (Fig. 19.1) indicate germination in 1937 and 1951. Flood debris contains artifacts dating from the mid-1930s to mid-1950s, and the driftwood is dominated by milled and cut wood. This terrace and related deposits formed during the larger floods of the 1920s to 1957–1958.

Alluvium of Post-dam Age

The post-dam alluvial deposits lie entirely within that part of the channel affected by operation of Glen Canyon Dam. Schmidt and Graf (1990) classify the post-dam alluvium as channel-margin deposits, reattachment bars, and separation bars. These deposits accumulate in areas of low current velocity related to changes in the width of the channel or in hydraulic roughness of the channel margin. They have formed at several topographic levels controlled by the discharge rate prevailing during deposition. River runners refer to all of these deposits as beaches, which are well-liked as camp sites for raft trips.

Reattachment bars result from recirculating flow that develops in the distinctive, arcuate topographic setting downstream of a large channel constriction, usually a debris fan. Downstream of the constriction, the flow separates from the main current and moves upstream, rejoining the main current at the head of the recirculation zone or eddy (Schmidt 1990). Several studies have addressed the sedimentology of reattachment bars and their relation to regulated streamflow (Schmidt 1990; Schmidt and Graf 1990; Rubin et al. 1990).

Aside from the deposits of the active channel, the most conspicuous post-dam deposits are those of the 1983 flood, a largely unplanned flood release, and those dating from 1984 to 1986. Earlier deposits dating from 1963 to 1983 were probably eroded during the floods and prolonged high releases of 1983 to 1986. The flood sand (unit fs of Fig. 19.2) is distinctively light-colored well-sorted very fine-grained sand with clay and silt content less than 5 percent. Because it is light-colored and relatively well sorted, the flood sand resembles eolian sand in the river corridor. The flood sand is usually about 1–2 m thick, and its distribution is locally spotty; where the sand is absent, a flood line showing evidence of scour is present. The scour line is marked by flood debris and an alignment of relatively small, submature saltcedar that germinated between 1970 and 1983. The flood debris contains vintage early 1980s material such as beer and soda-pop cans without disposable tabs and partly decomposed plastic artifacts. This sand and related floodmarks are below the level of the pre-dam alluvium.

The high-flow sand (unit hf in Fig. 19.2) is a very fine- to fine-grained sand with silt and clay content greater than 5 percent. Geomorphically, the high-flow sand is similar to the flood sand except that it is inset beneath the 1983 deposit. The high-flow sand was deposited during high flows from 1984 to 1986.

Sand deposited in the active channel below the 1986 high-flow level was partly reworked and covered during a controlled, experimental flood between March 22 and April 8, 1996, when flow rates were increased to 1270 m³/s (45,000 ft³/s). The stage of the experimental flood was near the upper limit of the high-flow sand. This controlled flood release was designed to reposition sand submerged in the channel onto subaerial river banks. Although we have not mapped these deposits, we anticipate that the experimental flood sand (unif efs of Fig. 19.2) was deposited just below the level of the high-flow sand.

Generally, the post-dam alluvial deposits record the depositional activity of the Colorado River since 1983. The largest flows of the post-dam era (Hyatt 1990) produced the 1983 flood sand in June–August of 1983, when peak discharge was 2700 m³/s (96,000 ft³/s) and sustained flows were above 1400 m³/s (50,000 ft³/s). During May–June of 1984–1986, sustained daily releases were the second highest of the post-dam era, ranging from about 900 to 1400 m³/s (32,000 to 50,000 ft³/s); this flow regimen resulted in the high-flow sand. Thus, a broad pattern of erosion followed by deposition at progressively lower levels produced the present configuration of the post-dam channel.

Sand Dunes

Eolian deposits forming dunes ranging in height from a few meters up to 10–20 m are widespread along the Colorado River in Grand Canyon. Grain-size analyzes show that the sand is moderately well-sorted to moderately sorted very fine- to fine-grained sand. Average silt and clay content is 5 percent with a range of 2–9 percent; this contrasts with pre-dam alluvium, which usually has average silt and clay content of 7 percent with a range of 2–16 percent. The eolian sand forms dunes, sand sheets, and other dune-like features that blanket older deposits. For the most part, these are coppice dunes or nabkhas, the terms applied to sand hummocks or mounds that develop around plants, which partially anchor the wind-blown sand (McKee 1982b, pp. 48–49; Cooke et al. 1993, pp. 356–358). In the river corridor, mesquite is usually associated with the coppice dunes.

The dunes are classified as active, moderately active to inactive, and largely inactive (units ea, ey, and eo of Fig. 19.2). The degree of eolian activity coincides with vegetation cover and topographic relief of the dune. Active dunes are sparsely vegetated and have relatively high relief. Vegetation is more abundant and relief is relatively subdued on the less active dunes (ey and eo).

The alluvial deposits of the river corridor are the immediate source of wind-blown sand. Eolian deposits typically occur downwind of sand flats formed on gravel bars, downwind of a terrace rise, or parallel with a terrace rise. Eolian erosion is enhanced where alluvial deposits are directly exposed to the wind, either by having large surface area or by exposure of sand in steep banks. The sand flats provide the fetch necessary for wind erosion, and a terrace rise exposes sand along the steep bank.

Holocene Tributary Debris Fans

Large boulders of locally derived bedrock are deposited in the river corridor by debris flows (Webb et al. 1988, 1989). We classify debris flows with colluvium (Fig. 19.2) to emphasize their close association with mass movement (Costa 1984; Hooke 1987). These impressive, potentially dangerous events result from mass wasting of the steep, talus-covered slopes of Grand Canyon following intense rainfall. For the most part, a mixture of water and sediment is driven downslope by the weight of the sediment, without water as the transporting medium. At the base of the slope, the debris enters a tributary channel and flows to the river. There, the deposits accumulate on debris fans, roughly fan-shaped landforms at the mouths of most tributaries (Hamblin and Rigby 1968; Dolan et al. 1978). The river is unable to move the debris except during relatively large floods (Graf 1980; Kieffer 1985; Webb et al. 1996). As a result, these deposits determine the course of the river between bedrock walls, the location and severity of rapids, and sites of alluvial deposition (Graf 1979; Howard and Dolan 1981; Kieffer 1985). In map view, these alluvial deposits encircle the debris fans outlining the fan-like shape (Hereford 1996; Hereford et al. 1997, 2001, 2001a).

Surficial Geology and Geomorphology of Large Tributary Debris Fans

The geomorphology of the large debris fans discussed here (Fig. 19.1, nos. 1–26) includes two elements (Hereford et al. 1996a). The larger element is the primary fan surface, which is segmented into several surfaces of different ages. Segmentation of fan surfaces is discussed by Cooke et al. (1993, pp. 19–184). The larger debris fans are evidently similar to Type IB alluvial fans of Blair and McPherson (1994, pp. 394–395), except that fan margins are extensively eroded by the Colorado River. These fans have a stratigraphic record of multiple, clast-rich debris flows and minor interbedded fluvial gravels. In the Grand Canyon, the deposits underlying the main fan surfaces are referred to as the fan-forming debris-flow deposits, because the primary surfaces are developed on these deposits.

The smaller element is an active debris-flow channel with a small debris fan at the river. The active channel is usually entrenched 2–5 m below the segmented surface, a notable exception is the deeply entrenched channel of the Prospect Canyon fan (Webb et al. 1996). Deposits transported down the active channel spread into the channel of the Colorado River, forming a small debris fan at the margin of the primary fan. Deposits of the entrenched channel and related fan are referred to as the channelized debris-flow deposits. The channelized deposits are identical to the historic-age debris flows or inset debris-flow surfaces discussed by Griffiths and others (Chapter 20, this volume). In contrast, the fan-forming deposits are mainly of prehistoric age, and they do not have a historic-age counterpart at the debris fans we studied (Fig. 19.1, nos. 1–26).

Both the fan-forming and channelized debris-flow deposits consist primarily of clast-supported angular to subangular boulders of local bedrock ranging in size from granules to large boulders as large as 3–5 m on an edge. The boulders are a mixture of resistant Paleozoic sandstone, limestone, and dolomite, which are present in the walls of the canyon and steep talus slopes beneath the cliffs (Huntoon et al. 1986). The debris-flow matrix is a poorly sorted mixture of clay to coarse sand and granule gravel. Large, angular boulders up to 1–3 m on an edge occur near the apex of many fans, although boulders this large are also present at the margin of the fans. Broadly speaking, the degree of rounding and maximum size of boulders are related to the size of the drainage basin (Hereford et al. 1997). Small basins with relatively short, steep channels produce large, angular boulders, whereas larger basins with relatively long channels and low gradients produce smaller subrounded to subangular boulders. Fluvial gravel of tributary origin is locally interbedded with the debris-flow sediment. The fluvial gravel typically consists of subrounded weakly imbricated pebble to small boulder-size clasts with minor coarse sand matrix; these gravels are distinctly finer grained and relatively well sorted compared with debris-flow gravel.

Large-scale topographic maps (Lucchitta 1991; Hereford et al. 1993, 1997, 2001; Hereford 1996; Webb et al. 1996) show that the area of the fan-forming deposits is substantially larger than the area of the entrenched channel and its associated fan. Typically, the area of the primary fan surface is about six times larger than the area of the active debris-flow channel and related fan, based on the median ratio of fan to channel area (Hereford et al. 1996a). In eastern Grand Canyon, the total area of the debris fans we studied ranges from 1 ha (2.47 ac) to 22 ha (54.3 ac) at upper Palisades Creek and Unkar Creek, respectively (Fig. 19.1, nos. 6 and 13).

The channelized and fan-forming debris-flow deposits are further subdivided according to age and topographic expression. Where we have mapped them (Fig. 19.1), the channelized debris-flow deposits consist of a younger unit forming the active debris-flow channel and an older unit forming one or more poorly to well-developed surfaces in the entrenched channel (units dcy and dco of Fig. 19.2, respectively). Both units were deposited by at least two debris flows in the past 100 years, based on the relation of the deposits to Colorado River alluvium of known age (Hereford et al. 1997, 2001; Hereford, 1996).

Between one and six fan-forming debris-flow deposits form the primary surface of the typical segmented fan in Grand Canyon (Fig. 19.1, nos. 1–26). These surfaces parallel the underlying deposits and are contemporaneous with deposition—as opposed to cross-cutting surfaces, which are younger than the underlying deposit. The deposits and surfaces are grouped into three main units: the older, intermediate, and younger fan-forming debris flows (units dfo, dfi, and dfy of Fig. 19.2), which are further subdivided where necessary. The surfaces are distinguished from each other by relative topographic position and surface-weathering characteristics (Table 19.1). The older surfaces have the highest elevation and are farthest from the river. In addition, the degree of surface weathering increases with distance from an elevation above the river.

Erosion of Prehistoric Debris Fans by the Colorado River

Geologic and large-scale topographic maps of several prehistoric debris fans reveal the extent of erosion by the Colorado River, which in turn affects how the fans aggraded. Most of the prehistoric debris fans are eroded and truncated at their margins. The rate and processes of debris-fan erosion by floods during historic time were discussed by Webb et al. (1996, pp. 84–87). This study showed that relatively small fans from channelized debris flows were largely removed from the active channel in only 1–2 years before regulation of streamflow and in 3–7 years following regulated streamflow. However, if fan-forming debris flows had substantially larger volume than channelized debris flows, as suggested by Hereford et al. (1993, 1996a), erosion of the larger deposit should require more than a few years and exposure to numerous large floods, given similar geomorphic setting.

Here we discuss several examples of prehistoric, partly eroded debris fans in eastern Grand Canyon (Hereford et al. 1996a). The distribution of the fan-forming deposits and the shape of the upper part of the Palisades Creek fan (Fig. 19.1, no. 6) resemble the radiating pattern normally associated with alluvial fans (Hooke 1987). An erosional scarp at the toe of the fan (Fig. 19.5a), however, is evidence that the Colorado River eroded and partly truncated the fan. The minimum height of the scarp ranges from 2 to 3 m. Downslope projection of the intermediate-age surface suggests that the fan probably extended across the river at Lava Canyon Rapids (Fig. 19.5a). The scarp truncates the older and intermediate-age debris-flow deposits, and it is overlapped by alluvium that is younger than 1200 A.D. (Hereford 1996). Therefore, erosion of the scarp occurred between 550 A.D. and 1200 A.D. Much of the fan remains unaffected by erosion because most of it lies well above the level of prehistoric and historic floods. Prehistoric alluvial deposits range in elevation from 823 m to 825 m around the margin of the fan, which is well below the fan apex at 845 m. The fan lacks well-developed inset relations among the surfaces, and it appears to have aggraded vertically (Fig. 19.5a).

Unlike the Palisades Creek fan, the lower Tanner Canyon fan (Fig. 19.1, no. 10) retains little of the original fan shape (Fig. 19.5b). Most of the lower Tanner Canyon fan lies within the range of prehistoric floods. Prehistoric flood-related alluvial deposits are downstream of the fan at elevations of 808–812 m; the elevation of the two fan segments range from 810 m to 815 m, which is well within the range of prehistoric floods. The intermediate and older debris-flow deposits are both truncated along two distinct west-trending scarps with relief of 3 m on the south scarp and 5 m on the north scarp (Fig. 19.5b). The erosional scarps were produced by at least two mainstem floods. The first flood or floods truncated the older debris flow. Truncation of the fan margin steepened the fan gradient, causing entrenchment of a channel. The intermediate-age flow was subsequently deposited in this channel, 2–3 m below the surface of the older deposit, and extended beyond the scarp in the older deposits. The second flood truncated the fan margin formed by the intermediate-age deposits (Fig. 19.5b). Finally, the fan has not aggraded vertically in the usual meaning of the term; instead, it developed mainly by aggradation of inset segments. The inset segments result from one or more floods that cut into the medial portion of the fan. This lowered the base level of the fan leading to entrenchment of the surface and emplacement of subsequent debris flows at a lower level.

Detailed topographic maps of Holocene fans in eastern Grand Canyon (Lucchitta 1991; Hereford et al. 1993) show that the Unkar Creek, Cardenas Creek, Comanche Creek, Espejo Creek, and the upper Palisades Creek fans (Fig. 19.1, nos. 13, 12, 8, 7, and 6, respectively) are moderately truncated, because they retain an almost complete fan shape. In contrast, the Basalt Creek and Tanner Canyon fans (Fig. 19.1, nos. 11 and 9) are severely truncated, as indicated by steep scarps, with up to 5 m of relief, that surround the margin of each fan. These different morphologies are probably related to the course of the river around the fan, the elevation of the fan relative to flood levels (which is related to particle size of the deposits), and the age of the fan.

The course of the channel relative to the debris fan differs between the moderately and severely truncated fans. At moderately truncated fans, the river is relatively straight and essentially parallel with the distal-fan margin, whereas at the severely truncated fans the river flows into and around the fan margin (Fig. 19.1, nos. 9 and 10, for example) or it flows entirely around the fan margin (Basalt Canyon; Fig. 19.1, no. 11). In these situations, the river is particularly effective at eroding the fan because it flows directly into the upstream margin of the fan, or the river completely surrounds the fan margin. A severely truncated fan could be older than a moderately truncated fan, because the older fan has been subjected to a larger number of mainstem floods. Large debris fans in eastern Grand Canyon and elsewhere in Grand Canyon (Fig. 19.1), however, differ in age by only a few thousand years.

Finally, the elevation of the fan relative to the river also controls the extent of erosional modification. The apex of the moderately truncated Palisades Creek fan (Fig. 19.1, no. 6) lies well above flood levels, suggesting that the volume and particle size of sediment was larger than the river was able to remove during recent millennia. On the other hand, most of the severely truncated lower Tanner Canyon fan (Fig. 19.1, no. 10) lies within the level of prehistoric floods, and the fan was relatively low and easily eroded. This difference between high and low elevation fans is evidently related to the slope or gradient of the fan, which in turn is a function of basin relief, lithology, and area (Cooke et al. 1993, pp. 177–178; Webb et al. 1996).

TABLE 19.1. Surface Weathering Characteristics of Fan-Forming Debris Flows, Grand Canyon, Arizona

	Debris-Flow Age Category		
Characteristic	Younger	Intermediate	Older
Carbonate coatings underside of boulders	None	Incipient Stage I, discontinuous, thin, <0.1 mm	Stage I, continuous, thin <0.5 mm
Splitting, spalling, and granular disintegration of sandstone boulders	None	Slight to common	Common
Average pit depth carbonate boulders (mm)	0 to ~1.5	>1.5 to ~4.7	>4.7 to ~17.5
Rilling of carbonate boulders	None	None	Present on 5 percent of boulders
Rock varnish, sandstone boulders dark	Absent to incipient on 50 percent of boulders	Present on all boulders, brown to dark brown	Well-developed on all boulders, brown to black

Ages and Correlation of Fan Surfaces

The relative ages of fan surfaces or segments is established by the degree of surface weathering (Table 19.1). The younger surface has weakly developed or no discernible rock varnish and the carbonate boulders appear fresh with little surface roughness. On the intermediate-age surface, rock varnish coats 50–100 percent of the sandstone boulders and carbonate boulders are distinctly roughened with dissolution pits. The undersides of boulders on the surface have a thin, very light-gray to white discontinuous coating of calcium carbonate, which is weakly developed Stage I soil carbonate morphology (Machette 1985). On the older surface, up to 100 percent of the sandstone boulders have well-developed rock varnish; carbonate boulders are distinctly and deeply pitted; and the undersides of boulders have a thin, mostly continuous coating of calcium carbonate. On the oldest surfaces, rillenkarren (Table 19.1), a distinctive pattern of shallow grooves or rills, is on up to 5 percent of the carbonate boulders.

Although relative ages are readily determined using weathering criteria (Table 19.1), finding the numerical age of fan segments is difficult. The relation of debris fans to dated Holocene alluvial deposits places the age of the larger fans in the late Holocene (Hereford et al. 1996a). Dated archeologic sites are present on fan surfaces locally, but sites provide only minimum ages. Briefly, fan surfaces are difficult to date directly because organic material is extremely rare in the coarse-grained deposits, which severely limits the use of radiocarbon dating. However, the depth of dissolution pits on carbonate boulders increases systematically with the relative age of the surface as shown independently by surface weathering such as degree of patination and soil development (Hereford et al. 1996a). Dissolution pits result from weakly acidified rainfall when atmospheric or metabolic CO_2 combines with water to form carbonic acid (H_2CO_3), which slowly

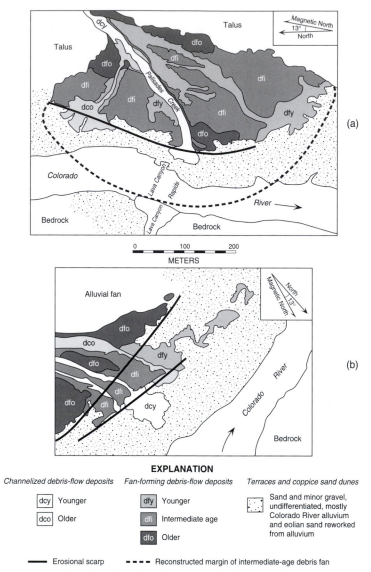

FIGURE 19.5. Generalized surficial geologic maps of eroded debris fans. (a) Moderately truncated Palisades Creek debris fan (Fig. 19.1, no. 6; Hereford, 1996) showing radiating pattern of fan forming debris flow deposits, reconstructed outline of intermediate-age debris-flow deposits, and trace of erosional scarp at base of fan. (b) Severely truncated lower Tanner Canyon fan (Fig. 19.1, no. 10) showing two erosional scarps at margin of older and intermediate-age deposits, respectively. (Modified from Hereford et al. 1996, Fig. 19.6.)

etches the surfaces of carbonate boulders. Thus, the depth of dissolution pits is related to the numerical age of a particular fan surface. The problem with using dissolution pits to estimate exposure time is that the rate of pit deepening must be established from independently dated surfaces, which are few in number.

Using the average depth of dissolution pits measured on four independently dated surfaces, the average rate of pit deepening was estimated to be 2.60 ± 0.12 mm/ka for debris fans in Grand Canyon. The procedure used to estimate deepening rate, methods of measuring and sampling pit depth, and statistical treatment of the data are in Hereford et al. (1997a). The above deepening rate was recalculated from the data in Hereford et al. (1997) using an improved computational procedure.

Limestone weathering is a linear function of time for at least the past 3 ka, as reported by numerous empirical studies of archeologic sites and gravestones [Hereford et al. (1997a) summarize the literature]. The physical processes of limestone weathering proceed at a constant rate—unlike weathering of silicate rocks, which decreases with time (Colman 1981). It is important to realize that the dissolution process is cumulative (Lipfert 1989). The influence on dissolution rate of increases and decreases in precipitation, therefore, averages out over time, maintaining an essentially constant long-term average rate.

Briefly, 6883 individual pit measurements were made on 618 boulders of Redwall Limestone on 71 fan surfaces at 26 tributary debris fans (Fig. 19.1). The measurements are summarized in Figure 19.6 by average depth and the corresponding 95 percent confidence interval for each of the 71 surfaces. The results are consistent with field relations indicating the relative age of fan segments at a particular fan. In other words, surfaces with deep dissolution pits appear older based on stratigraphic position and surface weathering criteria without considering pit depth (Table 19.1).

Ages estimated with the 2.60 mm/ka deepening rate range from 500 to about 7000 cal yr B.P. (calibrated or calendar years before present, i.e., 2001). The oldest surface on a debris fan, river mile 124.2L (Fig. 19.1, no. 17), has dissolution pits averaging 12.22 mm deep. This corresponds to about 4700 cal yr B.P. Even older surfaces are located at the mouths of Forster and Fossil Canyons and in the tributary canyon at river mile 220R. The three surfaces have dissolution pits averaging 17.29, 17.41, and 17.42 mm, respectively; this is about 6700 cal yr B.P. These surfaces are debris-flow levees preserved in the mouths of the tributary canyon, where they are protected from the erosional effects of the Colorado River.

The youngest surfaces datable by this method are at 209 Mile Canyon, Granite Park Wash, Palisades Creek, and Nankoweap Canyon (Figs. 19.1 and 19.9). Dissolution pits on the four surfaces average 1.2 to 1.47 mm deep, which is probably between 500 and 600 cal yr B.P. An average depth of around 1.2 mm is probably the minimum amount of surface pitting detectable by this method; carbonate boulders on fan-forming surfaces younger than this do not have measurable dissolution pits (Table 19.1).

The typical time between fan-forming debris flows at a particular fan is estimated from the differences in pit depth calculated for tributary fans with more than one surface (23 of the studied fans). The median difference ($n = 45$) in average depth of dissolution pits on the 23 fans is 2.13 mm, and the interquartile range is 1.32 to 3.22 mm. This suggests that surfaces of the large fans studied here differ in age by about 820 years within a range of 500 to 1200 years. This 820-year average interval is much longer than the 10- to 50-year average recurrence interval (Chapter 20, this volume) of channelized debris flows during historic time.

Several widely spaced tributaries have debris-fan surfaces with exposure times that cannot be shown to be different based on average pit depth; these are connected by the five horizontal lines in Figure 19.6. The surfaces are evidently time correlative within the limitations of this method, suggesting that conditions leading to formation of debris-fan surfaces influence the entire Grand

Canyon region, although these conditions do not necessarily trigger debris flows in every tributary.

Based on average pit depth, most of the surfaces fall into five broadly defined age clusters between about 800 and 4000 cal yr B.P. Five episodes of widespread debris-flow activity are indicated by our mapping, which delineates five to six surfaces of different age canyon-wide (Hereford et al. 1997, 2001). The surfaces in each cluster are shown by the pattern of alternating solid and open symbols with heavy lines in Figure 19.6. Surfaces were assigned to a particular cluster using k-means cluster analysis (Davis 1986, pp. 513–514). Surfaces at the younger and older limits of a cluster are not clearly separated from adjoining clusters. However, in most cases, surfaces within about ± 0.5 mm of the cluster centroid (vertical bars on horizontal lines in Fig. 19.9) are clearly separated from those in nearby clusters, based on nonoverlapping confidence intervals.

The five clusters are around 2-, 3.9-, 5.5-, 7.5-, and 10.4-mm average pit depth, which corresponds to about 800, 1500, 2100, 2900, and 4000 cal yr B.P., respectively. Thus, five broadly defined episodes of heightened debris-flow activity are discernible (Fig. 19.6), although the occurrence of fan-forming debris flows was highly variable during the late Holocene. These in turn may correspond to episodes of increased precipitation with each episode lasting perhaps several centuries.

A striking feature of the correlation chart is that most of the fan surfaces are younger than about 4.7 ka. The surfaces are mostly late Holocene, and 75 percent of them are younger than 2.8 ka. Preservation of prehistoric debris fans is linked to interaction between the frequency of fan-forming debris flows and removal of the deposits by the Colorado River. This interaction is probably controlled by the climate of the Grand Canyon region through its effect on local debris-flow activity and by the climate of the Colorado River drainage basin, which controls the size and frequency of mainstem floods. Only fans deposited in the last 3 to 4 ka are preserved (Fig. 19.6), suggesting that gradual, flood-related shifting of the main channel eventually removed early and mid-Holocene deposits.

An alternative explanation for the lack of early and mid-Holocene deposits in the river corridor is that they are present, but are buried by net aggradation of the fans. This seems unlikely, however, given that older surfaces have the highest elevation relative to younger surfaces on any particular fan. Early Holocene debris-flow deposits are present (Fig. 19.6) in protected sites in the mouths of tributary canyons. However, the surfaces of these deposits project downstream to the debris fan where they are topographically above younger surfaces. Thus, it seems likely that the absence of early and mid-Holocene debris fans resulted from lack of preservation in flood prone parts of the river corridor.

SUMMARY

The Holocene surficial geology of the Colorado River in Grand Canyon consists mainly of terraces and related alluvium, tributary debris fans, and sand dunes. These deposits are well-developed where the river corridor is at least 200–400 m wide. The deposits are mainly late Holocene; mid-Holocene deposits are uncommon, and early Holocene deposits are as yet unknown. The absence of older Holocene alluvium and debris fans probably results from poor long-term preservation along the narrow, flood-dominated channel of the Colorado River.

Perhaps the most important deposits are those forming prehistoric tributary debris fans, because they control the position of the channel and sites of high-level alluvial deposition. These bouldery, roughly fan-shaped deposits resist erosion, producing rapids and forcing the river to flow around the fans. Areas of low-current velocity develop around debris fans enhancing formation of high-level terraces. These alluvial deposits are in turn reworked by wind, forming dunes, sheets, and mounds of wind-blown sand.

Large debris fans of late Holocene age with multiple surfaces are present throughout Grand Canyon. These are composed of sediment derived from tributaries that was deposited in the river corridor by debris flow. The deposits are very poorly sorted and extremely coarse grained, boulders up to several meters in maximum diameter are locally abundant. The geomorphology of the fans consists of the inactive-primary fan surface and the entrenched, active debris-flow channel. Deposits related to the primary fan surfaces are termed fan-forming debris-flow deposits, and those in the channel are referred to as channelized debris-flow deposits. The fan-forming deposits are mainly of prehistoric age, although four of them were deposited as recently as 90–200 cal yr B.P. In almost every case, the channelized deposits are of historic age, having been deposited within the past 5–100 cal yr B.P.

The fan-forming deposits are divided into three relative age categories based on the degree of surface weathering. The youngest category is essentially unweathered with only incipient rock varnish and no subsurface soil development. Intermediate-age surfaces have rock varnish on most sandstone boulders, boulders that are weakly to moderately disintegrated, weakly developed Stage I soil-carbonate morphology, and carbonate boulders with relatively shallow dissolution pits. The older category has sandstone boulders with relatively dark patination, moderately developed Stage I soil carbonate, heavily disintegrated boulders, and carbonate boulders with relatively deep dissolution pits.

The numerical ages of fan surfaces are difficult to determine using standard dating methods, because organic material is extremely rare in the coarse-grained deposits. Several numerical dates, however, have been obtained from debris-flow deposits. The depth of dissolution pits on carbonate boulders is observed to increase systematically with degree of surface weathering. Using the average pit depth on carbonate boulders of dated fan surfaces, the rate of pit deepening is estimated to be 2.60 ± 0.12 mm/ka, and the rate is independent of time through at least the late Holocene. Using this rate, we determined the ages of 71 primary fan surfaces at 26 debris fans throughout Grand Canyon.

The oldest surface on a debris fan is 4700 cal yr B.P. Three debris-flow levees in protected sites just within the mouth of tributary canyons are the oldest dated surfaces at about 6700 cal yr B.P. The youngest surfaces datable by this method (several centuries are required for dissolution pits to develop) formed between 500 and 600 cal yr B.P. For the most part, 75 percent of the dated surfaces are younger than 2.8 ka. At a particular fan, the average time between fan-forming debris flows is about 820 years, indicating that the primary fan surfaces are relatively dormant compared with debris-flow activity in the entrenched channel. We believe that a primary fan surface results from an unusually high-volume debris flow or from a large number of small debris flows. Comparable in volume to historic-age flows, frequent relatively low-volume debris flows could eventually overtop the entrenched channel and spread over the primary surface. In either case, conditions leading to deposition on the primary fan surface are probably unusual as they occur infrequently.

Sandy alluvial deposits associated with large tributary debris fans range in age from 2500–1300 B.C. to the present. Five deposits form terraces that record the depositional activity of the Colorado River before completion of Glen Canyon Dam in 1963. These are informally named the striped alluvium (2500–1300 B.C. to 300 A.D.), alluvium of Pueblo II age (700–1200 A.D.), upper mesquite terrace (1400–1880 A.D.), lower mesquite terrace (1884 A.D. to early 1920s A.D.), and the pre-dam alluvium (early 1920s A.D. to 1957–1958 A.D.). A number of important archeologic sites are associated with the alluvium of Pueblo II age, which was deposited for the most part during Anasazi occupation of eastern Grand Canyon.

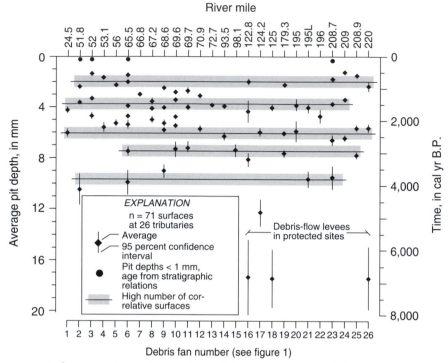

FIGURE 19.6. Age and correlation chart of debris-fan surfaces inferred from average depth of dissolution pits. From top to bottom, alternating solid and open symbols with heavy lines are surfaces clustered about the five horizontal lines, respectively. (Modified from Hereford et al. 1997a.)

Two major periods of erosion and nondeposition separate the late Holocene alluvia. The oldest, between A.D. 300–700, intervened between deposition of the striped alluvium and alluvium of Pueblo II age. The youngest, between 1200 A.D. and 1400 A.D. separates the alluvium of Pueblo II age from the alluvium of the upper mesquite terrace.

Broadly speaking, the alluvial deposits resulted from flood-related aggradation of the river banks; in places the terraces are extensive and may have been floodplains. Indeed, evidence suggests the floodplain of the alluvium of Pueblo II age was locally cultivated by the Anasazi. The erosional episodes isolated the floodplains from the active channel, which formed the terraces. The causes of alternating deposition and erosion are poorly understood, although climate, through its control on flood regimen and sediment load, is probably a first-order factor. Indeed, the broad similarity between alluvial chronologies of southern Colorado Plateau streams and the alluvial history of the Colorado River suggests a common causal mechanism.

Finally, the channel system is out-of-balance with regard to fan-forming debris flows. Over the life span of Glen Canyon Dam, one or more fan-forming events are expected to occur at any major tributary. In the recent past, when such debris flows entered the channel, perhaps forming a waterfall-like obstruction or greatly increasing the severity of a rapid, the debris was eventually removed by large mainstem floods. In the future, however, such channel obstructions will be difficult to remove given the limitations imposed by Glen Canyon Dam.

• 20 •

Debris Flows and the Colorado River

Robert H. Webb, Theodore S. Melis, and Peter G. Griffiths

INTRODUCTION

Debris flows are a type of flash flood that occur throughout Grand Canyon National Park (Webb et al. 1988, 1989, 1996) and elsewhere in the United States where unconsolidated deposits reside on steep terrain (Sharp and Nobels 1953; Glancy 1969; Williams and Guy 1971; Johnson and Rodine 1984; Osterkamp et al. 1986; Kochel 1987; Wells 1987; Wieczorek et al. 1989; Wohl and Pearthree 1991; Whipple and Dunne 1992; Hammack and Wohl 1996). These slurries of water, mud, and rock move with spectacular swiftness through the tributary canyons of the Colorado River. The colorful, sculptured cliffs enclosing Grand Canyon attract millions of tourists every year; these same cliffs, when subjected to intense summer thunderstorms, are the reason that debris flows are common here. The steep exposures of unstable bedrock, containing abundant clay, are ideal for initiation of debris flows, and debris-flow deposition is the primary reason for rapids in the Colorado River.

Debris flow is a generic term for a fluid with a sediment concentration of about 80 percent or greater (Pierson and Costa 1987). *Mudflows* contain more clay and fewer boulders, and other terms, such as *debris torrent* and *mudslides*, have special connotations. Debris flows in the southwestern United States were first described by Blackwelder (1928). Péwé (1968) briefly described a 1967 debris flow at Lees Ferry. The first scientific descriptions of debris flows in Grand Canyon resulted from the flood of December 1966 in Crystal and Lava-Chuar Creeks (Cooley et al. 1977). Ford et al. (1974) report several small debris flows near the mouth of the Little Colorado River in. More than 25 debris flows occurred in Grand Canyon between 1983 and 1996 (Melis et al. 1994; Webb et al. 1996).

Most debris flows are related to fault uplift or landscape disturbances such as volcanic eruptions, forest fires, or poor land-use practices (Pierson 1985; Gallino and Pierson 1985; Meyer et al. 1995). In Grand Canyon, as well as in nearby river canyons, debris flows occur because the Colorado River has rapidly downcut through bedrock, leaving ideal conditions for debris-flow initiation. In addition, the arid climate minimizes erosion of the limestone cliffs, thereby creating steep topography; summer thunderstorms provide energetic rainfall that is required for debris-flow initiation. Debris flows transport boulders and finer sediment relatively long distances because the bedrock channels of most tributary canyons confine the flow to minimize losses to deposition.

Many large debris flows have swept down drainages in Grand Canyon during the last century (Table 20.1). Boulders deposited during the 1939 debris flow,

TABLE 20.1. Major Debris Flows in Grand Canyon Between 1890 and 1997 that Deposited Sediment in the Colorado River

Tributary	Rapid	Miles	Side[a]	Year or Range of Largest Flow	Effects
Badger Creek	Badger Creek	7.9	R	1897–1909	Changed right side of rapid
Soap Creek	Soap Creek	11.2	R	1923–1934	Made navigation easier
Rider Canyon	House Rock	16.8	R	About 1966	Increased constriction
22-Mile Wash	22-Mile	21.4	L	Unknown	Unknown
Unnamed canyon	24-Mile	24.1	L	1989	Changed rapid completely
Tiger Wash	Tiger Wash	26.6	L	Unknown	Increased constriction
Unnamed canyon	None	30.2	R	1987?	Deposited new debris fan
South Canyon	Unnamed	31.6	R	1940–1965	Completely changed fan
Unnamed canyon	President Harding	43.7	L	1984	Changed left side
Unnamed canyon	New	62.5	R	1990	Created new rapid
Lava Creek	Lava Canyon	65.5	R	1966	Unknown
Palisades Creek	Lava Canyon	65.5	L	1990	Increased constriction
75-Mile Creek	Nevills	75.5	L	1990	Deposition on debris fan
Hance Creek	Sockdolager	78.7	L	Unknown	Deposition on left side
Monument Creek	Granite	93.5	L	1984	Increased constriction
Hermit Creek	Hermit	95.5	L	1996	Increased constriction
Boucher Creek	Boucher	96.7	L	1950s	Changed rapid
Crystal Creek	Crystal	98.3	R	1966	Changed rapid
Waltenberg Canyon	Waltenberg	112.2	R	1940s	Increased constriction
Unnamed canyon	New	127.6	L	1989	Created new rapid
128-Mile Creek	128-Mile	128.5	R	1890–1923	Increased constriction
Specter Chasm	Specter	129.0	L	1989	Increased rapid severity
Bedrock Canyon	Bedrock	130.5	R	1989	Increased constriction
Unnamed canyon	Unnamed	133.0	L	Unknown	Deposition on debris fan
Kanab Canyon	Kanab	143.5	R	1923–1942	Increased constriction
Prospect Canyon	Lava Falls	179.3	L	1939	Changed rapid
205-Mile Canyon	205-Mile	205.4	L	1937–1956	Changed rapid
Diamond Creek	Diamond Creek	225.5	L	1984	Increased constriction
231-Mile Canyon	231-Mile	230.8	R	Unknown	Increased constriction

[a]L, left side of the river when viewed from upstream; R, right side of the river.

Source: Modified from Webb (1996).

the largest historic event, completely changed Lava Falls Rapid (Webb et al. 1996, 1997). The better-known 1966 debris flow in Crystal Creek changed a benign rapid into challenging whitewater (Cooley et al. 1977; Kieffer 1985; Webb et al. 1989). No one saw these debris flows, but two others were witnessed. The eye-witness accounts vividly describe this fascinating type of flash flood.

South Canyon, July 1889

In the summer of 1889, Robert Brewster Stanton's expedition to determine the feasibility of a railroad at river level through Grand Canyon had failed, with three men lost to drowning. On July 18, Stanton and his remaining crew stashed their remaining gear, released their boats, and hiked up South Canyon (mile 31.5). A mid-morning thunderstorm overtook them and, according to Stanton

As the rain commenced to fall we heard some rocks roll down the slope behind us, when we looked up, and it seemed as if the whole slopes of the gorge had begun to move at the top. Little streams of water came over the top, and in a moment they changed into streams of mud; and as they came down they gathered strength and turned to streams of mud and rock, undermining larger rocks; and starting them they plunged ahead, and in a few moments the whole sides of the canyon seemed to be moving down upon us with a roar and awful rumbling noise; and as the larger rocks plunged ahead of the streams, they crashed against other rocks, breaking into pieces; and the fragments flew in to the air in every direction, hundreds of feet above our heads; and as these came nearer the bottom where we were, it seemed as if we were to be buried in an avalanche of rock and mud (Smith and Crampton, 1987).

The debris flow Stanton described is similar to ones that occur every year somewhere in Grand Canyon. Intense rainfall dislodges rocks and unconsolidated sediment from the steep slopes and cliffs, which mixes with storm runoff and forming a slurry that typically travels less than a kilometer. These small debris flows do not have much energy and usually stop on gentle slopes, loading channels with debris that can then fuel larger debris flows in the future.

Diamond Creek, July 1984

On July 20, 1984, two river-rafting companies completed their Grand Canyon trips at Diamond Creek in western Grand Canyon. At 3:30 P.M., the guides began driving two trucks up the road, which is little more than a channel bed smoothed by bulldozers in Diamond Creek. A flood the day before had washed out the road. The guides saw no sign of the severe thunderstorm that had just drenched the headwaters of this large drainage. About a kilometer up Diamond Creek, one of the trucks stalled in the middle of the channel. What happened next was described in detail by Ghiglieri (1992) and Webb (1996).

A guide walking up the road for help watched as a "wall of water" spread over a wide section of Diamond Creek, knocking trees over like they were matchsticks. In the constricted reach downstream where the trucks were stalled, the other guides heard the roar of the approaching flood, which to them sounded like a bulldozer. As the flood bore down on the trucks, the guides scrambled up nearby slopes to safety. The debris-flow front, nearly 2-m high, was described as a dark slurry containing wood but no visible boulders. The lead truck rose and pirouetted into the flow, accompanied by the sound of breaking glass. The second truck was quickly swept away by the flood, and both vehicles were last seen floating upside down with their wheels out of the slurry. At the Colorado River, other witnesses saw a wave 5–6 m high form when the debris flow entered the rapid, and the vehicles rose on the wave. The vehicles were found a few months later at the bottom of Diamond Creek Rapid (Ghiglieri 1992, p. 276).

The Diamond Creek flood demonstrated the awesome transport power of debris flows. The front that hit the vehicles, called the *snout*, was a mixture of wood, mud, and boulders; the larger particles were hidden behind a curtain of mud and floating wood. The trucks became particles entrained in the debris flow, floating like corks on the surface of the flow despite their weight. The slurry pushed the Colorado River to the side, changing the left side of Diamond Creek Rapid. Ultimately, the river reclaimed most of its channel, moving the trucks and other particles to the foot of the rapid.

THE PROCESS OF DEBRIS FLOW

Debris flows in Grand Canyon have three distinct phases: initiation, transport, and deposition. These phases are well illustrated by the 1984 debris flow in Monument Creek (Webb et al. 1988). Intense rain pelted a slope of Hermit Shale, and slurries of mud and colluvium from the overlying units moved quickly down the steep slopes. A large block of Esplanade Sandstone gave way as an avalanche and mixed with the slurries as they plunged over a 300-m cliff of Redwall Limestone. At the base of the cliff the mixture of boulders, ground-up rock, and muddy water had enough energy to flow down the lower gradient stream channel, leaving behind a deposit of the avalanche material. Boulders and finer sediment from the channel margins were entrained into the debris, causing it to "bulk up." As the debris flow approached the Colorado River, the slurry no longer was confined by bedrock. Most of the sediment and water decelerated, spread out over the debris fan, and stopped. The snout had enough momentum to enter the Colorado River, and deposition of large boulders diverted river flow from the left side of the rapid. The entire process, from initiation to deposition, was extremely fast; Webb et al. (1988) estimated that the main debris-flow pulse lasted only 3 minutes.

Initiation

Initiation of debris flows in Grand Canyon requires intense rainfall on steep slopes resulting in a mass movement, or slope failure, in consolidated or unconsolidated sediment. The intense rainfall and height of the initiation point above the river are the sources of energy. Slope failures supply the bulk sediment and plunge pools at the bases of cliffs, or long chutes through colluvial wedges, are where the mixing of sediment and water takes place. Although debris flows can be mobilized in any type of sediment, the ones that travel the longest distances are those derived from certain shales and their associated colluvial wedges.

The intense, sometimes protracted thunderstorms of July through September initiate most of the debris flows in Grand Canyon. Thunderstorms are either (a) widespread, affecting numerous tributaries, or (b) concentrated over one tributary. Few rain gages are present in areas where debris flows begin (Griffiths et al. 1997). Rainfall intensity during historical debris flows has ranged from 10 to 40 mm/hr at remote climate stations, and rainfall may last several hours. Debris flows do not necessarily occur during above-average rainfall seasons, although precipitation during the preceding weeks or month is typically well above normal (Melis et al. 1994; Webb et al. 1996; Griffiths et al. 1997).

In addition to summer thunderstorms, certain types of regional storms also cause debris flows. Thunderstorms from a dissipating tropical cyclone caused debris flows in Prospect Canyon in 1939 (Webb et al. 1996). The Crystal Creek flood of December 1966 resulted from a warm winter storm, with rainfall falling on a preexisting snowpack (Cooley et al. 1977). Historically, this type of storm spawned debris flows only in 1966 and 1995. Winter storms that have initiated debris flows typically have had lower intensities, and debris-flow initiation is likely a function of both rainfall intensity and duration (Wilson and Wieczorek 1995). In the Grand Canyon region, a closing period of higher intensity precipitation defines storms that produce debris flows; otherwise, streamflow floods result (Webb et al. 1996).

TABLE 20.2. Clay Mineralogy of Shales, Colluvium, and Debris-Flow Deposits in Grand Canyon

Formation	Member[a]	Depositional Environment	Illite[b] (%)	Kaolinite[b] (%)	Smectite[b] (%)	Other[b] (%)
Shales not Producing Debris Flows						
Chinle Formation	Petrified Forest	Lacustrine	41	11	42	6
Hakatai Shale[c]	Middle	Near-shore	52	15	2	31
Tropic Shale		Marine	16	19	57	8
mean ± standard deviation			36 ± 18	15 ± 4	34 ± 28	15 ± 14
Shales Producing Debris Flows						
Bright Angel Shale			68	22	0	10
Chinle Formation	Monitor Butte?	Fluvial	32	44	7	17
Dox Formation	Lower Middle	Fluvial	65	14	0	21
Esplanade Sandstone	Basal shale	Fluvial	50	40	2	8
Galeros Formation	Carbon Canyon	Lacustrine	63	23	0	14
Galeros Formation	Jupiter	Lacustrine	35	42	2	8
Hermit Shale		Fluvial	54	41	0	5
mean ± standard deviation			52 ± 14	32 ± 12	1 ± 3	14 ± 7
Colluvial Wedges						
Mean ± standard deviation ($n = 16$)			34 ± 13	43 ± 19	5 ± 5	18 ± 8
Debris Flow Deposits						
Mean ± standard deviation ($n = 10$)			41 ± 15	28 ± 9	6 ± 11	25 ± 11

[a]Nomenclature follows summary in Elston et al. (1989).
[b]Minerals identified by semiquantitative x-ray diffraction. Margin of error ± 20% by weight (Starkey et al. 1984).
[c]Typically not high enough on cliffs to produce debris flows.

If initiation is to occur, intense rainfall must trigger slope failures in bedrock or loose colluvium. Shales form unstable slopes, whereas sandstones and limestones form cliffs. Massive, uniform rock layers such as the Redwall Limestone are very stable, whereas units comprised of alternating limestone, sandstone, and shale layers, such as strata of the Supai Group, are unstable and susceptible to failure. Susceptibility has less to do with "hard" and "soft" units and more to do with thickness of strata and presence of underlying shales. In other words, sandstones from the Supai Group may be as resistant or more so than the Redwall Limestone, but stratification in the Supai Group produces the critical instability of thin, resistant layers being undercut by erosion of the weaker layers.

Certain types of clay minerals increase the propensity of bedrock failure and debris-flow mobilization. A mixture of sand, silt, clay, and water fills the interstices between larger particles. This fluid has a higher density and apparent viscosity than water. Bedrock units that contain significant amounts of clay are the most important units for the initiation of debris flows. Failures in the Hermit Shale, for example, have contributed to most of the largest historical debris flows in Grand Canyon. The Muav Limestone, a silty dolomite that grades downward into the Bright Angel Shale, is another important stratum producing slope failures. Many units of the Precambrian Grand Canyon Series, particularly the Dox Sandstone, contain enough fine particles to produce debris flows.

Not all shale units present in the Grand Canyon region produce debris flows. The clay mineralogy of shales greatly affect their stability (Trask 1959; Hampton 1975; Griffiths et al. 1996). Shales that produce debris flows are composed primarily of illite and kaolinite and are of terrestrial origin (Table 20.2). These single-layer, nonswelling clays allow deep penetration of rainfall to failure surfaces, whereas the cracks in multilayer, swelling clays, such as smectites, quickly close after wetting, preventing deep percolation. The only terrestrial shales that do not produce debris flows are the lacustrine strata in the Chinle Formation. These strata are diagenetically altered volcanoclastic sediment, which results in a high content of smectites. Marine units, such as the Tropic Shale in the vicinity of Lake Powell, are high in smectites (Table 20.2). Colluvial wedges that produce debris flows, and the debris-flow deposits themselves, have clay mineralogies similar to the terrestrial shales.

Major cation chemistry of bedrock units affects their proclivity for failure (Hampton 1975; Pierson and Costa 1987). Sodium causes clay minerals to disperse, sealing cracks and inhibiting deep percolation. Shales produced in a marine environment typically are high in sodium. Terrestrial shales typically are low in sodium and high in magnesium. All shales in the Grand Canyon region are high in calcium, and the presence of dispersed gypsum greatly increases the potential for slope failures. For example, the Fossil Mountain Member of the Kaibab Formation in western Grand Canyon does not contain significant shale layers, but the Harrisburg Member, a prominent gypsum-bearing unit, produces small debris flows.

Even the most failure-prone lithologic units do not produce debris flows when they are close to river level. The longitudinal profile of the Colorado River in upper Marble Canyon (Fig. 20.1) illustrates the relation between height of

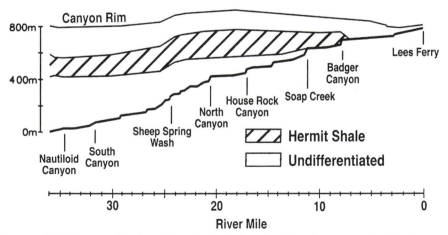

Figure 20.1. Longitudinal profile of the Colorado River in upper Marble Canyon showing the relation between the height of the Hermit Shale and rapids.

(a)

FIGURE 20.2. (a) The initiation area for debris flows in Monument Creek (Mile 93.5) consists of the interbedded Supai Group, primarily sandstones, siltstones, and shales (middle of the photograph), overlying the near-vertical Redwall Limestone and underlying the Hermit Shale. The 1984 debris flow was initiated in the Hermit Shale and in the Esplanade Sandstone Member of the Supai Group (Webb et al. 1988; photo by R. Webb). (b) Schematic diagram illustrating the initiation of debris flows by the failure of bedrock—usually the Hermit Shale and (or) members of the Supai Group—during intense rainfall.

377

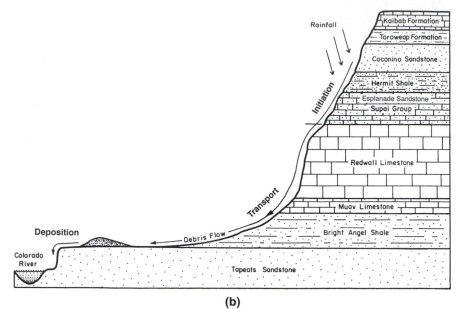

(b)

FIGURE 20.2. (*Continued*)

shales above the river and the presence of rapids, an indicator of debris flows. In the first 20 miles of river downstream from Lees Ferry, only five rapids are present. Only two of these—Badger and Soap Creek Rapids—are significant navigational hazards. The debris flows that created these rapids came from shales, which were deposited in a fluvial environment, of the Chinle Formation at the base of the Vermillion Cliffs. Downstream from about Colorado River mile 15, the Chinle Formation is not close enough to the river corridor to produce debris flows that reach the Colorado River. By mile 20, the Hermit Shale, which first appeared at river level at about mile 6, has risen over 300 m above the river. With higher elevation comes the potential energy necessary to mobilize falling debris into slurries. The result is more debris flows: Seven rapids and numerous riffles are in the Colorado River between mile 20 and 27, and the slopes above the river are strewn with loose boulders.

Bedrock failures (Fig. 20.2) are associated with 13% of historic debris flows in Grand Canyon. These failures are responsible for many of the largest flows of the last century (Melis et al. 1994; Griffiths et al. 1996, 1997). In December 1966, 11 slope failures in the Hermit Shale, Supai Group, and Muav Limestone contributed to the debris flow in Crystal Creek (Cooley et al. 1977). These failures occurred 1400–2000 m above and 20 km from the Colorado River. Of 93 slope failures throughout Grand Canyon that resulted from the storm in December 1966, 70 percent were in the Hermit Shale and Supai Group (Webb et al. 1989, p. 24). The 1984 debris flow in Monument Creek (Fig. 20.2) began as an avalanche from the Esplanade Sandstone of the Supai Group and flowed 5 km in its 1000-m descent (Webb et al. 1988).

Most debris flows in Grand Canyon result from failure of colluvial wedges at the base of Redwall Limestone cliffs (Fig. 20.3). The boulders and cobbles in these colluvial wedges are mostly derived from the Redwall Limestone, Kaibab Formation, and Supai Group sandstones; sand, silt, and clay are contributed by the Hermit Shale and fine-grained strata of the Supai Group. This poorly sorted mixture makes an ideal source for debris flows. Two types of failures begin in

(a)

FIGURE 20.3. (a) Failure scars caused by the "firehose effect" along gullies through colluvial wedges at Mile 62.5. The vertical cliffs at the top of the photograph are Redwall Limestone; the colluvial wedges overlie Muav Limestone and Bright Angel Shale. (Photo by T. Melis.) (b) Schematic diagram illustrating the initiation of debris flow by colluvial wedge failure during intense rainfall.

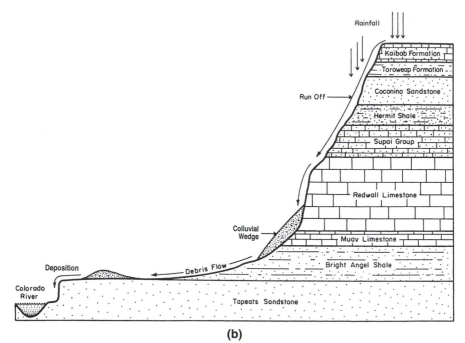

(b)

FIGURE 20.3. (*Continued*)

colluvial wedges. Runoff from intense rainfall erodes gullies into the colluvial wedges, where more and more material is entrained until a slurry forms. This type of direct failure has contributed to 20% of historical debris flows in Grand Canyon. The firehose effect (Johnson and Rodine, 1984) involves water pouring off cliffs onto colluvial wedges, causing failures that contributed to 37% of historical debris flows in Grand Canyon (Griffiths et al. 1997).

Debris flows initiated in colluvial wedges are small and flow only short distances, except in the case of failures caused by the firehose effect. The 1987 flow in 18-Mile Wash was caused by a flash flood cascading over a 100-m cliff in the Kaibab Formation and falling onto an unstable slope of Hermit Shale and strata of the Supai Group. In 1990, runoff cascading over the Redwall Limestone onto colluvial wedges overlying slopes of Muav Limestone caused several debris flows between mile 62 and 64 (Melis et al. 1994). Flood waters in Prospect Canyon, a large drainage that meets the Colorado River at Lava Falls Rapid (mile 179.4), must pour over a 325-m cliff onto a scree slope of loose basalt boulders and other colluvium, which then can mobilize and flow a short 1.6 km to the Colorado River. Prospect Canyon yielded 6 debris flows between 1939 and 1995, all initiated by the firehose effect (Webb et al. 1996).

Transport

In the vicinity of the South Rim, the source areas for debris flows are 1000–1300 m above and many kilometers south of the Colorado River. Resistant bedrock units, such as the Redwall Limestone, form bedrock-floored canyons that act as confining chutes for the moving debris, allowing long-distant transport with minimal loss of mass by deposition. Most side canyons preserve little evidence of past debris flows except an occasional lateral levee in a wide section of the tributary canyon or boulder fields deposited where smaller debris flows had stopped short of the river. Debris flows in Grand Canyon carry a relatively small volume of material delivered in a large pulse; without confining channels, debris flows would lose most of their volume to levee deposition.

Debris flows can move long distances in Grand Canyon. The 1966 debris flow in Crystal Creek flowed 21 km from its initiation points to the Colorado River (Cooley et al. 1977; Webb et al. 1989). Prehistoric debris flows, which have left depositional evidence in Shinumo and Kanab Creeks, could have flowed as far as 40 km. Most debris flows travel shorter distances, primarily because few tributary canyons are longer than several kilometers. More typical debris flows, like the 1984 flood in Monument Creek and the 1987 flood in 75-Mile Canyon (Melis et al. 1994), travel 3–6 km. The exact mechanism of how debris flows can travel such distances is not well understood (Johnson and Rodine 1984; Costa 1984) but is currently under intense study.

Three types of debris flows occur in Grand Canyon. Melis et al. (1997) reported hypothetical hydrographs developed from the combination of stratigraphic analyses of debris-fan and channel-margin deposits and reconstruction of peak discharges. Type I flows consist of a single unsustained pulse of debris flow followed by recessional streamflow (Fig. 20.4). Type I flows often occur in Grand Canyon and typically are the smallest debris flows. The 1990 debris flows at mile 62.5 and 62.6 are examples of this type of runoff (Melis and Webb 1993; Melis

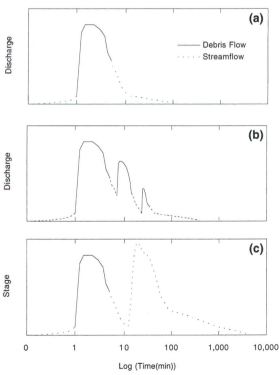

Figure 20.4. Generalized hydrographs of three types of debris flows in Grand Canyon reconstructed from depositional evidence in channels and on debris fans. (a) Type I flow. (b) Type II flow. (c) Type III flow. (From Melis et al. 1997.)

et al. 1994). Type II flows consist of a main debris-flow pulse, followed by alternating pulses of streamflow and debris flow. Depositional evidence on debris
fans provides the basis for this type of hydrograph; the 1987 debris flow at 18-
Mile Wash and the 1989 debris flow in a left tributary at 127.6 mile are excellent examples. Type III flows consist of a debris-flow pulse followed by a higher-
stage streamflow that removes most of the debris-flow evidence (Fig. 20.4). The
streamflow discharge is not necessarily higher than the debris-flow discharge,
because debris-flow deposition may raise the channel bed significantly, allowing smaller streamflow discharges to erase the evidence of larger debris flows
(Melis et al. 1997). The 1989 floods in Fossil and Forster Canyons are good examples of type III flows (Melis et al. 1994).

Peak discharges for debris flows in Grand Canyon range from 100 to 1000
m³/s (Melis et al. 1994). These discharges are well within the known realm of
debris-flow size (Pierson and Costa 1987). Using the total depositional volume,
Webb et al. (1988, 1989) and Melis et al. (1994) showed that the duration of debris-flow pulses ranges from 30 seconds to 3 minutes. Peak velocities for debris
flows range from 2 to 10 m/s (Melis et al. 1994), and channel slope is not the
most important factor. Higher velocity does not result in higher discharge. For
example, the Crystal Creek debris flow of 1966, which had a discharge of about
310 m³/s, had velocities that ranged from 3 to 6 m/s; other debris flows in steep-
angled chutes with discharges of about 100 m³/s had velocities closer to 7 m/s.

Peak discharges of debris flows are similar to those of post-dam flows in
the Colorado River except that debris flows typically contain 70–90% sediment.
Water content of the 1966 debris flow in Crystal Creek was about 30 percent by
weight; therefore, the peak discharge contained about 220 m³/s of sediment.
Melis et al. (1994) found that the water content of 20 debris flows reconstituted
in the laboratory ranged from 5 percent to nearly 40 percent, depending on the
amount of silt and clay carried in the debris flow. The variability in water content for a given flow is 1–5 percent; in other words, water content varies considerably among debris flows, but for any one debris flow the water content falls
within a fairly precise range.

The particle-size distribution of sediment carried in debris flows shows a
very wide range of sizes (Fig. 20.5). Gravel and cobbles (2- to 256-mm diameter) constitute the largest amount of debris flows (50–90 percent); 10–25 percent
of a typical debris flow is sand (Melis et al. 1994). The clay content, which is
critical for long-distant transport, ranges from <1% to 5%. Although boulders typically comprise <20% of debris flows, they are the most visible and important
aspect of debris-flow transport (Beatty 1989; Rodine and Johnson 1976), particularly with respect to Colorado River geomorphology.

As yet, the exact mechanism of boulder transport in debris flows is unknown,
but several hypotheses have been advanced. A large part of the mechanism is
buoyancy, or more specifically the difference in density between fluid and rock.
The density of sandstones, limestones, and basalts in Grand Canyon ranges from
2600 to 2700 kg/m³, and debris flows have densities between 2200 and 2400
kg/m³. The small density difference indicates that buoyancy forces should provide considerable lift to boulders in a moving debris flow. Despite this, boulders
still should sink, albeit slowly. Another upward force on boulders is created by
particle collisions in the slurry. In the jostling among all those buoyant boulders,
collisions may cause sinking particles to move upward and (or) ascending particles to sink. In pure water, forces are transmitted only in the fluid. In debris flows,
forces are transmitted both in the fluid and by particle-to-particle collisions.

Early models treated debris flows as deformable plastic moving as plug flow
with little internal deformation (Johnson and Rodine 1984). Plug flow models,

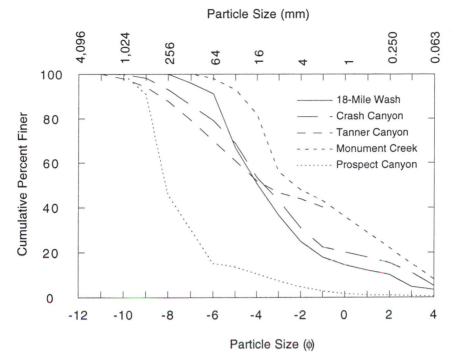

FIGURE 20.5. Particle-size distributions for selected debris flows in Grand Canyon. 18-Mile Wash, the 1987 debris flow at 18-Mile Wash; Crash Canyon, the 1990 debris flow at Mile 62.6; Tanner Canyon, the 1993 debris flow at Tanner Canyon, Mile 68.5; Monument Creek, the 1984 debris flow in Monument Creek (Mile 93.5); and Prospect Canyon, the 1995 debris flow in Prospect Canyon (Mile 179.4).

also known as Bingham models, require the assumption that below a certain point, called the yield stress, debris flows could withstand pressures without deforming. One indication of yield stress is critical thickness; when a debris flow stops, it has a finite thickness that is a function of yield stress. When slurries move in straight channels, most of the mass flows as a plug with little internal deformation (shearing). The only shearing in the flow is assumed to occur very close to the sides and bottoms of the flow. Using the plug-flow model, with no internal shear and a finite yield stress, rocks could bob along in the flow primarily as a result of buoyancy forces.

The problem is that debris flows moving in natural channels have considerable internal shear, although not necessarily as much as turbulent water flow. Something else is helping to support those boulders. One suggestion of what that force might be came from experiments in which landslides were created in a flume (Iverson and LaHusen 1989). High-frequency fluctuations in pressure, on the order of one cycle per second, occurred in the interstitial fluid at the base of the landslide. These pressures could have occurred for many reasons; they have resulted from flow over a rough bed like the bottom of a typical Colorado River tributary. The pressures generated were of sufficient magnitude to support the media; in other words, transient pressures generated within the fluid could provide the supporting forces for boulders.

Deposition

Debris fans at the mouths of tributaries provide the depositional site for most debris flows and a setting for interaction between fluvial and hillslope processes. Debris fans create most Grand Canyon rapids (Howard and Dolan 1981); only two rapids were created by rockfalls (Webb et al. 1988). Melis (1997) described 444 debris fans in Grand Canyon that periodically have debris-flow deposition. The tributary channel across most debris fans is confined, not by bedrock, but by levees of mud and boulders left by past debris flows. Terraces on the higher parts of debris fans, untouched by the Colorado River in recent years, are clearly of debris-flow origin, indicating earlier Holocene debris-flow activity (Hereford et al. 1997a).

Some debris flows lack sufficient mass and energy to reach the Colorado River. Their mass and energy are expended during deposition of the confining levees on the debris fan. Type I debris flows typically create a uniform depositional area on part of the debris fan, whereas type II flows create a complex array of debris-flow and streamflow deposits (Melis et al. 1997). The surfaces of recent deposits are rough with pressure ridges, which reflect pulses in debris flow (Major 1997), and levees. Boulder trains, which are reworked levees, are also common. Only residual boulder trains (some very large) and streamflow deposits mark deposition by type III debris flows.

Most debris flows reach and constrict the river to some extent, although the effect of debris flows on the mainstem varies with reach morphology, particularly channel width (Melis 1997). No twentieth century debris flow has fully dammed the river or flowed downstream; the 1939 debris flow from Prospect Canyon constricted Lava Falls Rapid by about 80 percent, whereas the 1966 debris flow from Crystal Creek had a similar effect on Crystal Rapid (Webb et al. 1997). Debris flows lose too much energy and mass on debris fans; in addition, the turbulent water in rapids dilutes the snout, further reducing its momentum. In most Grand Canyon rapids, the largest boulders are at the head, which is also the closest point to the mouth of the tributary canyon. Boulders near the bottom of rapids are generally pushed there by the force of the Colorado River.

Some of the boulders deposited in the Colorado River are enormous. At mile 62.5, one boulder transported into the river in 1990 weighs 280 Mg. Boulders transported in some debris flows that occurred after 1983 range between 70 and 100 Mg (Melis et al. 1994). In Crystal Creek, a boulder weighing 45 Mg did not reach the Colorado River in 1966, and larger ones deposited in the river are what increased the severity of Crystal Rapid. The size of boulders transported is not necessarily related to the discharge of debris flows; boulder size varies tremendously among debris flows of about the same size (Webb et al. 1989).

FREQUENCY OF DEBRIS FLOWS

Estimating the frequency of debris flows requires an approach that is considerably different from traditional flood-frequency analysis. Until the last few decades, no one recorded debris flows in most of Grand Canyon. The history of debris flows must be reconstructed from their deposits or their effects on the Colorado River. Several methods are available, but most have limitations. For example, radiocarbon dating of organic materials is the standard way to determine the age of sediments deposited during the last 40,000 years, but very little organic material was deposited with the mud and boulders. Under certain circumstances, such as association with archaeological sites, debris flows can be dated using ra-

(a)

FIGURE 20.6. Repeat photographs of Boucher Rapid (mile 96.7). (a) (February 8, 1890; R.B. Stanton). Historical evidence of debris flows in Grand Canyon largely comes from repeat photography (Webb 1996). Boucher Rapid is a moderate-sized rapid in the Inner Gorge that has had several debris flows in the last century. In this 1890 view, fresh-looking gravel in the channel of Boucher Creek suggests that a flash flood, but probably not a debris flow, had occurred a short time before this photograph was taken. (b) (February 18, 1992; R. Webb). Boucher Rapid is not considered a very formidable reach of whitewater today. The debris flow of 1966 at Crystal Rapid, only 2 km downstream, raised the river level sufficiently to drown out the tailwaves of Boucher Rapid (Péwé 1968). Similarly, Boucher Creek had a debris flow in the early 1950s that drowned out the tailwaves of Hermit Rapid, about 2 km upstream. The debris flow in Boucher Creek caused deposition of boulders over the entire debris fan. Boulders 1–2 m in diameter were displaced, and new ones of about the same size were deposited.

diocarbon analysis (Hereford et al. 1996). Using twigs extracted from mud plastered against an overhanging wall, Melis et al. (1994) dated a 5400 yr B.P. debris flow at river mile 63.3. Limestone particles dissolve with time, and the depth of pitting in limestone boulders can be used to date prehistoric debris flows (Hereford et al. 1997a). However, development of measurable pitting requires several centuries, rendering the technique useless for twentieth century debris flows.

Repeat photography is the best technique for documenting historical debris flows (Webb 1996). Before closure of Glen Canyon Dam in 1963, floods on the Colorado River eroded away most newly deposited debris within five years. Residual accumulations of boulders remain that are obvious in repeat photographs (Fig. 20.6). Repeat photography can also reveal the number of debris flows from a tributary in the last century, but only under certain conditions. A sequence of historical photographs can be used to reconstruct a history of debris-flow activ-

(b)

FIGURE 20.6. (*Continued*)

ity, such as in Prospect Canyon at Lava Falls Rapid (Webb et al. 1996, 1997). A single historical view can also be used if it was taken from a vantage point that looks down on a debris fan. In this case, changes in the depositional patterns on debris fans may indicate many past debris flows.

Melis et al. (1994) discriminated 525 tributaries of the Colorado River between Lees Ferry (mile 0) and Diamond Creek (mile 225) that periodically yield debris flows; Griffiths et al. (1996) extended the definition downstream to mile 246 to include 600 tributaries. The debris fans of 164 of these tributaries are visible in 483 photographs taken after 1872 (Webb 1996; Griffiths et al. 1996). By examining changes in the arrays of boulders on the debris fans, we found 96 tributaries that had one or more debris flows in the last century, or 56 percent of all debris fans recorded in the repeat photography. One half of the 63 largest tributaries (>10 km^2) had at least one debris flow in the last century. We also observed debris flows in 17 small chutes or gullies that otherwise are insignificant.

Using logistic-regression analysis of the data obtained from the repeat photography, Griffiths et al. (1996) estimated the probability of debris-flow occurrence throughout Grand Canyon (Fig. 20.7). Because the physiography of western and eastern Grand Canyon are considerably different, Griffiths et al. (1996) divided the canyon at Hermit Rapid (mile 95.0) and modeled each half separately. A calibration model was developed using the repeat photography information from 164 tributaries. A verification model was also developed from 214 tributaries that included those used in the calibration model combined with additional tributaries with frequency information obtained by methods other than

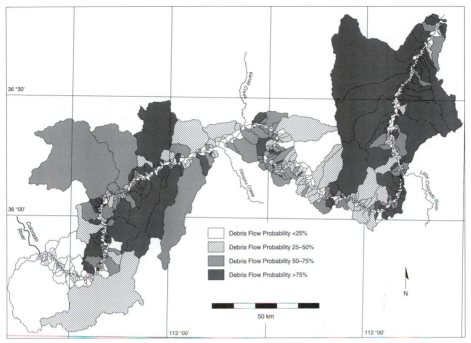

FIGURE 20.7. Map showing four probability classes of debris-flow frequency during the last century in 600 tributaries of the Colorado River in Grand Canyon. (From Griffiths et al. 1996.)

repeat photography. Griffiths et al. (1996) used 21 morphometric and lithologic variables that were suspected of influencing debris-flow frequency.

Based on the calibration data, the logistic-regression model of Griffiths and others (1997) had an accuracy of about 70% in eastern Grand Canyon and 74% in western Grand Canyon. Of the morphometric variables, tributary drainage area and channel gradient, the aspect of the river corridor at the mouth of the tributary, and the channel gradients to the Hermit Shale and Muav Limestone were statistically significant and were used to estimate frequency. The resultant map of probabilities of debris flows during the last century in Grand Canyon (Fig. 20.8) indicates that debris flows are more frequent in eastern than in western Grand Canyon. The highest frequency of debris flows has occurred between the Little Colorado River and Hance Rapid (mile 61.5 to 77). The reach between Havasu Creek (river mile 157) and Lower Granite Gorge (river mile 213) had the lowest frequency of debris flows. Debris-flow frequency is clustered in distinct reaches along the Colorado River, particularly where the river's course is southwesterly and where Hermit Shale is between 100 and 1000 m above and relatively close to the river.

For individual tributaries, debris flows recur every 10–50 years, on average, in 60 percent of the tributaries of the Colorado River. Some tributaries had no debris flows during the last century; Prospect Canyon has had six debris flows clustered between 1940 and 1995, but none in the preceding 50 years (Webb et al. 1996, 1997). A large flow may destabilize enough sediment to supply additional debris flows until the supply is exhausted. Another major slope failure oc-

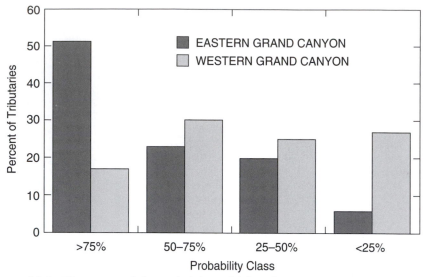

Figure 20.8. Histograms of the probability of debris-flow occurrence in eastern and western Grand Canyon. (From Griffiths et al. 1996.)

curs—perhaps a half century later, perhaps centuries later—and the process repeats itself. Some tributaries, such as 217-Mile Canyon, do not appear to have experienced a debris flow for perhaps several centuries. Whereas these tributaries have small floods on a frequent basis, debris flows are rare. The notable exception is Prospect Canyon, one of the most active debris-flow producers in Grand Canyon (Webb et al. 1996, 1997).

THE GEOMORPHIC FRAMEWORK OF THE COLORADO RIVER

River runners and scientists have long recognized parts of the association between tributary canyons, whitewater, and sand bars (Howard and Dolan 1981; Schmidt and Graf 1990). Major John Wesley Powell first recognized that the boulders in rapids were deposited during tributary floods (Powell 1876); other early explorers also noticed the close association of tributaries and rapids. But Powell believed that differential erosion of bedrock somehow affected the river; his crew learned to dread the appearance of granite and limestone along the river because it indicated, so they thought, intense rapids ahead. They were wrong. The lithology at river level affects mostly channel width, not slope (Hamblin and Rigby 1968; Howard and Dolan 1981). However, channel width dramatically affects debris fan stability and evolution, including fan shape, size, spacing, and composition (Melis 1997).

Piles of boulders in the river channel create hydraulic controls on the Colorado River. Debris flows transport boulders from the cliffs to the river; river floods transport much of the newly deposited material downstream where it accumulates in islands, rock gardens, or debris bars (Webb et al. 1989, 1997a; Melis 1997). This process is called reworking. What remains after repeated debris flows from the tributaries and reworking by the Colorado River is a debris fan, typically comprised of boulders on the surface (Melis and Webb 1993); a rapid full of large boulders that form various waves and holes; and an island or other deposit of well-sorted cobbles or boulders just downstream. Mantled around the debris fan on its upstream and downstream sides are sand bars, typically the best camping beaches in Grand Canyon. Debris-flow deposits, rapids, and sand bars make up the geomorphic framework of the Colorado River.

Rapids exist because the Colorado River lacks sufficient stream power to remove all boulders deposited from debris flows. Most of the sediment deposited by debris flows is removed; typically, the river will erode all mud and sand whenever the debris fan is inundated, and most of the cobbles and small boulders in the river are transported away. Whereas clay, silt, and sand may be transported a long distance from the source debris fan, the eroded cobbles and boulders are transported only a short distance downstream. Just below the rapid, the cobbles and small boulders accumulate in neatly sorted piles. These deposits, properly termed debris bars and informally called "islands" and "rock gardens" (Kieffer 1985, 1987), form secondary riffles and rapids. These stretches of whitewater are known informally as the "Son of . . ." rapids (e.g., Son of Hance) in recognition of their genetic link to the parent rapid. Howard and Dolan (1981) refer to well-sorted, coarse-grained deposits as cobble bars. Cobbles, which have diameters between 64 and 256 millimeters, are not the only constituents of these bars; sand and boulders also are present. Gravel bars are common between river mile 65 and 72, and the only exposed rock garden is downstream from Crystal Rapid (Kieffer 1985). The majority of reworked debris fans in Grand Canyon consist of Redwall and Muav Limestone boulders (Melis 1997).

Debris fans, rapids, and sand bars are all controlled, directly or indirectly, by the process of debris flows. Debris fans that are reworked remain relatively stable until modified by additional debris flows (Melis 1997). In terms of navigation, rapids change with each change in water level of the river, but the underlying configuration of boulders that create the waves and holes change little until the next debris flow. Rock gardens and islands remain until the next debris flow and reworking flood in the Colorado River. Finally, the sand bar, the least stable element of this geomorphic framework, can change with fluctuations in pre-dam flow or the pattern of flow release from Glen Canyon Dam. But the locations of most sand bars are fixed relative to piles of boulders, and the general configuration of sand bars is dependent on when the last debris flow occurred.

The presence of debris fans and rapids has broad implications as to whether a bedrock canyon is downcutting or not. The rapids account for about 10 percent of the length while causing 50 percent of the fall in the Colorado River through Grand Canyon (Leopold 1969). In the relatively warm and dry climate of the present, the Colorado River likely is expending its erosive energy on removing boulders from the rapids instead of eroding downward through bedrock.

Since completion of Glen Canyon Dam in 1963, the delicate balance between tributaries and river has shifted in favor of aggrading debris fans and narrower rapids. Concern about the decreased competence of the regulated Colorado River was raised shortly after the dam was completed (Péwé 1968). Graf (1980) concluded that debris fans were aggrading in bedrock canyons of the Green River in Utah. Reportedly, 25 percent of debris fans in Grand Canyon aggraded between 1965 and 1973 (Howard and Dolan 1981). Our observations and repeat photography (Melis et al. 1994; Webb 1996) suggest that this is too high a percentage for such a short period of time, but debris fans are aggrading at even the largest rapids (Webb et al. 1996).

The depth of the Colorado River is not controlled by resistant bedrock (at least not during the Holocene). Instead, the 444 debris fans (not all of the 525 tributaries have debris fans at their mouths) along the length of the river corridor inhibit further downcutting. The energy of flowing water is expended on removing all the massive boulders thrown into it by minuscule tributaries, not eroding the underlying bedrock. It is a titanic struggle between small tributaries and a large river: Tributaries push boulders in, and the river tries to transport them downstream or dissolve them. The tributaries currently are winning, because rapids are present along the river corridor. Now, operation of Glen Canyon Dam is accelerating that aggradation of boulders and cobbles (Webb et al. 1997a).

SUMMARY

Debris flows occur in 525 tributaries of the Colorado River in Grand Canyon between Lees Ferry and Diamond Creek (river miles 0 to 225). An episodic type of flash flood, debris flows transport poorly sorted sediment ranging in size from clay to boulders into the Colorado River. Debris flows in Grand Canyon are initiated by slope failures that occur during intense rainfall. Failures in weathered bedrock, particularly in the Hermit Shale and Supai Group, have initiated many historic debris flows in Grand Canyon. A second mechanism, termed the "fire-hose effect," occurs when runoff pours over cliffs onto unconsolidated colluvial wedges, triggering a failure.

Interpretation of repeat photographs spanning 125 years yielded information on the frequency of debris flows in 168 tributaries. Of these, 96 contain evidence of debris flows that have occurred since 1872, whereas 72 tributaries have not had a debris flow during the last century. The oldest, dated debris flow occurred 5400 ^{14}C years ago at mile 63.3. The frequency of debris flows ranges from one every 10 to 15 years in certain eastern tributaries, to less than one per century in 40 percent of the tributaries. Debris flows are more frequent in Marble Canyon and eastern Grand Canyon than in western Grand Canyon.

Debris flows in Grand Canyon have three types of hydrographs. Type I events consist of a single, unsustained pulse of debris flow, whereas type II events have multiple debris-flow pulses. Type III events consist of a debris flow followed by a streamflow flood of larger stage. Although peak discharges of most debris flows range from 100 to 300 m^3/s, the largest debris flow in Grand Canyon during the last century—the 1939 debris flow in Prospect Canyon—had a peak discharge of about 1000 m^3/s. The water content of debris flows ranges from 10 to 25 percent.

Debris flows create and maintain debris fans and the hundreds of associated riffles and rapids that partly control the geomorphic framework of the Colorado River downstream from Glen Canyon Dam. Before regulation, debris fans aggraded by debris flows were periodically reworked by large river floods. Operations of Glen Canyon Dam have reduced flood frequency in the Colorado River, which has limited reworking of recently aggraded debris fans. The presence of rapids indicates that the bed of the Colorado River is aggrading, not eroding through bedrock, in the Holocene.

• 21 •

SIDE CANYONS OF THE COLORADO RIVER IN GRAND CANYON

Andrew R. Potochnik and Stephen J. Reynolds

INTRODUCTION

The majesty of the Grand Canyon emanates from the magnitude of its impressive dimensions. The Colorado River has cut an exquisite gorge; its tributaries are no less spectacular. Trails from the rim are few, the hike is lengthy, and access to much of the canyon from above is limited; however, a river trip through the Grand Canyon, with ample time for hiking, enables one to explore many side canyons with relative ease. Unlike a view from the rim, a view from the river provides an inside-out perspective (Fig. 21.1). The rim view paints an overall picture of the grand-scale geological scene. In the side canyons, it is possible to examine the details of geological features and relationships that tell a more complete story. The sequence of rocks in the Grand Canyon can be divided into four groups, each separated by a major unconformity (Fig. 21.2). The oldest group includes Proterozoic igneous and metamorphic rocks (e.g., Vishnu Group) that were formed during a major episode of deformation and metamorphism about 1.7 billion years ago (Ilg et al. 1996). These rocks are overlain by the Grand Canyon Supergroup, a series of tilted sedimentary rocks and interlayered mafic flows and sills that formed between 1.3 and 0.8 billion years ago (Larson et al. 1994; Elston and McKee 1982). Extensive erosion of these Proterozoic rocks formed a conspicuous, regionally planar unconformity on which the canyon's characteristic, horizontally stratified Paleozoic rocks were deposited between 550 and 250 million years ago. The fourth group of rocks in the canyon includes a veneer of late Cenozoic sediments and volcanic rocks that were deposited mostly since six million years ago, during and after the main period of canyon cutting. The classic physiography of the canyon and surrounding Colorado Plateau is largely a signature of Cenozoic erosion in response to post-mid-Cretaceous uplift of the region to a height of about two miles (3.2 km) above sea level.

From its source in north-central Colorado to its mouth at the Gulf of California, the Colorado River crosses four major physiographic provinces: the Rocky Mountains, the Colorado Plateau, the Transition Zone, and the Basin and Range Province. On the final leg of its long journey across the Colorado Plateau, the river follows a sinuous course in northern Arizona through the Grand Canyon. The Grand Canyon, the Transition Zone, and the Colorado Plateau end abruptly at the Grand Wash Cliffs, where the river flows into the Basin and Range Province

FIGURE 21.1. The Colorado River within Marble Canyon, as seen from Saddle Canyon. (Photograph by S. Reynolds.)

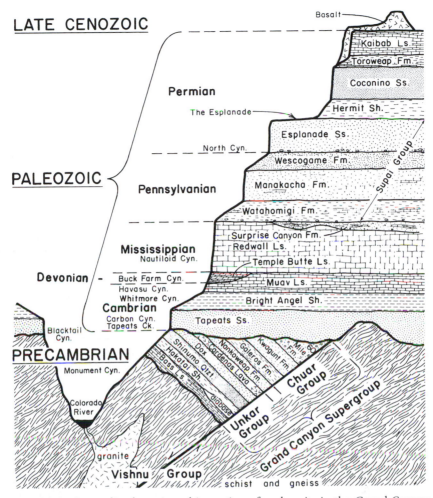

FIGURE 21.2. Generalized stratigraphic section of rock units in the Grand Canyon.

at Lake Mead. The canyon is 277 miles (444 km) long, up to 18 miles (30 km) wide, and nearly one mile (1.6 km) deep. Although many rivers in the world follow the trend of preexisting structural weaknesses, a peculiar feature of the Colorado River in the Grand Canyon is that it generally cross-cuts major north–south faults and folds and flows against the regional dip of strata through most of its length. It parallels the major structures only twice and for a mere one-sixth of its length (Fig. 21.3).

In contrast, side canyons are influenced more strongly by structural and geomorphic controls such as faults, regional dip of strata, and rock hardness. Below, we summarize these controls and show how they have resulted in the formation of segments of the Grand Canyon, each having a distinctive side-canyon morphology. Finally, we discuss geological features of some of the more interesting side canyons. This discussion is, in part, modified from Potochnik and Reynolds (1986).

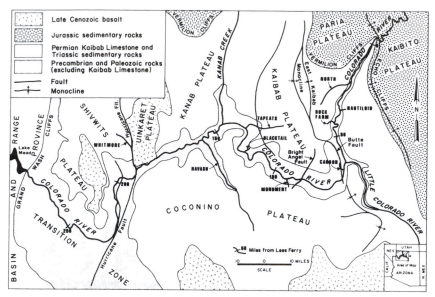

FIGURE 21.3. Simplified geologic map of the Grand Canyon region.

CONTROLS ON THE MORPHOLOGY OF SIDE CANYONS

The morphology of side canyons reflects various combinations of stratigraphic and structural controls, especially the regional dip of strata, folds and faults, and differential erosional susceptibility of the various rock units. Other factors influencing side-canyon morphology include climate and amount of time since the beginning of side-canyon incision.

A first-order control of side-canyon morphology is the regional or subregional dip of Paleozoic strata. Side canyons that flow down the regional dip tend to be longer and wider than those draining into the river against the dip. This is exemplified along the north rim of the canyon near Bright Angel Creek. Here, we see a more branched network of longer streams flowing with the southward dip, compared to those flowing against it from the south rim (Fig. 21.4).

A more local control of side-canyon morphology is displayed along folds and faults. Fault-controlled canyons, such as Bright Angel Creek, are recognized by distinct linear trends and longer-than-normal lengths (Fig. 21.4), attributes that reflect the easily eroded character of faulted and brecciated rock along the faults. Regional folds, such as the structurally low sag along Havasu Creek (Fig. 21.4), control the general location of some major tributaries. More structurally abrupt folds, such as the East Kaibab monocline (Fig. 21.3), cause rapid downstream changes in the morphology of a single side canyon as the stream cuts across a wide variety of rock units.

The character of side canyons is influenced strongly by differences in the erosional susceptibility of different rock units. Canyons cut in hard, resistant rocks, such as the Proterozoic Vishnu Group or the Redwall Limestone, have steep walls and narrow, V-shaped or vertical profiles. In contrast, canyons cut entirely in soft rocks, especially the easily eroded Dox Sandstone and Chuar Group, are broad and bowl-shaped, with rounded, receding walls. Most side canyons display both characteristics because of the successive alternation of rel-

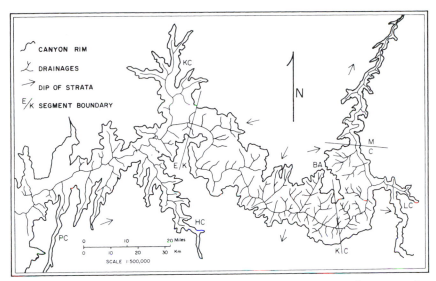

FIGURE 21.4. Simplified drainage-network map of the eastern Grand Canyon, showing the four segments of the canyon discussed in the text. Segments are as follows: M, Marble Canyon; C, Chuar; K, Kaibab; and E, Esplanade. Labeled tributaries are as follows: BA, Bright Angel Creek; HC, Havasu Creek; KC, Kanab Creek; LC, Little Colorado River; and PC, Prospect Canyon.

atively hard and soft rocks. A rapid lateral retreat of resistant cliffs probably begins as the underlying soft rock is exposed. As a consequence, side canyons commonly become wider once the stream has incised down through a cliff-former and into an underlying weak unit.

Climatic variation also causes different side-canyon morphologies. Precipitation varies considerably, according to changes in elevation across the region. Greater precipitation in uplifted regions, such as the Kaibab Plateau, provides more erosive power to streams originating in those areas. Frost action and chemical weathering also are greater in those same regions. Climatic control is best illustrated by comparing the relatively unincised Desert Facade [6000 ft (1829 m) elevation] near the mouth of the Little Colorado River to the topographically higher [7300 ft (2225 m) elevation] and more incised, pine-covered, south rim to the west. Climatic changes with time also have affected the types and magnitudes of processes responsible for incising and enlarging side canyons. For example, a change from a warmer and more humid climate in the Eocene to a colder and drier one in the Oligocene (Frakes and Kemp 1973) probably affected the rate of denudation of Mesozoic rocks from the plateau prior to any canyon cutting. Pleistocene climatic fluctuations likewise may have affected the recurrence interval of high-discharge flow events and the rates at which lava dams were scoured from the river.

Base-level control is of fundamental importance when considering the causative factors in canyon development. The local base level that controls Grand Canyon tributaries is the Colorado River. An interesting example of this effect is the lower gorge of the Little Colorado River. Upstream from the gorge, the Little Colorado River meanders across northeastern Arizona in a broad, open valley cut in Mesozoic strata. In its last 30 miles (48 km), however, the river gradient increases by about 500 percent and cuts a gorge through the entire 2900-foot (884-m)-thick Paleozoic section. Such an abrupt change in gradient probably was initiated by integration of the Little Colorado River into the main Colorado River system. Regional considerations indicate that this occurred about four million years ago, when Lake Bidahochi, centered in the present Little Colorado River Valley, was drained (Scarborough 1989). Through the process of headwater erosion, the ephemeral Little Colorado will be adjusting its gradient to this change for thousands of years to come.

MORPHOLOGICAL DIVISIONS OF THE GRAND CANYON

The Grand Canyon can be subdivided into four segments based on the general morphology of side canyons: (1) Marble Canyon, (2) Chuar, (3) Kaibab Plateau, and (4) Esplanade. The differences between these four segments can be observed readily on the Geologic Map of Arizona (Reynolds 1988) and in Fig. 21.4.

Marble Canyon Segment

This segment includes all canyons cut into the Marble Platform from below Lees Ferry to Little Nankoweap Creek (mile 52). It is characterized by fairly short, narrow, relatively unbranched gorges, such as North and Nautiloid canyons. These streams flow in accord with the regional northeast dip, while the river flows against it, resulting in distinctively barbed tributaries (Fig. 21.4). Despite fairly large watersheds, the canyons do not incise a large area of the Marble Platform and have the appearance of being newly formed.

Chuar Segment

From Little Nankoweap Creek to the beginning of the Upper Granite Gorge, canyon development is influenced strongly by the East Kaibab monocline and the Butte fault. The most distinctive feature in this segment is the abrupt appearance of enormous tributaries from the west and the virtual absence of lateral gorges from the east (Fig. 21.4). The river flows south, paralleling the base of the East Kaibab monocline, then swings broadly westward as it cuts across the monocline, deep into the core of the Kaibab Plateau. This segment is the structurally and topographically deepest part of the Grand Canyon. The large tributaries from the Kaibab uplift owe their size to the greater runoff and more significant relief of the lofty upland region. Moreover, broad, open valleys in their upper reaches are a consequence of incision into the soft Chuar Group shales and the Dox Formation. On the east rim, the absence of canyons along the Desert Facade is the result of low rainfall, resistant rock, and the regional dip of Paleozoic strata away from the river. The Little Colorado River is the only tributary from the east along this entire 48-mile (80-km)-long segment, yet it is the largest of the tributaries entering the Grand Canyon. Not far below its confluence with the Little Colorado River, the main Colorado River crosses a splay of the East Kaibab monocline, where the easily eroded Upper Proterozoic shales are exposed on both sides of the river. Here, small canyons begin to appear on the east side because of the weak shales, despite the continued dip of strata away from the river.

Kaibab Segment

This segment includes all canyons controlled by the Kaibab upwarp between Red Canyon (mile 77) and Kanab Creek (mile 143) (Fig. 21.4). Through the uplifted Kaibab Plateau, the river cuts to its deepest stratigraphic level, nearly 1800 feet (547 m) into the Proterozoic crystalline basement. Through most of this distance, the main river channel is incised in a narrow, steep-walled canyon called the Upper Granite Gorge. Snowmelt from the Kaibab Plateau produces several large, perennial streams that flow radially toward the canyon off the southern and western flanks of the plateau. The north rim tributaries commonly are long and fault-controlled and display a well-developed dendritic pattern. In contrast, southern tributaries flow against the dip of the strata and form short, steep, and generally unbranched canyons (Fig. 21.4). Outwash from these small, south-side tributaries produces many of the Grand Canyon's most tyrannical rapids (e.g., Hance, Sockdolager, Horn Creek, and Granite Falls).

Esplanade Segment

This segment is named for the characteristically broad topographic bench formed on the Esplanade Formation by the erosional retreat of the Kaibab–Coconino cliffs. It includes the entire western Grand Canyon downstream from Kanab Creek (mile 143) (Fig. 21.4). The Esplanade surface actually becomes a prominent feature rather abruptly at the lower end of the Upper Granite Gorge (mile 114) but does not become fully developed until the river leaves the Kaibab upwarp near Kanab Creek. Many side canyons in this segment are uncommonly long and linear, reflecting a strong control by north-trending faults that break the western Colorado Plateau into a series of fault blocks. Such faults control the position and trend of Prospect Creek, Peach Springs Canyon, and other side canyons, in addition to the overall course of the main Colorado River along the Hurricane fault (Fig. 21.3). The great width of these canyons is confined to the portion above the level of the Esplanade. Much narrower gorges are cut in the underlying Pennsylvanian–Cambrian rocks. The tenfold increase in width of the Esplanade west of the Kaibab Plateau suggests that canyon-cutting into the Kaibab Limestone began much earlier here than in areas to the east.

The Esplanade segment also includes two major tributaries—Kanab Creek and Havasu Creek—and side canyons with long and complex geological histories. For example, the present mouth of Prospect Canyon is a mere remnant of a wider, deeper canyon that was filled with basalt flows after its incision (Hamblin 1976). Other canyons, such as Milkweed and Peach Springs canyons, originally were formed by north-northeast-flowing drainages that predated cutting of the present Grand Canyon; the walls of these present-day side canyons contain remnants of sedimentary and volcanic rocks deposited during this earlier drainage regime.

GEOLOGICAL HIGHLIGHTS OF
SELECTED SIDE CANYONS

A variety of factors have influenced the size, orientation, and morphology of side canyons. Some of these factors are local features, such as faults, whereas others are regional in extent, resulting in the four canyon segments discussed above. To illustrate the attributes of side canyons and the processes that control them, the geological highlights of nine side canyons (Fig. 21.3) are presented below. These nine canyons have been chosen for discussion because they display a wide range of geological features in a variety of settings. We have included brief descriptions of the aesthetics of most canyons in an attempt to convey the natural wonder of the Grand Canyon as viewed from river level and from the vantage point of short hikes up the side canyons.

North Canyon

In the first 20 miles (32 km) below Glen Canyon Dam, the Colorado River cuts through the red-colored Mesozoic rocks of the Vermilion and Echo cliffs into the underlying, light-colored Permian rocks that form the familiar rim of the Grand Canyon. By North Canyon (mile 20 from Lees Ferry), the river also has deeply incised the underlying, rust-colored Permian Hermit Shale and the Pennsylvanian–Permian Supai Group. North Canyon, typical of the Marble Canyon segment, is a narrow cleft cut in sculptured formations of the Supai Group, including a gray conglomerate that marks the base of the Esplanade Sandstone. The canyon walls are steep and high, with the warm colors, soft textures, and acoustics of a cathedral. Parts of the canyon have the feeling of a sanctuary because of smooth, concave walls formed by dramatic, curved fractures in the sandstone (Fig. 21.5). The unusual fractures presumably reflect exfoliation of the sandstone walls that was caused by canyon incision.

Nautiloid Canyon

Not far downstream from North Canyon, gray- and cream-colored strata appear at river level and form a small cliff along the river's shores. Through the next 12 miles (19 km), the river's course becomes confined within the towering walls of the canyon's single most formidable barrier to river-rim hiking namely, the Redwall Limestone. This fossiliferous, carbonate rock was deposited in a vast, shallow sea that inundated much of western North America during Mississippian time, more than 300 million years ago. In 1869, John Wesley Powell named Marble Canyon for the beautiful, marblelike polish of the Redwall Limestone flanking the river. The limestone's true ivory and gray colors, exposed by running water or recent rockfalls, usually are concealed by a red iron-oxide stain derived from the overlying Supai Group and Hermit Shale.

The 500-foot (152-m) vertical walls of nearly pure limestone and dolomite are highly resistant to erosion in the semiarid climate of the inner canyon. Numerous solution caverns, however, were formed by groundwater dissolution of the limestone during the geological period between deposition of the Redwall and the overlying Supai Group (McKee 1976). Many large caverns developed along vertical joints or cracks in the limestone and are conspicuous—particularly near mile 35, where a large set of joints bisects the canyon.

In this vicinity, a narrow cleft called Nautiloid Canyon enters from the east (Fig. 21.3). A short climb into this canyon reveals numerous fossilized remains of the chambered nautiloid (Fig. 21.6). These fossils were the first of their kind to be found in the Grand Canyon (Breed 1968). The polished limestone floor of Nautiloid Canyon bears many cross sections of these ancient, cone-shaped creatures, whose size can be up to 20 inches (55 cm) in length. The tentacle-like appendages on some specimens may be relics of soft parts. These fossils, found in the cherty Thunder Springs Member of the Redwall Limestone, are but one of several types of preserved marine life.

Buck Farm Canyon

Small rapids and riffles occasionally punctuate the shady serenity of Marble Canyon as the river cuts into older Paleozoic strata downstream from Nautiloid Canyon. Only the keenest observer will notice where the river intersects the inconspicuous horizontal contact between the Mississippian Redwall Limestone and the underlying Cambrian Muav Limestone. The parallel layering and similar ap-

FIGURE 21.5. North Canyon. Curved fractures and bedding surfaces occur in the Permian Esplanade Sandstone of the Supai Group. (Photograph by A. Potochnik.)

pearance of these two red-stained limestones belies the staggering difference in their ages. Rocks representing Late Cambrian through Early Mississippian time (nearly 180 million years of earth history) are missing at this stratigraphic boundary. A clue to this unrecorded period is evident in the canyon walls upstream from Buck Farm Canyon (mile 41), where purplish lenses of Devonian Temple Butte Limestone 50 feet (15 m) high and hundreds of feet long are present between the Redwall and Muav limestones (Fig. 21.7). Especially well-exposed examples of these lenses are present in Buck Farm Canyon.

The bowl-shaped cross section of these lenses suggests the following evolutionary scenario. After deposition of the Cambrian Muav Limestone, the continent emerged above sea level, and streams carved channels into the landscape. Subsequent transgression of a Devonian sea filled these channels with impure limestone, which accumulated to even greater thicknesses in the western Grand Canyon (Beus 1987). Both Temple Butte Limestone and Muav Limestone then

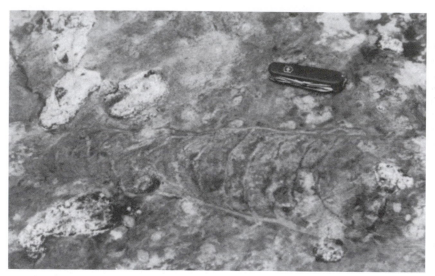

FIGURE 21.6. Fossil nautiloid on stream-polished outcrop of Mississippian Redwall Limestone. (Photograph by S. Reynolds.)

were eroded down to a peneplain as the landmass once again emerged from the sea in Late Devonian time. This second erosional period so thoroughly leveled the landscape that the Temple Butte Limestone in the eastern Grand Canyon was preserved only at the bottom of channels in which it first had accumulated. A third marine advance from the west in middle Mississippian time blanketed the region with the Redwall Limestone. This activity preserved the channel infillings in the geological record. The relatively recent cutting of the Grand Canyon affords cross-sectional views of these channels.

FIGURE 21.7. Lens of Devonian Temple Butte Limestone in a channel cut into the underlying Cambrian Muav Limestone. The lens is overlain by Mississippian Redwall Limestone. (Photograph by S. Reynolds.)

How did the landscape appear during the erosional intervals, and what lived in the Devonian sea? The flat-lying, undisturbed nature of the Paleozoic rocks tells us that dynamic crustal activity, such as mountain building, faulting, and volcanism, was absent. During the first erosional episode, the area was a bleak, featureless landscape of Muav Limestone that was incised by local stream channels. Probably no plants or animals lived on land, but early, bony-plated fishes first made their appearance during the transgression of the Devonian sea. The landmass subsequently reemerged, and the earliest land plants took root. The subsequent inundation by the Mississippian sea brought a much greater abundance and diversity of marine life—including corals, sponges, shellfish, echinoderms, and nautiloids.

Carbon Canyon

The Marble Canyon segment of Grand Canyon ends about 11 miles (18 km) downstream of Buck Farm Canyon, where the river enters the Chuar segment. The configurations of the side canyons change dramatically in response to the uplift of the broad Kaibab Plateau. Carbon Canyon, a lateral gorge entering the river from the west at mile 65, provides a fascinating structural and geomorphical perspective of the eastern boundary of the uplifted Kaibab Plateau.

The canyon walls in the lower portion of Carbon Canyon consist of purple sandstone and siltstone of the Proterozoic Dox Formation and buff-colored, coarse-grained arkose of the overlying Cambrian Tapeats Sandstone. The latter exhibits striking examples of honeycomb-weathering and colorful Liesegang banding. The purple- and rust-colored, parallel banding forms graceful, curving patterns throughout the rock (Fig. 21.8). It was caused by precipitation of iron oxides as groundwater migrated through the saturated matrix of the porous sandstone.

Evidence of the forces that uplifted the Kaibab Plateau relative to areas to the east is displayed dramatically a short distance up Carbon Canyon. Here, bedding planes in the Tapeats Sandstone become gently tilted upward as one walks up the canyon until a place is reached where the beds abruptly bend vertically. At this point, the narrow gorge opens into a wide valley of rolling hills with many small tributaries that drain the soft shales of the Proterozoic Chuar Group. The Kaibab Plateau, underlain by the entire Paleozoic sequence, is visible on the western skyline 5 miles (8 km) away and 2000 feet (610 m) higher than the elevation of the same formations along the river. The sharp upturn in the strata in Carbon Canyon is a local fold caused by the Butte fault, which parallels the East Kaibab monocline, the eastern boundary of the Kaibab Plateau. The Butte fault was a normal fault (west side down) in the Proterozoic but was reactivated as a reverse fault (west side up) during the Late Cretaceous to early Tertiary uplift of the Kaibab Plateau. Cenozoic erosion has breached the monocline, causing removal of the entire Paleozoic sequence and exposure of the underlying Chuar Group shales.

Monument Creek

Near mile 93, within the Kaibab segment, Monument Creek enters the Colorado River from the south side in the deepest section of the Upper Granite Gorge. Within the walls of this side canyon are exposures of Early Proterozoic metamorphic and granitic rocks, whose resistance to erosion is responsible for the steep-walled, "V" shape of the inner gorge (Fig. 21.9). The steep metamorphic layering and numerous convoluted folds within the rocks were formed during

FIGURE 21.8. Liesegang banding in Cambrian Tapeats Sandstone, Carbon Canyon. Bedding dips to the left. (Photograph by S. Reynolds.)

metamorphism and deformation that occurred during amalgamation and suturing of terranes along a continental margin to the southeast (Anderson 1986; Ilg et al. 1996). The mountains that formed during this tectonic episode were eroded away before the Cambrian sea encroached on the landscape and buried it beneath the beach sands of the Tapeats Sandstone. The "Great Unconformity" between the steeply dipping metamorphic rocks and the overlying, gently inclined Tapeats Sandstone is well-exposed and represents over one billion years of time.

In the more recent geological past, boulders carried down the present canyon of Monument Creek and deposited at its mouth have partially dammed the Colorado River, creating the major rapid of Granite Falls (Webb et al. 1989). Monument Canyon, which contains remnants of debris deposited in 1984, has the characteristics that a side canyon needs to form a large rapid in the river. These include a steep stream gradient, a narrow canyon with a flat floor, a sufficient supply of large boulders, and a source of fine-grained material to generate mudflows or debris flows (large boulders are more easily transported in a medium that is thicker than water). Monument Creek does not have a large drainage area, compared to other canyons; apparently, this factor is of secondary importance in creating a large rapid (Griffiths et al. 1996).

Blacktail Canyon

The Upper Granite Gorge ends in the vicinity of Blacktail Canyon (mile 120), where the regional westward dip of the strata causes the Tapeats Sandstone to descend to river level. Blacktail Canyon is a narrow, somewhat tubelike notch cut along the Great Unconformity between the sandstone and underlying schist of the Proterozoic Vishnu Group. The details of the unconformity are incredibly well exposed along the polished walls of the canyon (Fig. 21.10). The vertical metamorphic layering in the schist is overlain by sandstone and conglomerate derived from weathering and erosion of the schist. Although thin, vertical quartz veins in the schist were somewhat resistant to weathering, they finally were eroded into the small quartz pebbles now found in the basal sandstone. When

FIGURE 21.9. Upper Granite Gorge. Dark-colored walls are Proterozoic metamorphic and igneous rocks. (Photograph by S. Reynolds.)

standing here, it is easy to imagine the waves of the Cambrian sea 550 million years ago crashing onto jagged hills of schist and churning the metamorphic rock into sandy beaches as the sea advanced across the barren landscape.

Tapeats Creek

A few miles downstream, near mile 134, a cold and clear-flowing perennial stream called Tapeats Creek joins the Colorado River. The lowest formations of the Middle Proterozoic Unkar Group are exposed near river level (Fig. 21.11). These formations have a characteristic tilt that readily distinguishes these rocks from the more flat-lying Paleozoic rocks. A creek-side path through ancient ruins and garden sites of the Anasazi Indians traverses upward through the Unkar Group into the overlying Paleozoic rocks. Here, Thunder Springs bursts from a cavern high in the Muav Limestone wall. The Bass Formation, the oldest unit in the Unkar Group, contains wavy, fossilized algal mats, the oldest preserved evidence of life revealed in the Grand Canyon. Bright, reddish orange shales and siltstones of the overlying Hakatai Shale contain ripplemarks and mudcracks, features that suggest deposition in a tidal-flat environment. A gradational contact between these shales and the underlying Bass Limestone indicates that Hakatai mudflats gradually displaced the algal marine environment as the Bass sea retreated from the area. The tidal flats, in turn, were covered by a thick sequence of sand deposited near the shoreline of a sea. Consolidation of these sands formed the overlying Shinumo Quartzite, a cliff-forming unit that constitutes the steep-walled, narrow canyon of upper Tapeats Creek. A sill of dark-colored diabase approximately one billion years old occurs within the Bass Formation. As the layers in

FIGURE 21.10. The Great Unconformity between Tapeats Sandstone and the under-lying Proterozoic Vishnu Group, Blacktail Canyon. Material from the light-colored layer within the schist has been incorporated into the thin, basal conglomerate of the overlying sandstone. The unconformity represents more than one b.y. missing from the geologic record. (Photograph by A. Potochnik.)

FIGURE 21.11. Proterozoic diabase (dark ledge at river level) and slope-forming units of the Proterozoic Unkar Group along the Colorado River near Tapeats Creek. Tilted units of Unkar Group are overlain unconformably by Cambrian Tapeats Sandstone. (Photograph by J. Blaustein.)

the Bass Formation were pushed apart to accommodate the magma, they reacted with this magma to form thin layers of green serpentine and fibrous, chrysotile asbestos.

An up-canyon view from high on the Thunder River switchbacks reveals the angular unconformity between the Shinumo Quartzite and the overlying Cambrian rocks (Fig. 21.12). The Tapeats Sandstone, a beach sand of the advancing Cambrian sea, was deposited on the shore of a Shinumo Quartzite island that stood as a large remnant of late Proterozoic erosion. As the sea deepened, the island became submerged, and offshore muds of the Bright Angel Shale lapped across the top of the former island.

Havasu Creek

Downstream from Tapeats Creek, the river turns west and quickly passes out of the Kaibab segment and into the Esplanade segment. The regional tilt of the rock layers causes the cliff-forming Paleozoic limestones, once again, to appear at river level. The confluence of Havasu Creek with the Colorado River near mile 157 is easily missed. A narrow Muav Limestone gorge obscures the enormity of this large tributary, second only to the Little Colorado River in size. Havasu Creek is known for its spectacular waterfalls. The verdant banks of this perennial, aqua-blue stream are lined with velvet ash, cottonwood, and wild grape.

FIGURE 21.12. Tilter Proterozoic Shinumo Quartzite (Unkar Group) overlain unconformably by flat-lying Cambrian strata, Tapeats Creek. The resistant quartzite was once a craggy island. Along its flanks, Cambrian seas deposited the sands of the Tapeats, which forms the dark ledge shown in the left center of the photograph. The island later was buried by marine muds, which were compacted into the overlying, slope-forming units of the Cambrian Bright Angel Shale. (Photograph by S. Reynolds.)

Travertine deposits are perhaps the most fascinating geological feature of Havasu Creek. The travertine is formed by the precipitation of calcium carbonate as the creek waters warm and evaporate during the long flow to the Colorado. The travertine tends to encrust the surface and to take the form of any object over which the water passes.

Distinctive features of travertine cementation are the flat-topped and sinuous dams so commonly seen in the creek. These dams form by a self-enhancing process. An obstruction tends to catch sticks and leaves, which become encrusted with calcium carbonate, thereby increasing the size of the obstruction. When the obstruction becomes large enough, mosses colonize it and provide an additional substrate that increases its width and size. Eventually, a dam forms across the channel with perhaps one or two spillways through which the stream flows. The process becomes self-restricting in the spillways because water velocity is sufficient to prevent accumulation of debris, growth of moss, and precipitation of travertine.

Whitmore Wash

Near mile 188, below the notorious Lava Falls rapids, Whitmore Wash preserves evidence of a time even more tumultuous than that experienced by the river traveler while navigating the rapids. Whitmore Wash and the area around Lava Falls contain remnants of dark, basalt lava flows that once filled the Grand Canyon to a depth of more than 1400 feet (427 m). The lava erupted from volcanic vents, such as Vulcan's Throne, that pierced the Esplanade surface approximately 3000 feet (910 m) above the canyon bottom (Hamblin 1976 and Chapter 17, this volume).

A number of volcanic vents occur near the Hurricane fault, a major, recently active, north–south fault that may have served as a conduit for the ascending lavas. Upon eruption, the flows of molten lava cascaded over the walls of the Grand Canyon into the Colorado River 3000 feet (910 m) below, creating enormous clouds of steam and filling tributary canyons that drained into the Grand Canyon prior to volcanism. A cross-sectional view of the lava-filled canyon of "old" Whitmore Wash is visible from river level where the Whitmore Trail climbs the north wall of the Grand Canyon into Whitmore Valley. Less than one-half mile downstream from the trail is the "new" Whitmore Wash, a narrow side canyon that drains the same extensive watershed as the former wash. The new wash, however, has cut into the Paleozoic limestones instead of excavating the more erosionally resistant lava that fills the old channel. In the main Grand Canyon, the dams formed by lava flows were more transient, probably surviving less than 10,000 years (McKee et al. 1968; Damon et al. 1967; Hamblin 1976).

SUMMARY

The canyons described above represent just a small sample of the geological features and natural beauty found within side canyons of the Colorado River. Each canyon is unique, both in scenary and in the array of exposed geological features. Short hikes within these canyons complement the river running, which alternates between the relaxing tranquility of long, slow-moving stretches and the burst of apprehension, excitement, and chaos within the rapids. The entire experience is difficult to describe, but impossible to forget.

BIBLIOGRAPHY

Adams, D.C., and G.R. Keller. 1996. Precambrian Basement Geology of the Permian Basin Region of West Texas and Eastern New Mexico: A Geophysical Perspective. *AAPG Bulletin*, vol. 80, no. 3, pp. 410–431.

Adams, J.E., and M.L. Rhodes. 1960. Dolomitization by Seepage Refluxion. *American Association of Petroleum Geologists Bulletin*, vol. 44, pp. 1912–1921.

Ahlbrandt, T.S., S.S. Andrews, and D.T. Gwynne. 1978. Bioturbation in Eolian Deposits. *Journal of Sedimentary Petrology*, vol. 48, pp. 839–848.

Ahnert. 1960. The Influence of Pleistocene Climates Upon the Morphology of Cuesta Scarps on the Colorado Plateau. Association of American Geographers *Annals*, vol. 50, pp. 139–156.

Aitken, J.D. 1967. Classification and Significance of Cryptalgal Limestones and Dolomites, with Illustrations from the Cambrian and Ordovician of Southwestern Alberta. *Journal of Sedimentary Petrology*, vol. 37, pp. 1163–1178.

Aitken, J.D. 1978. Revised Models for Depositional Grand Cycles, Cambrian of the Southern Rocky Mountains, Canada. *Bulletin of Canadian Petroleum Geology*, vol. 26, pp. 515–542.

Akers, J.P., J.H. Irwin, P.R. Stevens, and N.E. McClymonds. 1962. Geology of the Cameron Quadrangle, Arizona. *U.S. Geological Survey Map* GQ 162.

Aldrich, L.T., G.W. Wetherill, and G.L. Davis. 1957. Occurrence of 1350 Million-Year-Old Granitic Rocks in the Western United States. *Geological Society of America Bulletin*, vol. 68, pp. 655–656.

Aleinikoff, J.N. 1996. Shrimp U-Pb Ages of Felsic Igneous Rocks, Belt Supergroup, Western Montana: Geological Society of America. *Abstracts with Programs* (Rocky Mountain Section), vol. 28 (7), p. 376.

Alf, R.M. 1968. A Spider Trackway From the Coconino Formation, Seligman, Arizona. *Southern California Academy of Sciences Bulletin*, vol. 67 (2), pp. 125–128.

Allmendinger, R.W., et al. 1987. Overview of the COCORP 40°N Transect, Western United States, the Fabric of an Orogenic Belt, *Geological Society of America Bulletin*, vol. 98, pp. 308–319.

Altany, R.M. 1979. Facies of the Hurricane Cliffs Tongue of the Toroweap Formation, Northewestern Arizona. *In:* Baars, D.L., ed., *Permianland. Four Corners Geological Society Guidebook*, 9th Field Conference, pp. 101–104.

Altschul, J.H., and H.C. Fairley. 1989. Man, Models, and Managment: An Overview of the Archaeology of the Arizona Strip and the Managment of its Cultural Resources: Report prepared for U.S. Department of Agriculture Forest Service and U.S. Department of Interior Bureau of Land Managment Contract Number 53-8371-5-0054, 410 pp.

Anderson, J.J., P.D. Rowley, R.J. Fleck, and A.E.M. Nairn. 1975. Cenozoic Geology of the Southwestern High Plateaus of Utah. *Geological Society of America Special Paper* 160, 88 pp.

Anderson, Phillip. 1986. Summary of the Proterozoic Plate Tectonic Evolution of Arizona from 1900 to 1600 m.y.a. *In:* Beatty, B. and P.A.K. Wilkinson, eds., Frontiers in Geology and Ore Deposits of Arizona and the Southwest. *Arizona Geological Society Digest*, vol. 16, pp. 5–11.

Anderson, R.E. 1978. Geologic Map of the Black Canyon 15-minute Quadrangle, Mohave County, Arizona, and Clark County, Nevada. *U.S. Geological Survey Quadrangle Map* GQ 1394.

Anderson, R.E., and P.W. Huntoon. 1979. Holocene Faulting in the Western Grand Canyon, Arizona, Discussion and Reply. *Geological Society of America Bulletin*, vol. 90, pp. 221–224.

Anderson, R.Y. 1982. A Long Geoclimatic Record from the Permian. *Journal of Geophysical Research*, vol. 8, pp. 7285–7294.

Anderson, R.Y., W.E. Dean, D.W. Kirkland, and H.I. Snider. 1972. Permian Castile Varved Anhydrite Sequence, West Texas and New Mexico. *Geological Society of America Bulletin*, vol. 83, pp. 59–86.

Andrews, E.D. 1991. Sediment Transport in the Colorado River Basin. *In: Colorado River Ecology and Dam Managment*. Washington, DC: National Academy of Science Press, pp. 54–74.

Armstrong, A.K., and J.E. Repetski. 1980. The Mississippian System of New Mexico and Southern Arizona. *In:* Fouch, T.D., and E.R. Magathan, eds., *Paleogeography of West Central United States, Rocky Mountain Paleogeography Symposium I*, Rocky Mountain Section. Denver: Society of Economic Paleontologists and Mineralogists, pp. 82–100.

Babcock, R.S. 1990. Precambrian Crystalline Core. *In:* Beus, S.S., and Morales, M., eds., Grand Canyon Geology. New York: Oxford University Press, pp. 11–28.

Babcock, R.S., E.H. Brown, M.D. Clark, and D.E. Livingston. 1979. Geology of the Older Precambrian Rocks of the Upper Granite Gorge of the Grand Canyon. Part II. The Zoroaster Plutonic-Complex and Related Rocks: *Precambrian Research*, vol. 8, pp. 243–275.

Babenroth, D.L., and A.N. Strahler. 1945. Geomorphology and Structure of the East Kaibab Monocline, Arizona and Utah. *Geological Society of America Bulletin*, vol. 56, pp. 107–150.

Badiozamani, K. 1973. The Dorag Dolomitization Model—Application to the Middle Ordovician of Wisconsin. *Journal of Sedimentary Petrology*, vol. 43, pp. 965–984.

Baeseman, J.F., and H.R. Lane. 1985. Taxonomy and Evolution of the Genus *Rhachistognathus* Dunn (Conodonta: Late Mississippian to Early Middle Pennsylvanian): Courier Forshungsinstitut Senckenberg, 74, pp. 93–166.

Bagnold, R.A. 1966. An Approach to the Sediment Transport Problem from General Physics. U.S. Geological Survey Professional Paper 422-I, 37 pp.

Bagnold, R.A. 1980. An Empirical Correlation of Bedload Transport Rates in Flumes and Natural Rivers. Royal Society of London, *Royal Society of London Proceedings*, Series A, vol. 372, (1751), pp. 453–473.

Baird, Donald. 1952. Revision of the Pennsylvanian and Permian Footprints *Limnopus, Allopus,* and *Baropus. Journal of Paleontology,* vol. 26, pp. 832–840.

Baker, V.R. 1973. Paleohydrology and Sedimentology of Lake Missoula Flooding in Eastern Washington. *Geological Society of America Special Paper* 144, 69 pp.

Baker, V.R. 1984. Flood Sedimentation in Bedrock Fluvial Systems. *In:* Koster, E.H., and R.J. Steel, R.J., eds., *Sedimentology of Gravels and Conglomerates,* Canadian Society Petroleum Geologists, Memoir 10, pp. 87–98.

Baker, V.R., and G. Pickup. 1987. Flood Geomorphology of the Katherine Gorge, Northern Territory, Australia. *Geological Society of America Bulletin,* vol. 98, pp. 635–646.

Bakhmeteff (Bakhmetev), B.A. 1932. *Hydraulics of Open Channels.* New York: McGraw-Hill, 329 pp.

Barnes, C.W. 1987. Geology of the Gray Mountain Area, Arizona. *Geological Society of American Centennial Field Guide,* Rocky Mountain Section, pp. 379–384.

Bassler, R.S. 1941. A Supposed Jellyfish From the Precambrian of the Grand Canyon. *U.S. National Museum Proceedings,* vol. 89, pp 519–522.

Basu, A. 1981. Weathering Before the Advent of Land Plants: Evidence from Unaltered Detrital K-feldspars in Cambro-Ordovician Arenites. *Geology,* vol. 9, pp. 132–133.

Batten, R.L. 1964. Some Permian Gastropoda from Eastern Arizona. *American Museum Novitates,* No. 2165, 16 pp.

Beatty, C.B. 1989. Great Boulders I Have Known. *Geology,* vol. 17, pp. 349–352.

Beer, B. 1988. We Swam the Grand Canyon. *The Mountaineers,* Seattle, 169 pp.

Belden, W. 1954. The Stratigraphy of the Toroweap Formation, Aubrey Cliffs, Coconino County, Arizona. M.S. thesis, University of Arizona, Tucson, 87 pp.

Bennett, V.C., and D.J. DePaolo, 1987. Proterozoic Crustal History of the Western United States as Determined by Neodymium Isotopic Mapping. *Geological Society of American Bulletin,* vol. 99, pp. 674–685.

Bernaski, G.E. 1985. *Compressional Laramide Deformation in the Southwestern Uinta Mountains, Northwestern Colorado, and Northwestern Utah.* University of Wyoming Master of Science Thesis. 154 pp.

Best, M.G., and W.H. Brimhall. 1970. Late Cenozoic Basalt Types in the Western Grand Canyon Region. *In:* Hamblin, W.K., and M.G. Best, eds., *The Western Grand Canyon District: Guidebook to the Geology of Utah.* Utah Geological Society, pp. 57–74.

Best, M.G., and W.K. Hamblin. 1978. Origin of the Northern Basin and Range Province: Implications From the Geology of its Eastern Boundary. *In:* Smith, R.B., and G.P. Eaton, eds., *Cenozoic Tectonics and Regional Geophysics of the Western Cordillera.* Geological Society of America, Memoir 152, pp. 313–340.

Beus, S.S. 1964. Fossils From the Kaibab Formation at Bee Springs, Arizona. Contributions to the Geology of Northern Arizona. *MNA Bulletin* No. 40, pp 59–64.

Beus, S.S. 1965. Permian Fossils from the Kaibab Formation at Flagstaff, Arizona. *Plateau,* vol. 38, 1–5.

Beus, S.S. 1978. Late Devonian (Frasnian) Invertebrate Fossils from Jerome Member of the Martin Formation, Verde Valley, Arizona. *Journal of Paleontology,* vol. 52, pp. 40–54.

Beus, S.S. 1980. Late Devonian (Frasnian) Paleography and Paleoenvironments in Northern Arizona. *In:* Fouch, T.O., and R. Magathan, eds., *Paleozoic Paleogeography of West-Central United States, Rocky Mountain Paleography Symposium 1.* Rocky Mountain Section, Society of Economic Paleontologists and Mineralogists, pp. 55–69.

Beus, S.S. 1980. Devonian Serpulid Biotherms in Arizona. *Journal of Paleontology,* vol. 54 (5), pp. 1125–1128.

Beus, S.S. 1986. A Geologic Surprise in the Grand Canyon. Arizona Bureau of Geology and Mineral Technology, *Fieldnotes,* vol. 16 (3), pp. 1–4.

Beus, S.S. 1987. Devonian and Mississippian Geology of Arizona. *In:* Jenney, J.P. and S.J. Reynolds, eds., *Geologic Evolution of Arizona. Arizona Geological Society Digest,* vol. 17.

Beus, S.S. 1989. Devonian and Mississippian Geology of Arizona. *In:* Jenney, J.P., and S.J. Reynolds, eds., *Geologic Evolution of Arizona. Arizona Geological Society Digest,* vol. 17.

Beus, S.S. 1990. The Distribution and Orientation of Shells from Brachiopod Population, Permian Kaibab Formation, Northern Arizona, USA. *In:* MacKinnon, D.I., Lee, D.E., and Campbell, J. D., eds., *Brachiopods Through Time, Proceedings of the 2nd International Brachiopod Congress,* University of Otago, Dunedin, New Zealand. Rotterdam: Balkema, pp. 233–239.

Beus, S.S. 1995. *Paleontology of the Surprise Canyon Formation (Mississippian) in Grand Canyon, Arizona. Fossils of Arizona,* vol. III, Proceedings 1995, Southwest Paleontological Society and Mesa Southwest Museum, Mesa, Arizona, pp. 25–36.

Beus, S.S., and W. Breed. 1968. A New Nautiloid Species from the Toroweap Formation in Arizona. *Plateau,* vol. 40 (4), pp. 128–135.

Beus, S.S., R.R. Rawson, R.O. Dalton, Jr., G.M. Stevenson, V.S. Reed, and T.M. Daneker. 1974. Preliminary Report on the Unkar Group (Precambrian) in Grand Canyon, Arizona. *In:* Karlstrom, T.N.V., G.A. Swann, and R.L. Eastwood, eds., *Geology of Northern Arizona, part 1—Regional Studies.* Geological Society of America Field Guide, Rocky Mountain Section, Flagstaff, Arizona, pp. 34–64.

Beverage, J.P., and J.K. Culbertson. 1964. Hyperconcreations of Suspended Sediment: *American Society of Civil Engineers, Journal of the Hydraulics Division*, vol. 90, p. 117–126.

Billingsley, G.H. 1979. Preliminary Report of Buried Valleys at the Mississippian-Pennsylvanian Boundary in Western Grand Canyon. *In:* Beus, S.S., and Rawson, R.R, eds., *Carboniferous Stratigraphy in Grand Canyon Country, Northern Arizona and Southern Nevada*, 9th International Congress on Carboniferous Stratigraphy and Geology, Urbana, Illinois, *American Geological Institute Selected Guidebook Series* no. 2, pp. 115–117.

Billingsley, G.H. 1989. Mesozoic Strata at Lees Ferry, Arizona. *In:* S.Beus, ed., *Centennial Field Guide Volume 2.* Boulder: Rocky Mountain Section of the Geological Society of America, pp. 67–71.

Billingsley, G.H. 2000. Volcanic events of the Past 20 Ma in the Western Grand Canyon Region, Significance to Grand Canyon Erosion. *In:* Young, R.A., ed., Abstracts for a working Conference on the Cenozoic Geological Evolution of the Colorado River System and the Erosional Chronology of the Grand Canyon Region. *Grand Canyon Association* (Grand Canyon, Arizona), pp. 1–3.

Billingsley, G.H., L.S. Beard, S.S. Priest, in press. Geological Map of the Grand Wash Cliffs and vicinity, Mohave County, northwestern Arizona: *U.S. Geological Survey Miscellaneous Field Studies Map.*

Billingsley, G.H., and S.S. Beus. 1985. The Surprise Canyon Formation, an Upper Mississippian and Lower Pennsylvanian(?) Rock Unit in the Grand Canyon, Arizona. *U.S. Geological Survey Bulletin*, vol. 1605A, pp. A27–A33.

Billingsley, G.H., and J.D. Hendricks. 1989. Physiographic Features of Northwestern Arizona. *In:* T.L. Smiley, J.D. Nations, T.L. Pewe, and J.P. Schafer, eds., *Landscapes of Arizona: The Geological Story.* Lantham: University Press of America, pp. 67–71.

Billingsley, G.H., and P.W. Huntoon. 1983. Geologic Map of Vulcan's Throne and Vicinity, Western Grand Canyon, Arizona. Grand Canyon Natural History Association.

Billingsley, G.H., and E.D. McKee. 1982. Pre-Supai Buried Valleys. *In:* McKee, E.D., ed., *The Supai Group of Grand Canyon.* U.S. Geological Survey Professional Paper 1173, pp. 137–154.

Bissell, H.J. 1969. Permian and Lower Triassic Transition From the Shelf to Basin (Grand Canyon, Arizona to Spring Mountains, Nevada). *In:* Geology and Natural History of the Grand Canyon Region. *Four Corners Geological Society Guidebook*, pp. 135–169.

Blackwelder, E. 1928. Mudflow as a Geological Agent in Semiarid Mountains. *Geological Society of America Bulletin*, vol. 39, pp. 465–494.

Blackwelder, E. 1934. Origin of the Colorado River. *Geological Society of America Bulletin*, vol. 45, pp. 551–566.

Blackwell, D.D. 1978. Heat Flow and Energy Loss in the Western United States. *In:* Smith, R.B. and G.P. Eaton, eds., *Cenozoic Tectonics and Regional Geophysics of the Western Cordillera.* Geological Society of America Memoir 152, pp. 175–208.

Blair, T.C., and J.G. McPherson. 1994. Alluvail Fan Processes and Forms. *In:* Abrahams, A.D., and A.J. Parsons. eds., *Geomorphology of Desert Enviroments*: New York: Chapman and Hall, pp. 354–402.

Blair, W.N. 1978. Gulf of California in Lake Mead Area of Arizona and Nevada During Late Miocene Time. *The American Association of Petroleum Geologists Bulletin*, vol. 62, (7), pp. 1159–1170.

Blakey, R.C. 1979. Oil Impregnated Carbonate Rocks of the Timpoweap Member, Moenkopi Formation, Hurricane Cliffs Area, Utah and Arizona. *Utah Geology*, vol. 6 (1), pp. 45–54.

Blakey, R.C. 1980. Pennsylvanian and Early Permian Paleogeography, Southern Colorado Plateau and Vicinity. *In:* Fouch, T.D., and E.R. Magathan, eds., *Paleozoic Paleogeography of West-Central United States.* Rocky Mountain Section, SEPM, Denver, pp. 239–257.

Blakey, R.C. 1988. Basin Tectonics and Erg Response. *Sedimentary Geology*, vol. 56, (1–4), pp. 127–152.

Blakey, R.C. 1990. Stratigraphy and Geological History of Pennsylvanian and Permian Rocks, Mogollon Rim Region, Central Arizona and Vicinity: *Geological Society of America Bulletin*, vol. 102, pp. 1189–1217.

Blakey, R.C. 1996. Permian Eolian Deposits, Sequences, and Sequence Boundaries, Colorado Plateau. *In:* Longman, M.W., and Sonnenfield, M.D., eds., *Paleozoic Systems of the Rocky Mountain Region.* Rocky Mountain Section, SEPM (Society for Sedimentary Geology), pp. 405–426.

Blakey, R.C., R. Knepp, and L.T. Middleton. 1983. Permian Shoreline Eolian Complex in Central Arizona: Dune Changes in Response to Cyclic Sea Level Changes. *In:* Brookfield, M.E., and T.S. Ahlbrandt, eds., *Eolian Sediments and Processes.* Amsterdam: Elsevier Science Publishers, pp. 551–581.

Blakey, R.C., and R. Knepp. 1988. Pennsylvanian and Permian Geology of Arizona. *In:* Jenney, J.P., and S.J. Reynolds, eds., *Geologic Evolution of Arizona, Arizona Geological Society Digest*, vol. 17.

Blakey, R.C., and L.T. Middleton. 1998. Permian Rocks in North-Central Arizona: A Comparison of the Sections at Grand Canyon and Sedona. *In:* Duebendorfer, E.M., ed., *Geological Excursions in Northern and Central Arizona.* Geological Society of America, Rocky Mountain Section, *Field Trip Guidebook*, pp. 97–125.

Blakey, R.C., F. Peterson, and G. Kocurek. 1988. Synthesis of Late Paleozoic and Mesozoic Eolian Deposits of the Western Interior of the United States. *Sedimentary Geology*, vol. 56 (1–4), pp. 3–126.

Bloeser, B. 1985. Melanocyrillium, a New Genus of Structurally Complex Late Proterozoic Microfossils from the Kwagunt Formation (Chuar Group), Grand Canyon, Arizona. *Journal of Paleontology*, vol. 59, pp. 741–765.

Bloeser, B., J.W. Schopf, R.J. Horodyski, and W.J. Breed. 1977. Chitinozoans from the Late Precambrian Chuar Group of the Grand Canyon, Arizona. *Science*, vol. 195, pp. 676–679.

Bohannon, R.G. 1984. Nonmarine Sedimentary Rocks of Tertiary Age in the Lake Mead Region, Southeastern Nevada and Northwestern Arizona. *U.S. Geological Survey Professional Paper* 1259, 72 pp.

Bowers, W.E. 1972. The Canaan Peak, Pine Hollow, and Wasatch Formations in the Table Cliff Region, Garfield County, Utah. U.S. Geological Survey Bulletin, vol. 1331B, pp. 707–730.

Bowring, S.A., and K.E. Karlstrom. 1990. Growth and Stabilization of Proterozic Continental Lithosphere in the Southwestern United States. Geology, vol. 18, pp. 1203–1206.

Bowring, S.A., K.V. Hodges, D.P. Hawkins, D.S. Coleman, K.L. Davidek, and K.E. Karlstrom. 1996. Thermochronology of Proterozoic Middle Crust, Southwestern U.S.: Implications for Models of Lithospheric Evolution: GSA Abstracts with Programs, vol. 28 (7), p. 452.

Boyce, J.M. 1972. The Structure and Petrology of the Older Precambrian Crystalline Rocks, Bright Angel Canyon, Grand Canyon, Arizona. M.S. thesis, Northern Arizona University, Flagstaff, 88 pp.

Bradley, W.C. 1963. Large Scale Exfoliation in Massive Sandstones of the Colorado Plateau. Geological Society of America Bulletin, vol. 74 (5), pp. 519–528.

Brady, L.F. 1939. Tracks in the Coconino Sandstone Compared with Those of Small Living Arthropods. Plateau, vol. 12, pp. 32–34.

Brady, L.F. 1947. Invertebrate Tracks from the Coconino Sandstone of Northern Arizona. Journal of Paleontology, vol. 32, pp. 466–472.

Brady, L.F. 1949. Onischoidichnus, A New Name for Isopodichnus Brady 1947 not Bornemann 1889. Journal of Paleontology, vol. 23, p. 573.

Brady, L.F. 1955. Possible Nautilod Mandibles from the Permian of Arizona. Journal of Paleontology, vol. 29 (1): 102–104.

Brady, L.F. 1959. A New Area of Kaibab Formation Limestone with Silicified Fossils. Plateau, vol. 31: 86–89.

Brady, L.F. 1961. A New Species of Paleohelcura Gilmore from the Permian of Northern Arizona. Journal of Paleontology, vol. 35, pp. 201–202.

Brady, L.F. 1962. Note on the "Alpha" member of the Kaibab Formation. In: Weber, R. H., and Peirce, H.W., eds., New Mexico Geological Society Guidebook of the Mogollon Rim Region, East-Central Arizona, 13th Field Conference, p. 92.

Brand, L.R. 1978. Footprints in the Grand Canyon; Origins, vol. 5, pp. 64–82.

Brand, L.R. 1979. Field and Laboratory Studies on the Coconino Sandstone (Permian) Vertebrate Footprints and Their Paleoecological Implications. Palaeogeography, Palaeoclimatology, Palaeoecology, vol. 28, pp. 25–38.

Brathovde, J.R. 1986. Stratigraphy of the Grand Wash Dolomite (Upper? Cambrian), Western Grand Canyon, Mohave County, Arizona. M.S. thesis. Northern Arizona University, Flagstaff, 140 pp.

Breed, W.S. 1968. The Discovery of Orthocone Nautiloids in the Redwall Limestone–Marble Canyon, Arizona. Four Corners Geological Society, 5th Field Conference, p. 134.

Bremner, J.A.C. 1986. Microfacies Analysis of Depositional and Diagenetic History of the Redwall Limestone in the Chino and Verde Valleys. M.S. thesis, Northern Arizona University, Flagstaff, 161 pp.

Brenckle, P.L. 1973. Smaller Mississippian and Lower Pennsylvanian Calcareous Foraminifers from Nevada. Cushman Foundation for Foraminiferal Research, Special Publication, vol. 11, pp. 1–82.

Brezinski, D.K. 1991. Permian Trilobites from the San Andres Formation, New Mexico, and Their Relationship to Species from the Kaibab Formation of Arizona. Journal of Paleontology, vol. 65 (3); pp. 480–484.

Briggs, D.E.G., A.G. Plint, and R.K. Pickerill. 1984. Arthropleura trails from the Westphalian of Eastern Canada. Paleontology, vol. 27, p. 843–855.

Brookfield, M.E. 1977. The Origin of Bounding Surfaces in Ancient Eolian Sediments. Sedimentology, vol. 24, pp. 303–332.

Brown, E.H., R.S. Babcock, M.D. Clark, and D.E. Livingston. 1979. Geology of the Older Precambrian Rocks of the Grand Canyon. Part I. Petrology and Structure of the Vishnu Complex. Precambrian Research, vol. 8, pp. 219–241.

Brown, S.C., and R.E. Lauth. 1958. Oil and Gas Potentialities of Northern Arizona. In: Roger Anderson and J. Harshbarger, eds., Guide Book of the Black Mesa Basin. Albuquerque: New Mexico Geological Society, pp. 153–160.

Brumbaugh, D.S. 1987. A Tectonic Boundary for the Southern Colorado Plateau. Tectonophysics, vol. 136, pp. 125–136.

Burchfiel, B.C., D.S. Cowan, and G.A. Davis. 1992. Tectonic Overview of the Cordilleran Orogen in the Western United States: In: Burchfiel, B.C., P.W. Lipman, and M.L. Zoback, eds., DNAG, The Cordilleran Orogen: Conterminous U.S., vol. G-3, pp. 407–479.

Campbell, I., and J.H. Maxson. 1938. Geological Studies of the Archean Rocks at the Grand Canyon: Carnegie Institution of Washington Year Book vol. 37, pp. 359–364.

Carothers, S.W., and B.T. Brown. 1991. The Colorado River through the Grand Canyon: Natural History and Human Change. Tucson, University of Arizona Press, 235 pp.

Carson, M., and M. Kirkby. 1972. Hillslope Form and Process. Cambridge: Cambridge University Press, pp. 448 and 449.

Chan, M., and G. Kocurek. 1988. Complexities in Eolian and Marine Interactions: Processes and Eustatic Controls on Erg Development. Sedimentary Geology, vol. 56 (1–4), pp. 283–300.

Chapin, C.E. 1983. An Overview of Laramide Wrench Faulting in the Southern Rocky Mountains with Emphasis on Petroleum Exploration. In: Lowell, J.D., ed., Rocky Mountain Foreland Basins and Uplifts. Rocky Mountain Association of Geologists, pp.169–179.

Chapin, C.E., and S.M. Cather. 1983. Eocene Tectonics and Sedimentation in the Colorado Plateau-Rocky Mountain Area. In: Lowell, J.D., ed., Rocky Mountain Foreland Basins and Uplifts. Rocky Mountain Association of Geologists, pp. 33–56.

Cheevers, C.W. 1980. Stratigraphic Analysis of the Kaibab Formation in Nothern Arizona, Southern Utah, and Southern Nevada. M.S. thesis, Northern Arizona University, Flagstaff, 144 pp.

Cheevers, C.W., and R.R. Rawson. 1979. Facies Analysis of the Kaibab Formation in Northern Arizona, Southern Utah, and Southern Nevada. *Permianland: Four Corners Geological Society Source Book,* pp. 105–133.

Chow, V.T. 1959. *Open-Channel Hydraulics.* New York: McGraw-Hill, 680 pp.

Chronic, H. 1952. Molluscan Fauna From the Permian Kaibab Formation, Walnut Canyon, Arizona. *Geological Society of America Bulletin,* vol. 63, pp. 95–166.

Chronic, H., 1953. Molluscan Fauna from the Permian Kaibab Formation, Walnut Canyon, Arizona. Unpublished M.S. thesis, Columbia University, 88 pp.

Chronic, H. 1983. *Roadside Geology of Arizona.* Missoula: Mountain Press Publishing Co., 314 pp.

Cisne, J.L. 1971. Paleoecology of Trilobites of the Kaibab Limestone (Permian) in Arizona, Utah, and Neveda. *Journal of Paleontology* vol. 45 (3), pp. 525–533.

Cisne, J.L., 1977. Middle Permian Trilobites of the Kaibab Formation in Arizona, Utah, and Nevada. Unpublished manuscript, Yale University Press, 37 pp.

Clark, M.D. 1976. The Geology and Petrochemistry of the Precambrian Metamorphic Rocks of the Grand Canyon, Arizona. Ph.D. thesis, University of Leicester, Leicester, England, 216 pp.

Clark, M.D. 1979. Geology of the Older Precambrian Rocks of the Grand Canyon, part III. Petrology of the Mafic Schists and Amphibolites. *Precambrian Research,* vol. 8, pp. 277–302.

Clark, R. 1980. Stratigraphy, Depositional Environments, and Petrology of the Toroweap and Kaibab Formations (lower Permian), Grand Canyon Region, Arizona.

Clemmensen, L.B., H. Olsen, and R. Blakey. 1989. Erg-margin Deposits in the Lower Jurassic Moenave Formation and Wingate Sandstone, Southern Utah. *Geological Society of America Bulletin,* vol. 101, pp. 759–773.

Cloud, P.E. 1960. Gas as a Sedimentary and Diagenetic Agent. *American Journal of Science,* vol. 258A, pp. 35–45.

Cloud, P.E. 1968. PreMetazoan Evolution and the Origins of the Metazoa. *In:* E.T. Drake, ed., *Evolution and the Environment.* New Haven: Yale University Press, pp. 1–72.

Cloud, P.E. and M.S. Semikhatov. 1969. Proterozoic Stromatolite Zonation. *American Journal of Science,* vol. 267, pp. 1017–1061.

Colman, S.M., 1981. Rock-Weathering Rates as Functions of Time. *Quarternary Research,* vol. 15, pp. 250–264.

Condie, K.C. 1928. Plate-Tectonics Model for Proterozoic Continental Accretion in the Southwestern United States. *Geology,* vol. 10, pp. 37–42.

Condra, G.E., and M.K. Elias. 1945a. *Bicorbula,* a New Permian Bryozoan–Probably a Bryozoan-Algal Consortium. *Journal of Paleontology,* vol. 19, pp. 116–125.

Condra, G.E., and M.K. Elias. 1945b. *Biscrbis arizonica* Condra and Elias, New Name for *Bicorbula arizonica. Journal of Paleontology,* vol. 19, 411.

Coney, P.J. 1981. Accretionary Tectonics in Western North America. *Arizona Geological Society Digest,* vol. 14, pp. 23–37.

Cook, D.A., 1991. Sedimentology and Shale Petrology of the Upper Proterozoic Walcott Member, Kwagunt Formation, Chua Group, Grand Canyon, Arizona. M.S. thesis, Northern Arizona University, Flagstaff, Arizona, 158 pp.

Cooke, R.U., A. Warren, and A.S. Goudie. 1993. *Desert Geomorphology.* London, UCL Press, 526 pp.

Cooley, M.E. 1962. Late Pleistocene and Recent Erosion and Alluviation in Parts of the Colorado River Sytem, Arizona, and Utah. *U.S. Geological Survey Professional Paper* 450–B, pp. B48–B50.

Cooley, M.E., J.W. Harshbarger, J.P. Akers, and W.F.Hardt. 1969. Regional Hydrolgeology of the Navaho and Hopi Indian Reservations, Arizona, New Mexico, and Utah. *U.S. Geological Survey Professional Paper* 521A, 61 pp. plus maps.

Cooley, M.E., B.N. Aldridge, and R.C. Euler. 1977. Effect of the Catastrophic Flood of December 1966, North Rim Area, Eastern Grand Canyon, Arizona. *U.S. Geological Survey Professional Paper* 980, 43 pp.

Cooper, G.A., and R.E. Grant. 1972. Permian Brachiopods of West Texas. *Smithsonian Contributions to Paleobiology,* vol. 14, pp. 1–23.

Cooper, G.A., and N.D. Newell. 1948. Key Permian Section, Confusion Range, Western Utah. *Geological Society of America Bulletin,* vol. 59, no. 10, pp. 1053–1058.

Costa, J.E. 1984. Physical Geomorphology of Debris Flows. *In:* Costa, J.E., and P.J. Fleisher, *eds.* Developments and Applications of Geomorphology. Berlin: Springer-Verlag, pp. 268–317.

Cotter, E. 1978. The Evolution of Fluvial Style, with Special Reference to the Central Appalachian Paleozoic. *In:* Miall, A.E., ed., *Fluvial Sedimentology.* Canadian Society of Petroleum Geologists Memoir 5, pp. 361–384.

Crimes, T.P. 1970. The Significance of Trace Fossils in Sedimentology, Stratigraphy, and Paleoecology with Examples From Lower Paleozoic Strata. *In:* Crimes, T.P. and J.C. Harper, eds., *Trace Fossils. Geological Journal* Special Issue 3, pp. 101–126.

Crimes, T.P. 1975. The Stratigraphic Significance of Trace Fossils. *In:* Frey, R.N., ed., *The Study of Trace Fossils.* New York: Springer-Verlag, pp. 109–130.

Cudzil, M.R., and S.G. Driese. 1987. Fluvial, Tidal, and Storm Sedimentation in the Chilhowee Group (lower Cambrian), Northeastern Tennessee, U.S.A. *Sedimentology,* vol. 34, pp. 861–883.

Curray, J.R. 1964. Transgressions and regressions. *In:* R.L. Miller, ed. *Paper in Marine Geology.* Macmillan. pp. 175–203.

Dalrymple, G., Brent, and W.K. Hamblin. 1998. *Proc. Natl. Acad. Sci,* vol. 95.

Dalton, R.O., Jr. 1972. Stratigraphy of the Bass Formation. M.S. thesis, Northern Arizona University, Flagstaff, 140 pp.

Dalziel, I.W.D. 1991. Pacific Margins of Laurentia and East Antartica–Australia as a Conjugate Rift Pair: Evidence and Implications for an Eocambrian Supercontinent. *Geology*, vol. 19, pp. 598–601.

Dalziel, I.W.D. 1997. Neoproterozoic-Paleozoic Geography and Tectonics: Review, Hypothesis, Environmental Speculation, *GSA Bulletin*, vol. 109 (1), pp. 16–42.

Damon, P.E. 1968. Application to the Dating of Igneous and Metamorphic Rocks with the Basin Ranges. *In:* Hamilton, E.I. and F.B. Farquhar, eds., *Radiometric Dating for Geologists.* New York: Interscience, pp. 36–44.

Damon, P.E., M. Shafiqullah, and R.B. Scarborough. 1978. Revised Chronology for Critical Stages in the Evolution of the Lower Colorado River [Abstract]. *Geological Society of America Abstracts with Programs*, vol. 10 (3), p. 101.

Damon, P.E., et al. 1967. Correlation and Chronology of Ore Deposits and Volcanic Rocks. U.S. Atomic Energy Commission Annual Progress Report No. C00-689-76 (Contract AT(11-1)-689.

Daneker, T.M. 1974. Sedimentology of the Precambrian Shinumo Quartzite, Grand Canyon, Arizona [abstract]. Geological Society of America. *Abstracts with Programs*, vol. 6 (5), p. 438.

Darton, N.H. 1910. A Reconnaissance of Parts of Northwestern New Mexico and Northern Arizona. *U.S. Geological Survey Bulletin*, vol. 435, p. 27.

David, L.R. 1944. A Permian Shark from the Grand Canyon. *Journal of Paleontology*, vol. 18 (1): 90–93.

Davidson, E.S. 1967. Geology of the Circle Cliffs Area, Garfield and Kane Counties, Utah. *U.S. Geological Survey Bulletin*, vol. 1229, 140 pp.

Davis, G.A., J.L. Anderson, E.G. Frost, and T.J. Schackelford. 1980. Mylonitization and Detachment Faulting in the Whipple–Buckstin–Rawhide Mountains Terrane, Southeastern California and Western Arizona. *In:* Crittenden, M.D., P.J. Coney, and G.H. Davis, eds., *Cordilleran Metamorphic Core Complexes.* Geological Society of America Memoir 153, pp. 79–129.

Davis, G.H. 1978. Monocline Fold Pattern of the Colorado Plateau. *In:* Matthews, vol., *ed.*, Laramide Folding Associated with Basement Block Faulting in the Western United States. *Geological Society of American Memoir,* pp. 215–233.

Davis, G.H., and S.E. Tindall. 1996. Discovery of Major Right-Handed Laramide Strike-Slip Faulting Along the Eastern Margin of the Kaibab Uplift, Colorado Plateau, Utah. *EOS, Transactions of the American Geophysical Union,* pp. F641–F642.

Davis, J.C. 1986. *Statistics and Data Analysis in Geology.* New York: John Wiley and Sons, 646 pp.

Davis, W.M. 1901. An Excursion to the Grand Canyon of the Colorado. *Bulletin of the Museum of Comparative Zoology* (Harvard), vol. 38, pp. 107–201.

Davis, W.M. 1903. An Excursion to the Plateau Province of Utah and Arizona. *Harvard College Museum of Comparative Zoology Bulletin*, vol. 42, pp. 1–50.

Dawson, W. 1897. Note on Cryptozoon and Other Ancient Fossils. *Canadian Records of Science*, vol. 8, p. 208.

Dean, J.S. 1988. Dendrochronology and Paleoenvironmental Reconstruction on the Colorado Plateau, *In:* Gumerman, G.J., ed., *The Anasazi in a Changing Environmental.* New York: Cambridge University Press, pp. 119–167.

DeCourten, F.L. 1976. Trace Fossils of the Kaibab Formation (Permian) of Northern Arizona. Unpublished M.S. thesis, University of California, Riverside, 72 pp.

Dehler, C.M., 1998. Facies Analysis and Environmental Interpretation of the Middle Chuar Group (Proterozoic): Implications for the Timing of Rodinian Breakup. American Association of Petroleum Geologists. *Abstracts with Programs* (Annual Meeting), p.12.

Dehler, C.M., and J.M. Timmons. 1998. Unpublished data. Dissertation and Masters thesis in progress, respectively. University of New Mexico, Albuquerque.

Denison, R.G. 1951. Late Devonian Freshwater Fishes from the Western United States. Chicago Natural History Museum, *Fieldiana. Geology*, vol. 11 (5), pp. 221–261.

Dew, E.A. 1985. Sedimentology and Petrology of the DuNoir Limestone (upper Cambrian) Southern Wind River Range, Wyoming. M.S. thesis, Northern Arizona University, Flagstaff, 193 pp.

Dickinson, W.R. 1981. Plate Tectonic Evolution of the Southern Cordillera. *Arizona Geological Society Digest*, vol. 14, p. 113–135.

Dickinson, W.R., et al. 1987. Laramide Tectonics and Paleogeography Inferred from Sedimentary Record in Laramide Basins of Central Rocky Mountain Region. *Geological Society America Abstracts with Programs*, vol. 19 (5), p. 271.

Dolan, R.A., Howard, and D. Trimble. 1978. Structural Control of the Rapids and Pools of the Colorado River in the Grand Canyon. *Science*, vol. 202, pp. 629–631.

Dott, R.H. 1974. Cambrian Tropical Storm Waves in Wisconsin. *Geology*, vol. 2, pp. 243–246.

Douglas, R.J., W.H. Gabrielse, J.O. Wheeler, D.F. Stott, and H.R. Belyea. 1970. Geology of Western Canada. *In:* R.J.W. Douglas, ed., *Geology and Mineral Resources of Canada: Geological Survey of Canada, Economic Geology Report No. 1,* pp. 365–488.

DuBois, S.M., A.W. Smith, N.K. Nye, and T.A. Nowak. 1982. Arizona Earthquakes, 1776–1980. *Arizona Bureau of Geology and Mineralogy, Geological Survey Branch Bulletin*, vol. 193, 456 pp.

Dutton, C.E. 1882. Tertiary History of the Grand Canyon District. *U.S. Geological Survey Monograph 2*, 264 pp.

Edzwald, J.K., and C.R. O'Melia. 1975. Clay Distributions in Recent Estuarine Sediments. *Clays and Clay Minerals*, vol. 23, pp. 39–44.

Ekdale, A.A., and M.D. Picard. 1985. Trace Fossils in a Jurassic Eolianite, Entrada Sandstone, Utah, U.S.A. *In:* Curran, H.A., ed., *Biogenic Structures: Their Use in Interpreting Depositional Environments.* SEPM Special Publication no. 35, pp. 3–12.

Elliott, D.K. 1987. *Chancelloria,* an Enigmatic Fossil from the Bright Angel Shale (Cambrian) of Grand Canyon, Arizona. *Journal of the Arizona-Nevada Academy of Science,* vol. 21, pp. 67–72.

Elliott, D.K., and D.L. Martin. 1987. A New Trace Fossil from the Cambrian Bright Angel Shale, Grand Canyon, Arizona. *Journal of Paleontology,* vol. 61, pp. 641–648.

Elston, D.P. 1979. Late Precambrian Sixty Mile Formation and Orogeny at the Top of the Grand Canyon Supergroup, Northern Arizona. *U.S. Geological Survey Professional Paper* 1092, 20 pp.

Elston, D.P. 1986. Magnetostratigraphy of Late Proterozoic Chuar Group and Sixtymile Formation, Grand Canyon Supergroup, Northern Arizona: Correlation with Other Proterozoic Strata of North America [Abstract]. Geological Society of America. *Abstracts with Programs* 1986 (Rocky Mountain Section), vol. 18, p. 353.

Elston, D.P. 1989a. Grand Canyon Supergroup: Northern Arizona. Stratigraphic Summary and Preliminary Paleomagnetic Correlations with Parts of Other North American Proterozoic Successions. *In:* Jenney, J.P., and S.J. Reynolds, eds., *Geologic Evolution of Arizona. Geological Society Digest,* vol. 17.

Elston, D.P. 1989b. Middle and Late Proterozoic Grand Canyon Supergroup. *In:* Elston, D.P., G.H. Billingsley, and R.A. Young, eds., *Geology of the Grand Canyon, Northern Arizona (28th International Geological Congress, Field Trip Guidebook T115/315).* Washington DC: America Geophysical Union, pp. 94–105.

Elston, D.P., and S.L. Bressler. 1977. Paleomagnetic Poles and Polarity Zonation from Cambrian and Devonian Strata of Arizona. *Earth and Planetary Science Letters,* vol. 36, pp. 423–433.

Elston, D.P., and C.S. Grommé. 1974. Precambrian Polar Wandering from Unkar Group and Nankoweap Formation, Eastern Grand Canyon, Arizona. Geological Society of America. *Abstracts with Programs,* vol. 6, pp. 440–441.

Elston, D.P., and E.H. McKee. 1982. Age and Correlation of the Late Proterozoic Grand Canyon Disturbance, Northern Arizona. *Geological Society of America Bulletin,* vol. 93, pp. 681–699.

Elston, D.P., and G.B. Scott. 1973. Paleomagnetism of Some Precambrian Basaltic Lava Flows and Red Beds, Eastern Grand Canyon, Arizona. *Earth and Planetary Science Letters,* vol. 28, pp. 235–265.

Elston, D.P., and G.B. Scott. 1976. Unconformity of the Cardenas–Nankoweap Contact (Precambrian), Grand Canyon Supergroup, Northern Arizona. *Geological Society of America Bulletin,* vol. 87, pp. 1763–1772.

Elston, D.P., and R.A. Young. 1989. Development of Cenozoic Landscape of Central and Northern Arizona: Cutting of Grand Canyon. *In:* Elston, D., G. Billingsley, and R. Young, eds., *Geology of Grand Canyon, Northern Arizona with Colorado River Guides.* Washington, DC: American Geophysical Union, pp. 145–153.

Elston, D.P., P.K. Link, D. Winston, and R.J. Horodyski. 1993. Correlations of Middle and Late Proterozoic Succesions. *In:* Reed, J.C., M.E. Bickford, R.S. Houston, P.K. Link, D.W. Rankin, P.K. Sims, and W.R. Van Schumus, eds., DNAG Precambrian: Conterminous U.S.: *GSA, The Geology of North America,* vol. C-2, p. 468–487.

Fairley, H.C., P.W. Bungart, C.M. Coder, Jim Huffman, T.L. Samples, and J.R. Balsom. 1994. The Grand Canyon River Corridor Survey Project: *Archaeological Survey Along the Colorado River Between Glen Canyon Dam and Seperation Canyon:* Report prepared for Grand Canyon National Park Service, Cooperative Agreement No. 9AA-40-07920, 276 pp.

Faulds, J.E., C. Schreiber, S.J. Reynolds, L.A. Gonzales, and D. Okaya. 1997. Origin and Paleogeography of an Immense, Nonmarine Miocene Salt Deposit in the Basin Range (Western USA). *Journal of Geology,* vol. 105. pp. 19–36.

Fenton, C.R. 1998. *Cosmogenic 3-Helium Dating of Lava Dam Outburst—Floods in Western Grand Canyon.* University of Utah, M.S. thesis. 76 pp.

Finks, R.M. 1960. Late Paleozoic Sponge Faunas of the Texas Region: The Siliceous Sponges, *American Museum of Natural History Bulletin,* vol. 120, pp. 1–160.

Finks, R.M., E.L. Yochelson, and R.P. Sheldon. 1961. Stratigraphic Implications of a Permian Sponge Occurrence in the Park City Formation of Western Wyoming. *Journal of Paleontology,* vol. 35, pp. 564–568.

Finnell, T.L. 1962. Recurrent Movement on the Canyon Creek Fault, Navajo County, Arizona. *In: Short Papers in Geology, Hydrology, and Topography. U.S. Geological Survey Professional Paper* 450-D, pp. D80–D81.

Finnell, T.L. 1966. Geologic Map of the Cibecue Quadrangle, Navajo County, Arizona. *U.S. Geological Survey Geologic Quadrangle Map* GQ-545.

Fisher, W.L. 1961. Upper Paleozoic and Lower Mesozoic Stratigraphy of Parashant and Andrus Canyons, Mohave County, Northwestern Arizona. Unpublished Ph.D. thesis, University of Kansas, 345 pp.

Ford, T.D., and W.J. Breed. 1973a. Late Precambrian Chuar Group, Grand Canyon, Arizona: *GSA Bulletin,* vol. 84, p. 1243–1260.

Ford, T.D., and W.J. Breed. 1973b. The Problematical Precambrian Fossil Chuaria. *Paleontology,* vol. 16 (3), pp. 535–550.

Ford, T.D., W.J. Breed, and J.S. Mitchell. 1969. Preliminary Geologic Report of the Chuar Group, Grand Canyon (with an appendix on palynology by C. Downie). *In: Geology and Natural History of the Grand Canyon Region.* Four Corners Geological Society 5th Field Conference Handbook, pp. 114–122.

Ford, T.D., W.J. Breed, and J.S. Mitchell. 1970. Carbon Butte, An Unusual Landslide. *In: Plateau,* vol. 43, pp. 9–15.

Ford, T.D., W.J. Breed, and J.S. Mitchell. 1972. Name and Age of Upper Precambrian Basalts in the Eastern Grand Canyon. *Geological Society of America Bulletin,* vol. 83 (1), pp. 223–226.

Ford, T.D., W.J. Breed, and J.S. Mitchell. 1972a. The Problematical Precambrian Fossil *Chuaria.* 24th International Geological Congress (Montreal), Proceedings, Section I, *Precambrian Geology,* pp. 11–18.

Ford, T.D., W.J. Breed, and J.S. Mitchell. 1972b. The Chuar Group of the Proterozoic, Grand Canyon, Arizona. 24th International Geological Congress (Montreal), Proceedings, Section I, *Precambrian Geology*, pp. 3–10.

Ford, T.D., P.W. Huntoon, G.H. Billingsley, and W.J. Breed. 1974. Rock Movement and Mass Wastage in the Grand Canyon. *In:* Breed, W.J., and E. Roat, eds., *Geology of the Grand Canyon.* Flagstaff: Museum of Northern Arizona Press, pp. 116–128.

Frakes, L.A., and E.M. Kemp. 1973. Paleogene Continental Positions and Evolution of Climate. *In:* Tarling, D.H. and S.K. Runcorn, eds., *Implications of Continental Drift to the Earth Sciences.* London: Academic Press, vol. 1, pp. 539–558.

Fryberger, S.G., A.M. Al-Sari, and T.J. Clisham. 1983. Eolian Dune, Interdune, Sand Sheet, and Siliciclastic Sabkha Sediments of an Offshore Prograding Sand Sea, Dhahran Area, Saudi Arabia. *American Association of Petroleum Geologists*, vol. 67, pp. 280–312.

Gallino, G.L., and T.C. Pierson. 1985. Polallie Creek Debris Flow and Subsequent Dam-Break Flood of 1980, East Fork Hood River Basin, Oregon: *U.S. Geological Survey Water Supply Paper* 2273, 22 pp.

Ghiglieri, M.P., 1992. *Canyon:* Tuscon, Arizona, University of Arizona Press, 311 pp.

Gilbert, G.K. 1874. One the Age of the Tonto Sandstone [Abstract]. *Philosophical Society of Washington*, Bulletin 1, p. 109.

Gilbert, G.K. 1875. Report on the Geology of Portions of Nevada, Utah, California, and Arizona. *U.S. Geographical and Geological Survey West of the 100th Meridian* (Wheeler), vol. 3 (1), pp. 17–187 and pp. 503–567.

Giletti, B.J., and P.E. Damon. 1961. Rubidium-strontium Ages of Some Basement Rocks From Arizona and Northwestern Mexico. *Geological Society of America Bulletin*, vol. 72, pp. 639–644.

Gilmore, C.W. 1926. Fossil Footprints from the Grand Canyon. *Smithsonian Miscellaneous Collections*, vol. 77 (9), 41 pp.

Gilmore, C.W. 1927. Fossil Footprints From the Grand Canyon; Second Contribution. *Smithsonian Miscellaneous Collections*, vol. 80 (3), 78 pp.

Gilmore, C.W. 1928. Fossil Footprints from the Grand Canyon; Third Contribution. *Smithsonian Miscellaneous Collections*, vol. 80 (8), 16 pp.

Gilmour, E.H., and I.D. Vogel. 1978. Bryozoans of the Toroweap Formation, Southern Nevada. Geological Society of America. *Abstracts with Programs*, vol. 10 (3), p. 107.

Glaessner, M.F. 1966. Precambrian Palaeontology. *Earth Science Reviews*, vol. 1, pp. 29–50.

Glaessner, M.F. 1969. Trace-fossils from the Precambrian and Basal Cambrian. *Lethania*, vol. 2, pp. 369–393.

Glaessner, M.F. 1984. *The Dawn of Animal Life.* Cambridge: Cambridge University Press, 244 pp.

Glancy, P.A., 1969. A Mudflow in the Second Creek Drainage, Lake Tahoe Basin, Nevada, and Its Relation to Sedimentation and Urbanization: *U.S. Geological Survey Professional Paper* 650C, pp. 129–136.

Glazner, A.F., J.E. Nielson, K.A. Howard, and D.M. Miller, 1986. Correlation of the Peach Springs Tuff, a Large-Volume Miocene Ignimbrite Sheet in California and Arizona. *Geology*, vol. 14 (10), pp. 840–843.

Gordon, M., Jr. 1984. Biostratigraphy of the Chainman Shale. *In:* Lints, J., Jr., ed., Western Geology Excursions. *Geological Society of America Annual Meeting Guidebook*, vol. 1, Department of Geological Sciences, Mackay School of Mines, Reno, Nevada, pp. 74–77.

Graf, W.L. 1979. Rapids in Canyon Rivers. *Journal of Geology*, vol. 87, pp. 533–551.

Graf, W.L. 1980. The Effect of Dam Closure on Down Stream Rapids. *Water Resources Research*, vol. 16, pp. 129–136.

Gregory, H.E. 1950. Geology and Geography of the Zion Park Region, Utah and Arizona. *U.S. Geological Survey Professional Paper* 220, 200 pp.

Griffen, L.R. 1966. *Actinocoelia maendria* Finks, from the Kaibab Limestone of Northern Arizona. *Brigham Young University Geology Studies*, vol. 13, pp. 105–108.

Griffiths, P.G., R.H. Webb, and T.S. Melis. 1996. Initiation and Frequency of Debris Flows in Grand Canyon, Arizona. *U.S. Geological Survey Open-File Report* 96-491, W.K. 1994. Late Cenozoic Lava Dams in the Western Grand Canyon. *Geological Society of Americal Memoir 183*, 144 pp.

Griffiths, P.G., R.H. Webb, and T.S. Melis. 1997. Initiation of Debris Flows in Tributaries of the Colorado River in Grand Canyon, Arizona, *In:* Cheng, Cheng-lung, *Debris-Flow Hazards Mitigation: Mechanics, Prediction, and Assessment.* New York: American Society of Civil Engineers, pp. 12–20.

Grover, P.W. 1987. Stratigraphy and Depositional Environment of the Surprise Canyon Formation, an Upper Mississippian Carbonate-Clastic Estuarine Deposit, Grand Canyon, Arizona. M.S. thesis, Northern Arizona University, Flagstaff, 166 pp.

Gutschick, R.D. 1943. The Redwall Limestone (Mississippian) of Yavapai County, Arizona. *Plateau*, vol. 16 (1), pp. 1–11.

Hack, J.T., 1942. *The Changing Physical Environment of the Hopi Indians of Arizona.* Cambridge, MA: Harvard University Press, *Papers of the Peabody Museum of American Archaeology* vol. 35, 85 pp.

Hamblin, W.K. 1965. Origin of Reverse Drag on the Downthrown Side of Normal Faults. *Geological Society of America Bulletin*, vol. 76, pp. 1145–1164.

Hamblin, W.K. 1976. Late Cenozoic Volcanism in the Western Grand Canyon. *In:* Breed, W.S. and E. Roat, eds., *Geology of the Grand Canyon.* Museum of Northern Arizona Press, pp. 142–169.

Hamblin, W.K. 1994. Late Cenozoic Lava Dams in the Western Grand Canyon. *Geological Society of Memoir* 183, 139 pp.

Hamblin, W.K. 1994a. Rates of Erosion by the Colorado River in the Grand Canyon, Arizona: *Proceedings of the 29th International Geology Congress*, Part B, pp. 211–218.

Hamblin, W.K., and J.K. Rigby. 1968. *Guidebook to the Colorado River Part1: Lee's Ferry to Phantom Ranch in Grand Canyon National Park.* Provo, UT: *Brigham Young University Geology Studies*, vol. 15 (5), pp. 84, 125.

Hamilton, W. 1982. Structural Evolution of the Big Maria Mountains, Northewastern Riverside County, Southeastern California. *In:* Forst, E.G., and D.L. Martin, eds., *Mesozoic–Cenozoic Tectonic Evolution of the Colorado River Region, California, Arizona, and Nevada.* Cordilleran Publishers, pp. 1–27.

Hammark, L., and E. Wohl. 1996. Debris Fan Formation and Modification at Warm Springs Rapid, Yampa River, Colorado. *Journal of Geology,* vol. 104, pp. 729–740.

Hampton, M.A. 1975. Competence of Fine-Grained Debris Flows. *Journal of Sedimentary Petrology,* vol. 45, p. 834–844.

Hansen, M.C. 1978. A Presumed Lower Dentition and a Spine of a Permian Petalodontiform Chondrichthyan, *Megactenopetalus kaibabanus. Journal of Paleontology,* vol. 52 (1), pp. 55–60.

Hantzschel, W. 1962. Trace Fossils and Problematica. *In:* R.C. Moore, ed., *Treatise on Invertebrate Paleontology, Part W, Miscellanea.* Geological Society of America and University of Kansas Press, 229 pp.

Hantzschel. W. 1975. Trace Fossils and Problematica. *In:* Teichert, C., ed., *Treatise on Invertebrate Paleontology, Part W, Miscellanea, Supplement 1.* Geological Society of America and University of Kansas Press, 269 pp.

Haq, B.U., and F.W.B. Van Eysingar. 1987. *Geologic Time Table,* 4th edition. Amsterdam: Elsevier Science Publishers.

Hardie, L.A. 1987. Dolomitization—A Critical View of Some Current Views. *Journal of Sedimentary Petrology,* vol. 57, pp. 166–183.

Harland. W.B. et al., eds. 1982. *A Geologic Time Scale.* Cambridge Earth Science Series, Cambridge, England: University Press, 131 pp.

Harrison, J.E., and Z.E. Peterman. 1971. Windermere Rocks and Their Correlatives in the Western United States. Geological Society of America. *Abstracts with Programs,* vol. 2 (7), pp. 592–593.

Harshbarger, J.W., C.A. Repenning, and J.H. Irwin. 1957. Stratigraphy of the Uppermost Triassic and the Jurassic Rocks of the Navaho Country. *U.S. Geological Survey Professional Paper* 291, 74 pp.

Hasiotis, S.T., and T.M. Brown. 1992. Invertebrate Trace Fossils: The Backbone of Continental Ichnology. *In:* Maples, C.G., and R.R. West, eds., Trace Fossils, The Paleontological Society, *Short Courses in Paleontology,* no. 5, pp. 64–104.

Haubold, Harmut. 1984. *Saurier Fahrten.* A. Ziemsen Verlag (Wittenberg Luthekalt, East Germany), 231 pp.

Havholm, K.G., and G. Kocurek. 1994. Factors Controlling Aeolian Sequence Stratigraphy: Clues from Super Bonding Surface Features in the Middle Jurassic Page Sandstone. *Sedimentology,* vol. 41, pp. 913–934.

Hawkins, D. P., 1996. U-Pb Geochronological Constraints in the Tectonic and Thermal Evolution of Paleoproterozoic Crust in the Grand Canyon, Arizona. Ph.D. thesis, Cambridge, Massachusetts Institute of Technology, 321 pp.

Hawkins, D.P., and S.A. Bowring, in press. U-Pb Monazite, Xenotime, and Titanite Geochronological Constraints on the Prograde to Post-Peak Metamorphic Thermal History of Paleoproterozoic Migmatites from the Grand Canyon, Arizona. *Contributions to Mineralogy and Petrology.*

Hawkins, D.P., S.A. Bowring, B.R. Ilg, K.E. Karlstrom, and M.L. Williams. 1996. U-Pb Geochronological Constraints on Paleoproterozoic Crustal Evolution, Upper Granite Gorge, Grand Canyon, Arizona. *Geological Society of America Bulletin,* vol. 108, pp. 1167–1181.

Hayes, P.T., and G.C. Cone. 1975. Cambrian and Ordovician Rocks of Southern Arizona and New Mexico and Westernmost Texas. *U.S. Geological Survey Professional Paper* 873, 98 pp.

Helley, E.J. 1969. Field Measurement of the Initiation of Large Bed Particle Motion in Blue Creek Near Klamath, California. *U.S. Geological Survey Professional Paper* 562-G. 19 pp.

Hendricks, J.D., and I. Lucchitta. 1974. Upper Precambrian Igneous Rocks of the Grand Canyon, Arizona, *In:* Karlstrom, T.N.V., G.A. Swann, and R.L. Eastwood, eds., *Geology of Northern Arizona,* part 1—*Regional Studies.* Flagstaff: Geological Society of America Field Guide, Rocky Mountain section, pp. 65–86.

Hereford, R. 1975. Chino Valley Formation (Cambrian?) in Northwestern Arizona. *Geological Society of America Bulletin,* vol. 86, pp. 677–682.

Hereford, R. 1977. Deposition of the Tapeats Sandstone (Cambrian) in Central Arizona. *Geological Society of America Bulletin,* vol. 88, pp. 199–211.

Hereford, R. 1984. Driftwood in Stanton's Cave: The Case for Temporary Damming of the Colorado River at Nankoweap Creek in Marble Canyon, Grand Canyon National Park, Arizona. *In:* Euler, R.C., ed., *The Archaeology, Geology, and Paleobiology of Stanton's Cave.* Grand Canyon Natural History Association, Monograph 6, pp. 99–106.

Hereford, R. 1996. Map Showing Surficial Geology and Geomorphology of the Palisades Creek Area, Grand Canyon National Park, Arizona: *U.S. Geological Miscellaneous Investigations Series Map* I-2449, scale 1:2,000 (with discussion).

Hereford, R., H.C. Fairley, K.S. Thompson, and J.R. Balsom. 1993. Surficial Geology, Geomorphology, and Erosion of Archeological Sites Along the Colorado River, Eastern Grand Canyon, Grand Canyon, Grand Canyon National Park, Arizona: *U.S. Geological Survey Open-File Report* 93-517, 45 pp., 4 plates.

Hereford, R., G.C. Jacoby, and V.A.S. McCord. 1996. Late Holocene Alluvial Geomorphology of the Virgin River in the Zion National Park Area, Southwest Utah. *Geological Society of America Special Paper* 310, 46 p.

Hereford, R., K.S. Thompson, K.J. Burke, and H.C. Fairley. 1996a. Tributary Debris Fans and the Late Holocene Alluvial Chronology of the Colorado River, Eastern Grand Canyon, Arizona: *Geological Society of America Bulletin,* vol. 108, p. 3–19.

Hereford, R., K.J. Burke, and K.S. Thompson. 1997. Map Showing Quaternary Geology and Geomorphology of the Nankoweap Rapids Area, Marble Canyon, Arizona: *U.S. Geological Survey Miscellaneous Investigations Series Map* I-2608, scale 1:2,000 (with discussion).

Hereford, R., K.J. Burke, and K.S. Thompson. 1997a. Dating Prehistoric Tributary Debris Fans, Colorado River, Grand

Canyon National Park, Arizona, with Implications for Channel Evolution and River Navigability. *U.S. Geological Survey Open-File Report* 97–167, 17 pp.

Hereford, R., K.J. Burke, and K.S. Thompson. 2001. Map Showing Quarternary Geology and Geomorphology of the Granite Park Area, Grand Canyon, Arizona. *U.S. Geological Survey Miscellaneous Investigations Series* I-(number not yet assigned), scale 1:2,000 (with discussion).

Hereford, R., K.J. Burke, and K.S. Thompson. 2001a. Map Showing Quarternary Geology and Geomorphology of the Lee's Ferry Area, Arizona. *U.S. Geological Survey Miscellaneous Investigations Series* I-(number not yet assisigned), scale 1:2,000 (with discussion).

Herries, R.D. 1993. Contrasting Styles of Fluvial–Aeolian Interactions at a Downwind Erg Margin: Jurassic Kayenta–Navajo Transition, Northeastern Arizona, USA. *In:* North, C.P., and D.J. Prosser, eds., *Characterization of Fluvial and Aeolian Reservoirs.* Geological Society of Special Publication, no. 73, pp. 199–128.

Hinds, N.E.A. 1938. An Algonkian Jellyfish from the Grand Canyon of the Colorado. *Science,* vol. 88, pp. 186–187.

Hintze, L.F. 1988. *Geologic History of Utah.* Geology Studies Special Publication 7. Provo: Brigham Young University, 202 pp.

Hjulstrom, E.J. 1935. Studies of the Morphological Activity of Rivers as Illustrated by the River Fyris. *University of Uppsala [Sweden] Geological Institute Bulletin,* vol. 25, pp. 221–527.

Hoffman, P.F. 1977. The Problematical Fossil *Chuaria* from the the Late Precambrian Uinta Mountain Group, Utah. *Precambrian Research,* vol. 4, pp. 1–11.

Hoffman, P.F. 1989. Speculations on Laurentia's First Gigayear (2.0 to 1.0 Ga). *Geology,* vol. 17, pp. 135–138.

Hoffman, P.F. 1991. Did the Breakout of Laurentia Turn Gondwanaland Inside-out?: *Science,* vol. 252, pp. 1409–1412.

Hoffman, P.F. 1988. United Plates of America, the Birth of a Craton: Early Proterozoic Assembly and Growth of Laurentia: Annual Review of Earth and Planetary Science, vol. 16, p. 543–603.

Hofmann, H.J., and J.D. Aitken. 1979. Precambrian Biota from the Little Dal Group, Mackenzie Mountains, Northwestern Canada. *Canadian Journal of Earth Sciences,* vol. 16, pp. 150–166.

Holm, R.F. 1987. San Francisco Mountain: a Late Cenozoic Composite Volcano in Northern Arizona. *In:* Beus S., ed., *Centennial Field Guide Volume 2.* Boulder: Rocky Mountain Section of the Geological Society of America, pp. 389–392.

Holm, R.F., and R.B. Moore. 1987. Holocene Scoria Cone and Lava Flows at Sunset Crater, Northern Arizona. *In:* Beus, S., ed., *Centennial Field Guide Volume 2.* Boulder: Rocky Mountain Section of the Geological Society of America, pp. 393–397.

Homma, M., and S. Shima. 1952. On the Flow in a Gradually Diverged Open Channel. *Japan Science Review,* vol. 2 (3), pp. 253–260.

Hooke, R., Le B., 1987. Mass Movement in Semi-Arid Environments and the Morphology of Alluvial Fans, *In:* Anderson, M.G., and Richards, K.S., eds., *Slope Stability.* New York: John Wiley and Sons, pp. 505–529.

Hopkins, R.L. 1986. Depositional Environments and Diagenesis of the Fossil Mountain Member of the Kaibab Formation (Permian), Grand Canyon, Arizona. M.S. thesis, Northern Arizona University, Flagstaff, 244 pp.

Horodyski, R.J. 1986. Paleontology of the Late Precambrian Chuar Group, Grand Canyon, Arizona. Geological Society of America. *Abstracts with Programs* (Rocky Mountain section), vol. 18, p. 362.

Howard, A., and R. Dolan. 1976. Changes in Fluvial Deposits of the Colorado River in the Grand Canyon Caused by Glen Canyon Dam. *In:* Lin, R.M., ed., *First Conference on Scientific Research in the National Parks, Transactions and Proceedings,* No. 5, New Orleans, Nov. 9–12, 1975, U.S. National Park Service, pp. 845–851.

Howard, A., and R. Dolan. 1981. Geomorphology of the Colorado River in the Grand Canyon. *Journal of Geology,* vol. 89 (3), pp. 259–298.

Hubbert, M.K. 1951. Mechanical Basis for Certain Familiar Geologic Structures. *Geological Society of America Bulletin,* vol. 62, pp. 355–372.

Hunt, C.B. 1969. Geological History of the Colorado River. *In:* The Colorado River Region and John Wesley Powell. *U.S. Geological Survey Professional Paper* 669, 145 pp.

Hunter, R.E. 1977. Basic Types of Stratification in Small Eolian Dunes. *Sedimentology,* vol. 24, pp. 361–388.

Huntoon, P.W. 1973. High Angle Gravity Faulting in the Eastern Grand Canyon, Arizona. *Plateau,* vol. 45 (3) pp. 117–127.

Huntoon, P.W. 1975. The Surprise Valley Landslide and Widening of the Grand Canyon. *Plateau,* vol. 48 (1 and 2), pp. 1–12.

Huntoon, P.W. 1981. Grand Canyon Monoclines, Vertical Uplift or Horizontal Compression? *Contributions to Geology,* vol. 19, pp. 127–134.

Huntoon, P.W., 1993. Influence of Inherited Precambrian Basement Structure on the Localization and Form of Laramide Monoclines, Grand Canyon, Arizona. *In:* Schimdt, C.J., R.B. Chase, and E.A. Erslev, eds., Laramide Basement Deformation in the Rocky Mountain Foreland of the Western United States. *Geological Society of America Special Paper* 280. pp. 243–256.

Huntoon, P.W., and D.P. Elston. 1980. Origin of the River Anticlines, Central Grand Canyon, Arizona. U.S. *Geological Survey Professional Paper* 1126A, 9 pp.

Huntoon, P.W., and J.W. Sears. 1975. Bright Angel and Eminence Faults, Eastern Grand Canyon, Arizona. *Geological Society of America Bulletin,* vol. 86, pp. 465–472.

Huntoon, P.W., G.H. Billingsley, and M.D. Clark. 1981. *Geologic Map of the Hurricane Fault Zone and Vicinity, Western Grand Canyon, Arizona.* Grand Canyon Natural History Association.

Huntoon, P.W., G.H. Billingsley, and M.D. Clark. 1982. Geologic Map of the Lower Granite Gorge and Vicinity, Western Grand Canyon, Arizona. Grand Canyon Natural History Association.

Huntoon, P.W., et al. 1996. *Geological Map of the Eastern Part of the Grand Canyon National Park, Arizona*, 1996 edition. Grand Canyon Association.

Hutchison, J.L., 1996. Relative Timing of Porphyroblast Gowth and Peak Metamorphism in the Lower Granie Gorge, Grand Canyon, Arizona. M.S. thesis, Albuquerque, University of New Mexico, 59 pp.

Hyatt, M.L., 1990. Graphic Detail of Historic Streamflows, Water Releases, and Reservoir Storage for Glen Canyon Dam and Lake Powell: *Baseline Reference Data for the Glen Canyon Dam EIS*: Denver, Colorado, U.S. Bureau of Reclamation, Water Management Section, 300 figs.

Ilg, B.R. 1992. Early Proterozoic Structural Geology of Upper Granite Gorge, Grand Canyon, Arizona. M.S. thesis, Flagstaff, Northern Arizona University, 59 pp.

Ilg, B.R. 1996. Tectonic Evolution of Paleoproterozoic Rocks in the Grand Canyon: Insights into Middle Crustal Processes. Unpublished Ph.D. thesis, Albuquerque, University of New Mexico, 97 pp.

Ilg, B.R., and Karlstrom, K.E. 1996. Older Precambrian Geology of the Eastern Part of the Grand Canyon. *In:* Huntoon P.W., ed., Geology of the Eastern Part of the Grand Canyon, Grand Canyon, Grand Canyon Natural History Association.

Ilg, B.R., Karlstrom, K.E., Williams, M.L., and Hawkins, D.P., 1996. Tectonic Evolution of Paleoproterozoic Rocks in the Grand Canyon: Insights into Middle Crustal Processes, *Geological Society of America Bulletin*, vol. 108, p. 1149–1166.

Illing, L.V., A.J. Wells, and J.C.M. Taylor. 1965. Pennecontemporaneous Dolomite in the Persian Gulf. *In:* Dolomitization and Limestone Diagenesis—A Symposium. *Society of Economic Paleotologists and Mineralogists Special Paper* 13, pp. 89–111.

Ippen, A.T. 1951. Mechanics of Supercritical Flow. *Transactions American Society of Civil Engineers*, vol. 116, pp. 268–295.

Ippen, A.T., and J.H. Dawson. 1951. Design of Channel Contractions. *Transactions American Society of Civil Engineers*, vol. 116, pp. 328–346.

Irwin, C.D, 1971. Stratigraphic Analysis of Upper Permian and Lower Triassic Strata in Southern Utah. *American Association of Petroleum Geologists Bulletin*, vol. 55, pp. 1976–2007.

Irwin, M.L. 1965. General Theory of Epeiric Clear Water Sedimentation. *American Association of Petroleum Geologists Bulletin*, vol. 49, pp. 224–459.

Iverson, R.M., and R.G. LaHusen. 1989. Dynamic Pore-Pressure Flucuations in Rapidly Shearing Granular Materials: *Science*, vol. 246, p. 796–799.

Jackson, G., 1990. Tectonic Geomorphology of the Toroweap Fault, western Grand Canyon. Arizona: Implicatons for Transgression of Faulting on the Colorado Plateau. *Arizona Geological Survey Open File Report* 90–4, 66pp.

Johnson, A.M., and J.R. Rodine. 1984. Debris Flow. *In:* Brunsden, D., and D.B. Prior, edrs., *Slope Instability*. New York: John Wiley and Sons, pp. 257–361.

Johnson, H.D. 1977. Shallow Marine Sandbar Sequences: An Example from the Late Precambrian of North Norway. Sedimentology, vol. 24, pp. 245–270.

Johnson, R.R., 1991. Historic Changes in Vegetation Along the Colorado River in the Grand Canyon, *In: Colorado River Ecology and Dam Managment:* Washinton, DC: National Academy Press, pp. 178–205.

Jones, A.T. 1986. A Cross Section of Grand Canyon Archeology: Excavations at Five Sites Along the Colorado River: Tuscon, Arizona, *National Park Service Publications in Anthropology*, No. 28, 357 pp.

Karlstrom, K.E., and S.A. Bowring. 1988. Early Proterozoic Assembly of Tectonostratigraphic Terranes in Southwestern North America. *Journal of Geology*, vol. 96, pp. 561–576.

Karlstrom, K.E., and S.A. Bowring. 1993. Proterozoic Orogenic History in Arizona. *In:* Van Schmus, W.R., and Brickford, M.E., eds. *Transcontinental Proterozoic Provinces, The Geology of North America* vol. c-2. Precambrian of the Conterminous U.S. Geological Society of America (DNAG), pp. 188–228.

Karlstrom, K.E., and M.L. Williams. 1995. The Case for Simultaneous Deformation, Metamorphism and Plutonism: An Example from Proterozoic Rocks in Central Arizona. *Journal of Structural Geology*, vol. 17, pp. 59–81.

Karlstrom, K.E., and M.L. Williams. 1998. Heterogeneity of the Middle Crust: Implications for Strength of Continental Lithosphere. *Geology*.

Karlstrom, T.N.V. 1988. Alluvial Chronology and Hydrological Change of Black Mesa and Nearby Regions, *In:* Gumerman, G.J., ed., *The Anasazi in a Changing Environment*. New York: Cambridge University Press, pp. 45–91.

Kauffman, E.G. and J.R. Steidtmann. 1981. Are These the Oldest Metazoan Trace Fossils? *Journal of Paleontology*, vol. 55, pp. 923–947.

Keller, G.R., R.B. Smith, and L.W. Braile. 1975. Crustal Structure Along the Great Basin–Colorado Plateau Transition from Seismic Reflection Studies. *Journal of Geophysical Research*, vol. 80, pp. 1093–1098.

Kendall, C.G. St. C., and W. Schlager. 1981. Carbonates and Relative Changes in Sea Level, *Marine Geology*, vol. 44, pp. 181–212.

Kent, W.N., and R.R. Rawson. 1980. Depositional Environments of the Mississippian Redwall Limestone in Northeastern Arizona. *In:* Fouch, T.D., and E.R. Magathon, eds., *Paleozoic Paleogeography Symposium 1, Rocky Mountain Section*. Denver: Society of Economic Paleontologists and Mineralogists. pp. 101–109.

Keppel, D. 1932. A Study of the Bryozoan Collection for the Museum of the National Park Service at Grand Canyon, Arizona. Unpublished M.A. thesis, Columbia University, 27 pp.

Keyes, C.R. 1938. Basement Complex of the Grand Canyon, *Pan-American Geologist*, Vol. 70 (2), pp. 91–116.

Kieffer, S.W. 1985. The 1983 Hydraulic Jump in Crystal Rapid: Implications for River-Running and Geomorphic Evolution in the Grand Canyon. *Journal of Geology*, vol. 93, pp. 385–406.

Kieffer, S.W. 1986. The Rapids of the Colorado River (a 20-minute VHS video). U.S. Geological Survey Open File Report 86–503.

Kieffer, S.W. 1987. The Rapids and Waves of the Colorado River, Grand Canyon, Arizona. U.S. Geological Survey Open-File Report 87-096, 97 pp.

Kieffer, S.W. 1988. Hydraulic Maps of Major Rapids, Grand Canyon, Arizona. U.S. Geological Survey Miscellaneous Investigations Map Series I-1897A-J.

Kieffer, S.W., J.B. Graf, and J.C. Schmidt. 1989. Hydraulics and Sediment Transport of the Colorado River. *In:* Elston, D.P., G.H. Billingsley, and R.A. Young, eds., *Geology of the Grand Canyon, Northern Arizona (with Colorado River Guides)*, Field Trip Guidebook T115/315 for the 28th International Geological Congress, 239 pp.

King, P.B. 1977. *The Evolution of North America.* Princeton: Princeton University Press, 197 pp.

Kirkland, P. 1962. Permian Stratigraphy and Stratigraphic Paleontology of the Colorado Plateau. M.S. thesis. University of New Mexico, Albuquerque, 245 pp.

Knight, R.L., and J.R. Cooper. 1955. Suggested Changes in Devonian Terminology in the Four Corners Area: *Four Corners Geological Society First Field Conference Guidebook*, pp. 56–58.

Kochel, R.C., 1987. Holocene Debris Flows in Central Virginia: *Reviews in Engineering Geology*, vol. 7, p. 139–155.

Kocurek, G. 1981. Significance of Interdune Deposits and Bounding Surfaces in Aeolian Dune Sands. *Sedimentology*, vol. 28, pp. 753–780.

Kocurek, G. 1986. Origins of Low-Angle Stratification in Aeolian Deposits. *In:* Nickling, W.G., ed., *Aeolian Geomorphology.* Proceedings of the 17th Annual Binghamton Symposium, pp. 177–193.

Kocurek, G., and Nielson, J., 1986. Conditions Favorable for the Formation of Warm-Climate Aeolian Sand Sheets. *Sedimentology*, vol. 33, pp. 795–816.

Koons, E.D. 1945. Geology of the Uinkaret Plateau, Northern Arizona. *Geological Society of America Bulletin*, vol. 56, pp. 151–180.

Koons, E.D. 1948. High-Level Gravels of Western Grand Canyon. *Science*, vol. 197, pp. 475–476.

Koons, E.D. 1955. Cliff Retreat in the Southwestern United States. *American Journal of Science*, vol. 253, pp. 44–52.

Korolev, V.S., and S.M. Rowland. 1993. Correlation of Unclassified Cambrian Dolomites of the Grand Canyon. Journal of the Arizona–Nevada Academy of Science, Proceedings Supplement, vol. 28, p. 42.

Kruger-Kneupfer, J.L., M.L. Sbar, and R.M. Richardson. 1985. Microseismicity of the Kaibab Plateau, Northern Arizona, and Its Tectonic Implications. *Bulletin of the Seismological Society of America*, vol. 75 (2), pp. 491–506.

Kues, B.S., and Lucas, S.G. 1989. Stratigraphy and Paleontology of a San Andres Formation (Permian, Leonardian) outlier, Zuni Indian Reservation, New Mexico. *In:* Anderson, O.J., S.G. Lucas, D.W. Love, and S.M. Cather, eds., *New Mexico Geological Society Guidebook*, 40th Field Conference, Southeastern Colorado Plateau, pp. 167–176.

Lancaster, N., 1984. Late Cenozoic Fluvial Deposits of the Tsonab Valley, Central Namib Desert. *Madoqua*, vol. 13, pp. 257–269.

Lancaster, N. 1989. The Namib Sand Sea: *Dune Forms, Processes, and Sediments.* Rotterdam: Balkema, A.A., 200 pp.

Lancaster, N. 1995. *Geomorphology of Desert Dunes.* London: Routledge, 290 pp.

Landau, L.D., and E.M. Lifshitz. 1959. *Fluid Mechanics.* Oxford: Pergamon Press, 536 pp.

Langford, R.P., 1989. Fluvial–Aeolian Interactions, Part I. Modern Systems. *Sedimentology*, vol. 36, pp. 1023–1035.

Langford, R.P., and M.A. Chan. 1989. Fluvial–Aeolian Interactions. Part II. Ancient Systems. *Sedimentology*, vol. 36, pp. 1037–1051.

Lapinski, P.W. 1976. The Gamma Member of the Kaibab Formation (Permian) in Northern Arizona. M.S. thesis, University of Arizona, Tucson, 138 pp.

Larson, A.A., P.E. Patterson, and F.E. Mutschler. 1994. Lithology, Chemistry, Age, and Origin of the Proterozoic Cardenas Basalt, Grand Canyon, Arizona. *Precambrian Research*, vol. 65, pp. 255–276.

Larson, E.E., P.E. Patterson, M.H. Amini, and J. Rosenbaum. 1986. Petrology, Chemistry and Revised Age of the Late Precambrian Cardenas Lava, Grand Canyon, Arizona. Geological Society of America. *Abstracts with Programs* (Rocky Mountain section), vol. 18, p. 353.

Leopold, L. 1969. The Rapids and the Pools—Grand Canyon. *U.S. Geological Survey Professional Paper 669*, pp. 131–145.

Levy, M., and N. Christie-Blick. 1991. Tectonic Subsidence of the Early Paleozoic Passive Continental Margin in Eastern California and Southern Nevada. *GSA Bulletin*, vol. 103, pp. 1590–1606.

Lingley, W.S. 1973. Geology of the Older Precambrian Rocks in theVicinity of Cleark Creek and Zoroaster Canyon, Grand Canyon, Arizona. M.S. thesis, Western Washington University, Bellingham, 78 pp.

Lingley, W.S. 1976. Some Structures in Older Precambrian Rocks of Clear Creek—Cremation Creek Area, Grand Canyon National Park, Arizona. *Arizona Geological Society Digest*, vol. 10, pp 27–37.

Link, P.K. (editor), N. Christie-Blick, W.J. Devlin, D.P. Elston, R.J. Horodyski, M. Levy, J.M.G. Miller, R.C. Pearson, A. Prave, J.H. Stuart, D. Winston, L.A. Wright, and C.T. Wrucke. 1993. Middle and Late Proterozoic Stratified Rocks of the Western U.S. Cordillera, Colorado Plateau, and Basin and Range Province. *In:* Reed, J.C., M.E. Bickford, R.S. Houston, P.K. Link, D.W. Randall, P.K. Sims, and W.R. Van Schumus, eds., DNAG *Precambrian: Counterminous U.S.*; GSA, *The Geology of North America*, vol. C-2, pp. 463–428.

Lipfert, F.W. 1989. Atmospheric Damage to Calcareous Stones: Comparison and Reconciliation of Recent Experimental Findings: *Atmospheric Environment*, vol. 23, pp. 415–429.

Lochman-Balk, C. 1970. Upper Cambrian Faunal Patterns on the Craton. *Geological Society of America Bulletin*, vol. 81, pp. 3197–3224.

Lochman-Balk, C. 1971. The Cambrian of the Craton of the United States. *In:* Holland, C.H., ed., *Cambrian of the New World.* New York: John Wiley and Sons, pp. 79–167.

Longwell, C.R. 1921. Geology of the Muddy Mountains, Nevada, with a Section to the Grand Wash Cliffs in Western Arizona. *American Journal of Science*, vol. 1, pp. 39–62.

Longwell, C.R. 1936. Geology of the Boulder Reservoir Floor, Arizona–Nevada. *Geological Society of America Bulletin*, vol. 47, pp. 1393–1476.

Longwell, C.R. 1946. How Old is the Colorado River? *American Journal of Science*, vol. 244 (12), pp. 817–835.

Longwell, C.R., E.H. Pampeyan, B. Bower, and R.J. Roberts. 1965. Geology and Mineral Deposits of Clark County, Nevada. *Nevada Bureau of Mines Bulletin*, vol. 62, 218 pp.

Loope, D.B. 1985. Episodic Deposition and Preservation of Eolian Sands: A Late Peolozoic Example from Southeastern Utah. *Geology*, vol. 13, pp. 73–76.

Loughlin, W.D. and P.W. Huntoon, 1983. *Compilation of available ground water quality data for sources within the Grand Canyon of Arizona.* Department of Geology and Geophysics, University of Wyoming (Laramie), 9 pp. plus appendices.

Lovejoy, E.M.P. 1980. The Muddy Creek Formation at the Colorado River in Grand Wash—The Dilemma of the Immovable Object. *Arizona Geological Society Digest*, vol. 12, pp. 177–192.

Lucchitta, I. 1967. Cenozoic Geology of the Upper Lake Mead Area Adjacent to the Grand Wash Cliffs, Arizona. Ph.D. dissertation, Pennsylvania State University, 218 pp.

Lucchitta, I., 1970. Late Cenozoic uplift of the southwestern Colorado Plateau and adjacent lower Colorado River region. *Tectonophysics*, vol.61, pp. 53–95.

Lucchitta, I. 1972. Early History of the Colorado River in the Basin and Range Province. *Geological Society of America Bulletin*, vol. 83, pp. 1933–1948.

Lucchitta, I. 1975. The Shivwits Plateau. *In:* Application of ERTS Images and Image Processing to Regional Geologic Problems and Geologic Mapping in Northern Arizona. *Jet Propulsion Laboratory Technical Report* 32-1597, pp. 41–72.

Lucchitta, I. 1979. Late Cenozoic Uplift of the Southwestern Colorado River Region. *Tectonophysics*, vol. 61, pp. 53–95.

Lucchitta, I. 1984. Development of the Landscape in Northwest Arizona—The Country of Plateaus and Canyons. *In:* Smiley, T.L., et al., eds., *Landscapes of Arizona—The Geological Story.* University Press of America, pp. 269–302.

Lucchitta, I., 1991. Topographic Maps of the Palisades–Unkar Area, Grand Canyon, Arizona. *U.S. Geological Survey Open-File Report* 91-535, scale 1:5,000.

Lucchitta, I., and S.S. Beus. 1987. Field Trip Guide for Marble Canyon and Eastern Grand Canyon. *In:* Davis, G.H., and G.M. VandenDolder, eds., *Geologic Diversity of Arizona and Its Margins: Excursions to Choice Areas.* Arizona Bureau of Geology and Mineral Technology, Special Paper 5, pp. 3–19.

Lucchitta, I., and J.D. Hendricks. 1983. Characteristics, Depostional Environment, and Tectonic Interpretations of the Proterozoic Cardenas Lavas, Eastern Grand Canyon, Arizona. *Geology*, vol. 11 (3), pp. 77–181.

Lucchitta, I., and E.H. McKee. 1975. New Chronological Constraints on the History of the Colorado River and the Grand Canyon [Abstract]. Geological Society of America. *Abstracts with Programs*, vol. 9, pp. 746–747.

Lucchitta, I., and R.A. Young. 1986. Structure and Geomorphic Character of Western Colorado Plateau in the Grand Canyon-Lake Mead Region. *In:* Nations, J.D., C.M. Conway, and G.A. Swann, eds., *Geology of Central and Northern Arizona.* Geological Society of America, Rocky Mountain Section, Field Trip Guidebook, pp. 159–176.

Lucchitta, I., C.M. Dehler, M.E. Davis, K.J. Burke, and P.O. Basdekas. 1995. Quarternary Geological Map of the Palisades Creek-Comanche Creek Area, Eastern Grand Canyon, Arizona: *U.S. Geological Survey Open-File Report* 95-832, scale 1:5,000 (with discussion).

Lull, R.S. 1918. Fossil Footprints from the Grand Canyon of the Colorado. *Smithsonian Miscellaneous Collections*, vol. 80, pp. 1–16.

Machette, M.N. 1985. Calcic Soils of the Southwestern United States. *Geological Society of America Special Paper 203*, p. 1–21.

Machette, M.N., and J.N. Rosholt. 1991. Quaternary Geology of the Grand Canyon. *In:* Morrison, R.B., ed., The Geology of North America, V K-2, Quaternary Nonglacial Geology of the Conterminous United States. *Geological Society of America*, pp. 397–401.

Major, J.J. 1997. Depositional Processes in Large-Scale Debris-Flow Experiments. *Journal of Geology*, vol. 105, p. 345–366.

Major, J.J., and T.C. Pierson. 1992. Debris Flow Rheology: Experimental Analysis of Fine-Grained Slurries. *Water Resources Research*, vol. 28, p. 841–857.

Mamet, B.L., and B. Skipp. 1970. Preliminary Foraminiferal Correlations of Early Carboniferous Strata in the North American Cordilleran. *Les Congres et Colloques di l'Universite di Liege*, vol. 55, pp. 237–348.

Marcou, J. 1856. Resume and Field Notes: United States Pacific Railroad Exploration, *Geological Report*, vol. 3 (4), pp. 165–171.

Martin, D.L. 1985. Depositional Systems and Ichnology of the Bright Angel Shale (Cambrian), Eastern Grand Canyon, Arizona. M.S. thesis, Northern Arizona University, Flagstaff, 365 pp.

Martin, D.L., L.T. Middleton, and D.K. Elliott. 1986. Depositional Systems of the Middle Cambrian Bright Angel Shale, Grand Canyon, Arizona. Geological Society of America. *Abstracts with Programs*, vol. 18, p. 394.

Martin, H., 1992. Conodont Biostratigraphy and Paleoenvironment of the Suprise Canyon Formation (Late Mississippian), Grand Canyon, Arizona. M.S. thesis, Northern Arizona University, Flagstaff, 365 pp.

Mather, T. 1970. Stratigraphy and Paleontology of Permian Kaibab Formation, Mogollon Rim Region, Arizona. Unpublished Ph.D. thesis, University of Colorado, Boulder, 164 pp.

Matthews, J.J. 1961. *Geologic Map of the Bright Angel Quadrangle, Grand Canyon National Park, Arizona.* Grand Canyon Natural History Association.

Maxson, J.H. 1968. *Geologic Map of the Bright Angel Quadrangle, Grand Canyon National Park, Arizona*: Grand Canyon, Arizona, Grand Canyon Natural History Association, scale 1:48,000.

McKee, E.D. 1932. Some Fucoids from the Grand Canyon. *Grand Canyon Nature Notes*, vol. 7, pp. 77–81.

McKee, E.D. 1933a. Landslides and Their Part in Widening the Grand Canyon. *Grand Canyon Nature Notes*, vol. 8, pp. 58–161.

McKee, E.D. 1933b. The Coconino Sandstone—Its History and Origin. Washington, DC: Carnegie Institute, *Publication 440*, pp. 77–115.

McKee, E.D. 1937. Research on Paleozoic Stratigraphy in Western Grand Canyon. Washington, DC: Carnegie Institute, *Yearbook 36*, pp. 340–343.

McKee, E.D. 1938. The Environment and History of the Toroweap and Kaibab Formations on Northern Arizona and Southern Utah. Washington, DC: Carnegie Institute, Publication 492, 268 pp.

McKee, E.D. 1939. Studies on the History of Grand Canyon Paleozoic Formations. Washington, DC: Carnegie Institute, *Yearbook 38*, pp. 313–314.

McKee, E.D. 1944. Tracks That Go Uphill in Coconino Sandstone, Grand Canyon, Arizona. *Plateau*, vol. 16, pp. 61–72.

McKee, E.D. 1945. Small-Scale Structures in Coconino Sandstone of Northern Arizona. *Journal of Geology*, vol. 53, pp. 313–325.

McKee, E.D. 1947. Experiments on the Development of Tracks in Fine Cross-bedded Sand. *Journal of Sedimentary Petrology*, vol. 17, pp. 23–28.

McKee, E.D. 1963. Lithologic Subdivisions of the Redwall Limestone in Northern Arizona: Their Paleogeographic and Economic Significance. *U.S. Geological Survey Professional Paper* 400-B.

McKee, E.D. 1974. Paleozoic Rocks of Grand Canyon. *In:* Breed, W.J., and E. Roat, eds., *Geology of the Grand Canyon.* Museum of Northern Arizona Press, pp. 42–75.

McKee, E.D. 1975. The Supai Group, Subdivision and Nomenclature. *U.S. Geological Survey Bulletin* 1395-J, 11 pp.

McKee, E.D. 1976. Paleozoic Rocks of the Grand Canyon. *In:* Breed, W.J., and E. Roat, eds., *Geology of the Grand Canyon.* Museum of Northern Arizona Press, pp. 42–64.

McKee, E.D. 1979. Ancient Sandstone Considered to be Eolian. *In:* McKee, E.D., ed., A Study of Global Sand Seas. *U.S. Geological Survey Professional Paper* 1052, pp. 187–238.

McKee, E.D. 1982a. The Supai Group of Grand Canyon. *U.S. Geological Survey Professional Paper* 1173, 504 pages.

McKee, E.D. 1982b. Sedimentary Structures in Dunes of the Nambia Desert, South West Africa: *Geological Society of America Special Paper* 188, 64 pages.

McKee, E.D., and J.J. Bigarella. 1972. Deformational Structures in Brazilian Coastal Dunes. *Journal of Sedimentary Petrology*, vol. 42, pp. 670–681.

McKee, E.D., and W. Breed. 1969. The Toroweap Formation and Kaibab Limestone. *In:* Summers, W.K., and F.E. Kottlowski, eds., The San Andres Limestone, a Resevoir for oil and water in New Mexico. *New Mexico Geological Society Special Publication* 3, pp. 12–26.

McKee, E.D., J.R. Douglas, and S. Rittenhouse. 1971. Deformation of Lee-Side Laminae in Eolian Dunes. *Geological Society of America Bulletin*, vol. 82, pp. 359–378.

McKee, E.D., and R.C. Gutschick. 1969. History of the Redwall Limestone of Northern Arizona. *Geological Society of America Memoir*, vol. 114, 726 pp.

McKee, E.D., W.K. Hamblin, and P.E. Damon. 1968. K–Ar Age of Lava Dam in Grand Canyon. *Geological Society of America Bulletin*, vol. 79 (1), pp. 133–136.

McKee, E.D., and E.H. McKee. 1972. Pliocene Uplift of the Grand Canyon Region: Time of Drainge Adjustment. *Geological Society of America Bulletin*, vol. 83 (7), pp. 1923–1932.

McKee, E.D., and R.J. Moiola. 1975. Geometry and Growth of the White Sands Dune Field, New Mexico. *U.S. Geological Survey, Journal of Research*, vol. 3, pp. 59–66.

McKee, E.D., and C.E. Resser. 1945. Cambrian History of the Grand Canyon Region. Washington DC: Carnegie Institute, Publication 563, 232 pp.

McKee, E.D., and E.T. Schenk. 1942. The Lower Canyon Lavas and Related Features at Toroweap in Grand Canyon. *Journal of Geomorphology*, vol. 5, pp. 245–273.

McKee, E.D., R.F. Wilson, W.J. Breed, and C.S. Breed. 1967. Evolution of the Colorado River in Arizona. *Museum of Northern Arizona Bulletin* 44, 67 pp.

McKee, E.H., and D.C. Noble. 1974. Radiometric Ages of Diabase Sills and Basaltic Lava Flows in the Unkar Group, Grand Canyon. Geological Society of America. *Abstracts with Programs*, vol. 6, p. 458.

McKee, E.H., and D.C. Noble. 1976. Age of the Cardenas Lavas, Grand Canyon, Arizona. *Geological Society of America Bulletin*, vol. 87, pp. 1188–1190.

McKinney, F.K. 1983. Ectoprocta (Bryozoa) from the Permian Kaibab Formation, Grand Canyon National Park, Arizona. *Fieldiana Geology*, n. s., no. 13, 17 pp.

McNair. 1951. Paleozoic Stratigraphy of Part of Northwestern Arizona. *American Association of Petroleum Geologists Bulletin* 35, pp. 503–541.

Melis, T.S. 1997. Geomorphology of Debris Flows and Alluvial Fans in Grand Canyon National Park and Their Influences on the Colorado River Below Glen Dam Canyon, Arizona. Ph.D. dissertation. Tuscon, University of Arizona. 495 pp.

Melis, T.S., and R.H. Webb. 1993. Debris Flows in Grand Canyon National Park, Arizona—Magnitude, Frequency, and Effects on the Colorado River. *In:* Shen, Hsieh Wen, S.T. Su, and Feng Wen, editors, *Hydraulic Engineering* 1993, vol. 2. New York, American Society of Civil Engineering, pp. 1290–1295.

Melis, T.S., R.H. Webb, P.G. Griffiths, and T.J. Wise. 1994. Magnitude and Frequency Data for Historic Debris Flows in Grand Canyon National Park and Vicinity, Arizona. *U.S. Geological Survey Water Resources Investigations Report* 94-4214, 285 pp.

Melis, T.S., R.H. Webb, and P.G. Griffiths. 1997. Debris Flows in Grand Canyon National Park: Peak Discharges, Flow Transformations, and Hydrographs, *In:* Chen, Cheng-lung, ed., *Debris-Flow Hazards Mitigation: Mechanics, Prediction, and Assessment.* New York: American Society of Civil Engineers, pp. 727–736.

Metzger, D.G. 1968. The Bouse Formation (Pliocene) of the Parker Blythe-Cibola Area, Arizona and California. *In:* Geological Survey Research 1968. *U.S. Geological Survey Professional Paper* 600-D, pp. D126–D136.

Meyer, G.A., S.G. Wells, and A.J.T. Jull. 1995. Fire and Alluvial Chronology in Yellowstone National Park: Climatic and Intrinsic Controls on Holocene Geomorphic Processes. *Geological Society of America Bulletin,* vol. 107, p. 1211–1230.

Middleton, L.T. 1988. Cambrian and Ordovician Depositional Systems in Arizona. *In:* Jenney, J.P., and S.J. Reynolds, eds., *Geological Evolution of Arizona. Arizona Geological Society Digest,* vol. 17, pp. 273–286.

Middleton, L.T., and R.C. Blakey. 1983. Processes and Controls on the Interonguing of the Kayenta and Navajo Formations, Northern Arizona. *In:* Brookfield, M.E., and Ahlbrandt, T.S., eds., *Eolian Sediments and Processes.* Amsterdam: Elsevier Science Publishers, pp. 613–634.

Middleton, L.T., and R. Hereford. 1981. Nature and Controls on Early Paleozoic Fluvial Sedimentation Along a Passive Continental Margin: Examples from the Middle Cambrian Flathead Sandstone (Wyoming) and Tapeats Sandstone (Arizona). International Association of Sedimentologists Special Congress, Keele, United Kingdom, Modern and Ancient Fluvial Systems. *Sedimentology and Processes,* p. 83.

Middleton, L.T., J.R. Steidtmann, and D. DeBoer. 1980. Stratigraphy and Depositional Setting of Some Middle and Upper Cambrian Rocks, Wyoming. *Wyoming Geological Association,* 32nd Annual Field Conference, pp. 23–35.

Miller, A.K., and A.G. Unklesbay. 1942. Permian Nautiloids from Western U.S. *Journal of Paleontology* 16(6), pp. 719–738.

Miller, H., and W. Breed. 1964. *Metococeras Bowmani,* A New Species of Nautiloid from the Toroweap Formation (Permian) of Arizona. *Journal of Paleontology,* vol. 38, pp. 877–880.

Moores, E.M. 1991. Southwest U.S.-East/Antartic (SWEAT) Connection: A Hypothesis. *Geology,* vol. 19, pp. 425–428.

Mullens, R. 1967. Stratigraphy and Environment of the Toroweap Formation (Permian) North of Ashfork, Arizona. M.S. thesis, University of Arizona, Tucson, 101 pp.

Naeser, C.W., et al. 1989. Fission-Track Dating, Ages for Cambrian Strata and Laramide and Psot-Middle Eocene Cooling Events from the Grand Canyon, Arizona. *In:* Eltson, D.P., G.H. Billingsley, and R.A. Young, eds., *Geology of Grand Canyon, Northern Arizona.* American Geophysical Union, International Geological Congress, 28th Guidebook T115/315. pp. 139–144.

National Research Council. 1987. *River and Dam Management:* A Review of the Bureau of Reclamation's Glen Canyon Environmental Studies. Washington, DC: National Academy Press, 203 pp.

Newell, N.D. 1948. Key Permian Section, Confusion Range, Western Utah. *Geological Society of America Bulletin,* vol. 59(10), 1053–1058.

Nicol, D. 1944. Paleoecology of Three Faunules in the Permian Kaibab Formation at Flagstaff, Arizona. *Journal of Paleontology,* vol. 18 (6), pp. 553–557.

Nielson, R.L. 1981. *Stratigraphy and Depositional Environments of the Toroweap and Kaibab Formations, Southwestern Utah,* Ph.D. dissertation, University of Utah, Salt Lake City, 1015 pp.

Nitecki, M.H. 1971. Pseudo-organic Structures From the Precambrian Bass Limestone in Arizona. *Geology,* vol. 23 (1), pp. 1–9.

Noble, L.F. 1914. The Shinumo Quadrangle, Grand Canyon District, Arizona. *U.S. Geological Survey Bulletin,* vol. 549, 100 pp.

Noble, L.F. 1922. A Section of Paleozoic Formations of the Grand Canyon at the Bass Trail. *U.S. Geological Survey Professional Paper* 131-B, pp. 23–73.

Noble, L.F. 1928. A Section of the Kaibab Limestone in Kaibab Gulch, Utah, *U.S. Geological Survey Professional Paper* 150, pp. 41–60.

Noble, L.F., and J.F. Hunter. 1916. Reconnaissance of the Archean Complex of the Granite Gorge, Grand Canyon, Arizona. *U.S. Geological Survey Professional Paper* 98-I, pp. 95–113.

Nyman, M.W., K.E. Karlstrom, E. Kirby, and C. Graubard. 1994. 1.4 Contractional Orogeny in Western North America: Evidence from Ca. 1.4 Ga Plutons. *Geology,* vol. 22, p. 901–904.

O'Conner, J.E., R.H. Webb, and V.R. Baker. 1986. Paleohydrology of Pool-and-Riffle Pattern Development: Boulder Creek, Utah. *Geological Society of America Bulletin,* vol. 98, pp. 410–420.

O'Conner, J.E., L.L. Ely, E.E. Whol, L.E. Stevens, T.S. Melis, S.K. Vishwas, and V.R. Baker. 1994. A 4500-Year Record of Large Floods on the Colorado River in the Grand Canyon, Arizona. *Journal of Geology*, vol. 102, pp. 1–9.

Ossian, C.R. 1976. Redescription of *Megactenopetalus kaibabanus* David 1944 (Chnodrichthyes: petalodontidae) with Comments on its Geographic and Stratigraphic Distribution. *Journal of Paleontology*, vol. 50 (3), pp. 392–397.

Osterkamp, W.R., C.R. Hupp, and J.C. Blodgett. 1986. Magnitude and Frequency of Debris Flows, and Areas of Hazard on Mount Shasta, Northern California. *U.S. Geological Survey Professional Paper* 1396-C, 21 p.

Pasteels, P., and L.T. Silver. 1965. Geochronological Investigations in the Crystalline Rocks of the Grand Canyon, Arizona [Abstract]. *Geological Society of America Annual Meeting Program*, Kansas City, 122 pp.

Peirce, H.W. 1984. The Mogollon Escarpment. *Arizona Bureau of Geology and Mineral Technology Fieldnotes*, vol. 14 (2), pp. 8–11.

Peirce, H.W., and J.D. Nations. 1986. Tectonic and Paleogeographic Significance of Tertiary rocks of the Southern Colorado Plateau and Transition Zone. *In:* Nationals, J.D., C.M. Conway, and G.A. Swann, eds., *Geology of Central and Northern Arizona*. Geological Society of America, Rocky Mountain Section, Field Trip Guidebook, pp. 159–176.

Peirce, H.W. M. Shafiqullah, and P.E. Damon. 1979. An Oligocene (?) Colorado Plateau Edge in Arizona. *Tectonophysics*, vol. 61, pp. 1–24.

Peterson, F. 1986. Jurassic Paleotectonics in the West-Central Part of the Colorado Plateau, Utah and Arizona. *In:* Peterson, J.A., ed., *Paleotectonics and Sedimentation in the Rocky Mountain Region, United States. American Association of Petroleum Geologists Memoir* 41, pp. 563–596.

Peterson, F. 1988. Pennsylvanian to Jurassic Eolian Transportation Systems in the Western United States, *Sedimentary Geology*, vol. 56 (1–4), pp. 207–260.

Péwé, T.L. 1968. Colorado River Guidebook: *A Geologic and Geographic Guide from Lee's Ferry to Phantom Ranch.* Tempe, Arizona, privately published, 78 pp.

Pierson, T.C. 1985. Initiation and Flow Behavior of the 1980 Pine Creek and Muddy River Lahars, Mount St. Helens, Washington: *Geological Society of America Bulletin*, vol. 96, p.1056–1069.

Pierson, T.C., and J.E. Costa. 1987. A Rheologic Classification of Subaerial Sediment-Water Flows: *Reviews in Engineering Geology*, vol. 7, pp. 1–12.

Podolny, W., Jr., and J.D. Cooper. 1974. Toward an Understanding of Earthquakes. *Highway Focus*, vol. 6 (1), 108 pp.

Potochnik, A.R., and S.J. Reynolds. 1986. Geology of Side Canyons of the Colorado, Grand Canyon National Park. Arizona Bureau of Geology and Mineral Technology. *Fieldnotes*, vol. 16 (1), pp. 1–8.

Powell, J.W. 1876. *Exploration of the Colorado River of the West*. Washington DC:, Smithsonian Institution, 291 pp.

Powell, J.W. 1895. Canyons of the Colorado: Flood and Vincent. Reprinted as *The Exploration of the Colorado River and Its Canyons*. New York: Dover, 400 pp.

Pratt, B.R., and N.P. James. 1986. The St. George Group (lower Ordovician) of Western Newfoundland: Tidal Flat Island Model for Carbonate Sedimentation in Shallow Epeiric Seas. *Sedimentology*, vol. 33, pp. 313–343.

Racey, J.S. 1974. Conodont Biostratigraphy of the Redwall Limestone at East-central Arizona. M.S. Thesis, Arizona State University, Tempe, 199 pp.

Ragan, D.M., and M.F. Sheridan. 1970. The Archean Rocks of the Grand Canyon, Arizona. Geological Society of America. *Abstracts with Programs*, part 2, pp. 132–133.

Ransome, F.L. 1904. The Geology and Ore Deposits of the Bisbee Quadrangle, Arizona. *U.S. Geological Survey Professional Paper* 21, 168 pp.

Rawson, R.R., and C.E. Turner-Peterson. 1974. The Toroweap Formation—A New Look. *In:* Swann, G., T. Karlstrom, and R. Eastwood, eds., *Guidebook to the Geology of Northern Arizona,* part I. Flagstaff Publishing Company, Flagstaff, Arizona, pp. 155–190.

Rawson, R.R., and C.E. Turner-Peterson. 1979. Marine-Carbonate, Sabkha, and Eolian Facies Transitions Within the Permian Toroweap Formation, Northern Arizona. *In:* Baars, D.L., ed., *Permianland, Four Corners Geological Society Guidebook*, 9th Field Conference, pp. 87–99.

Rawson, R.R., and C.E. Turner-Peterson. 1980. Paleogeography of Northern Arizona During the Deposition of the Permian Toroweap Formation. *In:* Fouch, T.D., and E.R. Magathan, eds., *Paleozoic Paleogeography of West Central United States*. Rocky Mountain Section, Society of Economic Paleontologists and Mineralogists, pp. 341–352.

Read, C.B., and A.A. Wanek. 1961. Stratigraphy of Outcropping Permian Rocks in Parts of Northeastern Arizona and Adjacent Areas. *U.S. Geological Survey Professional Paper* 374-H, pp. H1–H10.

Reches, Z. 1978a. Analysis of Faulting in Three Dimensional Strain Field. *Tectonophysics*, vol. 47, pp. 109–129.

Reches, Z. 1978b. Development of Monoclines, part I, Structure of the Palisades Creek Branch of the East Kaibab Monocline, Grand Canyon, Arizona. *Geological Society of America Memoir*, vol. 151, pp. 235–271.

Reches, Z., and A.M. Johnson. 1978. Development of Monoclines, part II, Theoretical Analysis of Monoclines. *Geological Society of America Memoir*, vol. 151, pp. 273–311.

Reed, V.S. 1974. Stratigraphy of the Hakatai Shale, Grand Canyon, Arizona [Abstract]. Geological Society of America. *Abstracts with Programs*, vol. 6 (5), p. 469.

Reeside, J.B., and H. Bassler. 1922. Stratigraphic Sections in Southwestern Utah and Northwestern Arizona, *U.S. Geological Survey Professional Paper*, vol. 129, pp. 53–77.

Reiche, P. 1938. An Analysis of Cross-lamination—The Coconino Sandstone. *Journal of Geology*, vol. 46, pp. 905–932.

Reineck, H.E., and F. Wunderlich. 1968. Classification and Origin of Flaser and Lenticular Bedding. *Sedimentology*, vol. 11, pp. 189–228.

Reynolds, S.J. 1988. Geologic Map of Arizona. *Arizona Geological Survey*, map 26, scale 1:1,000,000.

Reynolds, M.W., and D.P. Elston. 1986. Stratigraphy and Sedimentation of Part of the Proterozoic Chuar Group, Grand Canyon, Arizona. Geological Society of America. *Abstracts with Programs* (Rocky Mountain section), vol. 18, p. 405.

Rigby, J.K. 1976. Some Observations on Occurrences of Cambrian *Porifera* in Western North America and Their Evolution. *In:* Robinson, R.A., and A.J. Rowell, eds., *Paleontology and Depositional Environments: Cambrian of Western North America. Brigham Young University Geological Studies*, vol. 23, pp. 51–60.

Ritter, S.M. 1983. Conodont Biostratigraphy of Devonian–Pennsylvanian Rocks of Iceberg Ridge, Mojave County, Northwest Arizona. M.S. thesis, Brigham Young University, Provo, 54 pp.

Robinson, K., 1994. Metamorphic Petrology of the Lower Granite Gorge and the Significnace of the Gneiss Canyon Shear Zone, Grand Canyon. M.S. thesis, Amherst, University of Massachusetts, 197 pp.

Rodine, J.D., and A.R. Johnson. 1976. The Ability of Debris, Heavily Freighted with Coarse Clastic Materials, to Flow on Gentle Slopes: *Sedimentology*, vol. 23, pp. 213–234.

Rogers, J.D., and M.R. Pyles. 1979. Evidence of Catastrophic Erosional Events in the Grand Canyon of the Colorado River, Arizona. 2nd Conference on Scientific Research in the National Parks. 26–30 November 1979, San Francisco, California.

Rose, E., L. Middleton, and D. Elliott. 1998. Storm- and Fair-Weather Controls of Deposition of the Middle Cambrian Bright Angel Shale, Grand Canyon Arizona: Sedimentologic and Ichnologic Evidence. *Geological Society of America 50th. Annual Meeting*, Rocky Mountain Section, vol. 30, pp. 35.

Ross, G.M., M.E. McMechan, and F.J. Hein. 1989. Proterozoic History; The Birth of the Miogeoclin. *In:* Rickets, B.D. ed., *Western Canada Sedimentary Basin: A Case History*. Calgary: Canadian Society of Petroleum Geologists, pp. 79–99.

Rouse, H., B.V. Bhoota, and E. Hsu. 1951. Design of Channel Expansions. *Transactions. American Society of Civil Engineers*, vol. 116, pp. 347–363.

Rowley, P.D., T.A. Steven, and H.H. Mahnert. 1981. Origin and Structural Implications of Upper Miocene Rhyolites in Kingston Canyon, Piute County, Utah. *Geological Society of America Bulletin*, vol. 92 pp. 590–602.

Rubin, D.M., and R.E. Hunter. 1982. Bedform Climbing in Theory and Nature. *Sedimentology*, vol. 29, pp.121–138.

Rubin, D.M., and R.E. Hunter. 1983. Reconstructing Bedform Assemblages from Compound Crossbedding. *In:* Brookfield, M.E., and T.S. Ahlbrandts, eds., *Eolian Sediments and Processes*. Amsterdam: Elsevier Science Publishers, pp. 407–427.

Rubin, D.M., and R.E. Hunter. 1985. Why Deposits of Longitudinal Dunes Are Rarely Recognized in the Geologic Record. *Sedimentology*, vol. 32, pp. 147–157.

Rubin, D.M., J.C. Schmidt, and J.N. Moore. 1990. Origin, Structure, and Evolution of a Reattchment Bar, Colorado River, Grand Canyon, Arizona: *Journal of Sedimentary Petrology*, vol. 60, p. 982–991.

Sadler, C.J. 1993. Arthropod Trace Fossils from the Permian De Chelly Sandstone, Northeastern Arizona. *Journal of Paleontology*, vol. 67, pp. 240–249.

Scarborough, R.B. 1989. Cenozoic Erosion and Sedimentation in Arizona. *In:* Jenney, J.P., and S.J. Reynolds, eds., *Geologic Evolution of Arizona. Arizona Geological Society Digest*, vol. 17.

Scarborough, R.B., C.M. Menges, and P.A. Pearthree. 1986. *Map of Late Pliocene-Quaternary (post-4-my) Faults, Folds, and Volcanic Outcrops in Arizona*. Map 22. Arizona Bureau of Geology and Mineral Technology.

Schmidt, J.C. 1990. Recirculating Flow and Sedimentation in the Colorado River in Grand Canyon, Arizona. *Journal of Geology*, vol. 98, pp. 709–724.

Schmidt, J.C., and J.B. Graf. 1987. Aggradation and Degradation of Alluvial-Sand Deposits, 1965–1986, Colorado River, Grand Canyon National Park, Arizona. *U.S. Geological Survey Open-File Report* 87-555, approximately 100 pp.

Schmidt, J.C., and J.B. Graf. 1990. Recirculating Flow and Sedimentation in the Colorado River in Grand Canyon, Arizona. *Journal of Geology*, vol. 98, 74 pp.

Schopf, W.J., T.D. Ford, and W.J. Breed. 1973. Microorganisms from the Late Precambrian of the Grand Canyon, Arizona. *Science*, vol. 179, pp. 1319–1321.

Schreiber, B.C. 1986. Arid Shorelines and Evaporites. *In:* Reading, H.G., ed., *Sedimentary Environments and Facies*. London: Blackwell Scientific Publications, pp. 189–228.

Schumm, S.A., and R.J. Chorley. 1966. Talus Weathering and Scarp Recession in the Colorado Plateau. *Zeitschrift fur Geomorphologie*, vol. 10, pp. 11–36.

Schuster, R.L., and J.E. Costa. 1986. A Perspective on Landslide Dams. *In:* Schuster, R.L., ed., *Landslide Dams: Processes, Risk, and Mitigation, Proceedings*, Geotechnical Engineering Division of the American Society of Civil Engineers, Geotechnical Special Publication No. 3. New York: American Society of Civil Engineers, pp. 1–20.

Seaman, S.J., Karlstrom, K.E., Williams, M.L., and Petruski, A.J. 1997. Proterozoic Ultramafic Bodies in the Grand Canyon. Geological Society of America, Abstracts with Programs, vol. 29 (6), p. A-89.

Sears, J.W. 1973. Structural Geology of the Precambrian Grand Canyon Series. M.S. thesis, University of Wyoming, Laramie, 100 pp.

Sears, J.R., and Price, R.A. 1978. The Siberian Connection: A Case for the Precambrian Separation of the Siberian and North American Cratons. *Geology*, vol. 6, pp. 267–270.

Seilacher, A. 1970. Cruziana Stratigraphy of Non-fossiliferous Paleozoic Sandstones. *In:* Crimes, T.P., and J.C. Harper, eds., *Trace Fossils. Geological Journal* Special Issue, pp. 447–476.

Sepkoski, J.J. 1982. Flat-Pebble Conglomerates, Storm Deposits, and the Cambrian Bottom Fauna. *In:* Einsele, G., and A. Seilacher, eds., *Cyclic and Event Stratification.* New York: Springer-Verlag, pp. 371–385.

Sharp, R.P. 1940. Ep-Archean and Ep-Algonkian Erosion Surfaces, Grand Canyon, Arizona. *Geological Society of America Bulletin,* vol. 51, pp. 1235–1270.

Sharp, R.P., and L.H. Nobles. 1953. Mudflow of 1941 at Wrightwood, Southern California: *Geological Society of America,* vol. 64, p. 547–560.

Shinn, E.A. 1983. Tidal Flat Environment. *In:* Scholle, P.A., D.G. Debout, and C.H. Moore, eds., Carbonate Depositional Environments. *American Association of Petroleum Geologists Memoir,* vol. 33, pp. 171–210.

Shinn, W.A., R.N. Ginsburg, and R.M. Lloyd. 1965. Recent Supratidal Dolomite from Andros Island, Bahamas. *In: Dolomitization and Limestone Diagenesis—A Symposium. Society of Economic Paleontologists and Mineralogists Special Publication,* vol. 13, pp. 89–111.

Shirley, D.H. 1987. Geochemical Facies Analysis of the Surprise Canyon Formation in Fern Glen Channelway, Central Grand Canyon, Arizona. M.S. thesis, Northern Arizona University, Flagstaff, 208 pp.

Shoemaker, E.M., R.L. Squires, and M.J. Abrams. 1974. The Bright Angel and Mesa Butte Fault Systems of Northern Arizona. *In:* Karlstrom, T.N.V., G.A. Swann, and R.L. Eastwood, eds., *Geology of Northern Arizona,* part 1: *Regional Studies.* Rocky Mountain Section, Geological Society of America Annual Meeting, Flagstaff, pp. 255–291.

Shoemaker, E.M., R.L. Squires, and M.J. Abrams. 1975. The Bright Angel, Mesa Butte, and Related Fault Systems of Northern Arizona. *California Institute of Technology Jet Propulsion Laboratory Technical Report* 32-1597: 23–41. *Also in:* 1978. Bright Angel and Mesa Butte Fault Systems of Northern Arizona. *In:* Smith, R.B. and G.P. Eaton, eds., *Cenozoic Tectonics and Regional Geophysics of the Western Cordillera. Geological Society of America Memoir,* vol. 152, pp. 341–367.

Shride, A.F. 1967. Younger Precambrian Geology in Southern Arizona. *U.S. Geological Survey, Professional Paper* 566, 89 pp.

Shuey, R.T., D.K. Schellinger, E.G. Johnson, and L.B. Allen. 1973. Aeromagnetics and the Transition Between the Colorado Plateau and Basin Range Provinces. *Geology,* vol. 1, pp. 107–110.

Skipp, B. 1969. Foraminifera. *In:* McKee, E.D. and R.C. Gutschick, eds., *History of the Redwall Limestone of Northern Arizona. Geological Society of America Memoir,* vol. 114, pp. 173–195.

Skipp, B., 1979. Great Basin Region, Paleotectonic Investigations of the Mississippian System in the United States. *U.S. Geological Survey, Professional Paper* 1010, Part 1, Chapter P, pp. 273–382.

Smith, D.L., and C.G. Crampton. 1987. *The Colorado River Survey.* Salt Lake City: Howe Brothers, 305 pp.

Smith, P.B. 1970. New Evidence for Pliocene Marine Embayment Along the Lower Colorado River Area, California and Arizona. *Geological Society of American Bulletin,* vol. 81, pp. 1411–1420.

Smith, R.B. 1978. Seismicity, Crustal Structure, and Intraplate Tectonics of the Interior of the Western Cordillera. *In:* Smith, R.B., and G.P. Eaton, eds., Cenozoic Tectonics and Regional Geophysics of the Western Cordillera. *Geological Society of America Memoir,* vol. 152, pp. 111–144.

Smith, R.B., and M.L. Sbar. 1974. Contemporary Tectonics and Seismicity of the Western United States with Emphasis on the Intermountain Seismic Belt. *Geological Society of America Bulletin,* vol. 85, pp. 1205–1218.

Snow, J.I. 1945. Trilobites of the Middle Permian Kaibab Formation of Northern Arizona. *Plateau,* vol. 18 (2): 17–24.

Snyder, W.S., W.R. Dickinson, and M.L. Silberman. 1976. Tetonic Implications of Space–Time Patterns of Cenozoic Magmatism in the Western United States. *Earth and Planetary Science Letters,* vol. 32, pp. 91–106.

Sorauf, J.E. 1962. Structural Geology and Stratigraphy of the Whitmore Area, Mohave County, Arizona. Unpublished Ph.D. thesis, University of Kansas, 361 pp.

Sorauf, J.E., and G.H. Billingsley. 1991. Members of the Toroweap and Kaibab Formations, Lower Permian, Northern Arizona and Southwestern Utah. *The Mountain Geologist,* vol. 28 (1), pp. 9–24.

Spamer, E.E. 1984. Paleontology in the Grand Canyon of Arizona: 125 Years of Lessons and Enigmas from the Past to the Present. *The Mosasaur,* vol. 2, pp. 45–118.

Stevens, L. 1985. The 67 Elephant Theory or Learning to Boat Big Water Hydraulics. *River Runner,* vol. 5 (1), pp. 24–25.

Stevens, L., 1990. *The Colorado River in Grand Canyon, a Guide:* Flagstaff, Arizona, Red Lake Books, 110 pp.

Stevenson, G.M. 1973. Stratigraphy of the Dox Formation, Grand Canyon, Arizona. M.S. thesis, Northern Arizona University, Flagstaff, 225 pp.

Stevenson, G.M., and S.S. Beus. 1982. Stratigraphy and Depositional Setting of the Upper Precambrian Dox Formation in Grand Canyon. *Geological Society of America Bulletin,* vol. 93, pp. 163–173.

Stewart, J.H. 1972. Initial Deposits in the Cordilleran Geosyncline: Evidence of a Late Precambrian Continental Separation. *Geological Society of America Bulletin,* vol. 83, pp. 1345–1360.

Stewart, J.H., F.G. Poole, and R.F. Wilson. 1972. Stratigraphy and Origin of the Chinle Formation and Related Upper Triassic Strata in the Colorado Plateau Region. *U.S. Geological Survey Professional Paper* 690, 336 pp.

Stewart, J.H., and C.A. Suczek. 1977. Cambrian and Latest Precambrian Paleogeography and Tectonics in the Western United States. *In:* Stewart, J.H., C.H. Stevens, and A.E. Fritsche, eds., Paleozoic–Paleogeography of the Western United States. *Society of Economic Paleontologists and Mineralogists, Pacific Section, Paleogeography Symposium* 1, pp. 1–18.

Stokes, W.L. 1968. Multiple parallel-truncation Bedding Planes—A Feature of Wind-deposited Sandstone Formations. *Journal of Sedimentary Petrology,* vol. 38, pp. 510–515.

Stokes, W.L. 1986. *Geology of Utah.* Occasional Paper No. 6. Salt Lake City: Museum of Natural History, 310 pp.

Strahler, A.N. 1940. Landslides of the Vermilion and Echo Cliffs. *Journal of Geomorphology,* vol. 3, pp. 285–300.

Strahler, A.N. 1948. Geomorphology and Structure of the West Kaibab Fault Zone and Kaibab Plateau. *Geological Society of America Bulletin*, vol. 59, pp. 513–540.

Strand, R.I. 1986. Water Related Sediment Problems. Water Systems Management Workshop—1986, Session 4–1, U.S. Bureau of Reclamation, Denver, Colorado, pp. 1–30.

Sturgul, J.R. and Z. Grinshpan. 1975. Finite Element Model for Possible Isostatic Rebound in the Grand Canyon. *Geology*, vol. 3, pp. 169–171.

Sturgul, J.R., and T.D. Irwin. 1971. Earthquake History of Arizona. *Arizona Geological Society Digest*, vol. 9, pp. 1–23.

Talbot, M.R. 1985. Major Bounding Surfaces in Aeolian Sandstones—A Climatic Model. *Sedimentology*, vol. 32, pp. 257–265.

Teichert, C. 1965. Devonian Rocks and Paleogeography of Central Arizona. *U.S. Geological Survey Professional Paper* 464, 181 pp.

Thompson, K.L. 1995. Paleoecology and Biostratigraphy of the Fossil Mountain Member, Kaibab Formation, in Northwestern Arizona. Unpublished M.S. thesis, Northern Arizona University, 160 pp.

Thompson, P.A. 1972. *Compressible-Fluid Dynamics*. New York: McGraw-Hill, 665 pp.

Tidwell, W.D., J.R. Jennings, and S.S. Beus. 1992. A Carboniferous Flora from the Surprise Canyon Formation from the Grand Canyon, Arizona. *Journal of Paleontology*, vol. 66, pp. 1013–1021.

Timmons, J.M. 1998. Unpublished data. Masters Thesis in progress, University of New Mexico, Albuquerque.

Timmons, J.M., K.E. Karlstrom, and C.M. Delher. 1998. Structure and Sedimentary Tectonics of the Chuar Basin and Butte Fault, Grand Canyon: Evidence for a Neoproterozoic 'Growth' Fault. Geological Society of America. *Abstracts with Programs* (Rocky Mountain Section), vol. 30 (6), p. 38.

Trask, P.D. 1959. Effect of Grain Size on Strenght of Mixtures of Clay, Sand, and Water. *Bulletin of the Geological Society of America*, vol. 70, pp. 569–579.

Turner, C.E. 1974. Facies of the Toroweap Formation, Marble Canyon, Arizona. M.S. thesis, Northern Arizona University, Flagstaff, 120 pp.

Turnet, R.M., and M.M. Karpiscak. 1980. Recent Vegetation Changes Along the Colorado River Between Glen Canyon Dam and Lake Mead, Arizona: *U.S. Geological Survey Professional Paper* 1132, 125 pp.

Van Eysinga, F.S.B. 1975. *Geological Time Table*. Amsterdam: Elsevier Science Publishers.

Van Gundy, C.E. 1937. Jellyfish from the Grand Canyon Algonkian. *Science*, vol. 85, p. 314.

Van Gundy, C.E. 1951. Nankoweap Group of the Grand Canyon Algonkian of Arizona. *Geological Society of America Bulletin*, vol. 62, pp. 953–959.

Van Schumus, W.R., 1992. Tectonic Setting of the Midcontinent Rift System: Tectonophysics, vol. 213, p. 1–15.

Vidal, G. 1981. Aspects of Problematic Acid-Resistant, Organic-Walled Microfossils (Acritarchs) in the Proterozoic of the North Atlantic Region. *Precambrian Research*, vol. 15, pp. 9–23.

Vidal, G., and T.D. Ford. 1985. Microbiotas From the Late Proterozoic Chuar Group (Northern Arizona) and Uinta Mountain Group (Utah) and Their Chronostratigraphic Implications. *Precambrian Research*, vol. 28, pp. 349–389.

Walcott, C.D. 1880. The Permian and Other Paleozoic Groups of the Kanab Valley, Arizona. *American Journal of Science*, 3rd Series, vol. 20, pp. 221–225.

Walcott, C.D. 1883. Pre-Carboniferous Strata in the Grand Canyon of the Colorado, Arizona. *American Journal of Science*, 3rd Series, vol. 16, pp. 437–442; 484.

Walcott, C.D. 1889. A Study of a Line of Displacement in the Grand Canyon of the Colorado in Northern Arizona. *Geological Society of America Bulletin*, vol. 1, pp. 49–64.

Walcott, C.D. 1890. The Fauna of the Lower Cambrian or *Olenellus* Zone. *U.S. Geological Survey, 10th Annual Report*, pp. 509–760.

Walcott, C.D. 1894. Precambrian Igneous Rocks of the Unkar Terrane, Grand Canyon of the Colorado. *U.S. Geological Survey 14th Annual Report for 1892/3*, part 2, pp. 497–519.

Walcott, C.D. 1899. Precambrian Fossiliferous Formations. *Geological Society of America Bulletin*, vol. 10, pp. 199–244.

Walcott, C.D. 1910. Cambrian Geology and Paleontology II; Abrupt Appearance of the Cambrian Fauna on the North American Continent. *Smithsonian Miscellaneous Collection*, vol. 57, p.14.

Walcott, C.D. 1918. Cambrion Geology and Paleontology IV; Appendages of Trilobites. Smithsonian Miscellanious Collections, Vol. 67, No. 4, p. 115–216.

Walcott, C.D. 1920. Cambrian Geology and Paleontology, IV; Middle Cambrian Spongiae. *Smithsonian Miscellaneous Collection*, vol. 67, pp. 261–364.

Walen, M.B. 1973. Petrogenesis of the Granitic Rocks of Part of the Upper Granite Gorge, Grand Canyon, Arizona. M.S. thesis, Western Wshington University, Bellingham.

Wanless, H.R. 1973. Cambrian of the Grand Canyon: A Reevaluation of the Depositional Environment. Ph.D. dissertation, Johns Hopkins University, Baltimore, 128 pp.

Wanless, H.R. 1975. Carbonate Tital Flats of the Grand Canyon Cambrian. *In:* Ginsburg, R.N., ed., *Tidal Deposits: A Casebook of Recent Examples and Fossil Counterparts*. New York: Springer-Verlag, pp. 269–277.

Wardlaw, B.R. 1986. Paleontology and Deposition of the Phosphoria Formation. *In:* Boyd, D.W., and J.A. Liiegraven, eds., *Contributions to Geology*, Phosphoria Issue, University of Wyoming, vol. 24, pp. 107–142.

Wardlaw, B.R., and J.W. Collinson. 1978. Stratigraphic Relations of the Park City Group (Permian) in Eastern Nevada and Western Utah, *American Association of Petroleum Geologists Bulletin*, vol. 62 (7), pp. 1171–1184.

Webb, R.H. 1996. Grand Canyon, A Century of Change: *Rephotography of the 1889–1890 Stanton Expedition*. Tucson: University of Arizona Press, 290 pp.

Webb, R.H., P.T. Pringle, and G.R. Rink. 1987. Debris Flows from Tributaries of the Colorado River in Grand Canyon National Park, Arizona. *U.S. Geological Survey Open-File Report* 87–118, 64 pp.

Webb, R.H., P.T. Pringle, S.L. Reneau, and G.R. Rink. 1988. Monument Creek Debris Flow, 1984: Implications for Formation of Rapids on the Colorado River in Grand Canyon National Park. *Geology*, vol. 15, pp. 50–54.

Webb, R.H., P.T. Pringle, and G.R. Rink. 1989. Debris Flows from Tributaries of the Colordo River in Grand Canyon National Park, Arizona. *U.S. Geological Survey Professional Paper* 1492, 39 pp.

Webb, R.H., T.S. Melis, T.W. Wise, and J.G. Elliott. 1996. "The Great Cataract," Effects of Late Holocene Debris Flows on Lava Falls Rapid, Grand Canyon National Park and Hualapai Indian Reservation, Arizona. *U.S. Geological Survey Open-File Report* 96-460, 96 pp.

Webb, R.H., T.S. Melis, P.G. Griffens, J.G. Elliott, T.E. Cerling, R.J. Poreda, T.W. Wise and J.E. Pizzuto. 1997. Lava Falls Rapid in Grand Canyon, Effects of Late Holocene Debris Flows on the Colorado River. *U.S. Geological Survey Professional Paper* 1591, 90 pp.

Webb, R.H., T.S. Melis, P.G. Griffens, J.G. Elliott. 1997a. Reworking of Aggraded Debris Fans by the 1996 Controlled Flood on the Colorado River in Grand Canyon National Park, Arizona. *U.S. Geological Survey Open-File Report* 97–16, 36 pp.

Webster, G.D. 1969. *Chester Through Derry Conodonts and Stratigraphy of Northern Clark on Southern Lincoln Counties, Nevada.* University of California Publications in Geological Services, vol. 79, 119 pp.

Webster, G.D. 1984. Conodont Zonations Near the Mississipian–Pennsylvanian Boundary in Eastern Great Basin. *In:* Lintz, J., Jr., ed., *Western Geology Excursions (Geological Society of America Annual Meeting Guidebook)*. Department of Geological Sciences, Mackay School of Mines, Reno, Nevada, vol. 1, pp. 78–82.

Wells, W.G., II. 1987. The Effects of Fire on the Generation of Debris Flows in Southern California. *Reviews in Engineering Geology*, vol. 7.

Welsh, J.E., W.L. Stokes, and B.R. Wardlaw. 1979. Regional Stratigraphic Relationships of the Permian "Kaibab" or Black Box Dolomite of the Emery High, Central Utah. *In:* Baars, D.L., ed., *Permianland—A Field Symposium. Guidebook of the Four Corners Geological Society*, pp. 143–149.

Wenrich, K.J., G.H. Billingsley, and P.W. Huntoon. 1986. Breccia Pipe and Geologic Map of the Northeastern Hualapai Indian Reservation and Vicinity, Arizona. *U.S. Geological Survey Open-File Report* 86-458C, 29 pp.

Wenz. W. 1938. Gastropoda. *In:* O.H. Schindewolf, ed., *Handbuch der Paläozoologie,* vol. 6, no.1, 240 pp., Berlin Borntraeger.

Whipple, K.X., and T. Dunne. 1992. The Influence of Debris-Flow Rheology on Fan Morphology, Owens Valley, California. *Geological Society of America Bulletin*, vol. 104, p. 887–900.

White, D. 1929. Flora of the Hermit Shale, Grand Canyon, Arizona. Washington, DC: Carnegie Institute, *Publication 405*, pp. 1–221.

Wieczorek, G.F., E.W. Lips, and S.D. Ellen. 1989. Debris Flows and Hyperconcentrated Floods Along the Wasatch Front, Utah, 1983 and 1984: *Bulletin of the Association of Engineering Geologists*, vol. 236, p. 191–208.

Williams, G.P., and H.P. Guy. 1971. Debris Avalanches—A Geomorphic Hazard, *In:* Coates, D., ed., *Environmental Geomorphology.* Binghamton, State University of New York, pp. 25–46.

Williams, G.P., and M.G. Wolman. 1984. Downstream Effects of Dams on Alluvial Rivers. *U.S. Geological Survey Professional Paper* 1286, 83 pp.

Williams, M.L., and Karlstrom, K.E., 1996. Looping P-T Paths and High-t, Low-P Middle Crustal Metamorphism: Proterozoic evolution of the Southwestern United States: *Geology*, vol. 24, p. 1119–1122.

Wilson, E.D. 1962. Resume of the Geology of Arizona. *Arizona Bureau of Mines Bulletin* 171, 140 pp.

Wilson, E.D., R.T. Moore, and J.R. Cooper. 1969. Geologic Map of Arizona. Arizona Bureau of Mines. Scale 1:500,000.

Wilson, I.G. 1972a. Aeolian Bedforms: Their Development and Origins. *Sedimentology*, vol. 19, pp. 173–210.

Wilson, I.G. 1972b. Ergs. *Sedimentary Geology*, vol. 10, pp. 77–106.

Wilson, R.F. 1974. Mesozoic Stratigraphy of Northeastern Arizona. *In:* Karlstrom, T., G. Swann, and R. Eastwood, eds., *Geology of Northern Arizona.* Flagstaff: Geological Society of America, Rocky Mountain Section Meeting.

Wilson, R.C., and G.F. Wieczorek. 1995. Rainfall Thresholds for the Initiation of Debris Flows at La Honda, California: *Environmental and Engineering Geoscience*, vol. I, p. 11–27.

Wohl, E.E., and P.P. Pearthree. 1991. Debris Flows as Geomorphic Agents in the Huachuca Mountains of Southeastern Arizona. *Geomorphology*, vol. 4, p. 273–292.

Wood, W.H. 1956. The Cambrian and Devonian Carbonate Rocks at Yampai Cliffs, Mohave County, Arizona. Ph.D. dissertation, University of Arizona, Tucson, 228 pp.

Wooden, J.L., and Ed. Dewitt. 1991. Pb Isotopic Evidence for the Boundary Between the Early Proterozoic Mojave and Central Arizona Crustal Provinces in Western Arizona. *In:* Karlstrom, K.E., ed., *Early Proterozoic Geology and Ore Deposits of Arizona. Arizona Geological Society Digest*, vol. 19, pp. 27–50.

Woodward-Clyde Consultants. 1982. Geologic Characterization Report for the Paradox Basin Study Region, Utah Study Areas, vol. 1, Regional Overview. Battelle Memorial Institute Office of Nuclear Waste Isolation, ONWI-290.

Young, R.A. 1966. Cenozoic Geology Along the Edge of the Colorado Plateau in Northwestern Arizona. Ph.D. dissertation, Washington University, Washington, 167 pp.

Young, R.A. 1970. Geomorphological Implications of Pre-Colorado and Colorado Tributary Drainage in the Western Grand Canyon Region. *Plateau*, vol. 42, pp. 107–117.

Young, R.A. 1979. Laramide Deformation, Erosion and Plutonism Along the Southwestern Margin of the Colorado Plateau. *Tectonophysics*, vol. 61, pp. 25–47.

Young, R.A. 1989. Paleogene-Neogene Deposits of Western Grand Canyon, Arizona. *In:* Beaus, S., ed., *Centennial Field Guide*, Volume 2. Boulder: Rocky Mountain Section of The Geological Society of America.

Young, R.A. 1999. Nomenclature and Ages of Late Cretaceous(?)–Tertiary Strata in the Hualapi Plateau Region, Northwestern Arizona. *In:* Billingsley, G.H., K.J. Wenrich, P.W. Huntoon, and R.A. Young. *Breccia-Pipe and Geologic Map of the Southwestern part of the Hualapai Indian Reservation and Vicinity, Arizona.* U.S. Geological Survey Miscellaneous Investigations Series map I-2554, pp. 21–50.

Young, R.A., and W.J. Brennan. 1974. Peach Springs Tuff, Its Bearing on Structural Evolution of the Colorado Plateau and Development of Cenozoic Drainage in Mohave County, Arizona. *Geological Society of America Bulletin*, vol. 85, pp. 83–90.

Young, R.A., and E.D. McKee. 1978. Early and Middle Cenozoic Drainage and Erosion in West-Central Arizona. *Geological Society of America Bulletin*, vol. 89, pp. 1745–1750.

INDEX

accretion, of island arcs, 226–227
acritarchs, 68–70
Actinocoelia, 205–208
Adegnathus, 129
Adei Eechii Cliffs, 219
Adetognathus, 131–132
Agostopus, 170, 171
Albertella, 97
algal mats, 65, 122
Allopus, 171
alluvium: dating, 359–360; lower mesquite terrace, 357; post-dam, 360–361; pre-dam, 360; Pueblo II age, 357, 359–360; relative deposits, 355; striped, 357, 359–360; upper mesquite terrace, 357–359, 359–360
Alokistocare-Glossopheura, 95, 96
Amblyopus, 171
amphibolite facies, 19–20
Anasazi: evidence of, 352–353, 405
Angulichnus, 99–100
Antagmus, 97
anticline development, 255
Antler orogeny, 226
Apache Group, 51–52
aphanitic lime dolomite, 190–191
aphanitic lime mudstone, 190–191
arc plutons, 21
Arenicoloides, 99–100
Asterosoma, 56
Atokan Series, 161
Aubrey Limestone, 196
Awatubi Member, 57, 58, 64, 71

backwaters, 290
Baicalia, 62
Banded Mountain Member, 110
Baropezia, 171
Baropus, 170, 171
Barypodus, 167, 170, 171
basements, 18
Basin and Range Province, 242–244, 270
basins, 153
Bass Formation, 405
Bass Limestone, 42, 44
Bass Shear Zone, 12, 32–33
Bat Tower, 133
Bat Tower Section, 123
Bathyuriscus-Elrathina, 96
Beck Springs Dolomite, 74
Bellerophontid, 206
Belt Supergroup, 74
Bidahochi Formation, 216, 270
Bill's Cone, 342
Black Ledge Dam, 320
Black Mesa: area cross section, 220–221; physiographic map, 215; regional topography, 267; surrounding topography, 219–221
Blacktail Canyon, 403
Blue Springs Basin, 250
Bolaspis, 97
Bolsa Quartzite, 91
Bothreolepis, 112
bottle-green member, 49
Boucher Block, 15, 30–31

Boucher Pluton, 21, 22–23
Boucher Rapid, 385–386
Bouse Formation, 271, 273
Boxonia, 64
Brady Canyon, 180
Brady Canyon Member, 181, 184, 193, 194
Brahma Schist, 20, 15
Bright Angel Creek, 307, 394
Bright Angel Fault, 79–80, 251
Bright Angel Pluton, 22–23
Bright Angel Rapids, 283
Bright Angel Shale, 94–95, 98, 103–104, 256
Bright Angel Shear Zone, 13, 15, 30–31
Bright Angel system, 349–350
Brooksella canyonensis, 56
Buck Farm Canyon, 399–401
Buried Canyon Dam, 320, 330, 335–337
Butte Fault, 82–83, 401–402

Calamites, 129
Callipteris, 151
Cambrian deposits, 90–91
Cambrian-Devonian unconformity, 109–110
Carbon Butte, 258
Carbon Butte Member, 57, 58, 61, 62–63, 71
Carbon Canyon, 401–402
Carbon Canyon Member, 57, 58, 60–62, 71
carbonate lithofacies, 189
Cardenas Lava, 42, 43, 47, 48–49, 60
Carmel Formation, 216, 218, 221
Cataract Basin, 250
Cataract Creek, 269
Cataract Creek fault system, 350
Cavusgnathus, 129
Cedar Mesa, 219
Cedar Mesa Sandstone, 162
Cenozoic strata, 212–213.
Chainman Formation, 132
Chancelloria, 97
channels, 110, 125
chert, 202
Chinle Formation, 216, 218, 220, 228
Chocolate Cliffs, 215
Chuar Butte, 235
Chuar Canyon, 234
Chuar Group, 47, 53, 54, 57, 58, 66–71, 74–75, 82, 84–85, 88
Chuar Segment, 396–397
Chuar syncline, 82–83
Chuaria, 59, 64, 66–68, 69
cinder cones, 342–343
Claron Formation, 219
Clear Creek Block, 15, 30–31
Cock's Comb, 235
Coconino Plateau, 214–215, 219
Coconino Sandstone, 148, 162, 163–173, 227–228
Cogswell Butte, 257
colluvial wedges, 378–380
Colorado Plateau, 154, 244–245, 222–223, 266, 346–351
Colorado Plateau Physiographic Province, 4, 6, 212–213, 253, 257–258, 260–264, 266–274, 275, 276–279, 344, 354–356, 367, 372, 376, 386–388, 389–390, 391, 392
Comanche Point Member, 47, 48

Composita, 129
compositional zoning, 29
conodonts, 113–114
Conostichus, 127
contractional faults, 79–80
Cooperella, 207
Cordilleran miogeocline, 88
Corophioides, 99–100
Cottonwood Pegmatite Complex, 22–23
Cow Springs Sandstone Formation, 216, 221
Cremation Pegmatitie Complex, 22–23
Cruziana, 99–100, 103, 128
Crystal Creek, 378, 382
Crystal Pluton, 21, 22–23
Crystal Rapids, 290, 297–299, 304, 305–309
Crystal Shear Zone, 13, 30–31, 34–35

"D" Dam, 320, 340
Dakota Formation, 216, 218, 221
De Chelly Sandstone, 162, 169
debris fans, 285, 299–304, 310–312, 353, 362–367, 369–371, 388–389
debris flows, 372, 374–380, 382–383, 384–388
Defiance Arch, 154
Defiance Plateau, 137
Defiance Uplift, 227
deformations, 26–28, 29
deposition, 154, 354, 361, 384
Derbyia, 206
Des Moinesian Series, 161
dessication cracks, 191
Deubendorff Rapids, 276, 278
Diamond Canyon, 238, 246
Diamond Creek, 372–373
Diamond Creek Block, 19, 32–33
Diamond Creek Pluton, 17, 19, 21, 24–25
Diamond Peak, 248
dikes, 49–50, 253–254, 341–342
Diplichnites, 99–100
Diplocraterion, 99–100, 103
Diplodichnus, 167
Diplopodichnus, 168, 169
dissolution pits, 362, 368–371
Dolichopodus, 170, 171
dolomitization, 192–193
downcutting, 344, 389
Dox Formation, 42–43, 46–48, 58
drainage system development, 262–263
driftwood, 358
dunes, 152, 155, 174, 176–177, 355, 361–362
Duppa Member, 57, 58, 62, 71
Dutton, C. E., 6

earthquakes, 346–349
East Kaibab monocline, 233, 234, 235, 237–238
Echo Cliffs, 219
eddies, 296–297
Elbert Formation, 114
Elves Chasm Block, 14, 32–33
Elves Chasm Pluton, 18–19, 21, 24–25, 32–33
embayments, 153
Entrada Sandstone Formation, 216, 218, 221
Eocrinus, 97
eolian deposits, 151–152, 171
Eosigmoilina, 131
erosion, 299–304, 343–345, 364–367
Escabrosa Limestone, 122
Escalante Creek Member, 47
Esplanade Cascades, 314, 329, 341
Esplanade Dam, 320, 330, 334–335
Esplanade Sandstone, 145–147, 150, 155, 157, 162, 378

Esplanade Segment, 397
evaporite lithofacies, 185–187
extensional faults, 80–81
extensional sag, 251

Farallon Plate, 243
fault reactivation, 86
faults, 224–225, 237, 246–252, 255, 349–350
Fern Glen Valley, 134
flat-pebble conglomerates, 104–105
flood of 1983, 297–299
floods, 311, 352, 356–359
Flour Sack Member, 94
flow, lava. *See* lava flows
flow, water, 288, 290–293, 295, 296, 303, 309
Fort Apache Member, 151
Fossil Mountain Member, 197, 199, 201, 205–208, 209–210
Froude number, 288

Galeros Formation, 57, 59–62, 71
Garden Valley Formation, 209
Garnet Pegmatite Complex, 22–23
garnets, 29
geologic time scale, 7
Gilbert, G. K., 6
Glen Canyon Dam, 305–307, 389–390
Glen Canyon Group, 220
Glossopleura, 97
Gnathodus, 129
Gneiss Canyon Shear Zone, 16, 18, 34–35
gneisses, 18
Grand Canyon: cross sections, 4, 5, 220–221
    debris study sites, 353; description, 1–4;
    drainage-network map, 395; expeditions, 6;
    formations, geologic time; relation, 7; geologic
    column, 7; geologic map, regional, 394;
    inhabitation, 4, 6; maps of, 2–3, 4; monocline
    offset, 231; morphological divisions, 396–397;
    physiographic map, 214–215; profile equilibrium,
    345; rim elevations, 2–4, 5; rock strata, 391, 393;
    seismicity, regional, 346–351; slope retreat, 344;
    tectonics, 223, 229–230, 240, 242–254, 349–350;
    topography, regional, 269
Grand Canyon Embayment, 153–154
Grand Canyon National Park, 1
Grand Canyon Supergroup, 7, 76–78, 86–88
Grand Staircase, 213–219, 268
Grand Wash Cliffs, 270, 271
Grand Wash Dolomite, 96
Grand Wash Fault, 246–247, 252, 349
Grand Wash Trough, 271
Grandeur Member, 209
Grandview-Phantom system, 350
Grandview Trail, 120
Granite Gorge Metamorphic Suite, 15, 19–20, 26–29, 34–37, 79
Granite Park, 134, 360
Granite Park Block, 32–33
Granite Park Mafic Complex, 24–25
Granite Rapids, 279–280
granites, 21, 26
Grapevine Camp Pluton, 22–23
Grassy Mountain, 272
gravel, 354–356
gravity faults, 255
Gray Cliffs, 218
Gray Ledge Dam, 320
Gray Mountain, 219
Great Unconformity, 37, 60, 79, 402, 404
Grenville orogeny, 87–88
ground water, 176, 254

Hakatai Shale, 42, 45, 405
Hancock Knolls, 253
hanging valleys, 335, 337
Harrisburg Member, 197, 199–200, 202–205, 206, 208,
  211, 376
Hasavu Canyon, 334
Havasu Creek, 339, 406
Hermit Formation, 140, 141, 147–149, 150, 158–159
Hermit Rapids, 293
Hermit Shale, 376, 378
Hermit Trail, 139, 142, 145, 157
Holbrook Basin, 154
Horn Creek Pluton, 21, 22–23
Horn Creek Rapids, 304
Horseshoe Mesa Member, 117, 120, 122, 123
Hotauta Conglomerate Member, 44
House Rock Rapids, 281
Hualapai Canyon, 172
Hualapai Limeston, 270
Hualapai Trail, 175, 178–179
Hurricane Cliffs tongue, 187
Hurricane Fault, 17, 19, 225, 247, 248, 249, 251–252,
  349
hydraulic jumps, 289, 292, 307

Iceberg Canyon, 271
Iceberg Ridge, 109, 112, 116
Imperial Formation, 273
Indian Springs Formation, 132
Intermountain Seismic Belt, 349, 350
irregular-bedded sandstone, 187–189
island arcs, 9, 11
isotopic boundaries, 34–35

Jackass Canyon, 188
Jerome Member, 114
Jupiter Member, 57, 58, 59–60, 71

Kaibab Formation, 196–199, 200–205, 206, 208–211, 218
Kaibab Plateau, 264
Kaibab Segment, 397
Kaibab Trail, 144, 145, 148
Kaiparowits Formation, 216
Kanab Canyon, 110
Kanab Creek, 269
Kanab Point, 204
karsted topography, 123, 126, 398–399, 400
Kayenta Formation, 216, 218, 221
Kingston Peak Formation, 74
Kootenia, 97
Kwagunt Formation, 57, 62–65, 71

Lake Mead, 271
Laoporus, 167, 170, 171
Laramide Orogeny, 229–231, 230, 231–234, 236,
  239–240, 241–242
Lava Butte Dam, 320, 328–329
lava dams, 316–319, 320, 321–324
Lava Falls Dam, 320
lava flows, 20, 313–314, 315–316, 329, 333
Lava Pinnacle, 341
Layered Dbs Dam, 320
Leadville Limestone, 122
Leonardian Series, 162
Lepidodendron, 127, 129
Liesegang banding, 401, 402
Lingulella, 97
Lipalian interval, 92
lithosphere, 9
Little Colorado River, 267, 268
Little Dal Group, 74

Log Springs Formation, 132
Lone Mountain Monocline, 249
Lower Granite Gorges, 36
Lower Permian rocks, 137

Manakacha Formation, 140–143, 150, 154, 157
Mancos Shale Formation, 216, 221
Manning Canyon Formation, 132
Marble Canyon, 111, 136, 360, 376, 396
Markham Fault Zone, 250
Martin Formation, 107
Marvine, A. R., 6
Massive Dbs Dam, 320
McKee, Edwin D., 6–7
Meade Peak Phosphatic Shale Member, 209
Melanocyrillium, 69
Meriwhitica Monocline, 16, 238–239
Mesa Butte system, 350
Mesichnium, 168, 169
Mesoproterozoic rocks, 77
Mesozoic strata, 212–213
metamorphism, 35–37
microfossils, 68–70
Midcontinent Rift, 87
Middle Granite Gorge Block, 14, 32–33
Mineral Canyon Block, 15, 30–31, 37
Mississippian rock, 119
Missourian Series, 161
Moenave Formation, 216, 218, 221
Moenkopi Formation, 200, 215, 216, 218, 220, 267–268,
  272
Mogollon Escarpment, 240
Mogollon Highlands, 266
Mogollon Rim, 139, 144–145, 147, 154, 184, 198, 267
Mogollon Shelf, 154
Mojave province, 34
Monoclines, 231, 236, 236–238, 249
Monte Cristo Group, 122
Monument Canyon/Creek, 374, 377, 378, 402–403
Mooney Falls, 119, 339
Mooney Falls Member, 117, 119–120, 122–123
Moreyella, 207
Morrison Formation, 216, 221
Morrowan Series, 161
Muav Limestone, 95–96, 104–105, 376
Muddy Creek Formation, 270
Muddy Peak Limestone, 107
Music Mountain Formation, 241

Naco Formation, 137
Nankoweap Formation, 42, 47, 53, 54, 55–56, 58, 74–75,
  81
Nankoweap Rapids, 355
Nanopus, 170, 171
Nautiloid Canyon, 398–399, 400
nautiloids, 400
Navajo Sandstone Formation, 216, 218, 220, 221
Neostreptognathodus, 207, 208
Nisusia, 97
North American tectonic development, 9–11
North Canyon, 398, 399

Oak Creek Canyon, 155, 183, 184, 190
Ochoa Point Member, 47, 48
Octopodichnus, 168, 169
Olenellus, 97
Olenellus-Antagmus, 94
O'Niell Butte, 145
Oniscoidichnus, 168, 169
orogeny, 9–11, 226
Ouchita orogeny, 227

Pacific Plate, 243
Pahrump Group, 74
Painted Desert, 215, 219
Pakoon Limestone, 136, 150, 162
Palaeophycus, 99–100
Palaeopus, 170, 171
Paleohelcura, 167, 168, 169
paleontology, in Tonto Group, 96–97
Paleoproterozoic rocks, 12, 14, 77
Paleozoic rocks, 223
Palisades Creek debris fan, 363, 366, 367
Paradise Formation, 132
Paradox Basin, 154, 227
Parashant Canyon, 249
Park City Group, 208–209
Paterina, 97
Peach Springs Canyon, 241, 242
Peach Springs Tuff, 246–247, 266
pelletal wackestone, 190
Peniculauris, 200, 205–208
Pennsylvanian rocks, 137
Phantom Pluton, 22–23
Phycodes, 99–100
Pierce Canyon, 270
Pierce Ferry, 271
Pink Cliffs, 218–219
Pipe Creek Pluton, 21, 22–23
Pipe Creek Synform, 13, 15
Plagioglypta, 206
Plympton Formation, 209
Polygnathus, 113–114
Ponderosa Dam, 328–330
pool-and-rapid sequence, 276, 284–285
pools, 290
Pre-Redwall unconformity, 117
Precambrian rocks, 92, 224–225
pressure-temperature history, 29, 35–37
Prospect Canyon, 327, 329, 363
Prospect Dam, 320, 324–327
Prospect Dike, 342, 343
Prospect Lake, 327
pull-aparts, 173

Quartermaster Canyon, 112, 125, 128
Quartermaster Pluton, 16, 18, 24–25, 26
Quaternary system, 355
Queantoweap Sandstone, 147, 162

raindrop impressions, 173
rainstorms, 374
Rama Schist, 15, 19–20
rapids, 279–280, 282, 281, 285–287, 290–297, 300–301,
    305, 310, 372, 389
recreation, 353
Red Butte, 219
Red Pine Shale, 74
Redwall Limestone, 115–116, 117, 118–120, 122–123,
    126, 227
reverse drag, 251
Reynolds number, 288
Rhipidomella, 131
rifting, 86–88, 269–273
rim gravels, 266–267
Rio Grande River, 243
ripple marks, 172
rivers, 261, 265
Road Canyon Formation, 209
Roberts Roost Fanglomerate, 246
Robinson Wash, 204
Rock Canyon conglomerate, 200
rock gardens, 304, 389

Rodinia, 77, 88
Rose Well Camp, 250–251
rotational slides, 256–258
Ruby Block, 14, 30–31
Ruby Pluton, 21, 22–23
Rusophycus, 99–100, 103

San Andreas Fault, 243
San Andreas Limestone, 194
sand dunes, 355, 361–362
sand seas, 152
sand-volcanoes, 56
sandy dolomite, 190
Sapphire Pegmatite Complex, 22–23
Scalarituba, 99–100
scarps, development, 267–268, 272
Scenella, 97
Schizodus, 187, 193, 208
Schnebly Hill Formation, 137, 148, 162
Scolecocoprus, 167, 168, 169
Scolicia, 99–100
seas, 44–48, 51, 90, 96, 101, 122–123, 154, 159, 191,
    192, 194–195, 209, 211, 228–229
sediment, 275–276, 296–297, 352, 354, 382, 383
sedimentation, behind lava dams, 319
Sedona Arch, 153
seismicity, 347. *see also* earthquakes
Seligman Member, 181, 183–184, 185, 186
Separation Canyon, 116
Separation Pluton, 18, 24–25
Separation Shear Zone, 16
shear zones, 15
shelves, 153
Shinarump Cliffs, 215
Shinarump Conglomerate, 267–268, 272
Shinarump Member, 218
Shinumo Quartzite, 42–43, 45, 101, 405
Shivwits Plateau, 269, 271–272
side canyons morphology, 394–396
sills, 49–50
Sixtymile Formation, 53, 54, 57, 58, 72–73, 74–75, 85–86
skeletal packstone, 189–190
Skolithos, 99–100
slope retreat, 344–345
Sockdolager Antiform, 13, 15
Solomon Temple Member, 47–48
South Canyon, 183, 372–373
South Kaibab Trail, 116, 188, 197
South Rim, 219–221
Spencer Canyon Block, 18, 34–35
Spencer Pluton and dike complex, 24–25
Straight Cliffs Formation, 216, 218
Stratifera, 59
striped alluvium, 357, 359–360
stromatolites, 44–45, 61–62, 64–65, 66, 70, 128
Summerville Formation, 216
Supai Group, 7, 140–143, 145–149, 150–151, 152–160,
    375, 378
Surprise Canyon Formation, 123, 124–126, 127–128,
    129–135
Surprise Pluton, 18, 24–25
Surprise-Quartermaster Block, 18, 34–35
Surprise Valley, 258
Sweetina, 207
Sycamore Canyon, 184, 187, 189, 190
Syringopora, 118

Tanner Canyon, 358, 366, 367
Tanner Member, 57, 58, 59, 60, 71
Tanner Rapid, 233
Tapeats Canyon, 258

Tapeats Creek, 403–404
Tapeats Sandstone, 60, 92–94, 100–103, 256, 401, 402
tectonism, 225–229, 242–247, 249, 251–252, , 253–258, 265–274
Teichichnus, 99–100
temperature, 36
Temple Butte Formation, 107–109, 110–114
Temple Butte Limestone, 107, 400
Temple Cap Formation, 216, 218
terraces, 354–357
thermal history, Granite Gorge Metamorphic Suite, 37
Thunder River, 118
Thunder River Trail, 186
Thunder Springs, 405
Thunder Springs Member, 118, 122
tidal flats, 188
time scale, 7
Timpoweap Member, 200
Tonto Group, 7, 91, 92–96, 96–97, 100–103, 104–105
Tonto Platform, 90
Toreva Formation, 216, 221
Toroweap Canyon, 323
Toroweap Cascades, 314, 341
Toroweap Dam, 320, 330, 331–334
Toroweap Fault, 342, 349
Toroweap Formation, 164, 181–183, 185–187, 191–193, 194–195, 197
trace fossils, 150, 169–170
travertine dams, 406
Travertine Falls Pluton, 19, 24–25
Triavestiga, 168, 169
Trinity Block, 15, 30–31, 37
Trinity Pluton, 21, 22–23
Tropic Shale Formation, 216, 218
Troy Quartzite, 51–52
Tuckup Canyon, 111
Tuna Pluton, 22–23
turbulence, 288

Uinta Uplift, 230–231
Uncompahgre Uplift, 227
Unkar Group, 39, 40–41, 43–44, 46–48, 51–52, 54, 79–81
Upper Basalt Canyon, 60
Upper Granite Gorge, 12–13, 14, 36, 402

valley anticlines, 255
Vermilion Cliffs, 218
Virgilian Series, 161–162
Vishnu Schist, 15, 20, 37
Vishnu Shear Zone, 13, 15, 30–31
Visingso Group, 75
volcanic necks, 341–342
volcanism, 253–254
Vulcan's Forge, 341, 342
Vulcan's Throne, 314, 331, 342

Wahweap Formation, 217
Walcott Member, 57, 58, 64–65, 68, 71
Walhalla Plateau, 256–257
Waltenberg Block, 14, 32–33
Ward Terrace, 219
Watahomigi Formation, 138, 150, 161
waterfalls, 321–323
waves, 217, 288–297, 305–307
Wepo Formation, 221
Wescogame Formation, 143–145, 150, 155, 157, 161
West Kaibab fault system, 349–350
Whipple Mountains, 243
White Cliffs, 218
White Rim Sandstone, 199
White Rim Sandstone Member, 194
Whitmore Dam, 320, 337–339
Whitmore Wash, 118, 337, 338, 406–407
Whitmore Wash Member, 117, 118, 122
Willow Springs Formation, 270
Wingate Sandstone Formation, 217, 218, 221
Wolfcampian Series, 162
Woods Ranch Member, 180, 184, 186

Yale Point Sandstone Formation, 217, 221
Yavapai arcs, 35

Zacanthoides, 97
Zion Canyon, 218
Zoroaster Antiform, 13
Zoroaster Pluton, 18–19, 21, 22–23